Communications
in Computer and Information Science

1983

Rationale

The CCIS series is devoted to the publication of proceedings of computer science conferences. Its aim is to efficiently disseminate original research results in informatics in printed and electronic form. While the focus is on publication of peer-reviewed full papers presenting mature work, inclusion of reviewed short papers reporting on work in progress is welcome, too. Besides globally relevant meetings with internationally representative program committees guaranteeing a strict peer-reviewing and paper selection process, conferences run by societies or of high regional or national relevance are also considered for publication.

Topics

The topical scope of CCIS spans the entire spectrum of informatics ranging from foundational topics in the theory of computing to information and communications science and technology and a broad variety of interdisciplinary application fields.

Information for Volume Editors and Authors

Publication in CCIS is free of charge. No royalties are paid, however, we offer registered conference participants temporary free access to the online version of the conference proceedings on SpringerLink (http://link.springer.com) by means of an http referrer from the conference website and/or a number of complimentary printed copies, as specified in the official acceptance email of the event.

CCIS proceedings can be published in time for distribution at conferences or as post-proceedings, and delivered in the form of printed books and/or electronically as USBs and/or e-content licenses for accessing proceedings at SpringerLink. Furthermore, CCIS proceedings are included in the CCIS electronic book series hosted in the SpringerLink digital library at http://link.springer.com/bookseries/7899. Conferences publishing in CCIS are allowed to use Online Conference Service (OCS) for managing the whole proceedings lifecycle (from submission and reviewing to preparing for publication) free of charge.

Publication process

The language of publication is exclusively English. Authors publishing in CCIS have to sign the Springer CCIS copyright transfer form, however, they are free to use their material published in CCIS for substantially changed, more elaborate subsequent publications elsewhere. For the preparation of the camera-ready papers/files, authors have to strictly adhere to the Springer CCIS Authors' Instructions and are strongly encouraged to use the CCIS LaTeX style files or templates.

Abstracting/Indexing

CCIS is abstracted/indexed in DBLP, Google Scholar, EI-Compendex, Mathematical Reviews, SCImago, Scopus. CCIS volumes are also submitted for the inclusion in ISI Proceedings.

How to start

To start the evaluation of your proposal for inclusion in the CCIS series, please send an e-mail to ccis@springer.com.

Alessandro Ortis · Alaa Ali Hameed ·
Akhtar Jamil
Editors

Advanced Engineering, Technology and Applications

Second International Conference, ICAETA 2023
Istanbul, Turkey, March 10–11, 2023
Revised Selected Papers

 Springer

Editors
Alessandro Ortis
University of Catania
Catania, Italy

Alaa Ali Hameed
Istinye University
Istanbul, Türkiye

Akhtar Jamil
National University of Computer
and Emerging Sciences
Islamabad, Pakistan

ISSN 1865-0929   ISSN 1865-0937 (electronic)
Communications in Computer and Information Science
ISBN 978-3-031-50919-3   ISBN 978-3-031-50920-9 (eBook)
https://doi.org/10.1007/978-3-031-50920-9

This Springer imprint is published by the registered company Springer Nature Switzerland AG
The registered company address is: Gewerbestrasse 11, 6330 Cham, Switzerland

Paper in this product is recyclable.

Preface

This volume comprises the proceedings of the 2nd International Conference on Advanced Engineering, Technology and Applications (ICAETA 2023). The conference was co-organized by Istanbul University-Cerrahpaşa, Turkey and University of Catania, Italy, on March 10–11, 2023.

ICAETA offers a forum for researchers proposing solutions meeting the operational needs of industry. The conference is hence targeted both at researchers working on innovative technology and experts developing tools in the field. The goal of the conference is to attract papers investigating the use of technology outside the controlled environment of research laboratories.

The topics cover a range of areas related to engineering, technology, and applications. Main themes of the conference include, but are not limited to:

- Data Analysis, Visualization, and Applications
- Artificial Intelligence, Machine Learning, and Computer Vision
- Computer Communication and Networks
- Signal Processing and Applications
- Electronic Circuits, Devices, and Photonics
- Power Electronics and Energy Systems

We are delighted to announce the remarkable response to our Call for Papers for this conference, which exceeded our highest expectations. A total of 139 scholarly works spanning a multitude of domains were submitted, each subjected to rigorous Single-blind review by at least three experts in their respective fields. After initial screening, 126 papers were selected for further review. Following an intensive and highly competitive review process, we accepted 37 papers for presentation, culminating in an impressive acceptance rate of 26%. The selection of these papers was based solely on their technical excellence, profound significance, and impeccable clarity of presentation.

We extend our profound gratitude to the authors, presenters, and delegates whose presence and invaluable contributions were pivotal in making this event an outstanding success. Furthermore, we wish to express our heartfelt appreciation to our distinguished keynote speakers, session chairs, and diligent authors and reviewers.

Alessandro Ortis
Akhtar Jamil
Alaa Ali Hameed

Organization

General Chairs

Zeynep Orman   Istanbul University-Cerrahpaşa, Turkey
Alessandro Ortis   University of Catania, Italy

Steering Committee

Samee Khan   Mississippi State University, USA
Gui-Song Xia   Wuhan University, China
Marcello Pelillo   Ca' Foscari University of Venice, Italy
Fausto Pedro García Márquez   Castilla-La Mancha University, Spain
Nizamettin Aydin   Yildiz Technical University, Turkey
Shehzad Ashraf Chaudhry   Abu Dhabi University, UAE

Program Chairs

Francesco Rundo   ST Microelectronics, Italy
Alaa Ali Hameed   Istinye University, Turkey

Technical Program Chairs

Daniele Ravì   University of Hertfordshire, UK
Akhtar Jamil   National University of Computer and Emerging Sciences, Pakistan

Registration Committee

Ersin Onur Erdoğan   Istanbul University-Cerrahpaşa, Turkey
Doğukan Aksu   Istanbul University-Cerrahpaşa, Turkey
Davut Celik   Istanbul University-Cerrahpaşa, Turkey

Publicity Committee

Esra Tepe	Istanbul University-Cerrahpaşa, Turkey
Ümmet Ocak	Istanbul University-Cerrahpaşa, Turkey
Mehmet Yavuz Yağci	Istanbul University-Cerrahpaşa, Turkey

Technical Program Committee

Abbas Uğurenver	Istanbul Aydin University, Turkey
Ahmet Sertbaş	Istanbul University-Cerrahpaşa, Turkey
Akhan Akbulut	İstanbul Kültür University, Turkey
Aliyu Musa	Tampere University, Finland
Alladoumbaye Ngueilbaye	Shenzhen University, China
Atakan Kurt	Istanbul University-Cerrahpaşa, Turkey
Atta Ur Rehman	Ajman University, United Arab Emirates
Aya Magdy Fahmy Elfatyany	Menofia University, Egypt
Aysel Ersoy	Istanbul University-Cerrahpaşa, Turkey
Bahaa Alsheikh	Yarmouk University, Jordan
Berk Canberk	Istanbul Technical University, Turkey
Cengiz Polat Uzunoğlu	Istanbul University-Cerrahpaşa, Turkey
Chawki Djeddi	University of Rouen, France
Chiranji Lal Chowdhary	Vellore Institute of Technology, India
Costin Bădică	University of Craiova, Romania
Dario Allegra	University of Catania, Italy
Deepika Kumar	Bharati Vidyapeeth's College of Engineering, India
Derya Yiltaş Kaplan	Istanbul University-Cerrahpaşa, Turkey
Dharm Singh Jat	Namibia University, Namibia
Emel Arslan	Istanbul University-Cerrahpaşa, Turkey
Eylem Yücel Demirel	Istanbul University-Cerrahpaşa, Turkey
Fabrizio Messina	University of Catania, Italy
Faezeh Soleimani	Ball State University, USA
Fatih Keleş	Istanbul University-Cerrahpaşa, Turkey
Fatma Bozyiğit	Antwerp University, Belgium
Fatma Patlar Akbulut	İstanbul Kültür University, Turkey
Filippo Milotta	ST Microelectronics, Italy
Fırat Kaçar	Istanbul University-Cerrahpaşa, Turkey
Francesco Guarnera	University of Catania, Italy
Giovanni Puglisi	University of Cagliari, Italy
Gülsüm Zeynep Gürkaş Aydin	Istanbul University-Cerrahpaşa, Turkey
Hamza Issa	American University of the Middle East, Kuwait

Waleed Ead	Beni-Suef University, Egypt
Wing W. Y. Ng	South China University of Technology, China
Yusuf Sait Türkan	Istanbul University-Cerrahpaşa, Turkey
Zafar Ali	Southeast University, China

Contents

Multimodal Classifier for Disaster Response

Saed Alqaraleh[1]([✉]) [iD] and Hatice Sirin[2] [iD]

[1] Computer Engineering Department, Hasan Kalyoncu University, Gaziantep, Turkey
saed.alqaraleh@hku.edu.tr
[2] Software Engineering Department, Hasan Kalyoncu University, Gaziantep, Turkey

Abstract. Data obtained from social media has a massive effect on making correct decisions in time-critical situations and natural disasters. Social media content generally consists of messages, images, and videos. In situations of disasters, using multimedia files such as images can significantly help in understanding the damage caused by disasters compared to using text only. In other words, the exact situation and the effect of disaster are better understood using visual data.

So far, researchers widely use text datasets for building efficient disaster management systems, and a limited number of studies have focused on using other content, such as images and videos. This is due to the lack of available multimodal datasets. We addressed this limitation in this work by introducing a new Turkish multimodal dataset. This dataset was created by collecting disaster-related Turkish texts and their related images from Twitter. Then, by three evaluators and the majority voting, each sample was annotated as a disaster or not a disaster.

Next, multimodal classification studies were carried out with the late fusion technique. The BERT embedding approach and a pre-trained LSTM model are used to classify the text, and a pre-trained CNN model is used for the visual content (images). Overall, concatenating both inputs in a multimodal learning architecture using late fusion achieved an accuracy of 91.87% compared to early fusion, which achieved 86.72%.

Keywords: Multimodal Classifier · Disaster Management · Tweet Text Classification · Image Classification · Turkish language

1 Introduction

A vast number of images and texts captured during most of our daily events are uploaded to social media platforms worldwide. This large-scale data shared on social media can be classified using visual recognition and textual understanding. Since natural disasters are time-critical, a practical classification of data published on social networks is extremely useful for people in charge and humanitarian organizations to make plans and correct decisions on time. It is worth mentioning that, unfortunately, sometimes, during important events such as disasters, people share irrelevant information with disaster hashtags to ensure that more readers see their tweets.

Disasters are events that negatively affect people, the environment, and societies due to the life losses and damages that occur during disasters. Recently, messages and photographs have been highly used to describe the situations of people and the environment

© The Author(s), under exclusive license to Springer Nature Switzerland AG 2024
A. Ortis et al. (Eds.): ICAETA 2023, CCIS 1983, pp. 1–13, 2024.
https://doi.org/10.1007/978-3-031-50920-9_1

during natural disasters such as earthquakes, floods, fires, etc. Social media platforms, where information and news can be accessed and used in real-time, are considered one of the most widely used tools for communication and its purposes. However, in cases of misuse, it can create a chaotic environment and causes various harms. Hence, an automated system that can find the most valuable and relevant information before, during, or after disasters is vital.

Due to advances in deep learning, the performance of both text and image classification methods has increased significantly in recent years. This increases the interest in multimodal deep-learning classification systems. It demonstrates that deep representations of image and text data can be transferred to a new field by performing common deep-learning representations for different data types. However, most of these studies working on introducing efficient systems that automatically classify English only [1, 2]. In contrast, unfortunately, only a few works have been done in this field related to other languages, such as Turkish, and all of them focus only on the text [3–5]. In other words, no Turkish study uses the available text and images of disaster-related information to build an efficient multimodal classifier for disaster response.

Information on emergency management is typically time-sensitive, subject to constant change, and critical to society's readiness to respond to emergencies and disasters. Emergency managers are trying to allow people to report critical situations through all available channels, such as phones, TV, and the Internet (websites like social media). These channels also inform and guide the public before/during, and after disasters. For example, in 2011, a magnitude 5.8 earthquake occurred in the United States; authorities contacted the public via Twitter to report the disaster damage in their regions and inform them what to do. In this earthquake, where calls could not be made due to equipment disruptions, Twitter was used to reach people's relatives and get information from public institutions [6]. Also, during the Van Earthquake on October 23, 2011, in Turkey, people in Van and surrounding places organized social media campaigns to aid, support, and rescue activities. Also, it has been reported that some injured under the rubble of the Van Earthquake used social media to request help.

Overall, it is crystal clear that such platforms can effectively help crisis management. However, due to the immense amount of shared multimodal data, removing redundant and irrelevant information is essential to assist decision-makers in making the most suitable actions during such events.

The manuscript is organized as follows: The recent works and literature related to this study are summarized in Sect. 2. Section 3 describes our proposed method and architecture. Experimental setups, results, and analysis are presented in Sect. 4. Section 5 demonstrates the conclusions of this research.

2   Literature Review

The recent works and literature related to this study are summarized below. In [3], a new social media data analysis framework was proposed. This framework uses deep bidirectional neural networks trained on earthquakes, floods, and extreme flood datasets. It first works on learning from discrete handcrafted features and then fine-tuning the deep bidirectional transformer neural networks. Overall, the developed multiclass classifier integrated with support vector machines provides a precision of 0.83 and 0.79 for both random and original splits, respectively. While integrating Bernoulli naïve Bayes can achieve 0.59 and 0.76, multinomial naïve Bayes achieved 0.79 and 0.91.

The work presented in [4] aims to create a system that can detect crises that require assistance by making an effective classification for Turkish text tweets using KNN, SVM, and CNN algorithms. Their results indicated that the proposed model could achieve around 94% accuracy. Next, in [5], the work of [4] was further improved, and a text Turkish tweet dataset for crisis response and a deep-learning Turkish tweet classification system for crisis response was presented.

In [7], distinctive features were proposed to classify tweets as contextual and non-contextual. After classifying the tweets, situational tweets summarization techniques were used to convey awareness to government agencies. Then, the system of [7] was experimented with using the Uttarakhand floods and the Nepal earthquakes datasets, which contain tweets in English and Hindi. Results demonstrate that the domain-independent classifier outperforms the domain-dependent technique for English and Hindi tweets.

In [8], a method to classify tweets about damage detection that combines statistical and illuminating features was developed. This method uses Random Forest and AdaBoost classifiers. The results of experimental work in [8] showed that the proposed method outperformed the baseline SVM with the Bag-of-Words model.

Related to multimodal systems, there are two basic approaches to automatically merging different models, in our case, text and images, early fusion and late fusion [9]. Early fusion works on finding the best features from each multi-data (text and image), and the features of both text and image are combined in a single vector [10]. Then, a classifier is trained on top of the standardized vector. Related to the late fusion (voting): each modality, such as the image and text, is propagated through their classifier, and the output probabilities of both models are averaged, where the class is selected using the maximum value [11]. It is worth mentioning that a third type, known as Hybrid Fusion (GMU), exists and can support multiple languages (input text belongs to multiple languages). This type is usually trained using the best features per modality.

In the following, the recent studies related to multimodal systems are summarized. In [12], a simple feature augmentation approach that can leverage the text-image relationship to improve the classification of emergency tweets was presented. Also, a new multimodal dataset containing 4600 tweets (image + text) collected during the 2017 USA disasters and manually annotated was introduced. The model was tested on two categories, i.e., Humanitarian and Damage Assessment, and the observations indicated increased performance.

In [13], another multimodal classification model for mining social media disaster information was introduced. This model uses the Late Dirichlet Allocation (LDA) to identify subject information. Bert embedding and VGG-16 are used to analyze the multimodal data, where text and image data are classified separately. The Weibo data collected during the 2021 Henan heavy storm was used. Their results showed that an improvement of 12% can be achieved using the proposed model compared to "KGE-MMSLDA", a topic-based event classification model.

In [14], the integration of disaster data provided by text and image was investigated. Then, the attention mechanism was used to build a multimodal deep learning model called CAMM. This model was compared with "MUTAN' and "BLOCK" unimodal models and outperformed both by 6.31% and 5.91%, respectively.

Another multimodal disaster identification system was presented in [15]. This model combines the visual features with word features to classify each input tweet. In more detail, the visual features are extracted using a pre-trained convolutional neural network, while the textual features are extracted using a bidirectional long-term memory (BiL-STM) network with an attention mechanism. Next, a feature fusion and the Softmax classifier are used to decide on the class. Overall, the system of [15] outperformed some baselines unimodal and multimodal models by 1% and 7%, respectively.

Until this study, no multimodal crisis management systems were built specifically to classify Turkish language text and images. This work introduces the first-ever Turkish language multimodal crisis management system. We first collected Turkish text and visual tweets related to natural disasters to achieve our goal. Then, intensive experiments were performed to produce an efficient multimodal classification system that processes text and image data shared on Twitter before, during, and after natural disasters and informs authorities about relevant and critical disaster-related information.

3   The Proposed Multimodal Learning Approach

In this work, three evaluators annotated the collected samples as relevant to or irrelevant to natural disasters. Next, a new automated and multimodal classifier that supports the Turkish language was proposed.

As shown in Sect. 4, and based on intensive investigations, an LSTM was selected for text classification. At the same time, an in-depth feature extracted from a fully connected layer of AlexNet-based CNN achieved the best performance for image classification. In addition, after comparing the performance of BERT, Glove, and Word2Vec embedding systems, we selected BERT as it showed superior performance. Finally, we perform the late fusion of classification scores. As shown in Fig. 1, we verify the integration of visual and textual methods with multimodal techniques.

3.1   The Collected Sample

The collected Twitter data consisted of Turkish texts and their related images. The data will be used to analyze whether it is relevant to natural disasters or not. To collect the samples, deprem (earthquake), yangın (fire), trafik kazası (traffic accident), müsilaj (sea saliva), and sağanak (downpour) disaster-related keywords were used. Around 17 thousand samples belonging to five separate sub-datasets with text and image data are created. However, as data may contain meaningless words and symbols, the cleanup is performed afterward. Then, three annotators read each tweet, viewed the related image(s), and independently judged whether each sample was related to the disaster. The majority voting result was applied for the final labeling of each sample.

Note that, as mentioned before, sometimes people share irrelevant tweets during disasters with keywords and hashtags related to the disaster to increase the number of views and shares. Hence, it is crucial to distinguish tweets about the disaster. A sample of the collected relevant and irrelevant tweets (text and image) for different natural disasters are shown in Fig. 2. Overall, the used keywords and the total number of collected samples for each keyword are shown in Table 1.

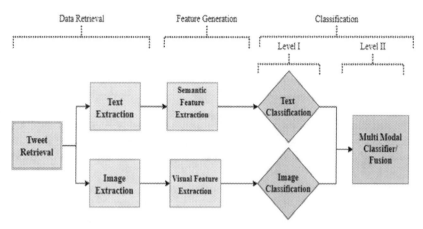

Fig. 1. The workflow of the proposed multi-modal classification system.

Fig. 2. Sample of the collected tweets (images along with their text).

Table 1. The number of samples for each sub-dataset.

Disaster	Datasets	# Relevant Tweets	#Irrelevant Tweets
Deprem	Dataset 1	2596	2400
Yangın	Dataset 2	1450	1448
Trafik Kazası	Dataset 3	1718	1692
Müsilaj	Dataset 4	800	780
Sağanak	Dataset 5	2029	2100
Total number of samples		8593	8420

3.2   The Proposed Text Classification Model

Text classification generally determines whether each sample belongs to one of the predefined classes. In other words, most candidate class is selected for each input. Until the last decade, the text classification process was challenging as computers can only process numbers. This process used traditional techniques, i.e., Bag of word approaches, such as TF-IDF. However, with the recent improvements in text's feature extraction and representation using word embeddings and also the impressive performance of deep learning, we can build not just a text classification system but also other text systems such as document summarization, customer relationship management, web mining, emotion analysis, etc. that can achieve human level. In the following, we summarize the steps of text classification.

Preprocessing
Significantly, the textual data may contain irrelevant and useless terms such as spaces, punctuation, stop words, and repetitive words. Such data will be removed. Also, Turkish has some specific characters, i.e., "ç, ğ, ı, ö, ş, ü" where users in general use the equivalent English character while writing, especially in informal writing. Note that it is an essential preprocessing step to convert back such character to its equivalent Turkish one (this process is named Deascification). Case folding is another preprocessing step. Here, all text is converted to lowercase. Other important text preprocessing are 1) Tokenize: which refers to dividing the string into tokens; 2) Stop word filtering works on eliminating common words. 3) Stemming filtering: Reduces each word to its root by removing prefixes or semed attachments.

Vector Representation of Texts
Fasttext, Word2vec and Glove are successful examples of the first generation of embedding approaches. Although they are easy to develop and use, their main weakness is that each word will always get the exact vector space representation. However, in real life, words can have multiple meanings and may have different meanings and contexts based on their surrounding words. The problem has been overcome by recently developed

embeddings such as BERT, ELMO, and XLNET. In this study, based on our prelimi-nary performance investigations, Word2vec, Glove, and BERT were selected, and their performance was investigated.

Text Classifier

Convolutional Neural Networks (CNN or ConvNet), a category of deep learning neural networks, have performed superhumanly in many areas, such as image recognition, object recognition, automatic video classification, and computer vision.

On the other hand, the recurrent neural network (RNN) is another well-known cate-gory of deep learning networks. The output of some of its nodes can affect the subsequent input to the same nodes through cycle connections between nodes. RNNs can use their memory (internal state) to process inputs with different length sequences. Overall, RNN is widely used in Natural language processing (NLP) and can predict the latter word from the former words given in a content. Like traditional neural networks, RNN uses a reverse propagation algorithm. During this backward spread, gradients prone to zero can occur, which is called the reset gradient problem. Long Short-Term Memory Net-works (LSTM) can be considered an improved RNN architecture introduced to solve the mentioned problem [16, 17]. LSTM also solves the problems of Memorization-overlap and reset gradient (vanishing gradient), two main problems in deep neural networks applications.

Figure 3 shows the architecture of the LSTM model utilized in this work, which is used to determine whether the text of the input tweet is related to a natural disaster or not.

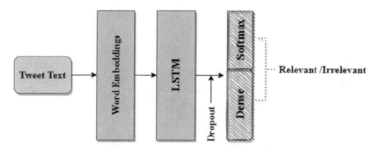

Fig. 3. LSTM network model architecture.

3.3 Image-Based Classification

When people look at an image(s), they can effortlessly differentiate between the color, size, similarity, types of items, etc. When it comes to computers, it is more challenging as computer process the numerical value of the image's pixels. However, recently CNN models have been able to achieve outstanding performance. In general, CNN processes these matrices using some hidden layers that detect the image properties/features to distinguish its objects. Thanks to the proposed multi-model dataset, where each sample was manually annotated to one of the predefined classes (disaster/not a disaster), we developed and trained a supervised CNN classification model. The main layers of the used model as summarized below.

Input Layer

This layer constitutes the first layer of CNN. While designing the model, selecting the correct input size for the images in this layer is critical and essential. For example, when the selected size is large, it requires more memory, training, and testing time. On the other hand, if the size is small, the network's performance may be low as we may lose the quality of the images while the memory requirement and training time are reduced. In other words, an appropriate input image size is needed for network success regarding network depth and hardware cost when performing image analysis.

Convolution Layer

In convolution, which is a customized linear process, the primary purpose is to extract designating properties for each input image. This layer performs convolution rather than matrix product [18]. In other words, the input image is represented by a matrix of its pixels read in the input layer. The convolution process is to scan this matrix using its filters to extract descriptive properties for images and texts.

Filters generate output data by implementing the convolution process on the previous layer's output. As a result of this convolution process, the activation map is created. Activation maps are regions where characteristics specific to each filter are found. During the training of CNNs, the coefficients of these filters modify with each learning loop in the training set. Thus, the network identifies which input regions are significant as designating properties.

Dropout Layer

CNN sometimes memorizes the data (samples). This layer prevents the network from memorizing and overfitting [19]. The basic logic implemented on this layer is the unloading of some nodes of the network. In other words, dropout works by temporarily ignoring some randomly chosen neurons' incoming and outgoing connections.

Activation Layer

This layer is also known as the rectifier unit (ReLU) layer. The ReLU function generally exchanges the negative input by 0 while it takes its value for positive entries. Hence, all negative values will be replaced by zero. Note that the output of the mathematical operations that are carried out on the convolution layer produces linear results, and this layer will make the deep network a nonlinear structure. With the use of this layer, the network learns faster.

Pooling Layer

Pooling, called "Down Sampling", is usually positioned after the ReLU layer. Its primary purpose is to reduce the size of ReLU output. The processes performed on this layer are working on representing the data in fewer values while still having efficient features. As a result, it will create less transactional load for the following layers and decrease the chance of the model overfitting (memorization).

Like the convolution process, certain filters are defined and applied here. These filters are routed around the image according to a particular step-by-step value and placed in the output matrix by taking the maximum values (maximum pooling) or the average of the values (average pooling). Based on our investigation, maximum pooling was selected as it outperformed others.

Fully Connected Layer
This layer connects all nodes of the previous layer. The entire matrix is given a single class vector with a size of 1 * 1 * 4096. This layer will be followed by another fully connected layer, which is explained below.

Classification Layer (Fully Connected Layer with Softmax)
Classification is carried out in this layer of deep learning models. Based on the previous layer's output, a weight matrix of 4096 * 2 is obtained for the classification layer. The output value of this layer is equal to the most candidate class of the two predefined classes, i.e., disaster and not a disaster. Different functions, such as Softmax and sigmoid, can be used to make the final decision, which is used in this layer. Here, Softmax was favored based on our observations.

4 Experiments

In this section, multiple investigational experiments that demonstrate the performance of proposed classification models vs. some state of art ones are done. First, we perform a comparison when models are used to classify image and text separately (we call it Unimodal Classification) and then when used to classify input samples with text and images (we call it Multimodal Classification). In other words, performance measurements were carried out with three separate classifications: tweet text, tweet image, and tweet text and image together. The performance of the trained models is measured by Accuracy, Precision, Recall, and F1 Score evaluation matrices.

To provide quality and robustness of the achieved results, all samples and sub-datasets presented in this paper were used in the experimental work.

Our experiments were conducted on a machine equipped with the Nvidia GTX1650 GPU with Intel(R) Core (TM) i7-10750H CPU and 16 GB of RAM running the Windows 10 Enterprise operating system. All codes were implemented using Python and its libraries. Google Colab environment is used for implementing the code.

4.1 Unimodal Classification

In this section, Unimodal image and text approaches were trained separately for each natural disaster (deprem-earthquake, yangın-fire, trafik kazası-traffic accident, müsilaj-mucilaj, sağanak-downpour).

Experiment 1. Performance of Image Classification
The main aim of this experiment is to choose the best architecture that will be later adapted and used for multimodal fusion. For the image unimodal experiment, we used

five state-of-the-art CNN architectures. Table 2 reports the results for the tested CNN models. As a result, although all models achieved almost similar performance, "Mouzannar's CNN model1" [20] performed marginally better than other models. It was also the fastest model for training and forecasting.

Table 2. Performance of the selected CNN models when used for image classification (unimodal).

Dataset #	Matrix	CNN Models				
		CNN_Model1 [20]	CNN_Model2 [1]	VGG19	ResNet50	Inception ResnetV2
1st	Acc	0.8548	0.8321	0.7183	0.7333	0.8313
	F1	0.852	0.8309	0.7169	0.7315	0.8316
2nd	Acc	0.8522	0.8295	0.717	0.7319	0.8291
	F1	0.8501	0.8287	0.7157	0.7302	0.8275
3rd	Acc	0.8578	0.8342	0.7219	0.7385	0.8342
	F1	0.8552	0.8321	0.7203	0.7364	0.8327
4th	Acc	0.8601	0.8344	0.7238	0.7407	0.8369
	F1	0.8579	0.8323	0.7217	0.7386	0.8342
5th	Acc	0.8512	0.8301	0.7141	0.7311	0.8301
	F1	0.8479	0.8279	0.7119	0.7381	0.8279

Experiment 2. Performance of Text Classification
In this section, we first compared the performance of the pre-trained Glove, Word2Vec, and BERT word embeddings using the proposed LSTM model and the CNN architecture of [21, 22]. Table 3 summarizes the performance results of CNN and an LSTM model. Overall, Word2Vec outperformed GloVe by around 1%, but BERT marginally outperformed both Word2Vec and GloVe.

Table 3. Performance of the classifiers using multiple embedding approaches.

Classifier	Word Embedding	Accuracy	Precision	Recall	F1-Score
Proposed Model_LSTM	Glove	0.8644	0.8581	0.8704	0.8642
	Word2Vec	0.8721	0.8643	0.8769	0.8701
	BERT	**0.8795**	**0.8702**	**0.8845**	**0.8772**
CNN [21, 22]	Glove	0.8498	0.8421	0.8546	0.8483
	Word2Vec	0.8547	0.8495	0.8592	0.8543
	BERT	**0.8576**	**0.8531**	**0.8607**	**0.8568**

4.2 Multimodal Classification

Experiment 3. Performance of the Late Fusion vs Early Fusion
Based on the results of the above experiments, the developed multimodal consisted of
BERT as a text feature extraction system, the LSTM used for text classification, and
the CNN model of [20] used for image classification. Then, the late fusion approach
is used to make the final decision on the class, whereas, for rule-based, the weighted
maximum decision rule is implemented. To finalize the experimental work, we have
compared the performance of early fusion and late fusion, as well as the text-only
unimodal and image-only unimodal. Overall, as shown in Table 4, it is obvious that late
fusion outperforms early fusion. Also, multimodal models provide a further performance
improvement compared to both Text-only and image-only unimodal.

Table 4. Performance comparison of different modalities.

Training Modal	Modality	Accuracy	Precision	Recall	F1 Score
Unimodal	Image	85.48	84.87	86.55	85.20
	Text	87.95	87.02	88.45	87.92
Multimodal (Text + Image)	Early Fusion	86.72	85.95	87.50	86.56
	Late Fusion	**91.87**	**90.34**	**92.25**	**91.28**

**Experiment 4. Performance of the Developed Model vs Some State of Art Deep
Learning Models**
In this experiment, the performance of the models of [20] and [23] and the proposed
model (LSTM(BERT)–CNN) were compared. Table 5 shows the performance comparisons of the mentioned models. Based on the accuracy, precision, recall, and F1 score,
as shown in Table 5, the proposed model has significantly outperformed the others.

Table 5. Performance comparisons of the state-of-art deep learning models.

Models	Datasets	Precision	Recall	F1 Score	Accuracy
CNN (Word2Vec) – CNN [23]	Dataset1	85.93	87.13	86.52	86.47
	Dataset2	84.42	85.82	85.11	85.12
	Dataset3	85.08	86.14	85.60	85.62
	Dataset4	84.63	85.94	85.27	85.44
	Dataset5	85.15	86.36	85.75	85.93
	Average	**85.04**	**86.28**	**85.65**	**85.72**

(*continued*)

Table 5. (*continued*)

Models	Datasets	Precision	Recall	F1 Score	Accuracy
LSTM (Word2Vec) – CNN [20]	Dataset1	90.72	91.95	91.33	91.46
	Dataset2	87.45	88.93	88.18	88.15
	Dataset3	89.77	91.02	90.39	90.38
	Dataset4	89.69	90.94	90.31	90.47
	Dataset5	90.29	91.71	90.99	91.07
	Average	**89.58**	**90.91**	**90.24**	**90.31**
LSTM (BERT) - CNN Proposed Model	Dataset1	90.34	92.25	91.28	91.87
	Dataset2	81.81	90.73	87.80	88.46
	Dataset3	89.12	91.16	88.89	90.87
	Dataset4	89.63	91.28	90.44	91.07
	Dataset5	90.03	92.11	91.05	91.23
	Average	**89.79**	**91.51**	**9.89**	**90.70**

5 Conclusion and Future Works

In this work, we have collected tweets (text) and their related images published before, during, or after some natural disasters. These samples were manually prepared and annotated using three evaluators. Then, a new multimodal classification system was presented after some intensive experimental work to ensure the efficiency and robustness of the proposed model. The late fusion was used to achieve multimodal classification; Also, a pre-trained BERT - LSTM model was used for processing text while a pre-trained CNN model was used for visual modal (images).

The experiment section indicated that the developed multimodal achieved an accuracy of 91.87%, while early fusion achieved 86.72%. Hence, such a model can improve disaster events' classification accuracy and help authorities make the most suitable timely decisions.

As future work, more intensive work on developing a new CNN model for images and using more advanced feature extraction methods for images can be applied. Another direction is to investigate the possibility of supporting multiple languages.

References

1. Alam, F., Ofli, F., Imran, M.: Descriptive and visual summaries of disaster events using artificial intelligence techniques: case studies of Hurricanes Harvey, Irma, and Maria. Behav. Inf. Technol. **39**(3), 288–318 (2020)
2. Ponce-López, V., Spataru, C.: Social media data analysis framework for disaster response. Discov. Artif. Intell. **2**(1), 1–14 (2022)
3. Tas, F., Cakir, M.: Nurses' knowledge levels and preparedness for disasters: a systematic review. Int. J. Disast. Risk Reduct. 103230 (2022)
4. Alqaraleh, S., Işik, M.: Efficient Turkish tweet classification system for crisis response. Turk. J. Electr. Eng. Comput. Sci. **28**(6), 3168–3182 (2020)

5. Alqaraleh, S.: Efficient Turkish text classification approach for crisis management systems. Gazi Univ. J. Sci. 1 (2021)
6. Soydan, E., Alpaslan, N.: Medyanin Doğal Afetlerdeki İşlevi. İstanbul J. Soc. Sci. Summer **7**, 53–64 (2014)
7. Rudra, K., Ganguly, N., Goyal, P., Ghosh, S.: Extracting and summarizing situational information from the Twitter social media during disasters. ACM Trans. Web (TWEB) **12**(3), 17 (2018)
8. Madichetty, S., Sridevi, M.: Disaster damage assessment from the tweets using the combination of statistical features and informative words. Soc. Netw. Anal. Min. **9**(1), 42 (2019)
9. Huang, F., Zhang, X., Zhao, Z., Xu, J., Li, Z.: Image–text sentiment analysis via deep multimodal attentive fusion. Knowl.-Based Syst. **167**, 26–37 (2019)
10. Yu, S., Cheng, Y., Xie, L., Luo, Z., Huang, M., Li, S.: A novel recurrent hybrid network for feature fusion in action recognition. J. Vis. Commun. Image Represent. **49**, 192–203 (2017)
11. Guo, D., Zhou, W., Li, H., Wang, M.: Online early-late fusion based on adaptive hmm for sign language recognition. ACM Trans. Multimed. Comput. Commun. Appl. (TOMM) **14**(1), 1–18 (2017)
12. Sosea, T., Sirbu, I., Caragea, C., Caragea, D., Rebedea, T.: Using the image-text relationship to improve multimodal disaster tweet classification. In: The 18th International Conference on Information Systems for Crisis Response and Management (ISCRAM 2021) (2021)
13. Zhang, M., Huang, Q., Liu, H.: A multimodal data analysis approach to social media during natural disasters. Sustainability **14**(9), 5536 (2022)
14. Khattar, A., Quadri, S.M.K.: CAMM: cross-attention multimodal classification of disaster-related tweets. IEEE Access **10**, 92889–92902 (2022)
15. Hossain, E., Hoque, M.M., Hoque, E., Islam, M.S.: A deep attentive multimodal learning approach for disaster identification from social media posts. IEEE Access **10**, 46538–46551 (2022)
16. Hochreiter, S., Schmidhuber, J.: Long short-term memory. Neural Comput. **9**, 1735–1780 (1997)
17. Srivastava, N., Hinton, G., Krizhevsky, A., Sutskever, I., Salakhutdinov, R.: Dropout: a simple way to prevent neural networks from overfitting. J. Mach. Learn. Res. **15**, 1929–1958 (2014)
18. Rodriguez, R., Gonzalez, C.I., Martinez, G.E., Melin, P.: An improved convolutional neural network based on a parameter modification of the convolution layer. In: Castillo, O., Melin, P. (eds.) Fuzzy Logic Hybrid Extensions of Neural and Optimization Algorithms: Theory and Applications. SCI, vol. 940, pp. 125–147. Springer, Cham (2021). https://doi.org/10.1007/978-3-030-68776-2_8
19. Loodos. loodos/bert-base-turkish-uncased hugging face (2022). https://github.com/Loodos/turkish-languagemodels
20. Mouzannar, H., Rizk, Y., Awad, M.: Damage identification in social media posts using multimodal deep learning. In: Proceedings of the 15th International Conference on Information Systems for Crisis Response and Management (ISCRAM), Rochester, pp. 529–543 (2018)
21. Chen, Y.: Convolutional neural network for sentence classification. Master's thesis, University of Waterloo (2015)
22. Yoon, K.: Convolutional neural networks for sentence classification [OL]. arXiv Preprint (2014)
23. Nguyen, D.T., Ofli, F., Imran, M., Mitra, P.: Damage assessment from social media imagery data during disasters. In: Proceedings of the 2017 IEEE/ACM International Conference on Advances in Social Networks Analysis and Mining 2017, pp. 569–576 (2017)

Image Encryption Using Spined Bit Plane Diffusion and Chaotic Permutation for Color Image Security

Renjith V. Ravi[1] , S. B. Goyal[2(✉)] , and Chawki Djeddi[3]

[1] Department of Electronics and Communication Engineering, M.E.A Engineering College, Malappuram, Kerala, India
[2] City University, Petaling Jaya, Malaysia
drsbgoyal@gmail.com
[3] Laboratoire de Vison et d'intelligence Artificielle, Université Larbi Tebessi, Tébessa, Algérie

Abstract. With the tremendous rise in multimedia output over the last ten years, image encryption has grown in importance as a component of information security. It is challenging to secure images using conventional encryption methods because of the intrinsic image properties that distinguish image cryptography from text cryptography. This study suggests employing binary key images for lossless encryption of color images. First, chaotic sequences from the 2D Henon map will be used to permute the locations of the pixels in the plaintext image. The key images, in this case, are bit planes created from an additional image. In order to improve the encryption image and make cracking more challenging, the bit plane will be further rotated. It concentrates on two methods for effective image encryption: bit plane slicing and bit plane spinning. In terms of a number of statistical tests, such as key experiments, information entropy experiments, and encryption quality testing, the simulation analysis also shows that the suggested technique is lossless, safe, and effective. Based on the findings, it can be shown that the new method takes less time to conceal the remaining intelligence than it does to encrypt the complete image.

Keywords: Bit-plane decomposition · Chaotic Permutation · Image Encryption · Image Cryptography · Bit plane Spinning

1 Introduction

Numerous digital photos may be quickly sent across a public network due to the rapid growth of computer networks and internet multimedia processing. Such as medicine, the military, business, and multi-media [1]. The security of images against unauthorized access has become a popular research and application topic. The inherent characteristics of a digital image, such as its large storage capacity and strong pixel correlation, may preclude the use of conventional data encryption algorithms for images [2, 3]. The aperiodicity, high sensitivity to beginning circumstances and factors, ergodicity, and pseudo-randomness of chaos have motivated several academics to develop chaos-based image cryptographic algorithms.

A. Ortis et al. (Eds.): ICAETA 2023, CCIS 1983, pp. 14–25, 2024.
https://doi.org/10.1007/978-3-031-50920-9_2

Cryptography is an effective means of transmitting confidential information safely. It encrypts images before transmitting them to change their structure. Due to this inability to easily acquire the original image, even the hacker will be in immune. The main objective is to better protect the original image. The principal use of slicing of bit-plane is the division of images into binary bit planes [4]. The intensity value of each pixel in an image is represented by a bit [5, 6]. The pixel intensities of an image are always used as the basis for image scrambling. To analyze the significance of each bit in an image, the image in digital format is separated into 8-bit planes.

The bit value of an image is simultaneously impacted by a little change in color. The image is broken up into 8-bit planes and is made up of several pixels [7]. It is used to define the significance of each bit in an image by representing the highest-order and lower-order bits. Compared to the other least important bit, the perceptional masking method, it produces superior image encryption. This step will be completed without affecting the overall quality of the image [8].

We present a technique for image encryption using chaotic maps and bit planes to circumvent the aforementioned issues. Bit-plane segmentation, chaotic series generation, and scrambling are the three adaptable components used. Because a digital computer can effectively execute bit-plane decomposition, the technique is computationally efficient in the lossless process [5]. The framework's use of permutation, which allows for the modification of pixel positions, raises the level of security of the algorithm [9]. Analysis of security and simulation results show that the method suggested works well and has a lot of potential [10]. The rest of the article is organized as follows, Sect. 2 discussed briefly about the materials and methods. All the results are discussed in Sect. 3 and finally the article is concluded in Sect. 4.

2 Materials and Methods

Think about a color image that is made up of many pixels. Bits are used to indicate each pixel. 8-bit planes [6, 11] from bit plane 1 to bit plane 8 make up the image. An image's lowest-order bits are found in plane 8, while all its higher-order bits are found in plane 1.

The slices of the eight planes contain information of the concerned image [6]. The output will get jumbled as we change the placements, values, and other factors.

The significant information in an image is shown on the bit plane. There are eight-bit planes in it. In the illustration mentioned above, all of the pixels in an image's pixels are represented by their lowest-order bits, or LSB bits, and their highest-order bits, or MSB bits. The goal of bit-plane slicing is to distinguish between noise and meaningful bit-plane data in an image. More distortion occurs when the MSB bit is substituted than when the LSB bit is. Less important bits are simple to encode, and keeping the most important bits is crucial.

2.1 The Henon Map

The Hénon Map (Eq. 1) is a two-dimensional, recursive map that was initially presented by Hénon (1976). It is among the best-known and most extensively researched instances of a nonlinear system with a weird attractor [12, 13]. It is a discrete-time process that employs the formula below to map a point (x, y) at time n.

$$x_{n+1} = 1 - a|x_n| + by_n$$
$$y_{n+1} = bx_n \tag{1}$$

where the process parameters are a and b.

2.2   Bit Plane Slicing

The image is separated into several bit planes using bit plane slicing [8]. Both higher-order and lower-order bits make up its structure. The bulk of the data in an image is contributed by MSB, which makes a considerable contribution to the overall image. Only a portion of an image's information is contributed by LSB. It is important for both image analysis and compression. The main contribution of this method is that it is used to provide critical information for each component of an image. Additionally, it makes sure that certain parts are beneficial. A bit plane primarily has many layers, each of which displays a different range of bit planes in an image. The number of bit planes used to break down the image also has an impact on the security level. The MSB is crucial in bit plane slicing since the loss of higher-order bit planes results in a considerably darker image. The quality of the image degrades more when the least important bit is encrypted than when the most important bit is encrypted. For each of the three distinct color planes, encryption is carried out using the bit plane slicing technique.

2.3   Bit Plane Spinning

The spinning (spinning) of bit planes is the basic technique for creating image encryption. Compared to the alternative bit-shifting approach, it is more efficient. Each bit plane is spinned at a different angle once the image has been broken up into bit planes [8]. At the receiving end, the bit plane of the original image is restored by reversing the spinning of the bit planes. Multiple angles, such as 90, 180, and 270 degrees, are applied to the spinning of a bit plane to make image encryption more complex. Without understanding the method used, an attacker would have a tough time obtaining the image. This involves the spinning of a tiny portion of a plane to alter its shape and produce an image. The crucial data is extracted from an image via bit-plane slicing.After bit plane slicing at different angles in 8-bit planes, bit spinning is performed. Each plane pixel is rotated at a new angle to produce an encrypted image. No bits are lost during plane spinning. By rotating bit 14 in high-dimensional space, this approach is accomplished. An original bit plane is created by combining all of the rotating bit planes. This technique prevents information loss when retrieving the actual image. This paper presents image encryption that acts on an image by using both the spinning of bits and the degree of spinning as a private key. By giving the 8-bit planes a spinning angle, this encryption operates. An encrypted image is produced by repeating these processes for each bit plane. The outcome shows that this encryption provides more confidentiality than the other methods.

2.4 Proposed Encryption Algorithm

The proposed encryption algorithm consists of a permutation of the position of pixels in the plaintext image and a two-level diffusion using bit planes. Here, there are two inputs one is a plaintext image, and the other is a key image. The key image is grayscale, and the plaintext image is a color image in RGB format. The color image will be separated into red, green and blue components for parallel processing. The position of pixels in each color image will be permuted according to the sequences from the 2D Henon map. Then the first stage of diffusion will be carried out using the keys generated from Bit planes 1, 2, and 3, and further, the second stage of diffusion will be performed using the keys generated from bit planes 4, 5 and 6. Finally, this processed R, G and B planes will be integrated to form the ciphertext image. We will do all of these processes in reverse order during decryption. Figure 1 shows the block diagram of proposed encryption algorithm.

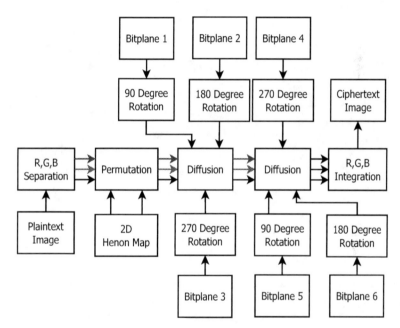

Fig. 1. Block diagram of the Proposed Encryption Algorithm

Key Generation

Here we are using a grayscale image as a key image. The key image will be decomposed into its six-bit planes for the diffusion process. Further, these bit planes will be rotated at 90, 180 and 270° to ensure more security. As we know, the bit planes consist of binary values, and those values cannot be used for XORing with the plaintext image. These values will be converted to gray values according to the following equation Eq. 2.

$$k_{new} = (k_{old} \times 10^3) \bmod 256 \tag{2}$$

3   Results and Discussion

For analyzing the performance of encryption and decryption algorithms, the test images *Lena*, *Sailboat* and *Baboon* are taken. The results of encryption and decryption are shown below in Fig. 2.

| (a) Sailbot -Plain | (b) Sailbot -Encrypted | (c) Sailbot -Decrypted |
| (d) Baboon - Plain | (e) Baboon -Encrypted | (f) Baboon Encrypted |

Fig. 2. Results obtained after encryption and decryption

3.1   Histogram Analysis

An essential statistical tool for determining the effectiveness of an image cryptosystem is histogram analysis. In Fig. 4, the histograms for the three colour channels of the original Lena and Baboon images and the corresponding created cypher images are shown [14]. The majority of the image data were obtained smoothly, and the histogram of the original image had a non-uniform distribution with a distinct and distinctive peak. The retrieved image, on the other hand, is a flawless reconstruction of the input images, whereas the cypher image is noise-like and unrelated to the initial plain image. The image that was retrieved is a flawless reconstruction of the input images. The histogram in the encrypted image is very consistent and smooth, demonstrating that no statistical information from the initial plain image was added. The suggested approach may therefore defend against cypher image and statistical assaults. The histograms of plaintext and ciphertext versions of the three-color components are depicted in Fig. 3.

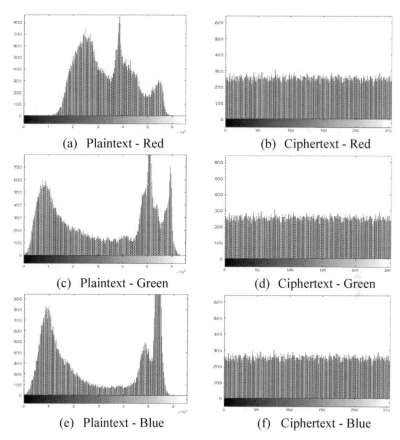

(a) Plaintext - Red

(b) Ciphertext - Red

(c) Plaintext - Green

(d) Ciphertext - Green

(e) Plaintext - Blue

(f) Ciphertext - Blue

Fig. 3. Histograms of Plain text and Ciphertext versions of the test image *Sailboat*

3.2 Correlation Factor Analysis

For determining two neighboring pixels in the horizontal, vertical, and diagonal orientations, the correlation factor (CF) is a crucial statistic. Two neighboring pixels, x and y, are among the N pairs of pixels that are chosen [14]. The formula for calculating the correlation factor may then be expressed by Eq. 3

$$CF = \frac{\sum_{i=1}^{N}(x_i - \bar{x})(y_i - \bar{y})}{\sqrt{\sum_{i=1}^{N}(x_i - \bar{x})^2}\sqrt{\sum_{i=1}^{N}(y_i - \bar{y})^2}} \tag{3}$$

Let N be 2000 now. Table 1 displays the findings of the correlation between the original and encrypted images. Because the correlation factor for adjacent pixels in the different directions is low, it may be assumed that the adjacent pixels are statistically independent. Each pair of horizontally adjacent pixels' correlation distribution is shown in Fig. 4. The source image's correlation factor is quite strong and very near to 1. The cypher image's correlation factor, on the other hand, is small and almost zero. The robustness of the suggested cryptosystem is thus confirmed by these findings. Table 2 shows the comparison of Correlation results with the literature.

Table 1. Correlation Coefficients

Image	Channel	Direction	Plaintext	Ciphertext
Lena	R	Horizontal	0.9897	0.0003
		Vertical	0.9796	−0.0017
		Diagonal	0.9665	0.0052
	G	Horizontal	0.9654	−0.0024
		Vertical	0.9567	0.0010
		Diagonal	0.9163	−0.0008
	B	Horizontal	0.9854	−0.0003
		Vertical	0.9657	0.0021
		Diagonal	0.9125	−0.0014
Sailboat	R	Horizontal	0.9867	0.0013
		Vertical	0.9853	−0.0008
		Diagonal	0.9635	0.0007
	G	Horizontal	0.9657	−0.0015
		Vertical	0.9524	−0.0013
		Diagonal	0.9737	−0.0017
	B	Horizontal	0.9428	0.0021
		Vertical	0.9754	−0.0020
		Diagonal	0.9542	0.0013

Table 2. Comparison of correlation values with the literature

Image	Reference	Vertical	Horizontal	Diagonal
Baboon	Proposed	−0.0002	0.0005	−0.0032
	Alghamdi et al. [15]	−0.0001	0.0006	−0.0021
	Alanezi et al. [16]	−0.0001	−0.0002	0.0011
	Arif et al. [17]	−0.0036	−0.0019	−0.0033
	Lu et al. [18]	−0.0004	0.0007	0.0029

3.3 Differential Attack

By changing a few of the original, unaltered pixels in the image, the adversary may get useful information. Evaluations of the encrypted plain image's resilience to differential raids often use the number of changing pixel rate (NPCR) and the unified averaged changed intensity (UACI). We used the described encryption method to encrypt P1 and P2 in order to acquire their respective encrypted images, denoted by C1 and C2, using a single secret key [14]. P2 is the result of changing one pixel from the initial plaintext image P1, which is represented by P1. After then, the NPCR and UACI might be administered by Eq. 4.

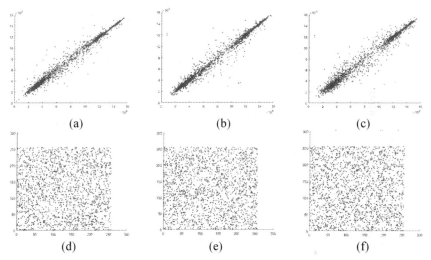

(a)   (b)   (c)

(d)   (e)   (f)

Fig. 4. Correlation plot in horizontal, vertical and diagonal direction. (a), (b) and (c) plaintext images. (d), (e) and (f) Ciphertext images

$$
\left.
\begin{aligned}
NPCR &= \frac{1}{MN} \sum_{i}^{M} \sum_{j}^{N} D(i,j) \cdot 100\% \\
D(i,j) &= \begin{cases} 1, & C_1(i,j) \neq C_2(i,j) \\ 0, & C_1(i,j) = C_2(i,j) \end{cases} \\
UACI &= \frac{1}{MN} \sum_{i}^{M} \sum_{j}^{N} \frac{|C_1(i,j) - C_2(i,j)|}{255} \cdot 100\%
\end{aligned}
\right\}
\tag{4}
$$

The values of NPCR and UACI obtained for the test image are shown in Table 3 and its comparison for the test image are shown in Table 4.

Table 3. NPCR and UACI

Image	Channel	Values	
		NPCR	UACI
Lena	R	99.4125	33.1252
	G	99.2564	31.2583
	B	99.3254	30.2473
Sailboat	R	99.8526	27.3251
	G	99.2748	33.3326
	B	99.2487	32.3245

Table 4. Comparison of NPCR and UACI Values with the literature

Image	Reference	NPCR	UACI
Baboon	Proposed	99.7123	33.5216
	Alghamdi et al. [15]	99.6153	33.4718
	Alanezi et al. [16]	99.6239	33.5615
	Arif et al. [17]	99.6059	33.4375
	Lu et al. [18]	99.5743	33.3941

3.4 Information Entropy

The matrix E(m) of the m-dimensional image data matrix served as a symbol for the Shannon entropy, which measures the security of the digital image [14].

$$E(m) = -\sum_{i=0}^{255} P_b(m_i)\log_2(P_b(m_i)) \tag{5}$$

A precise random image would provide values for all 256 pixels with the same probability, where Pb(mi) stands for the probability of m_i. Therefore, the information entropy's calculated value is closer to 8. *Lena* and *Sailbot* cypher images' entropy levels are shown in Table 3 to be quite near their theoretical values. In light of these findings, it can be concluded that the suggested cryptosystem effectively defends against information entropy attacks [19]. The values of information entropy obtained for various test images and its comparison with the literature are shown in Table 5 and Table 6 respectively.

Table 5. Information Entropy for Plaintext and Ciphertext

Image	Channel	Plaintext	Ciphertext
Lena	R	7.2543	7.9994
	G	7.3351	7.9995
	B	7.3254	7.9991
Sailboat	R	7.3452	7.9996
	G	7.6253	7.9999
	B	7.2581	7.9997

Table 6. Comparison of Entropy Values with the literature

Image	Reference	Entropy of Ciphertext
Baboon	Proposed	7.9998
	Alghamdi et al. [15]	7.9997
	Alanezi et al. [16]	7.9991
	Arif et al. [17]	7.9992
	Lu et al. [18]	7.9971

3.5   Encryption Time

The time taken for encryption using the proposed algorithm for the test image **Baboon** is shown in Table 7.

Table 7. Time Taken for Encryption

Reference	Time taken for Encryption
Proposed	0.07258 s
Alghamdi et al. [15]	0.09155 s
Alanezi et al. [16]	0.3033 s
Arif et al. [17]	1.28 s
Lu et al. [18]	1.243 s

4   Conclusion

In this study, a secure cryptographic algorithm that combines bit plane slicing and bit plane spinning is developed. In order to do this, we created a scrambling technique employing bit-plane spinning and bit-plane slicing. Histogram Analysis, Correlation,

Entropy, NPCR and UACI data are used to evaluate this method's effectiveness. An image's decryption is strong since no data is lost throughout the process. Security may be greatly increased using bit plane slicing and spinning for cryptography. The results demonstrate that it offers higher encryption without degrading the general image quality. Therefore, any image format may employ the mode, such as bit spinning, required for effective encryption. The testing findings demonstrate that the suggested approach improves security, which the Correlation, Entropy, NPCR and UACI, and histogram analysis support.

References

1. Dhane, H., Agarwal, P., Manikandan, V.M.: A novel high capacity reversible data hiding through encryption scheme by permuting encryption key and entropy analysis. In: 2022 4th International Conference on Energy, Power and Environment (ICEPE) (2022)
2. Manikandan, V.M., Zhang, Y.-D.: An adaptive pixel mapping based approach for reversible data hiding in encrypted images. Signal Process.: Image Commun. **105**, 116690 (2022)
3. Gopalakrishnan, T., Ramakrishnan, S., Balakumar, M.: An image encryption using chaotic permutation and diffusion. In: 2014 International conference on recent trends in information technology (2014)
4. Song, W., Fu, C., Zheng, Y., Tie, M., Liu, J., Chen, J.: A parallel image encryption algorithm using intra bitplane scrambling. Math. Comput. Simul. **204**, 71–88 (2023)
5. Wan, Y., Wang, S., Du, B.: A bit plane image encryption algorithm based on compound chaos. Multimed. Tools Appl. 1–19 (2022)
6. Gopinath, R., Sowjanya, M.: Image encryption for color images using bit plane and edge map cryptography algorithm. Int. J. Eng. Res. Technol. (IJERT) **1**, 4 (2012)
7. Xu, J., Zhao, B., Wu, Z.: Research on color image encryption algorithm based on bit-plane and Chen chaotic system. Entropy **24**, 186 (2022)
8. Vijayaraghavan, R., Sathya, S., Raajan, N.R.: Security for an image using bit-slice rotation method-image encryption. Indian J. Sci. Technol. **7**, 1 (2014)
9. Anwar, S., Meghana, S.: A pixel permutation based image encryption technique using chaotic map. Multimed. Tools Appl. **78**, 27569–27590 (2019)
10. Cao, W., Zhou, Y., Chen, C.P., Xia, L.: Medical image encryption using edge maps. Signal Process. **132**, 96–109 (2017)
11. Tripathi, M., Tomar, K., Chauhan, U., Chauhan, S.P.S.: Color images encryption by using bit plane crypt algorithm. In: 2022 2nd International Conference on Innovative Practices in Technology and Management (ICIPTM) (2022)
12. Moysis, L., Azar, A.T.: New discrete time 2D chaotic maps. Int. J. Syst. Dyn. Appl. **6**, 77–104 (2017)
13. Moysis, L., Azar, A.T., Tutueva, A., Butusov, D.N., Volos, C.: Discrete time chaotic maps with application to random bit generation. In: Azar, A.T., Kamal, N.A. (eds.) Advances in Systems Analysis, Software Engineering, and High Performance Computing), pp. 542–582. IGI Global (2021
14. Mohamed, H.G., ElKamchouchi, D.H., Moussa, K.H.: A novel color image encryption algorithm based on hyperchaotic maps and mitochondrial DNA sequences. Entropy **22**, 158 (2020)
15. Alghamdi, Y., Munir, A., Ahmad, J.: A lightweight image encryption algorithm based on chaotic map and random substitution. Entropy **24**, 1344 (2022)
16. Alanezi, A., et al.: Securing digital images through simple permutation-substitution mechanism in cloud-based smart city environment. Secur. Commun. Netw. **2021**, 1–17 (2021)

17. Arif, J., et al.: A novel chaotic permutation-substitution image encryption scheme based on logistic map and random substitution. IEEE Access **10**, 12966–12982 (2022)
18. Lu, Q., Zhu, C., Deng, X.: An efficient image encryption scheme based on the LSS chaotic map and single S-box. IEEE Access **8**, 25664–25678 (2020)
19. Manikandan, V.M., Masilamani, V.: A novel entropy-based reversible data hiding during encryption. In: 2019 IEEE 1st International Conference on Energy, Systems and Information Processing (ICESIP) (2019)

Hardware Implementation of MRO-ELM for Online Sequential Learning on FPGA

Önder Polat[✉] and Sema Koç Kayhan

Gaziantep University, 27310 Gaziantep, Turkey
{onderp,skoc}@gantep.edu.tr

Abstract. This paper presents a parallel hardware accelerator for an online variant of the extreme learning machine (ELM) algorithm, called mixed-norm regularized online ELM (MRO-ELM). ELM is a training algorithm for feedforward neural networks that has been widely adopted in the literature. The proposed parallel architecture is implemented on a field-programmable gate array (FPGA) and designed for classification tasks. It is designed to be scalable and reconfigurable for different problem sizes, and can be used for various numbers of hidden neurons. Among the existing studies, the proposed architecture is the first hardware implementation that has norm regularization with parallel processing capability. The implementation results for the proposed hardware accelerator are reported in terms of hardware efficiency and they show that the proposed design has lower resource utilization than existing parallel implementations.

Keywords: FPGA · online sequential ELM · hardware accelerator · regularization

1 Introduction

Artificial neural networks (ANN) are widely used computing systems for machine learning applications in various problem domains such as pattern recognition, computer vision, robotics, and medical diagnosis [1,7,12,21]. In the era of big data, most of these applications demand powerful computing platforms in terms of both speed and cost efficiency. ANNs that operate in real-time especially should exhibit high computational capability with low hardware complexity. ANNs are therefore widely implemented as custom digital hardware on field-programmable gate arrays (FPGAs) for their reconfigurability and flexibility in terms of area-speed trade-off.

ANNs are categorized into many different types based on their connection topologies and learning algorithms. One of the simplest type is the feedforward neural network, in which the input passes through the first, hidden, and last layers directly, in a forward manner. Extreme Learning Machine (ELM) [10] is a single-hidden layer feedforward network (SLFN) that belongs to the class of randomization-based networks [19]. ELM is trained by analytically finding the

A. Ortis et al. (Eds.): ICAETA 2023, CCIS 1983, pp. 26–37, 2024.
https://doi.org/10.1007/978-3-031-50920-9_3

weights for the output layer, given the training data and hidden layer weight and bias values. The weight and bias values for the hidden layer are randomly generated and does not require tuning, which simplifies the training process. Compared to the classical feedforward networks, which are usually trained using iterative gradient-descent based algorithms, ELM has the advantage of faster training speed [13] with sufficient generalization ability [9]. Owing to these properties, ELM and its variants are utilized for different application areas, such as robot learning [4], aircraft recognition [17], medical object detection [22], and electric load and price prediction [15].

Learning methods for ELM can be classified into offline and online learning, based on data availability. In offline (or batch) learning, whole data are available beforehand and used in the training process in a batch manner. Because the entire dataset is used for training, this method requires more resources and provides less flexibility for re-training the neural network. Online learning, however, involves training the neural network with each new data chunk continuously in an incremental manner. Re-training is performed rapidly on the fly without using or storing the previous data. Online learning is a natural choice for hardware implementation because it requires fixed and reduced memory resources, it trains faster for each data chunk, and it supports data streams for applications where the entire dataset might not be initially available.

In this paper, a reconfigurable and scalable hardware accelerator for an online variant of ELM called mixed-norm regularized online ELM (MRO-ELM) [16] is proposed for classification. The proposed accelerator architecture designed from start to provide parallel processing as much as possible and implemented on an FPGA. The main contribution of this study are that compared to the known existing literature; it is the first hardware implementation of an online ELM variant that has the norm regularization property with a parallel architecture and compared to an existing parallel architecture, it requires significantly lower resources.

The remainder of this paper is organized as follows. Section 2 briefly describes the MRO-ELM algorithm and Sect. 3 discusses the related work in the literature. The proposed hardware architecture is presented in Sect. 4. The results are presented and evaluated in Sect. 5 in terms of the hardware efficiency and the paper is concluded in Sect. 6.

2   The MRO-ELM Algorithm

The base ELM algorithm depends on two foundations. First, an SLFN with L neurons in the hidden layer and with an infinitely differentiable non-linear activation function can learn N different observations, given that the input and hidden layer weights and biases are assigned randomly [8, 11]. Second, after the initial random assignment of the previous layers, the output weights for the output layer can be determined by computing the inverse of the hidden-layer output matrix. Thus, iterative algorithms, such as gradient descent, are avoided [11].

A standard SLFN with activation function $g(x)$ and L hidden neurons is simply modelled as:

$$H\beta = T \tag{1}$$

$\mathbf{H} \in \mathcal{R}^{N \times L}$ is the hidden-layer output matrix such that:

$$\mathbf{H} = \begin{bmatrix} G(\mathbf{w}_1 \cdot \mathbf{x}_1^T + b_1) \dots G(\mathbf{w}_L \cdot \mathbf{x}_1^T + b_L) \\ \cdot \quad\quad\quad\quad\quad \cdot \\ \cdot \quad\quad\quad\quad\quad \cdot \\ \cdot \quad\quad\quad\quad\quad \cdot \\ G(\mathbf{w}_1 \cdot \mathbf{x}_N^T + b_1) \dots G(\mathbf{w}_L \cdot \mathbf{x}_N^T + b_L) \end{bmatrix}_{N \times L} \tag{2}$$

where each column corresponds to the output of the ith hidden neuron. $\beta \in \mathcal{R}^{L \times m}$ is the output weight vector and $\mathbf{T} \in \mathcal{R}^{N \times m}$ is the target vector that contains the labels for the training data for m classes. $\beta \in \mathcal{R}^{L \times m}$ and $\mathbf{T} \in \mathcal{R}^{N \times m}$ are given as:

$$\beta = \begin{bmatrix} \beta_1 \\ \cdot \\ \cdot \\ \cdot \\ \beta_L \end{bmatrix}_{L \times m} \quad \mathbf{T} = \begin{bmatrix} t_1 \\ \cdot \\ \cdot \\ \cdot \\ t_N \end{bmatrix}_{N \times m} \tag{3}$$

For an infinitely differentiable activation function $g(x)$, the solution to the system in Eq. 1 can be found by finding the pseudo-inverse of the matrix H such that:

$$\widetilde{\beta} = \mathbf{H}^{\dagger}\mathbf{T} \tag{4}$$

The base ELM algorithm described above learns from data using batch learning. Online sequential extreme learning machine (OS-ELM) [14] is the first ELM based algorithm proposed for online learning. It has the ability to learn the training data one-by-one or chunk-by-chunk. Processed training data is discarded, and OS-ELM has no prior knowledge as to how many training samples will be input into the system.

The OS-ELM algorithm first computes the initial β_0 for the initial chunk of data of size N_0. Since $N_0 \leq N$, initial output weight β_0 is:

$$\beta_0 = (H_0^T H_0)^{-1} H_0^T T_0 \tag{5}$$

When another chunk of data is presented to the network, OS-ELM updates β by using the recursive equation:

$$\beta_{k+1} = \beta_k + P_{k+1} H_{k+1}^T (T_{k+1} - H_{k+1}\beta_k) \tag{6}$$

where $K_0 = H_0^T H_0$, $K_{k+1} = K_k + H_{k+1}^T H_{k+1}$, and $P_k = K_k^{-1}$. P_k is also recursively updated with the following equation:

$$P_{k+1} = P_k - P_k H_{k+1}^T (I + H_{k+1} P_k H_{k+1}^T)^{-1} H_{k+1} P_k \tag{7}$$

Recursive least squares (RLS) has a similar derivation: when rank$(\mathbf{H}_0) = L$ (the number of hidden neurons), the learning outcome of OS-ELM is same as the base ELM algorithm. Thus, for OS-ELM, it is required that $L \leq N_0$ in initialization. A detailed explanation for the derivation of Eqs. (6) and (7) can be found in [14].

Regularization with norms constraints the target optimization problem to produce better results, such as reduced generalization error or sparse output. ANNs are typically regularized to prevent overfitting. The MRO-ELM algorithm is a mixed-norm and frobenius norm regularized online ELM algorithm where the mixed-norm induces row-wise sparsity for the output weight $\boldsymbol{\beta}$, whereas the frobenius norm provides stability. Row-wise sparsity eliminates some rows of $\boldsymbol{\beta}$ completely which corresponds to a neuron in ANN. This, in turn, reduces the network complexity, while reducing the generalization error.

MRO-ELM is based on the alternating direction method of multipliers (ADMM) [2] framework. ADMM is used to solve an optimization problem by splitting the variables and solving each of them in an alternating fashion. The ADMM iterations for MRO-ELM are derived in [16] as:

$$\boldsymbol{\beta}^{k+1} = (\boldsymbol{H}^T\boldsymbol{H} + \rho\boldsymbol{I})^{-1}(\boldsymbol{H}^T\boldsymbol{T} + \rho\boldsymbol{z}^k - \boldsymbol{y}^k) \tag{8a}$$

$$\boldsymbol{z}^{k+1} = rowsoft(\boldsymbol{\beta}^{k+1}, \boldsymbol{y}^k, \lambda_1, \lambda_2, \rho) \tag{8b}$$

$$\boldsymbol{y}^{k+1} = \boldsymbol{y}^k + \rho(\boldsymbol{\beta}^{k+1} - \boldsymbol{z}^{k+1}) \tag{8c}$$

$\mathbf{z}$ and $\mathbf{y}$ are ADMM variables that are initially set to zero and λ_1, λ_2, and ρ are hyperparameters used for regularization. The row-wise threshold function (rowsoft) for $\mathbf{z}$ is defined as [16]:

$$z_i = \begin{cases} 0, & \|\boldsymbol{w}\|_2 \leq \lambda_1 \\ \frac{\boldsymbol{w}(\|\boldsymbol{w}\|_2 - \lambda_1)}{(\lambda_2 + \rho)\|\boldsymbol{w}\|_2}, & \|\boldsymbol{w}\|_2 > \lambda_1 \end{cases} \tag{9}$$

where $\boldsymbol{w} = \boldsymbol{y}_i + \rho\boldsymbol{\beta}$. The recursive update formulas for MRO-ELM are similarly given as [16]:

$$C_k = (\boldsymbol{H}_k^T\boldsymbol{H}_k + \rho\boldsymbol{I})^{-1} \tag{10}$$

$$C_k = C_{k-1} - C_{k-1}\boldsymbol{h}_k^T(\boldsymbol{I} + \boldsymbol{h}_k C_{k-1}\boldsymbol{h}_k^T)^{-1}\boldsymbol{h}_k C_{k-1} \tag{11}$$

$$D_{k-1} = (\boldsymbol{H}_{k-1}^T\boldsymbol{T}_{k-1}) \tag{12}$$

$$D_k = D_{k-1} + \boldsymbol{h}_k^T\boldsymbol{t}_k; \tag{13}$$

Combined with the ADMM iterations in Eqs. (8a, 8b, and 8c), the MRO-ELM algorithm is presented in Algorithm 1.

Algorithm 1. The MRO-ELM algorithm

Input:

      Set the parameters: λ_1, λ_2, ρ, and L.

      N_0: The number of samples initially available

      T_0: $N_0 \times m$ dimensional initial target matrix of ELM

      H_0: $N_0 \times L$ dimensional initial hidden layer matrix.

Output:

      β: $L \times m$ dimensional output weight vector

      Initialization Stage:

      Compute H_0 as in (2)

      Initialize β_0, z_0, and y_0 to zero.

      $C_0 = (H_0^T H_0 + \rho I)^{-1}$,

      $D_0 = H_0^T T$

      Sequential Learning Stage:

1: **for** $k = 1, 2, 3, \dots$ **do**

2:    $C_k = C_{k-1} - C_{k-1} h_k^T (I + h_k C_{k-1} h_k^T)^{-1} h_k C_{k-1}$

3:    $D_k = D_{k-1} + h_k^T t_k$

4:    $\beta_k = C_k (D_k + z_{k-1} - y_{k-1})$

5:    $z_k = rowsoft(\beta_k, y_{k-1}, \lambda_1, \lambda_2, \rho)$

6:    $y_k = y_{k-1} + \rho(\beta_k - z_k)$

7: **end for**

8: **return** β_k

---

3 Related Work

In the literature, hardware implementations for both offline and online versions of ELM have been proposed. The most comprehensive work on offline ELM in the literature is presented in [5]. The authors of this work proposed three similar architectures for the basic ELM, which were implemented on a Virtex-6 FPGA. The architectures were named ELMv1, ELMv2, and ELMv3. In ELMv1, a serial computation approach was used to reduce resource usage at the cost of increased training time. ELMv2 and ELMv3 utilize parallel processing to decrease processing time. The difference between these two architectures is that one of them (ELMv3) applies matrix decomposition to the rectangular hidden layer matrix, whereas the other (ELMv2) applies it to a square matrix obtained from the hidden layer matrix. All architectures utilize QR decomposition to find the pseudo-inverse of the hidden layer matrix. The design proposed in [3] is implemented on a Virtex-6 FPGA device. However, hardware implementation part of this design only realizes a classifier. Thus, the learning procedure was performed externally and produced values were fed into the hardware classifier.

Because of its suitability for hardware implementation, various studies have been presented in the literature targeting OS-ELM [6,18,20]. The works described in [6] and [20] utilize one-by-one processing of the incoming data to reduce the hardware by complexity because this approach eliminates the matrix inversion process from the OS-ELM algorithm. The work in [6] uses serial computation, which is the online version of the work in [5]. It was implemented on a Virtex-7 FPGA as an intellectual property (IP) core. The work in [20] combines autoencoder with OS-ELM for anomaly detection application on a Virtex-7 FPGA. Matrix and vector products were fully parallelized for this architecture. The work in [18] is a hardware accelerator for OS-ELM that utilizes a system-on-chip (SoC) device that integrates a hard processor with programmable logic (FPGA). Contrary to the other OS-ELM implementations discussed previously, the study presented in [18] implemented the matrix inversion process using matrix decomposition.

4 Hardware Implementation

4.1 Adapting MRO-ELM for Hardware Implementation

The architecture proposed in this work utilize one-by-one processing of the incoming data to achieve reduced hardware complexity. The recursive equation in (11) can be reformulated with Sherman-Morrison formula such that $\mathbf{h} \in \mathcal{R}^{n \times L}$ and $n = 1$.

$$C_k = C_{k-1} - \frac{C_{k-1} h_k^T h_k C_{k-1}}{1 + h_k C_{k-1} h_k^T} \qquad (14)$$

This reformulation eliminates the costly matrix inversion in Eq. (11) that is usually solved with matrix decomposition methods. For each new data sample, $\mathbf{h}$ is the corresponding hidden layer vector of length L and $\boldsymbol{t}$ is the label vector of length m. The matrices $\mathbf{C} \in \mathcal{R}^{L \times L}$, and $\mathbf{D} \in \mathcal{R}^{L \times m}$ are recursively updated at each iteration and initialized according to Algorithm 1.

It can be deduced from the Algorithm 1 and Eq. (14) that majority of the operations are matrix-vector multiplications, element-wise product of matrices, and matrix addition. Except for $C_{k-1} h_k^T h_k C_{k-1}$ operation in Eq. (14), large scale matrix multiplications are not required. Because of these observations, in the proposed hardware, multiplications are performed by using parallel vector multipliers that can perform different operations such as multiply, multiply-accumulate (MAC), and multiply-add (MAD). Computation steps in Algorithm 1 are re-ordered and adapted to parallel vector multipliers to exploit this advantage and to reduce the number of operations and memory occupation.

4.2 Hardware Architecture

The proposed architecture is designed to be scalable and configurable for different numbers of hidden neurons, classes, and feature sizes. Scalability is achieved by defining the metrics such as the number of hidden neurons L, the number of classes m, and the hyperparameters λ_1, λ_2, ρ as configurable parameters to the design. These parameters are configured before the synthesis and implementation phase for the desired ANN size and hyperparameter values. The Verilog hardware description language (HDL) is used to design the proposed architecture and these parameters are integrated into HDL such that the tools will automatically expand the circuit for varying number of hidden neurons and classes. The hardware adapted MRO-ELM algorithm mainly consists of a parallel arithmetic unit (PAU) that is formed by inferring parallel DSP48 slices in Xilinx FPGAs. DSP48 slices are configurable and dedicated hardware that are efficient at digital signal processing applications. They contain multipliers, adders, registers, and multiplexers, and can provide many digital signal processing functions. PAU in the proposed architecture is configured to provide MAC and MAD functions in addition to basic multiplication and addition, depending on the algorithm step. A 32-bit fixed-point representation is used throughout the architecture. Each column of a matrix is stored in a different block RAM (BRAM) instance to provide parallel data read for PAU and parallel data write to BRAM.

In the sequential training phase, the incoming data is used to compute the vector h, and the operations in Algorithm 1 are then performed. Note that matrices C_0 and D_0 are computed externally, and the initial values are stored in the memory before the sequential learning phase. The hyperparameters λ_1, λ_2, and ρ are provided as constant values to the hardware. The proposed architecture can also classify the input data sample using the current β value. Classification is performed by multiplying the input data with β using PAU. Classification mode is enabled if the external mode control signal is asserted accordingly.

5 Experimental Results

In this section, the proposed architecture is evaluated for the hardware performance. The experimental setup consists of a computer with Xilinx Vivado implementation software version 2016.4 and a Xilinx Virtex-7 FPGA device. The proposed architecture was implemented using the Virtex-7 FPGA (XC7VX1140T). The Verilog HDL was used to manually design the proposed architecture, which was then synthesized and implemented using Vivado. Since the architecture is scalable and configurable, it is implemented for 10 classes and varying number of hidden neurons using fixed point Q12.20 format representation for data. The implemented design was evaluated for hardware performance metrics, such as the resource utilization and maximum clock frequency.

The resource utilization is the total number of resources that are utilized during the implementation phase of the Vivado software. These resources are

Table 1. Hardware Performance of the Implemented Hardware

Hidden Neurons	LUT	FF	DSP	BRAM	Max. Freq. (MHz)
10	18988	8823	54	35	118
20	38336	16460	94	70	110
30	55150	23859	134	105	105
40	73745	31356	174	140	100
50	91895	38691	214	150	96
100	187729	76160	414	350	84
200	384435	151128	814	700	70
300	561499	225995	1214	1050	50

look-up tables (LUTs), flip-flops (FFs), block RAMs (BRAMs) and digital signal processors (DSPs). LUTs are small RAMs that can implement any combina-
tional logic function depending on the number of inputs available. FFs are used
to implement the sequential logic where they act as simple memory elements.
BRAMs are internal memory blocks that is integrated to the FPGA chip for
faster access but their sizes are limited. DSP blocks are incorporated to modern
FPGAs for efficient digital signal processing operations such as multiplication.

The maximum clock frequency is obtained from the implementation software
based on the constraint given by the user. It is defined as the reciprocal of the
minimum clock period that the digital circuit supports without any error in its
operation. Its unit is hertz (Hz). Since faster clock frequency provides faster
data processing, it is important to carefully constraint the circuit for the fastest
achievable clock frequency. This metric is heavily dependent on the structure
of the architecture, amount of utilized resources, placement of the resources on
the device and the FPGA technology used. The data throughput is also related
to the maximum clock frequency and it is computed by dividing the maximum
clock frequency to the total number of clock cycles in each sequential iteration.

The resource utilization and maximum clock frequency of the proposed circuit are obtained by separately implementing it for various number of hidden
neurons. The values reported by the Vivado software are recorded and presented
in Table 1. Because of the parallel nature of the architecture, the DSP and BRAM
usage increases with the number of hidden neurons, whereas the max. frequency
decreases owing to increased complexity. Most DSP and LUT utilization are
directly caused by the 32-bit fixed-point number format.

A direct comparison with the existing works is not straightforward due
to differences in various aspects of the designs such as; FPGA platform,
design methodology (sequential, parallel), data input/output (I/O) interface and
latency, prediction type (regression, classification). The performance of the proposed design is nevertheless compared with the applicable reported values in the
existing works. Compared to the sequential design presented in [6], the resource

utilization for the proposed design is quite high (Table 2). However, this increase in resource utilization is an expected trade-off since the proposed design is a parallel architecture. The design in [6] has the advantage of lower DSP utilization due to fixed number of multipliers allocated for sequential operation. In a parallel architecture, however, the number of DSP blocks proportionally increases with the number of hidden neurons. On the other hand, the data throughput of the proposed design is significantly greater than the compared work. Thus the parallel architecture favors throughput whereas the sequential one favors resource utilization. The resource utilization and max. frequency is also compared with the parallel design proposed in [20] that is implemented on a Virtex-7 FPGA for 32 neurons, 10 classes, 784 features with 32-bit number format. The proposed design was implemented for the same values and the comparison is shown in Table 3. Compared to the parallel OS-ELM architecture in [20], the proposed design utilize much less resources while achieving slightly better clock frequency. Although the training data throughput for the proposed design is lower than the compared work, it is quite acceptable if the resource cost is taken into account. Furthermore, the proposed design has the advantage of better prediction accuracy since the proposed design implements an online ELM algorithm with regularization ability. Even though the detailed hardware architecture is not described in [20], the difference in resource utilization can be caused by the method of implementation (high level synthesis (HLS) versus manual HDL). HLS tools allow the developer to describe their design using a high level programming language such as C or Pyhton. While these tools are feasible for some applications, they do not always produce logic circuits that give the same performance as pure HDL designs. The reason for difference in the number of DSP blocks might be the inefficient mapping of vector multiplications. In the proposed design, all multiplications are performed using L multipliers and the number of DSP blocks are directly related to L, the number of hidden layer neurons.

Table 2. Comparison of hardware performance with the serial architecture presented in [6] for 100 hidden neurons

Implementation	LUT	FF	DSP	BRAM	Max. Freq. (MHz)	Throughput (KHz)
[6] (OS-ELM)[a]	43356	41196	41	162	192.3	≅3.8
Proposed[a]	187729	76169	414	350	84	≅50

a: The number of classes for [6] and the proposed work are 7 and 10, respectively.

Table 3. Comparison of hardware performance with the parallel architecture presented in [20] for 32 hidden neurons

Implementation	LUT	FF	DSP	BRAM	Max. Freq. (MHz)	Throughput (KHz)
[20] (OS-ELM)	330881	182825	3347	816	100	≅315
Proposed	56480	24996	142	112	105	≅74.5

6   Conclusion

In this study, an FPGA hardware accelerator architecture for the online sequential learning algorithm MRO-ELM was proposed. This is the first hardware implementation of an online ELM variant that supports regularization. The architecture was designed to provide parallel processing for higher training throughput. The experimental results demonstrate that the proposed accelerator architecture has significantly better resource utilization than the other existing parallel architecture. In future work, the parallel multiplier topology could be enhanced to lower hardware complexity for lower resource utilization and faster speed. In addition, an improved architecture may be developed to provide flexibility to the user to balance the design between sequential and parallel processing with configuration.

References

1. Abiodun, O.I., et al.: Comprehensive review of artificial neural network applications to pattern recognition. IEEE Access **7**, 158820–158846 (2019). https://doi.org/10.1109/ACCESS.2019.2945545, https://ieeexplore.ieee.org/document/8859190/
2. Boyd, S., Parikh, N., Chu, E.: Distributed Optimization and Statistical Learning via the Alternating Direction Method of Multipliers. Now Publishers Inc. (2011)
3. Decherchi, S., Gastaldo, P., Leoncini, A., Zunino, R.: Efficient digital implementation of extreme learning machines for classification. IEEE Trans. Circuits Syst. II Express Briefs **59**(8), 496–500 (2012). https://doi.org/10.1109/TCSII.2012.2204112
4. Duan, J., Ou, Y., Hu, J., Wang, Z., Jin, S., Xu, C.: Fast and stable learning of dynamical systems based on extreme learning machine. IEEE Trans. Syst. Man Cybern.: Syst. **49**(6), 1175–1185 (2019). https://doi.org/10.1109/TSMC.2017.2705279
5. Frances-Villora, J.V., Rosado-Muñoz, A., Martínez-Villena, J.M., Bataller-Mompean, M., Guerrero, J.F., Wegrzyn, M.: Hardware implementation of real-time extreme learning machine in FPGA: analysis of precision, resource occupation and performance. Comput. Electr. Eng. **51**, 139–156 (2016). https://doi.org/10.1016/j.compeleceng.2016.02.007, http://www.sciencedirect.com/science/article/pii/S0045790616300222
6. Frances-Villora, J., Rosado-Muñoz, A., Bataller-Mompean, M., Barrios-Aviles, J., Guerrero-Martinez, J.: Moving learning machine towards fast real-time applications: a high-speed FPGA-based implementation of the OS-ELM training algorithm. Electronics **7**(11), 308 (2018). https://doi.org/10.3390/electronics7110308, http://www.mdpi.com/2079-9292/7/11/308
7. Gao, M., Ding, L., Jin, X.: ELM-based adaptive faster fixed-time control of robotic manipulator systems. IEEE Trans. Neural Netw. Learn. Syst. 1–13 (2021). https://doi.org/10.1109/TNNLS.2021.3116958

8. Huang, G.-B., Babri, H.: Upper bounds on the number of hidden neurons in feedforward networks with arbitrary bounded nonlinear activation functions. IEEE Trans. Neural Netw. **9**(1), 224–229 (1998). https://doi.org/10.1109/72.655045, http://ieeexplore.ieee.org/document/655045/

9. Huang, G.B., Zhou, H., Ding, X., Zhang, R.: Extreme learning machine for regression and multiclass classification. IEEE Trans. Syst. Man Cybern. Part B (Cybern.) **42**(2), 513–529 (2012). https://doi.org/10.1109/TSMCB.2011.2168604

10. Huang, G., Huang, G.B., Song, S., You, K.: Trends in extreme learning machines: a review. Neural Netw. **61**, 32–48 (2015). https://doi.org/10.1016/j.neunet.2014.10.001, https://linkinghub.elsevier.com/retrieve/pii/S0893608014002214

11. Huang, G.B., Zhu, Q.Y., Siew, C.K.: Extreme learning machine: theory and applications. Neurocomputing **70**(1), 489–501 (2006). https://doi.org/10.1016/j.neucom.2005.12.126, http://www.sciencedirect.com/science/article/pii/S0925231206000385

12. Khan, M.A., et al.: Cucumber leaf diseases recognition using multi level deep entropy-ELM feature selection. Appl. Sci. **12**(2), 593 (2022). https://doi.org/10.3390/app12020593, https://www.mdpi.com/2076-3417/12/2/593

13. Markowska-Kaczmar, U., Kosturek, M.: Extreme learning machine versus classical feedforward network. Neural Comput. Appl. **33**(22), 15121–15144 (2021). https://doi.org/10.1007/s00521-021-06402-y

14. Liang, N.-Y., Huang, G.-B., Saratchandran, P., Sundararajan, N.: A fast and accurate online sequential learning algorithm for feedforward networks. IEEE Trans. Neural Netw. **17**(6), 1411–1423 (2006). https://doi.org/10.1109/TNN.2006.880583, http://ieeexplore.ieee.org/document/4012031/

15. Naz, A., Javed, M.U., Javaid, N., Saba, T., Alhussein, M., Aurangzeb, K.: Short-term electric load and price forecasting using enhanced extreme learning machine optimization in smart grids. Energies **12**(5), 866 (2019). https://doi.org/10.3390/en12050866, https://www.mdpi.com/1996-1073/12/5/866

16. Polat, O., Kayhan, S.K.: GPU-accelerated and mixed norm regularized online extreme learning machine. Concurr. Comput.: Pract. Exp. **34**(15), e6967 (2022). https://doi.org/10.1002/cpe.6967, https://onlinelibrary.wiley.com/doi/abs/10.1002/cpe.6967

17. Rong, H.J., Jia, Y.X., Zhao, G.S.: Aircraft recognition using modular extreme learning machine. Neurocomputing **128**, 166–174 (2014). https://doi.org/10.1016/j.neucom.2012.12.064, https://www.sciencedirect.com/science/article/pii/S0925231213010023

18. Safaei, A., Wu, Q.M.J., Akilan, T., Yang, Y.: System-on-a-chip (SoC)-based hardware acceleration for an online sequential extreme learning machine (OS-ELM). IEEE Trans. Comput.-Aided Design Integr. Circuits Syst. **38**(11), 2127–2138 (2019). https://doi.org/10.1109/TCAD.2018.2878162, https://ieeexplore.ieee.org/document/8509179/

19. Suganthan, P.N., Katuwal, R.: On the origins of randomization-based feedforward neural networks. Appl. Soft Comput. **105**, 107239 (2021).https://doi.org/10.1016/j.asoc.2021.107239, https://www.sciencedirect.com/science/article/pii/S1568494621001629

20. Tsukada, M., Kondo, M., Matsutani, H.: OS-ELM-FPGA: an FPGA-based online sequential unsupervised anomaly detector. In: Mencagli, G., et al. (eds.) Euro-Par 2018. LNCS, vol. 11339, pp. 518–529. Springer, Cham (2019). https://doi.org/10.1007/978-3-030-10549-5_41

21. Zhao, X., et al.: Chaos enhanced grey wolf optimization wrapped ELM for diagnosis of paraquat-poisoned patients. Comput. Biol. Chem. **78**, 481–490 (2019). https://doi.org/10.1016/j.compbiolchem.2018.11.017, https://www.sciencedirect.com/science/article/pii/S1476927118307965
22. Zhu, W., Huang, W., Lin, Z., Yang, Y., Huang, S., Zhou, J.: Data and feature mixed ensemble based extreme learning machine for medical object detection and segmentation. Multimed. Tools Appl. **75**(5), 2815–2837 (2016). https://doi.org/10.1007/s11042-015-2582-9

A Literature Survey on Event Detection for Indoor Environment Using Wireless Sensor Network

Lial Raja AlZabin[1]([⊠]), M. Jannathl Firdouse[2], and Baidaa Hamza Khudayer[3]

[1] College of Computer Science and Informatics, Amman Arab University, Amman, Jordan
l.alzabin@aau.edu.jo
[2] Al Zahra College for Women, Airport Heights, Muscat, Oman
[3] Al Buraimi University College, Al Buraimi, Oman

Abstract. Environmental incidents like fires, gas leaks and explosions that happen at random and unforeseen times might be greatly helped by event detection via a wireless sensor network. Determining the occurrence of a specific event is hence one of a system's most important tasks. The system must be able to gather and infer environmental data that will enable isolating and recognizing specific occurrences in order to do this. Several alternative event detection methods, such as those that can be handled by separate sensors, remote groups, or fusion centers, are employed in wireless sensor networks depending on how the environmental data is acquired. This paper provides an overview of event detection mechanisms and challenges, as well as the hypothesis, and covers the fundamental requirements for sensing systems. Furthermore, relevant research work on environmental event detection in current event detection approaches is reviewed and discussed. Thus, the purpose of this paper is to conduct a relative analysis of the event detection mechanism using fusion center-based fuzzy logic for improving the detection system's performance.

Keywords: Wireless Sensor Networks · event detection · data collection · challenges

1 Introduction

Wireless Sensor Networks (WSNs) are broadly used for many purposes, including ecological examination and event recognition scenarios. These nodes are randomly dispersed throughout a broad geographic area. These sensor nodes combine sensing, data communication, and processing components. These devices assist WSNs in communicating with other nodes, physically monitoring them, and exchanging the sensory data they collect. They also control regional processing and decide on observations [1]. As a result, the 'event detection' characteristic is used to describe both the outcome of event detection and anomalous data behaviors in a particular monitored environment. Many applications that use this characteristic of event detection in networks have focused a lot of attention on event detection. These applications include the identification and tracking of military targets, surveillance, disaster assistance, and medical care [2]. As a result, it is important to note that an event detection method must satisfy several conditions. The requirements include being on time, having a high rate of genuine detection, and having a low incidence of false alarms [3]. The background which includes event detection challenges and hypotheses is given in Sect. 2, related work that includes monitoring and reporting event strategies, and event detection approaches are explained in Sect. 3, followed by a discussion in Sect. 4 of this paper.

2 Background

The next subsections provide and discuss some background data about the challenges and hypotheses of an event detection system since this paper concentrates on event-based detection systems using Wireless Sensor Networks.

2.1 Event Detection Challenges

Limited sources, for example, battery life, processing power, and memory define sensor nodes. Therefore, coming up with a new, accurate event detection approach is a difficult endeavor given these constraints and difficulties. According to [3] based on [4], situational reliance, the criticality of an application, the presence of multiple and diverse data sources, and network topology are the most typical problems in an event detection scenario, as shown in Fig. 1.

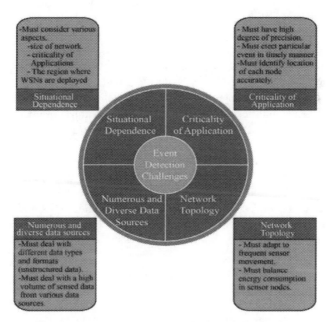

Fig. 1. WSNs Event Detection Challenges, Source: [4]

2.2 Event Detection Hypothesis

Four alternative optimization strategies can be used in a hypothesis-testing setting: distributed detection, Neyman-Pearson, Bayesian formulations, and fuzzy sets. These strategies are discussed in the section below.

- **Distributed Detection**

    In a traditional detection system, event detection and the related approximations such as position, velocity, etc. are basically done in a centralized way where many of the sensors are distributed in a particular region. The available environmental data from the sensing devices are communicated to a central processing center, which performs the estimation and best detection. Decentralized detection is allowed in situations where the application detects several events and those are irrelevant for the event detection. Different methods for data fusion and distributed detection can be used [5, 6] based on the network's characteristics, such as topology, type of sensors, etc. Decisions and local judgments can be made from each sensor using their own data and input from other sensors and submitted to the fusion center for final approval. Reduction of transmission bandwidth, lowering the cost, and increasing the lifespan of sensors is possible with distributed detection. There is performance loss even though the amount of information at the fusion center is minimal when compared to the centralized system.

- **Neyman-Pearson Formulation**

    Every sensor's binary hypothesis testing issue will be handled with the existing environmental information, which is useful to decide whether the event occurs or not. In this scenario, hypothesis H1 is related to the target presence and H0 is to the target absence. From each sensor, the data is received by the fusion center. If H1 is decided, then $x\_i = 1$ or $x\_i = 0$ if H0 is detected. At the fusion center, the related optimal fusion rules are applied to the existing information, which yields a global decision by keeping $x\_0 = 1$ if H1 or $x\_0 = 0$ if H0. Usually, the decision rules are based on finding the thresholds that satisfy the given situations, which is time-consuming and often a demanding task [7]. Determining the maximized global probability of detection using local and global optimal decision rules is done corresponding to the NP formulation.

- **Bayesian Formulation**

    Developing optimal fusion rules that reduce the Bayesian risk is considered the main goal of the Bayesian framework. Since they are in NP formulation, the binary test hypotheses are solved, and local decisions are taken in this situation. The event characteristics, for example, event power and network characteristics such as sensor position must be assumed or known prior in some of the systems. Other system characteristics such as probability density functions and the probability of sensor detection must be known or unknown too. Finding the optimal decision rule is a difficult task in distributed detection systems, which ensures the system's viability, and this can be described in [8–12]. Though this distributed detection finds the optimum answer in most situations, this has developed a famous studies topic in latest years.

- **Fuzzy Sets**

    Logic and fuzzy sets were first introduced in [13]. Since then, their qualities have been used in other fields. WSNs have been equipped with fuzzy logic to improve decision-making, conserve resources, and improve performance. Among the areas it has been applied to are cluster-head election [14, 15], security [16], data aggregation [17], routing [18], MAC protocols [19], and QoS [20]. Yet, there hasn't been much discussion regarding the application of fuzzy logic to event description and detection. [21] suggests using fuzzy logic with double-sliding window detection to improve the accuracy of event detection. Nevertheless, they don't look at how fuzzy logic functions on its own or how the geographical or temporal properties of the data affect categorization accuracy. Fuzzy logic is used in D-FLER [22] to combine local and personal data to predict the likelihood of an event. They come to the conclusion that fuzzy logic improves event detection precision. D-FLER can distinguish between actual fire data and nuisance testing by employing fuzzy values. On the other hand, temporal semantics are not used by D-method FLERs. Moreover, because all of the experiments only last for 60 s after the fire starts, the authors do not examine the number of false alarms generated by D-FLER. The organizational framework of the Fuzzy Logic System (FLS) is depicted in Fig. 2. The matching membership functions are used by the fuzzifier to transform the crisp input variables $x \in X$, where X is the set of potential input variables, into fuzzy language variables. "Variables whose values are words or sentences in a natural or artificial language, rather than integers," according to [23], are linguistic variables. Based on the membership degree, an input variable can be connected to one or more fuzzy sets.

Using if-then expressions, the fuzzified data are processed in line with a specified set of rules derived from the domain knowledge provided by experts. The inference technique now transforms the input fuzzy sets into the output fuzzy sets. The DE-fuzzifier then takes the fuzzy sets of rules and produces a precise result. The required control actions are represented by the crisp output value. The terms such as decision-making, defuzzification, and fuzzification are employed to characterize the three steps mentioned above.

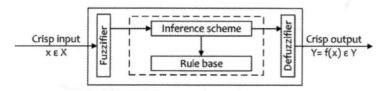

Fig. 2. The organizational framework of a Fuzzy Logic System

3 Related Work

Events typically occur rarely and produce themselves irregularly, leaving a large amount of tangible proof of their presence once they are over. They might happen briefly (transiently), in which case the detection probability should be increased, or they might happen repeatedly (persistently), in which case the detection latency should be reduced. Depending on the application domain, there may be a very minimal probability that an event will occur. Ephemeral and transient events are necessary for successful detection [24–26]. As a result, the objective of event detection is to identify events accurately by minimizing false alarms. Along with the detection method utilized to find the events depending on Wireless Sensor Networks, as shown in Fig. 3, this section also reviews associated academic works on event detection, including suitable monitoring and reporting procedures. Successful event identification in WSN monitoring applications depends on three crucial indicators, including:

Detection probability - describes the likelihood of finding an incident around the monitoring zone.

False alarm probability - describes the likelihood of a fake alarm occurring within the monitored area.

Detection delay - describes how long it takes for a message containing sensed data to transport from the sensing node to the sink node throughout the network. For event identification, several statistical and probabilistic methods have been developed [3, 27].

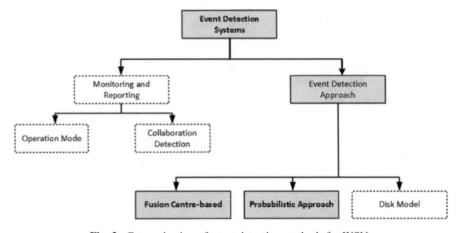

Fig. 3. Categorization of event detection methods for WSNs

3.1 Monitoring and Reporting Event Strategies

This section provides standard monitoring and reporting techniques as well as recent research. They are examined in terms of the metrics, and a determination of their purpose and impact is also made. Cooperative detection, periodic sleep scheduling, transceiver energy deactivation, node density, data filtering, and aggregation technique are some of these detection strategies.

- **Collaborative Detection**

    As described in [28], there is improved efficacy of energy and elevated detection accuracy when related to the centralized method. The expense of disseminating the details of the event detection from the sensors to reach the sink is considered substantial. The relationship that exists with neighboring nodes to decide about their existence with an adequate number leads to a change in the propagation of event detection reports as in [29]. The concept of using CH to forward the sensed data is applied in [30] to put forward the Probabilistic Event Monitoring Scheme (PEMS). In a multidimensional feature space, the support vector machine, K-Nearest Neighbour, and Gaussian Mixture Model were applicable. The proposed system uses the testbed system which is aiming to detect the leakages in the pipeline. An alarm application based on [31] was suggested in [32] to help firefighting operations.

In tough, hostile conditions, a system of temperature, light, and humidity sensors is used. The study team considered a dispersed WSN, which is made up of numerous isolated WSNs. Findings showed that fire breakout, where a quick response to such destructive events is primarily required, can benefit from more longevity (avoiding synchronization costs during the idle period). A mechanism for collaborative event detection, wherein sensing nodes collaborate to correctly conclude the incidence of an occasion as opposed to counting on a centralized BS for coordination or processing, was proposed [33] to recognize diverse instructions of software-precise occasions. It was discovered during the device evaluation using the WSN of 100 nodes installed at the fence of an actual worldwide production site that direct processing of the raw data at the nodes propagating only identified activities significantly reduces communication overhead. Data Service Middleware (DS Ware), which was earlier presented by [34], may swiftly offer event detection services by working together to correlate multiple sensor observations in accordance with the real-time characteristics of events. This method can tell the difference between events that happened and false alarms. As a result, collaborative detection can lead to a more complete and precise understanding of when and where events have occurred. It does not contribute to the detection of the outcome as a false positive. However, because of the delay in the delivery of event reports caused by exchanging collaboration information, the sink is not promptly notified of the event, affecting applications' reliability and effectiveness.

- **Periodic Sleep Scheduling**

For event-driven WSNs, periodic sleep scheduling works well because event detection is guaranteed and the network operation lasts long enough. Intermittent rest booking is exceptionally favored regarding energy streamlining and diminishing general energy utilization [35]. Sensor nodes are scheduled to turn off their power to extend the operational life of a WSN. In some event detection scenarios, the always-active approach described here does not satisfy the need for longevity. The PDC-SMAC protocol, which uses social sensing-based duty cycle management, was created in this context to decrease ineffective sensing i.e. when the node is effective but there is naught to detect [36]. It extends the lifetime of the network by minimizing ineffective sensing. According to the findings of the performance evaluation, PDC-SMAC reduced report delivery latency by 47% and 27%, respectively, for various rates of event occurrence and different intervals of packet generation. An algorithm for cooperatively determining node wakeups was proposed by [37] to identify the crucial trade-off between detection performance and system lifetime. When compared to random sleep scheduling, their extensive evaluation experiments of the proposed algorithm showed a reduction in detection latency of 31% and an increase in detection probability of 25%. They, therefore, suggest a stage-by-stage offset scheduling approach, like [38], which keeps the detection latency to 3D + 2L wherein D is the most number of hops to attain a relevant node, and L is the duration of the scheduled sleeping, in which the unit is the time slot size. An alarm is delivered to a central node after an event is detected, notifying all other nodes. It uses the FWI index. [39] uses a brand-new k-coverage method to find forest fires. With k or more sensor nodes monitoring each point, the K-coverage algorithm increases fault tolerance. To increase the network's lifespan, certain sensors might be placed in standby mode. It turned out that the k-coverage increased the network's lifespan. Finally,[40] proposed a DSR and DSP scheme for dynamic sleep scheduling with the goals of balancing energy consumption and lowering sleep latency. As a result of the computation of a potential increase in the residual energy based on the available harvesting opportunity, DSP enables each node to aggressively shorten its activation duration. Each DSR node changes its activation mode depending on how much energy is still available to it. When people in the monitoring area need to be alerted by the network, such as when a fire unexpectedly breaks out in a building or when mineworkers are exposed to a gas leak, the latency of the event detection and communication is crucial. This is because intermediary nodes with an unfavorable sleep cycle can have a considerable negative impact on the dissemination of information across a large multi-hop network. This strategy lowers the likelihood of detection. However, by increasing the number of nodes that detect the event, the detrimental effect on the detection probability can be reduced. Yet, because synchronization requires sending control messages, doing so increases network overhead.

- **Transceiver Power Deactivation**

Some other alternatives for full active/inactive node scheduling is to show off the wireless transceiver, which consumes the most energy. Media collisions and future transmission attempts, as well as idle listening times during which transceivers are powered on but no packets are received, all need the usage of energy, which is why they waste a lot of energy [41]. A generic Wake-Up Radio-based Medium Access Control (GWR-MAC) protocol for short-range communication WSNs evolved in order to decrease idle listening and improve power performance [42]. Deactivating the transmitter with the majority of WSNs, including implantable body sensor networks, can be quite beneficial in this situation (IBSN). Both source- and sink-initiated wake-up techniques are supported by GWR-MAC. GWR-MAC can be effective for detecting events that call for a sufficiently low detection delay, as shown by the analytical comparison with conventional periodic sleep scheduling MAC approaches for the purpose of enabling the right radio selection taking the application's situations into consideration. In [43] (where collision-free transmission is guaranteed) and [44], where communication and data transfer are carried out using two radios such as control radio and data radio respectively [45–47], transceivers are turned off when nodes are not required to send or receive reports. It is important to note that the energy needed to switch on the transceiver frequently equals the energy needed to transfer one data packet, which increases the amount of energy used overall [48]. The key sensor components are not turned off; hence the detection probability is unaffected. Deactivating the transceiver causes a necessary delay in the transmission of the reports to the targeted sink node, though. Additionally, turning on the transceiver adds to the network's overall energy usage.

- **Node Density**

  The periodic sleep scheduling approach, which was covered earlier, lengthens the life of WSNs. The possibility of detection is impacted by the reduced sensing coverage, though. A sufficient node density assists in preserving the sensing coverage in the monitoring region by enabling accurate synchronization of sleep scheduling among the network's nodes. By inserting redundant nodes inside the range of nearby sensing node pairs, it is possible to maintain excellent coverage while creating a fault-tolerant environment. [49] proposed the ESCARGO method, which provides fast event report transmission while continually maintaining network connectivity, for sleep coverage of unusual geographic occurrences. It has been demonstrated that ESCARGO can extend the life of a network without sacrificing detection likelihood or detection delay by adding redundant nodes to a WSN with low node density and a non-random deployment approach. [50] employed redundant nodes of their SENSLIDE system, which is recommended for anticipating a landslide, in order to recognize appropriate forested area fires. A dispensed sensor system with data collection and occasion detection is known as SENSLIDE. The system was created to effectively handle the challenges of a dispersed WSN environment with weak connectivity and scarce power. Energy-aware routing protocols were used to stop nodes' energy from running out too quickly. Due to the usage of redundant nodes, SENSLIDE was able to reach an acceptable level of fault tolerance. If redundant nodes are installed in the monitoring area, periodic sleep scheduling methods can maintain coverage. In this regard, [51] concluded that the dynamic topology change brought on by the sleep schedule has a detrimental impact on the operation and performance of the network and can be mitigated by increasing node deployment redundancy. To highlight the benefits of redundant nodes, the researchers examined two different sleep modes (random sleep and coordinated sleep) in the context of providing network coverage to give high detection probability. The results showed that coordinated sleep scheduling is substantially better than random sleep scheduling with the same amount of node redundancy while having an additional control overhead. The subject of an earlier study [46, 49–51] was node density. Nevertheless, node density can be increased to increase event detection probability because accurate on-time detection of transitory events necessitates a perfect detection probability. However, it complicates the network even more, which could cause the event notifications to spread more slowly.

- **Data Filtering and Aggregation**

  Several similar notification reports are issued by surrounding nodes when an event is identified by several of them. These reports are ultimately seen as copies, and it may be enough to advance some of them to confirm the event of the occasion [52, 53]. Swift Opportunistic Forwarding of Infrequent Events (SOFIE) was proposed in this context to guarantee the detection of events and the swift propagation of notification reports. It gives amazing region inclusion considering the mathematical properties that are normal to WSN. It was determined that in ideally and randomly distributed networks with different node densities, SOFIE provides event notification reports more quickly than traditional geographic forwarding techniques. Additionally, it was found that SOFIE kept a WSN's sensing coverage by periodically scheduling sleep, allowing primarily detecting nodes to sleep while other active nodes spread event notification reports [54]. A unique approach was proposed to achieve a speedy response to real-time event detection in a WSN. Only one representative node is required in the proposed method, which is based on clustering, to provide its data. As a result, data transfer requires fewer nodes, which causes event notification reports to propagate more quickly. To demonstrate its suitability for environmental uses like the detection of forest fires, the proposed method was subjected to an experimental evaluation. [55], which used the National Fire Danger Rating System (NFDRS) in their investigation, supports this.

For the rapid transmission of event notification reports, [56] also provided a distributed, geographical Correlation-based Collaborative Medium Access Control (CC-MAC) protocol. This protocol combines MACs relating to events and networks, often known as E-MAC and N-MAC, respectively. It is not necessary to broadcast all of the event notification reports from all of the nearby nodes that identified the occasion due to the use of the spatial correlation between nodes that identify an occasion. As a result, only one node is chosen to represent each of the nodes in the immediate area as well as those that are viewing the associated sensed data. You can ensure the set-off transmission of notification reports by propagating the created reviews from the consultant node to the appropriate sink node. As a result, considering that this approach deals with the transmission and spread of signals at some point in the network, it has no impact on the detection likelihood. The occasion notification reviews can be forwarded to the network sink or base station more excellent unexpectedly if the reproduction reviews are eliminated or silenced. This lessens the strain of useless network traffic, reducing the detection postponement (Table 1).

Table 1. Research studies of Event Detection Operation Mode based on WSNs.

Authors	Approach	Mechanism	Findings	Limitations
Misra et al., 2015	Periodic Sleep Scheduling	PDC-SMAC protocol being proposed Duty cycle management for unusual occasions using social sensing Reduce the instances where a node is active, but nothing can be detected	Reduced report delivery latency for intervals by 27 to 60% and for rates of event occurrence by 47%	Postpone in occasion detection
Karuppiah Ramachandran et al., 2016	Transceiver Power Deactivation	Reducing idle listening	Enhancing energy efficiency	Postpone transmitting reviews

(continued)

Table 1. (*continued*)

Authors	Approach	Mechanism	Findings	Limitations
Harrison et al., 2015	Node Density	ESCARGO for sleep coverage of uncommon geographical events Event reports are delivered on schedule while ensuring constant network connectivity	Broaden the network's lifespan without causing a delay in detection	More network complexity
Yang et al., 2012	Data Filtering and Aggregation	Real-time event detection in the WSN can be handled with quick response. Primarily based on clustering, simply one representative node desires to document its data	The reliability of event reports and energy efficiency are guaranteed	High rate of false alarms

3.2   Approaches for Event Detection

The common approach and current research are covered in the following subsection. Also, the remark in their result and purpose is given. These detection approaches consist of an event detection-based disk model, a probabilistic model, and a decision fusion model.

- **Event Detection-based Disk Model**

    In WSNs, event detection is a critical application. The vital topic of event detection has experienced a rise in interest among academics and scientists in recent years. The sensors' disk-shaped design was initially used extensively. Due to the sensor's detecting field's disk-like form, this technology is known as the "disk model". If the distance is less than the radius of the disk, the sensor can locate the target. [57] investigated network detection capabilities and search strategies. He talked about numerous target types and sensor models, such as mobile and stationary sensors, targets placed randomly or optimally, independent, or global coordination searches, as well as stealthy or obvious sensors. The Boolean sensing model with a fixed sensing radius was used in all the detection techniques. [58] also suggested a mathematical method for assessing WSN coverage. One might determine the number of active nodes required to provide the desired coverage by dividing the sensing range of a sensor node by the size of the overall deployment area. The disk model is really basic. An event will either be detected if it occurs within the sensor's sensing range or not. The disk model, however, ignores how the state of the surroundings and the strength of the signal being transmitted affect the function of the senses. The disk concept is overly utopian for the reasons listed previously. As a result, the signal detected by sensors continuously weakens with distance using the signal attenuation model, which also takes noise into account. [59] As a result, sensor signal readings rather than a straightforward binary value can be effectively depicted.

- **Probabilistic Model**

   Probability Detection is the main concept used in the event detection paradigm. The events are detected by the sensors with the calculation of the energy of signals generated by the occurrence. As they move farther from their source, most physical signals, like electromagnetic and acoustic signals, lose energy [2]. [60] investigate the possibility of detecting mobile targets in a scenario where several sensors are randomly positioned to observe a certain area of interest. The researchers discovered a connection between the detection issue and a line-set intersection issue. They concluded that the sensors' perimeters rather than their shapes determine the detection probability. As a result, the researchers put a new concept of an altered disk into practice. Based on this concept, the detecting range of the sensors can be adjusted in accordance with the perimeter's restrictions. Applications for energy-efficient event detection that consider physical factors are studied in [61]. These researchers initially developed an analytical framework to evaluate the metrics such as target detection missing probability and notification transmission latency. Their research provided guidance on how to set the system's parameters as efficiently as possible while still operating within realistic performance constraints. The sensors' sensing range can be changed to accommodate various environmental conditions. Researchers who employed a more realistic model based on sensor properties other than the disk model further examined how WSNs recognize events. [62] examined the coverage of target detection applications in WSNs. It is expected that sensors are disking in the sensing zones. However, using restrictions on false alarm probability and missing probability detection, the researchers were able to identify the sensing zone. As a result, a completely unique technique for cooperative detection was proposed. The "OR rule" turned into the final selection-fusion method selected by the researchers. In different phrases, the sink determines that the target is present if even one of the character sensors domestically identified it. Taking components in a similar context, cellular targets' detection with continuous movement becomes investigated [63]. These objectives had been divided into two groups: 1) the rational targets, which are aware of the competencies of the current sensors, and 2) the blind targets, which can only be in direct strains. The effectiveness of detection was studied by the researchers, who expected to identify only a few key locations for the deployment of more sensors and the possibility of some degree of freedom for the targets. They suggested a straightforward detection rule. A sensor movement scheduling technique was developed to attain almost ideal system detection performance within a specified detection latency bound.

- **Decision Fusion Model-Based Fuzzy Logic**

    A common technique for raising the performance of the detecting system is Data Fusion (DF) [64]. A group of strategies known as DF combines data from various sources, including sensors and CH, to draw more precise conclusions. There are now two different types of fusion centers: cluster-based fusion centers and non-cluster-based fusion centers.

i. *Cluster-based Fusion Centre*

It is advised [65] to employ a hierarchical decision fusion technique to increase the likelihood of network-wide detection. Each sensor first makes a local choice using a model of signal attenuation. A set of the nearest neighbour sensors is then used to create a cluster-level sensor fusion to produce new decision results. To reach the ultimate network-level decision, the sensors that provide a successful fusion outcome report their findings to the network fusion center. This fusion strategy is founded on several great decisions. These strategies have run into problems. The fact that there are consistently fewer or equal numbers of successful local decisions made by sensors than those upon which the final decision is based is one problem. So fewer wise local decisions would result in a better decisions all around. Nodes send local decision results to a fusion center for final decision-making under the "decision fusion" method of fusion. The "value fusion" fusion method, in which nodes report the fusion center of their raw energy measurements, must be kept in mind. Following that, a decision is made based on the measurements collected from various nodes. However, [66], quick sensor placement methods based on a probabilistic data fusion model were presented. Algorithms are used to position sensors in the best possible places to maximize detection performance. According to this method, sensors provide energy measurements to the CH, which compares the averages of all the data with a threshold. If a larger average than the threshold is attained, the CH determines if a target is present, or the CH determines that no target is there. To ensure the accuracy of sensor deployment, [66] investigated the issue of information coverage barriers. For intrusion detection, nearby sensors worked together and fused their information. They were designed with the intention of lowering the quantity of active sensors needed to cover a barrier. Considering this, a suitable technique to determine the coverage set of just a few active sensors was suggested. A virtual sensor incorporates the sensor readings to enable value-fusion-based decision-making. The process for choosing the threshold value is one of the issues that the value fusion approaches have come across. Most researchers who used these techniques haven't talked about potential issues. The threshold was instead left as an experienced 62 value.

ii. *Non-Cluster-based Fusion Centre*

Using a cluster-based network structure improved network performance and decreased communication costs in earlier fusion works. For efficient event identification, [67] proposes a collaborative fuzzy logic. Fuzzy values can increase the precision of detection. The improvement of the collaborative event detection method using clustered WSNs [68] culminates with some findings before being proposed in [69].

[22] explores the detection of house fires using temperature and smoke sensors and presents D-FLER, a distributed fuzzy inference engine for event detection. Using a distributed fuzzy logic engine, the proposed D-FLER combines local observation with data from individual sensors. In order to detect fire spreads as early as possible in open areas like towns and forests, [70] developed a sensor network technique that combines sensory data from a temperature sensor with a maximum likelihood algorithm. The system's architecture is made up of three subsystems: sensing, processing, and localized alerting. The findings suggested that using this technology in a situation requiring early fire detection would be successful. Some research on WSN-based event detection methods is described in Table 2 below.

Table 2. Some Research on WSN-based Event Detection Approaches

Authors	Approach	Mechanism	Findings	Limitations
(Wang et al., 2016) [59]	Disk Model	Application coverage for target detection in WSNs was investigated The sensing regions are thought to resemble disks	Increased accuracy by transferring all sensor data to the sink that will make the ultimate decision	False alarm possibility and missing probability
(Tian Wang et al., 2016) [2]	Probabilistic Model	A probabilistic decision model was combined with a realistic signal model to attain a final decision	High detection probability and few false alarms	The use of different values and units produced by a model is not permitted
(Liu et al., 2005) [1]	Disk Model	A mathematical model where the overall range of the deployment area is proportional to the range of each sensor node. It determines the number of active nodes need to achieve desired coverage	No need to know the location of sensors and improved energy consumption	Low-accurate detection with a high rate of false alarms

(*continued*)

Table 2. (*continued*)

Authors	Approach	Mechanism	Findings	Limitations
(Lazos et al., 2007) [60]	Probabilistic Model	Possibility of finding mobile targets A few sensors are stochastically placed to keep an area of interest under surveillance	The length of the sensor perimeters is used to determine the likelihood of detection, which is independent of the shapes of the sensors	Determining a few key locations for the placement of extra sensors only have a limited degree of flexibility
(Medagliani et al., 2012) [61]	Probabilistic Model	Analyze the parameters, including the likelihood of missing a target Delivery of notifications too slowly	The best method for setting system parameters within the boundaries of realistic performance	No realistic performance

4 Discussion

While taking into consideration the use of WSNs in early detection, to increase event detection accuracy and reduce the detection of false alarms, a number of event detection systems have been discussed and proposed. The most popular detection techniques and strategies were examined in this research together with their benefits and drawbacks. To improve detection accuracy and lower false alarms, various event detection methodologies can be applied. Each of these strategies is examined to suggest a mechanism that accounts for the deficiencies in earlier models. The followings are a few research gaps and promising solutions that can be addressed based on the literature:

4.1 Network Structure

Cluster-based networks and non-cluster-based networks are the two different types of network architectures used in event detection, according to linked research. Indeed, cluster network topology has several benefits over non-cluster networks. First, a cluster-based network offers scalability for numerous sensors and a large monitoring area where events can be detected in subareas effectively. Sensor nodes that are members of the same cluster will cooperate in detection rather than enlist the assistance of all the other sensors in the network. According to the literature, sensor nodes that are far from an event might not be able to detect it, and using them in detection might have an impact on how accurately an event is detected. Additionally, CH gathers data from its member sensors, decides, and sends this information to the fusion center. So, communication expenses could be decreased. Since there are several types of sensors being used to identify the current event, applicable approaches that aggregate multiple decisions to improve detection quality have limitations. The number of members will rise because of these sending their CH aggregate data. More memberships will make things more complicated, reduce detection precision, and raise the number of false alarms.

4.2 Sensing Data and Representation

There are often obstacles to using actual value: first, the cluster head and fusion center should manipulate more than a few sensor types (heterogeneous sensors). The design ought to be more complex if it must cope with a couple of units and values. The second difficulty is related to the sensor's set threshold. For example, if the temperature sensor's alarm threshold is fifty-five °C but the fire temperature is really 45 °C, the sensor might not sound an alarm till the temperature hits fifty-five °C, delaying detection and possibly leading to a false alarm. A probabilistic method, alternatively, has been efficiently used to lessen the reliance on a fixed threshold value for the representation of sensing data. Unluckily, the probabilistic method has not been used for diverse sensor types which can be represented with diverse units and value ranges.

4.3 Event Detection Using Fusion Centre-Based Fuzzy Logic

Most current event detection methods define event threshold values using crisp values. However, crisp numbers are unable to handle the various degrees of inaccurate sensor readings. In several literature studies, fuzzy logic was suggested for usage in fusion centers because of its advantages over other strategies like Bayesian theory. The brief list of these advantages is as follows: (i) Fuzzy logic is comparable to the logic used in human decision-making. (ii) Inaccurate sensor readings can be accommodated.

As an example, in an occasion defined by using more than a few temperatures above 55 °C and smoke obscuration stages above 15%, the event in the case of fire detection is detailed by way of a set value for temperature and smoke. (ii) Comparatively, fuzzy logic is more acceptable than other categorization methods. However, several factors that may increase complexity and reduce accuracy should be considered while designing the fuzzy logic model, such as the fuzzy logic's exponential growth as a function of the number of inputs and rule base. Thus, maintaining the number of inputs to a minimum can aid in lowering complexity, but it may impact the detecting system's accuracy. Fuzzy logic typically makes a trade-off between growing complexity and input complexity. As this study suggests, researchers should make efficient use of the fusion center to construct and determine these inputs to decrease the number of inputs. Using a fuzzy logic rule-based approach, the veracity of each cluster is evaluated. The fusion center analyzes each cluster's trustworthiness before deciding rather than treating each cluster equally. As a result, the event detection system's detection accuracy can be increased.

5 Conclusion

This report analyzed relevant background data and research on WSNs with feasibility and application fields. Likewise, the challenges in event detection and some of the event detection hypotheses were discussed to get an optimized outcome. Also discussed the value of environmental monitoring techniques like collaborative detection, periodic sleep scheduling, and transceiver power deactivation in this paper. This review examined popular methodologies and strategies created for event detection such as the event detection-based disk model, probabilistic model, and decision fusion model-based fuzzy logic. Similarly, these techniques point out their advantages and disadvantages by considering crucial metrics including false alarm likelihood, detection delay, and detection probability, and providing an inclusive observation appropriately. This paper also suggests the researchers use the fusion center with fuzzy logic to reduce the number of inputs and increase the detection accuracy with less false alarms.

References

1. Liu, M., Cao, J., Lou, W., Chen, L., Li, X.: Coverage analysis for wireless sensor networks. In: Jia, X., Wu, J., He, Y. (eds.) MSN 2005. LNCS, vol. 3794, pp. 711–720. Springer, Heidelberg (2005). https://doi.org/10.1007/11599463_69
2. Wang, T., et al.: Extracting target detection knowledge based on spatiotemporal information in wireless sensor networks. Int. J. Distrib. Sens. Netw. **12**(2), 5831471 (2016)
3. Nasridinov, A., Ihm, S.Y., Jeong, Y.S., Park, Y.H.: Event detection in wireless sensor networks: survey and challenges. In: Park, J., Adeli, H., Park, N., Woungang, I. (eds.) Mobile, Ubiquitous, and Intelligent Computing. Lecture Notes in Electrical Engineering, vol. 274, pp. 585–590. Springer, Heidelberg (2014). https://doi.org/10.1007/978-3-642-40675-1_87
4. Kerman, M.C., Jiang, W., Blumberg, A.F., Buttrey, S.E.: Event detection challenges, methods, and applications in natural and artificial systems. Lockheed Martin MS2 Moorestown NJ (2009)
5. Viswanathan, R., Varshney, P.K.: Distributed detection with multiple sensors part i. fundamentals. Proc. IEEE **85**(1), 54–63 (1997)
6. Blum, R.S., Kassam, S.A., Poor, H.V.: Distributed detection with multiple sensors II. Advanced topics. Proc. IEEE **85**(1), 64–79 (1997)
7. Acharya, S., Kam, M., Wang, J.: Distributed decision fusion using the Neyman-Pearson criterion performance analysis of simulated hard soft fusion view project distributed decision fusion using the Neyman-Pearson criterion. Department of Electrical and Computer Engineering, Drexel University (2014)
8. Gostar, A.K., Hoseinnezhad, R., Bab-Hadiashar, A.: Multi-Bernoulli sensor-selection for multi-target tracking with unknown clutter and detection profiles. Signal Process. **119**, 28–42 (2016)
9. Ciuonzo, D., Rossi, P.S.: Distributed detection of a non-cooperative target via generalized locally-optimum approaches. Inf. Fusion **36**(261–274), 2017 (2017)
10. Li, T., Corchado, J.M., Sun, S., Bajo, J.: Clustering for filtering: multi-object detection and estimation using multiple/massive sensors. Inf. Sci. **388**, 172–190 (2017)
11. Kaltiokallio, O., Yiğitler, H., Jäntti, R.: A three-state received signal strength model for device-free localization. IEEE Trans. Veh. Technol. **66**(10), 9226–9240 (2017)
12. Khaleghi, B., Khamis, A., Karray, F.O., Razavi, S.N.: Corrigendum: corrigendum to 'multi-sensor data fusion: a review of the state-of-the-art' [Information Fusion 14(1) (2013) 28–44]. Inf. Fusion **14**(4), 562 (2013)
13. Zadeh, L.A.: Fuzzy logic and its applications. New York, NY, USA (1965)
14. Gupta, I., Riordan, D., Sampalli, S.: Cluster-head election using fuzzy logic for wireless sensor networks. In: Proceedings of the 3rd Annual Communication Networks and Services Research Conference, pp. 255–260 (2005)

15. Kim, J.M., Park, S.H., Han, Y.J., Chung, T.M.: CHEF: cluster head election mechanism using Fuzzy logic in wireless sensor networks. In: International Conference on Advanced Communication Technology, ICACT, vol. 1, pp. 654–659 (2008)

16. Javanmardi, S.: A novel approach for faulty node detection with the aid of fuzzy theory and majority voting in wireless sensor networks. Int. J. Adv. Smart Sens. Netw. Syst. 2(4), 1–10 (2012)

17. Mahvy, M., Jahani, R., Shayanfar, H.A.: Using simulated annealing algorithm for optimal bidding strategy in electric markets. IJTPE J. 17–21 (2011)

18. Dastgheib, S.J.: An efficient approach to detect faulty readings in long-thin wireless sensor network using fuzzy logic. In: 2012 International Conference on Future Communication Networks, ICFCN 2012, vol. 4, no. 1, pp. 88–92 (2012)

19. Do, W.: Fuzzy logic-optimized secure media access control (FSMAC) protocol for wireless sensor networks Qingchun Ren and Qilian Liang, pp. 37–43. IEEE Xplore (2005)

20. Xia, F., Zhao, W., Sun, Y., Tian, Y.C.: Fuzzy logic control based QoS management in wireless sensor/actuator networks. Sensors 7(12), 3179–3191 (2007)

21. Liang, Q., Wang, L.: Event detection in wireless sensor networks using fuzzy logic system. In: Computational Intelligence for Homeland Security and Personal Safety (CIHSPS), pp. 52–55 (2005)

22. Marin-Perianu, M., Havinga, P.: D-FLER–a distributed fuzzy logic engine for rule-based wireless sensor networks. Ubiquit. Comput. Syst. 1, 86–101 (2007). https://doi.org/10.1007/978-3-540-76772-5_7

23. Zadeh, L.A.: Outline of a new approach to the analysis of complex systems and decision processes. IEEE Trans. Syst. Man Cybern. 1, 28–44 (1973)

24. L'ecuyer, P., Mandjes, M., Tuffin, B.: Importance sampling and rare event simulation. Rare event simulation using Monte Carlo methods, pp. 17–38 (2009)

25. Rubinstein, R.Y., Kroese, D.P.: Simulation and the Monte Carlo Method. Wiley, Hoboken (2016)

26. Nellore, K., Hancke, G.P.: A survey on urban traffic management system using wireless sensor networks. Sensors 16, 157 (2016)

27. Xiao, J., Zhou, Z., Yi, Y., Ni, L.M.: A survey on wireless indoor localization from the device perspective. ACM Comput. Surv. (CSUR) 49, 25 (2016)

28. Wittenburg, G., Dziengel, N., Adler, S., Kasmi, Z., Ziegert, M., Schiller, J.: Cooperative event detection in wireless sensor networks. IEEE Commun. Mag. 50, 124–131 (2012)

29. Wu, H., Cao, J., Fan, X.: Dynamic collaborative in-network event detection in wireless sensor networks. Telecommun. Syst. 62, 43–58 (2016)

30. Das, S.N., Misra, S.: Event-driven probabilistic topology management in sparse wireless sensor network. IET Wirel. Sens. Syst. 5, 210–217 (2015)

31. Rashid, S., Akram, U., Qaisar, S., Khan, S.A., Felemban, E.: Wireless sensor network for distributed event detection based on machine learning. In: Internet of Things (iThings), 2014 IEEE International Conference on, and Green Computing and Communications (GreenCom), IEEE and Cyber, Physical and Social Computing (CPSCom), pp. 540–545. IEEE (2014)

32. Bernardo, L., Oliveira, R., Tiago, R., Pinto, P.: A fire monitoring application for scattered wireless sensor networks. In: Proceedings of the International Conference on Wireless Information Networks and Systems, Barcelona, Spain (2007)

33. Wittenburg, G., Dziengel, N., Wartenburger, C., Schiller, J.: A system for distributed event detection in wireless sensor networks. In: Proceedings of the 9th ACM/IEEE International Conference on Information Processing in Sensor Networks, pp. 94–104. ACM (2010)

34. Li, S., Son, S.H., Stankovic, J.A.: Event detection services using data service middleware in distributed sensor networks. In: Zhao, F., Guibas, L. (eds.) IPSN 2003. LNCS, vol. 2634, pp. 502–517. Springer, Heidelberg (2003). https://doi.org/10.1007/3-540-36978-3_34

35. Rajesh, L., Reddy, C.B.: Efficient wireless sensor network using nodes sleep/active strategy. In: 2016 International Conference on Inventive Computation Technologies (ICICT), pp. 1–4. IEEE (2016)
36. Misra, S., Mishra, S., Khatua, M.: Social sensing-based duty cycle management for monitoring rare events in wireless sensor networks. IET Wirel. Sens. Syst. **5**, 68–75 (2015)
37. Kavitha, S., Lalitha, S.: Sleep scheduling for critical event monitoring in wireless sensor networks. Int. J. Adv. Res. Comput. Commun. Eng. **3** (2014)
38. Zhu, Y., Liu, Y., Ni, L.M.: Optimizing event detection in low duty-cycled sensor networks. Wirel. Netw. **18**, 241–255 (2012)
39. Guo, P., Jiang, T., Zhang, Q., Zhang, K.: Sleep scheduling for critical event monitoring in wireless sensor networks. IEEE Trans. Parallel Distrib. Syst. **23**, 345–352 (2012)
40. Yoo, H., Shim, M., Kim, D.: Dynamic duty-cycle scheduling schemes for energy-harvesting wireless sensor networks. IEEE Commun. Lett. **16**, 202–204 (2012)
41. Karuppiah Ramachandran, V.R., Ayele, E.D., Meratnia, N., Havinga, P.J.: Potential of wake-up radio-based mac protocols for implantable body sensor networks (ISBN)—a survey. Sensors **16**, 2012 (2016)
42. Karvonen, H., Petäjäjärvi, J., Iinatti, J., Hämäläinen, M., Pomalaza-Ráez, C.: A generic wake-up radio based MAC protocol for energy efficient short range communication. In: 2014 IEEE 25th Annual International Symposium on Personal, Indoor, and Mobile Radio Communication (PIMRC), pp. 2173–2177. IEEE (2014)
43. Rajendran, V., Obraczka, K., Garcia-Luna-Aceves, J.J.: Energy-efficient, collision-free medium access control for wireless sensor networks. Wirel. Netw. **12**, 63–78 (2006)
44. Feng, J., Potkonjak, M.: Power minimization by separation of control and data radios. In: 2002 IEEE CAS Workshop on Wireless Communication and Networking, pp. 112–121 (2002)
45. Olds, J.P., Seah, W.K.: Design of an active radio frequency powered multi-hop wireless sensor network. In: 2012 7th IEEE Conference on Industrial Electronics and Applications (ICIEA), pp. 1721–1726. IEEE (2012)
46. Vescoukis, V., Olma, T., Markatos, N.: Experience from a pilot implementation of an "in-situ" forest temperature measurement network. In: 2007 IEEE 18th International Symposium on Personal, Indoor and Mobile Radio Communications, PIMRC 2007. IEEE (2007)
47. Sheth, A., et al.: Senslide: a sensor network based landslide prediction system. In: 2005 Proceedings of the 3rd International Conference on Embedded Networked Sensor Systems, pp. 280–281. ACM (2005)
48. Hsin, C.-F., Liu, M.: Network coverage using low duty-cycled sensors: random & coordinated sleep algorithms. In:2004 Proceedings of the 3rd International Symposium on Information Processing in Sensor Networks, pp. 433–442. ACM (2004)
49. Mohamed, M.M.A., Khokhar, A., Trajcevski, G.: Energy efficient resource distribution for mobile wireless sensor networks. In: 2014 IEEE 15th International Conference on Mobile Data Management (MDM), pp. 49–54. IEEE (2014)
50. He, T., et al.: VigilNet: an integrated sensor network system for energy-efficient surveillance. ACM Trans. Sens. Netw. (TOSN) **2**, 1–38 (2006)
51. Tian, D., Georganas, N.D.: A coverage-preserving node scheduling scheme for large wireless sensor networks. In: Proceedings of the 1st ACM International Workshop on Wireless Sensor Networks and Applications, pp. 32–41. ACM (2002)
52. Yang, Z., Ren, K., Liu, C.: Efficient data collection with spatial clustering in time constraint WSN applications. In: Zu, Q., Hu, B., Elçi, A. (eds.) ICPCA/SWS 2012. LNCS, vol. 7719, pp. 728–742. Springer, Heidelberg (2013). https://doi.org/10.1007/978-3-642-37015-1_64
53. Pripužić, K., Belani, H., Vuković, M.: Early forest fire detection with sensor networks: sliding window skylines approach. In: Lovrek, I., Howlett, R.J., Jain, L.C. (eds.) KES 2008. LNCS (LNAI), vol. 5177, pp. 725–732. Springer, Heidelberg (2008). https://doi.org/10.1007/978-3-540-85563-7_91

54. Yu, L., Wang, N., Meng, X.: Real-time forest fire detection with wireless sensor networks. In: 2005 Proceedings of the International Conference on Wireless Communications, Networking and Mobile Computing. IEEE (2005)
55. Vuran, M.C., Akyildiz, I.F.: Spatial correlation-based collaborative medium access control in wireless sensor networks. IEEE/ACM Trans. Netw. (TON) **14**, 316–329 (2006)
56. Werner-Allen, G., et al.: Deploying a wireless sensor network on an active volcano. IEEE Internet Comput. **10**, 18–25 (2006)
57. Brass, P.: Bounds on coverage and target detection capabilities for models of networks of mobile sensors. ACM Trans. Sens. Netw. (TOSN) **3**(2), 9 (2007)
58. Liu, B., Dousse, O., Nain, P., Towsley, D.: Dynamic coverage of mobile sensor networks. IEEE Trans. Parallel Distrib. Syst. **24**(2), 301–311 (2013)
59. Wang, T., et al.: Extracting target detection knowledge based on spatiotemporal information in wireless sensor networks. Int. J. Distrib. Sens. Netw. (2016)
60. Lazos, L., Poovendran, R., Ritcey, J.A.: Probabilistic detection of mobile targets in heterogeneous sensor networks. In: 2007 6th International Symposium on Information Processing in Sensor Networks, IPSN 2007. IEEE (2007)
61. Medagliani, P., Leguay, J., Ferrari, G., Gay, V., Lopez-Ramos, M.: Energy-efficient mobile target detection in wireless sensor networks with random node deployment and partial coverage. Pervasive Mob. Comput. **8**(3), 429–447 (2012)
62. Zhou, J., Shi, J.: RFID localization algorithms and applications—a review. J. Intell. Manuf. **20**(6), 695 (2009)
63. Varshney, P.K.: Distributed Detection and Data Fusion. Springer, Heidelberg (2012)
64. Yi, S., Hao, Z., Qin, Z., Li, Q.: Fog computing: platform and applications. In: 2015 Third IEEE Workshop on Hot Topics in Web Systems and Technologies (HotWeb). IEEE (2015)
65. Tan, W., Wang, Q., Huang, H., Guo, Y., Zhang, G.: Mine fire detection system based on wireless sensor network. In: 2007 International Conference on Information Acquisition, ICIA 2007 (2007)
66. Yang, G., Qiao, D.: Barrier information coverage with wireless sensors. In: INFOCOM 2009. IEEE (2009)
67. Kapitanova, K., Son, S.H., Kang, K.-D.: Using fuzzy logic for robust event detection in wireless sensor networks. Ad Hoc Netw. **10**, 709–722 (2012)
68. Kieu-Xuan, T., Koo, I.: A cooperative spectrum sensing scheme using fuzzy logic for cognitive radio networks. KSII Trans. Internet Inf. Syst. **4**(3) (2010)
69. Thuc, K.-X., Insoo, K.: A collaborative event detection scheme using fuzzy logic in clustered wireless sensor networks. AEU-Int. J. Electron. Commun. **65**, 485–488 (2011)
70. Zervas, E., Sekkas, O., Hadjieftymiades, S., Anagnostopoulos, C.: Fire detection in the urban rural interface through fusion techniques. In: IEEE International Conference on Mobile Adhoc and Sensor Systems, MASS 2007 (2007)

A Computer Presentation of the Analytical and Numerical Study of Nonlinear Vibration Response for Porous Functionally Graded Cylindrical Panel

Ahmed Mouthanna[1,2(✉)], Sadeq H. Bakhy[2], and Muhannad Al-Waily[3]

[1] College of Engineering, University of Anbar, Ramadi, Anbar, Iraq
ahmed.mouthana@uoanbar.edu.iq
[2] Mechanical Engineering Department, University of Technology, Baghdad, Iraq
[3] Department of Mechanical Engineering, Faculty of Engineering, University of Kufa, Kufa, Iraq

Abstract. The current study uses a new analytical model and numerical method to present a study of free vibration carried out on a cylindrical shell panel that is simply supported and functionally graded. It is anticipated that the FG thickness attributes will be dependent on the porosity level and will change along the thickness axis in accordance with a distribution that follows a power-law. This work makes a contribution by analyzing the performance of porous FGMs, which are employed in a particularly wide variety of biomedical applications. For the purpose of determining the free vibration characteristics as well as the nonlinear vibration response, the governing equations are constructed on a first-order shear deformation theory by utilizing the Galerkin technique with the fourth-order Runge Kutta a close encounter with an incomplete FGM cylindrical shell panel and include different parameters. Parameters included are the power-law index, graded distributions of porosity, and FG thickness. With the help of both the ANSYS 2021-R1 software, a numerical investigation was carried out making use of the finite element approach, and a modal investigation was carried out. This was done in order to verify the analytical strategy.

Keywords: Porous · Materials with a Functional Grading · Theory of First-Order Shear Deformation · Analytical Investigation · Nonlinear Dynamic Response · Frequency

1 Introduction

In mechanical and construction engineering, investigating and developing new materials and structures is essential. Because of their lightweight nature and excellent mechanical properties, structures composed of advanced materials are often employed in numerous industries, including civil, mechanical, and aerospace engineering. The application in practice is normally in the shape as nanocomposite structures, Functionally Graded

Materials (FGM) structures, laminated, and sandwich structures [1]. In recent years, analyzing a new structure has become a popular study area for numerous material and engineering scientists. Many articles have been published about composite materials. Ranging from micro-scale (nanocomposite and FGM) to macro-scale (laminated and sandwich structures), the studies have been conducted using a variety of techniques, scales, and forms [2, 3]. In the dynamics case, numerous investigations have been done on the vibration analysis of structures [4–6]. Bagheri et al. [7] studied the geometrically nonlinear dynamic response of joined conical shells constructed of functionally graded material (FGM) subjected to thermal shock. Zhu et al. [8] employed Reddy's theory of Shear Deformation at Higher Orders with the geometric nonlinearity hypothesis (von Kármán) to investigate the relationship between nonlinear forced vibration properties and nonlinear free vibration properties of viscoelastic plates. Li and Liu [9] examined the thermal free vibration as well as buckling attitude of viscoelastic sandwich of FGM shallow shell having a core constructed of tunable auxetic honeycomb. Singh et al. [10] coupled piezoelectric sensors with time-dependent tri analytical solutions in order to investigate the free vibration of viscoelastic of orthotropic rectangular plates in-plane FG. Sahu et al. [11] presented the free vibration and damping investigation of the sandwich doubly-curved shallow shell with a viscoelastic-FGM layer as suggested by the theory of shear deformation of the first order. Moreover, there has been a significant increase in the number of investigations for shell structures constructed of FGM materials [26]. These published studies investigated the buckling properties as well as the linear and nonlinear vibration behavior in classical shell theory [12, 13]. Throughout the development of the material industry, FGM porous cores employed for lightweight structures are becoming a significant element in civil, mechanical, and aerospace engineering due to their electrical, mechanical, and thermal properties. In particular, the FG porous material has a high strength as well as an excellent energy absorption capability. Ghobadi et al. [14] investigated the influence of the various distributions of porosity on the static and dynamic behavior of sandwich FGM nanostructures under thermo-electro-elastic coupling. Esayas and Kattimani [15] studied the influence of porosity on the dynamic damping of geometrically nonlinear vibrations of an FG magneto-electro-elastic plate. Javid et al. [16] researched the free vibrational characteristics of porous FG micro-cylindrical shells with viscoelastic medium and two skins made of nanocomposite based on Biot's assumptions. According to third-order shear deformation theory, Keleshteri and Jelovica [17] sandwich panels with FG metal cores of foam were studied to determine their buckling and free vibrational behavior. Kumar H S et al. [18] approaching together transient responses and FG nonlinear free vibration of skew plate under the influence of porosity distribution. Srikarun et al. [19] researched the linear and nonlinear stability of sandwich beams with porous FG cores subjected to various types of distributed loads. Chan et al. [20] Using theory of first-order shear deformation shell, researchers were investigated free vibration with a nonlinear response of the dynamic properties of a porous functionally graded trimmed shell have aconical shap equipped with actuators of piezoelectric in warm settings. Dastjerdi and Behdinan [21] studied the free vibration characteristics of smart sandwich plates with carbon nanotube-reinforced and piezoelectric layers by employing Reddy's theory of third-order shear deformation. Yadav et al. [22] investigated statics of nonlinear of sandwich circular-cylindrical shells consisting of two carbon nanotube-reinforced face sheets and a porous FG core by utilizing higher-order shear deformation and thickness theory. Binh et al. [23] examined the nonlinear dynamic characteristics of a porous FG toroidal shell of variable thickness subjected to thermal loads and enclosed by a medium that has elasticity. Furthermore, numerous investigations displayed the advent of the porous core in strengthening structures [24, 25]. Based on 3D elasticity, the dynamic analyses and natural frequencies of porous FG cylindrical panels and annular sector plates were determined by Babaei et al. [26]. Li et al. [27] used the differential quadrature method to study the natural vibration properties of metal porous foam for conical shells consisting of two intriguing elastically restrained boundaries. Hung et al. [12] investigated the effect of the porous FG variable thickness for the toroidal shell on the nonlinear stability (buckling and post-buckling) under compressive loads and surrounded by the elastic foundation. Vinh and Huy [28] presented an inclusive analysis of the buckling, static bending and free vibration of the FG plates for sandwiches, including porosity distribution according to the finite element method with new hyperbolic shear deformation theory. Amir and Talha [29] employing finite element and high-order shear deformation technique, this research looked at the vibration of nonlinear thermo-elastic characteristics of FG porous double curve shallow shells. Keleshteri and Jelovica [30] employed high-order bidirectional porosity distributions to examine the free and forced vibration characteristics of FG porous beams. Nevertheless, there have only been a few studies on the FGM free vibration systems with porous metal formation. This investigation's objective is to carry out a study on the nonlinear analysis of free vibration of a simply-supported two-phase FGM cylindrical shell panel with porous as part of its research. In the present work, suppose that the FGM is constructed from ceramic and metal, mechanical characteristics are varied with reason disparate porosity distributions based on power-law distributions, with changes in the thickness direction. A novel model for the first-order shear deformation theory is formed to discover the nonlinear free vibration characteristics according to different FGM parameters. Utilizing the FEA strategy that is exemplified by ANSYS software, the results of mode and natural frequency forms of the FGM cylindrical shell with porous are provided here. The numerical findings for FG porous materials that are offered here are not found anywhere else in the literature, and as a consequence, should be of relevance to industrial applications. This study is organized into four parts. In the first section, theory of first-order shear deformation criteria is introduced, in addition to constitutive equations, features of FG porous structures, and an analytical vibration analysis of the porous cylindrical panel. The second section introduces numerical analysis and finite element simulation. Results and discussion are included in the third section. The last part includes study summaries and findings.

2   Models of Porous FGM Cylindrical Shell Panel

Consider a thick FGM cylindrical shell panel made of metal and ceramic, in which the lower surface is ceramic-rich and the upper surface is metal-rich, respectively. The FGM cylindrical panel is considered to carry porosities that distribute evenly and unevenly through the shell thickness direction (Fig. 1). The shell's thickness, Radius, and edges are represented by h, R, a, and b, respectively. To describe the shell's motion, a cartesian coordinate system (x, y, z) on the center surface of the shell is employed, where z identifies the out-of-plane coordinate, and x and y determine the shell's in-plane coordinates.

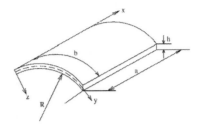

Fig. 1. The geometry of the cylindrical panel found on the FGM.

In addition, the power law, the sigmoid law, or the exponential law may be used to adequately represent the volume fraction of the FG cylindrical shell layers. Equation 1 makes the assumption that the distribution of the ceramic volume fraction Vc is governed by a power law [31]:

$$V_m + V_c = 1 V_c = V_c(z) = \left(\frac{2z + h}{2h}\right)^g \tag{1}$$

where, Vc and Vm are volume fractions of ceramic and metal, respectively. g is the power-law index. When the value of g is equal to infinity, it indicates a fully metallic shell, whereas when the value of g is equal to zero, it denotes a fully ceramic shell. The fundamental mechanical properties of the FGM cylindrical shell panels, with a porosity volume ratio of G (G < 1), adopt the modified form of Eq. 2, assuming that porosities disperse equally in the ceramic and metal phases

$$P(z) = P_m + (P_c - P_m)\left(\frac{2z + h}{2h}\right)^g - \frac{G}{2}(P_c - P_m) \tag{2}$$

Pm and Pc represented the values of the material properties of the metal and ceramic components of the FG shells, respectively. Young's modulus (E) and mass density (ρ) are taken to change in the thickness direction for our current formulations, while Poisson's ratio (v) will be assumed to remain constant for simplicity based on earlier research.

2.1 Fundamental Equations

This study takes into account thick porous FGM cylindrical shell panels subjected to external loading with varying boundary conditions. As a result, the system of governing equations is established, and the nonlinear vibration of the cylindrical shell is determined using first-order shear deformation plate theory [32]:

$$\begin{aligned}
\tilde{u}(x, y, z, t) &= u(x, y, t) + z\phi_x(x, y, t), \\
\tilde{v}(x, y, z, t) &= v(x, y, t) + z\phi_y(x, y, t), \\
\tilde{w}(x, y, z, t) &= w(x, y, t)
\end{aligned} \tag{3}$$

where, (ϕ_x, ϕ_y) describes the transverse normal slopes about x- and y-axes at $(z = 0)$. The following equations serve as the basis for the strain-displacement relationships of the cylindrical shell:

$$
\begin{Bmatrix} \varepsilon_x \\ \varepsilon_y \\ \gamma_{xy} \end{Bmatrix} = \begin{Bmatrix} \varepsilon_x^\circ \\ \varepsilon_y^\circ \\ \gamma_{xy}^\circ \end{Bmatrix} + z \begin{Bmatrix} \lambda_x \\ \lambda_y \\ \lambda_{xy} \end{Bmatrix}, \begin{Bmatrix} \gamma_{xz} \\ \gamma_{yz} \end{Bmatrix} = \begin{Bmatrix} \frac{\partial w}{\partial x} + \phi_x \\ \frac{\partial w}{\partial y} + \phi_y \end{Bmatrix} \tag{4}
$$

$$
\begin{Bmatrix} \varepsilon_x^\circ \\ \varepsilon_y^\circ \\ \gamma_{xy}^\circ \end{Bmatrix} = \begin{Bmatrix} \frac{\partial u}{\partial x} + \frac{1}{2}\left(\frac{\partial w}{\partial x}\right)^2 \\ \frac{\partial v}{\partial y} + \frac{1}{2}\left(\frac{\partial w}{\partial x}\right)^2 \\ \frac{\partial u}{\partial y} + \frac{\partial v}{\partial x} + \frac{\partial w}{\partial x}\frac{\partial w}{\partial y} \end{Bmatrix} \text{ and } \begin{Bmatrix} \lambda_x \\ \lambda_y \\ \lambda_{xy} \end{Bmatrix} = \begin{Bmatrix} \frac{\partial \phi_x}{\partial x} \\ \frac{\partial \phi_y}{\partial y} \\ \frac{\partial \phi_x}{\partial y} + \frac{\partial \phi_y}{\partial x} \end{Bmatrix}
$$

In the planes (xz, yz), the x designates the transverse shear strain components $(\gamma_{xz}, \gamma_{yz})$. The nonlinear stress-strain constitutive relations at a general point inside the skin of the cylindrical shell can be expressed as:

$$
\begin{Bmatrix} \sigma_x^{sh} \\ \sigma_y^{sh} \\ \tau_{xy}^{sh} \\ \tau_{xz}^{sh} \\ \tau_{yz}^{sh} \end{Bmatrix} = \begin{bmatrix} C_{11} & C_{12} & 0 & 0 & 0 \\ C_{12} & C_{22} & 0 & 0 & 0 \\ 0 & 0 & C_{44} & 0 & 0 \\ 0 & 0 & 0 & C_{55} & 0 \\ 0 & 0 & 0 & 0 & C_{66} \end{bmatrix} \begin{Bmatrix} \varepsilon_x \\ \varepsilon_y \\ \gamma_{xy} \\ \gamma_{xz} \\ \gamma_{yz} \end{Bmatrix} \tag{5}
$$

For, $C_{11} = C_{22} = \frac{E(z)}{1-v^2}$, $C_{12} = \frac{vE(z)}{1-v^2}$, $C_{55} = C_{66} = \frac{E(z)}{2(1+v)}$.

Numerous studies have found the shear correction factor for homogeneous structures. Furthermore, some researchers have given a thickness-shear vibration value of $\frac{\pi^2}{12}$, but this will result in a different value for the FGM structure due to the material properties continuously changing in the thickness direction. So, the shear correction factor is denoted by (K_s), and its value is offered $\left(K = \frac{5}{6}\right)$ as [33]. The stress and moment resultants of the FGM porous cylindrical shell panel can be represented as,

$$
N_x = \int_{-\frac{h}{2}}^{\frac{h}{2}} \sigma_x^{sh} dz, N_y = \int_{-\frac{h}{2}}^{\frac{h}{2}} \sigma_y^{sh} dz, N_{xy} = \int_{-\frac{h}{2}}^{\frac{h}{2}} \tau_{xy}^{sh} dz, (Q_x, Q_y) = \int_{-\frac{h}{2}}^{\frac{h}{2}} K_s \left(\tau_{xz}^{sh}, \tau_{yz}^{sh}\right) dz, M_x = \int_{-\frac{h}{2}}^{\frac{h}{2}} \sigma_x^{sh} z dz,
$$
$$
M_y = \int_{-\frac{h}{2}}^{\frac{h}{2}} \sigma_x^{sh} z dz, \ M_{xy} = \int_{-\frac{h}{2}}^{\frac{h}{2}} \tau_x^{sh} z dz \tag{6}
$$

where,

$$
N_x = I_{10}\varepsilon_x^\circ + I_{20}\varepsilon_y^\circ + I_{11}\frac{\partial \phi_x}{\partial x} + I_{21}\frac{\partial \phi_y}{\partial y}, \ N_y = I_{20}\varepsilon_x^\circ + I_{10}\varepsilon_y^\circ + I_{21}\frac{\partial \phi_x}{\partial x} + I_{11}\frac{\partial \phi_y}{\partial y}
$$

$$
N_{xy} = I_{30}\gamma_{xy}^\circ + 2I_{31}\left(\frac{\partial \phi_x}{\partial y} + \frac{\partial \phi_y}{\partial x}\right), \ M_x = I_{11}\varepsilon_x^\circ + I_{21}\varepsilon_y^\circ + I_{12}\frac{\partial \phi_x}{\partial x} + I_{22}\frac{\partial \phi_y}{\partial y}
$$

$$
M_y = I_{21}\varepsilon_x^\circ + I_{11}\varepsilon_y^\circ + I_{22}\frac{\partial \phi_x}{\partial x} + I_{12}\frac{\partial \phi_y}{\partial y}, \ M_{xy} = I_{31}\gamma_{xy}^\circ + I_{32}\left(\frac{\partial \phi_x}{\partial y} + \frac{\partial \phi_y}{\partial x}\right) \tag{7}
$$

$$
Q_x = K_s I_{30}\gamma_{xz}, Q_y = K_s I_{30}\gamma_{yz}.
$$

The nonlinear governing motion equations of a porous FGM cylindrical shell panel are given below,

$$\delta u : \frac{\partial N_x}{\partial x} + \frac{\partial N_{xy}}{\partial y} = I_0 \frac{\partial^2 u}{\partial t^2} + I_1 \frac{\partial^2 \phi_x}{\partial t^2}$$

$$\delta v : \frac{\partial N_{xy}}{\partial x} + \frac{\partial N_y}{\partial y} = I_0 \frac{\partial^2 v}{\partial t^2} + I_1 \frac{\partial^2 \phi_y}{\partial t^2}$$

$$\delta w : \frac{\partial Q_x}{\partial x} + \frac{\partial Q_y}{\partial y} + N_x \frac{\partial^2 w}{\partial x^2} + 2N_{xy} \frac{\partial^2 w}{\partial x \partial y}$$

$$+ N_y \frac{\partial^2 w}{\partial y^2} + q + \frac{N_y}{R} = I_0 \frac{\partial^2 w}{\partial t^2}$$

$$\delta \phi_x : \frac{\partial M_x}{\partial x} + \frac{\partial M_{xy}}{\partial y} - Q_x = I_2 \frac{\partial^2 \phi_x}{\partial t^2} + I_1 \frac{\partial^2 u}{\partial t^2}$$

$$\delta \phi_y : \frac{\partial M_{xy}}{\partial x} + \frac{\partial M_y}{\partial y} - Q_y = I_2 \frac{\partial^2 \phi_y}{\partial t^2} + I_1 \frac{\partial^2 v}{\partial t^2} \tag{8}$$

By submitting stress function f (x, y) as follow:

$$N_x = \frac{\partial^2 f}{\partial y^2}, \ N_y = \frac{\partial^2 f}{\partial x^2}, \ N_{xy} = -\frac{\partial^2 f}{\partial x \partial y} \tag{9}$$

Equation (9) is substituted into the first two equations of Eq. (8) to yield;

$$\frac{\partial^2 u}{\partial t^2} = -\frac{I_1}{I_0} \frac{\partial^2 \phi_x}{\partial t^2}, \quad \frac{\partial^2 v}{\partial t^2} = -\frac{I_1}{I_0} \frac{\partial^2 \phi_y}{\partial t^2} \tag{10}$$

Replacing (10) into the remaining three equations of Eq. (8), get,

$$\delta w : \frac{\partial Q_x}{\partial x} + \frac{\partial Q_y}{\partial y} + N_x \frac{\partial^2 w}{\partial x^2} + 2N_{xy} \frac{\partial^2 w}{\partial x \partial y} + N_y \frac{\partial^2 w}{\partial y^2} + q + \frac{N_y}{R} = I_0 \frac{\partial^2 w}{\partial t^2}$$

$$\delta \phi_x : \frac{\partial M_x}{\partial x} + \frac{\partial M_{xy}}{\partial y} - Q_x = \left(I_2 - \frac{I_1^2}{I_0} \right) \frac{\partial^2 \phi_x}{\partial t^2}$$

$$\delta \phi_y : \frac{\partial M_{xy}}{\partial x} + \frac{\partial M_y}{\partial y} - Q_y = \left(I_2 - \frac{I_1^2}{I_0} \right) \frac{\partial^2 \phi_y}{\partial t^2} \tag{11}$$

Through Eqs. (7) yield,

$$\varepsilon_x^o = A_{22}N_x - A_{12}N_y - B_{11}\frac{\partial \phi_x}{\partial x} - B_{12}\frac{\partial \phi_y}{\partial y}$$

$$\varepsilon_y^o = A_{11}N_y - A_{12}N_x - B_{21}\frac{\partial \phi_x}{\partial x} - B_{22}\frac{\partial \phi_y}{\partial y}$$

$$\gamma_{xy}^o = A_{66}N_{xy} - B_{66}\left(\frac{\partial \phi_x}{\partial y} + \frac{\partial \phi_y}{\partial x} \right) \tag{12}$$

The compatibility equation must be used to add an equation,

$$\frac{\partial^2 \varepsilon_x^{\,\circ}}{\partial y^2} + \frac{\partial^2 \varepsilon_y^{\,\circ}}{\partial x^2} - \frac{\partial^2 \gamma_{xy}^{\,\circ}}{\partial x \partial y} = \frac{\partial^2 w^2}{\partial x \partial y} - \frac{\partial^2 w}{\partial x^2}\frac{\partial^2 w}{\partial y^2} - \frac{1}{R}\frac{\partial^2 w}{\partial x^2} \tag{13}$$

Equation (12) is inserted into Eq. (7) and then into Eq. (11), which results in;

$$T_{11}(w) + T_{12}(\phi_x) + T_{13}(\phi_y) + S_1(w, f) + q = I_0 \frac{\partial^2 w}{\partial t^2},$$

$$T_{21}(w) + T_{22}(\phi_x) + T_{23}(\phi_y) + R_2(f) = \left(I_2 - \frac{I_1^2}{I_0}\right)\frac{\partial^2 \phi_x}{\partial t^2},$$

$$T_{31}(w) + T_{32}(\phi_x) + T_{33}(\phi_y) + R_3(f) = \left(I_2 - \frac{I_1^2}{I_0}\right)\frac{\partial^2 \phi_y}{\partial t^2} \tag{14}$$

When Eq. (12) is replaced with Eq. (13) and the stress functions, it provides the following formalization for the compatibility of FGM porous cylindrical shell panels

$$\left(\begin{array}{l} A_{11}\dfrac{\partial^4 f}{\partial x^4} + A_{22}\dfrac{\partial^4 f}{\partial y^4} + (A_{66} - 2A_{12})\dfrac{\partial^4 f}{\partial x^2 \partial y^2} - B_{21}\dfrac{\partial^3 \phi_x}{\partial x^3} - B_{12}\dfrac{\partial^3 \phi_y}{\partial y^3} + \\ (B_{66} - B_{11})\dfrac{\partial^3 \phi_x}{\partial x \partial y^2} + (B_{66} - B_{22})\dfrac{\partial^3 \phi_y}{\partial x^2 \partial y} - \left(\dfrac{\partial^2 w^2}{\partial x \partial y} - \dfrac{\partial^2 w}{\partial x^2}\dfrac{\partial^2 w}{\partial y^2} - \dfrac{1}{R}\dfrac{\partial^2 w}{\partial x^2}\right) \end{array}\right) = 0 \tag{15}$$

2.2 Nonlinear Vibration Analysis

In this paper, suppose that the porous FGM shell is subjected to impact of the uniformly distributed transverse load $q = Q \sin \Omega t$ with four edges simply supported boundary conditions:

$$w = N_{xy} = \phi_y = 0, \ \text{at} \ x = 0, a$$
$$w = N_{xy} = \phi_x = 0, \ \text{at} \ y = 0, b \tag{16}$$

The next equations are desired to apply to the displacements in the present cases that satisfy the assumed boundary conditions [34]:

$$w(x, y, t) = W(t) \sin \lambda_m x \sin \delta_n y$$
$$\phi_x(x, y, t) = \Phi_x(t) \cos \lambda_m x \sin \delta_n y$$
$$\phi_y(x, y, t) = \Phi_y(t) \sin \lambda_m x \cos \delta_n y \tag{17}$$

Replacing (17) with (15), gain;

$$f(x, y, t) = \tilde{A}_1(t) \cos 2\lambda_m x + \tilde{A}_2(t) \cos 2\delta_n y + \tilde{A}_3(t) \sin \lambda_m x \sin \delta_n y$$

$$\tilde{A}_1(t) = \frac{\delta_n^2}{32 A_{11} \lambda_m^2} W^2; \ \tilde{A}_2(t) = \frac{\lambda_m^2}{32 A_{22} \delta_n^2} W^2$$

$$\tilde{A}_3(t) = \frac{\left(\frac{W}{R}\right) + \left(B_{21}\lambda_m^3 + (B_{11} - B_{66})\lambda_m \delta_n^2\right)\Phi_x(t) + \left(\delta_n^3 B_{12} + (B_{22} - B_{66})\lambda_m^2 \delta_n\right)\Phi_y(t)}{\left(A_{11}\lambda_m^4 + A_{22}\delta_n^4 + (A_{66} - 2A_{12})\lambda_m^2 \delta_n^2\right)}$$

$$\tag{18}$$

After minor adjustments, the system of nonlinear motion equations in terms of displacements is derived by introducing the formula (17 and 18) into (14) and using the Galerkin method;

$$t_{11}W + t_{12}\Phi_x + t_{13}\Phi_y + t_{14}W\Phi_x + t_{15}W\Phi_y + t_{16}W + t_{17}W^2 + t_{18}W^3 + L_{32}q = I_0 \frac{d^2W}{dt^2}$$

$$t_{21}W + t_{22}\Phi_x + t_{23}\Phi_y + n_7 W + n_2 W^2 = \tilde{\rho}_1 \ddot{\Phi}_x,$$

$$t_{31}W + t_{32}\Phi_x + t_{33}\Phi_y + n_9 W + n_4 W^2 = \tilde{\rho}_1 \ddot{\Phi}_y, \tag{19}$$

The natural frequencies of the FGM porous cylindrical shell are obtained by solving Eq. (20), setting $q = 0$, and taking the linear components of Eq. (19):

$$\begin{vmatrix} t_{11} + t_{16} + I_0\omega^2 & t_{12} & t_{13} \\ t_{21} + n_1 & t_{22} + \tilde{\rho}_1\omega^2 & t_{23} \\ t_{31} + n_3 & t_{32} & t_{33} + \tilde{\rho}_1\omega^2 \end{vmatrix} = 0 \tag{20}$$

Three answers to Eq. (20) correspond to the axial, circumferential, and radial angular frequencies of the porosity FGM cylindrical shells. One with the lowest frequency is taken into account. The porosity FGM cylindrical shells panel subjected to orderly distributed load $q = Q\sin\Omega t$ is considered, Eq. (19) becomes:

$$I_0 \frac{d^2W}{dt^2} - t_{11}W - t_{12}\Phi_x - t_{13}\Phi_y - t_{14}W\Phi_x - t_{15}W\Phi_y - t_{16}W - t_{17}W^2 - t_{18}W^3 = L_{32}Q \sin \Omega t$$

$$t_{21}W + t_{22}\Phi_x + t_{23}\Phi_y + n_1 W + n_2 W^2 = \tilde{\rho}_1 \ddot{\Phi}_x$$

$$t_{31}W + t_{32}\Phi_x + t_{33}\Phi_y + n_3 W + n_4 W^2 = \tilde{\rho}_1 \ddot{\Phi}_y \tag{21}$$

The nonlinear dynamic responses of a porous FGM cylindrical shell panel can be calculated using the fourth-order Runge-Kutta technique by solving Eq. (21) with the following initial conditions: $W(0) = 0$. When the second and third equations relating to (Φ_x, Φ_y), are solved from Eq. (21), the results are then substituted into the first equation to yield:

$$I_0 \frac{d^2W}{dt^2} - (a_1 + a_2)W - (a_3 + a_4 + a_6 + r_{17})W^2 - (a_5 + r_{18})W^3 = L_{32}Q \sin \Omega t \tag{22}$$

The fundamental natural frequencies of the cylindrical shell may be calculated as follows:

$$\omega_{mn} = \sqrt{\frac{-(a_1 + a_2)}{I_0}} \tag{23}$$

3   Numerical Investigation

Using numerical methodologies is one way in which one may validate the precision of the analytical method that has been suggested. Problems may be solved using a wide variety of numerical methodologies, although the method of finite elements (FEM) [35] is considered to be the most accurate. The finite element method (FEM) that is outlined by the ANSYS software (Version 2020 R1) was used in this investigation. The construction of a three-dimensional prototype of the Functionally graded cylindrical shell panel, as shown in Fig. 2, is followed by the application of the matching shell's sides boundary conditions, which are subjected to modal study. In addition, as can be seen in Fig. 3, the prototype has been meshed with an 8-nodes SOLID186 slices type, which has resulted in a total of 19360 slices and 137922 nodes. This element type is a significant basic element that is used in the representation of structural models. As shown in Fig. 4 [36], the component consists of a higher-order solid element having 20 nodes that has quadratic spatial characteristics and three freedom degrees for translations along the normal axes. The equation that is used to determine the mechanical characteristics of the FGMs layers as follows: (2). The modal analysis for the selected models is carried out to determine the free vibration characteristics both natural frequencies and mode shapes, as shown in Fig. 5. This is done on the basis of the various factors that were described before.

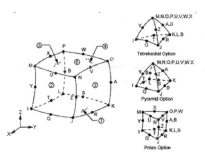

Fig. 2. FGM cylindrical shell panel

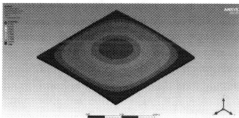

Fig. 3. Meshed Model

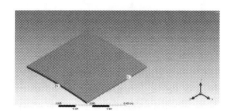

Fig. 4. Structural geometry of element type SOLID 186

Fig. 5. View of the modular analysis of the FGM cylindrical shell

4 Results and Discussion

In this study, a novel mathematical form was developed to analyze the natural frequencies and modes of vibration of FGM cylindrical shell panels supported by just a simply support with uniform porosity distribution based on power-law distribution. Influences of different properties on frequency parameters are analyzed. The material properties of numerous individual materials that were used in this investigation are presented in Table 1. The analytical solution was also verified using the (ANSYS 2020 R1) software that is available for purchase, and the results were tabulated and presented using a variety of curves. Assume that the dimensions of the cylindrical shell are R = 3 m, a = b = 0.5 m., porosity factor (0.1, 0.2, and 0.3), the power-law distribution g = 2, and FG thickness (h = 10, 14, and 16 mm).

Table 1. The material properties of many individual materials

Material Property	FG core	
	Aluminum (Al)	Ceramic (Al2O3)
Modulus of Elasticity, GPa	70	380
Mass density, Kg/m^3	2702	3800
Poisson's ratio	0.3	0.3

Table 2 demonstrates both analytical and numerical findings for the cylindrical panel's natural frequencies for a variety of porosity factors. According to the information shown in Table 2, the thickness of the FG has a considerable impact on the frequency characteristics. Good agreements are reached between analytical analysis tests and numerical tests when the percentage of difference between the two is less than 0.9%. When the porosity factor goes up, the natural frequencies go down because the material stiffness of the goes down. On the other hand, when the panel FG thickness goes up, the natural frequency goes up because the panels get better as the Functionally graded core thickness goes up. This can be seen as a consequence of the findings in Table 2, which show that the natural frequencies go down when the porosity factor goes up. Accordingly, the first six deflections of three-dimensional mode shapes are shown for FGM in Fig. 6 when the material is simply supported in the form of porous cylindrical shell panels with the following parameters: a gradient index of (g = 0.5), a porosity ratio of (g = 30%), a = b = 0.5 m, R = 3, m = n = 1, and a thickness of 10 mm for the FGM. In a like way, it's also possible to depict other 3D mode shapes that are supported by a variety of edge conditions.

Figures 7 and 8 illustrate, respectively, the impacts of the material gradient factor (g) on the dynamic response and natural frequency of cylindrical panels with three different porosity values. These panels were designed to test the effectiveness of the material gradient parameter. Due to a reduction in the material's rigidity, shown that the natural frequency decline whenever the gradient indicator (g) increases, and the dynamic of nonlinear response expands in three porosity magnitudes of (10, 20, and

30%). Additionally, according to Eq. 2, the amount of metal increases while a volume of ceramic decreases, which results in a decrease in shell stiffness. The influence that the porosity factor, denoted by the symbol G, has on the dynamic of nonlinear response is seen in Fig. 9. There are three different levels of porosity that are taken into consideration: $G = 10\%$, $G = 20\%$, and $G = 30\%$. As can be seen, increasing the pores results in a greater amplitude deflection of FGM cylindrical shells. This is something that can be noticed.

Analytical findings of the dynamic nonlinear response of the cylindrical shells with varying FG core thicknesses are shown in Fig. 10. (10, 15, and 20 mm). Figure 9 illustrates that the intensity of the dynamic of nonlinear response diminishes as the Functionally graded thickness of the shells grows. This is due to the fact that the stiffness of the FG panel increases as the Functionally graded thickness does.

Table 2. The FGM cylindrical panels' natural frequency with a power law index of $g = 2$.

Thickness	Porosity %	Analytical	Numerical	Discrepancy %
10	0.1	317.8349	315.35	0.63
	0.2	298.4475	296.66	0.67
	0.3	265.8474	265.61	0.075
14	0.1	400.8333	397.38	0.75
	0.2	372.2816	369.95	0.8
	0.3	322.9525	323.17	0.3
16	0.1	444.8098	440.7	0.9
	0.2	411.6792	408.92	0.72
	0.3	353.9510	354.32	0.28

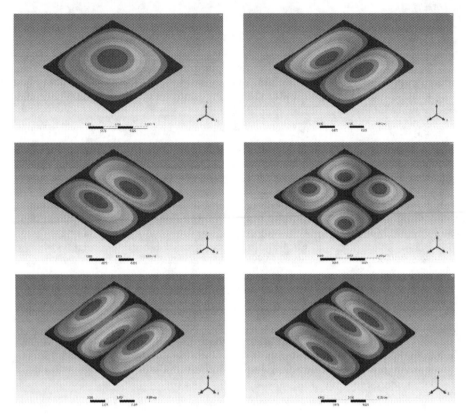

Fig. 6. The initial six mode shapes of Porous FGM simply supported cylindrical panels at G = 0.3, g = 2.

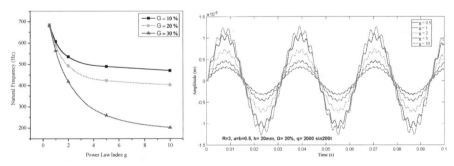

Fig. 7. The result of applying the power law index on porous FGM cylindrical panel was the following natural frequency results

Fig. 8. A look at how the gradient index of a material affects the dynamic response of porous cylindrical shells made from fibrous graphene (FGM).

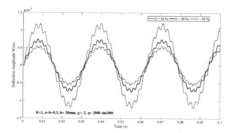

Fig. 9. The influence of porosity on the dynamic response of cylindrical panels

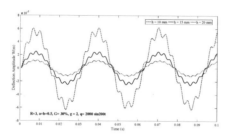

Fig. 10. Cylindrical shells made with porous FGM vibrate with varying FGM thicknesses.

5 Conclusion

In the current work, the nonlinear free vibration characteristics of an FG two-phase porous cylindrical shell with a simply supported boundary condition based on the FSDT are developed. A simply supported cylindrical shell's analytical formulation is presented in order to calculate the nonlinear free dynamic behavior. In order to validate the findings of an analytical solution, a numerical analysis was carried out with the assistance of ANSYS 2021 R1. Also, on deflection-time curve and natural frequency, the research findings for material gradient, porous parameter, and FG thickness are displayed. Based on the findings, it can be deduced that the porosity parameter does, in fact, have some kind of influence on the essential natural frequency of the FG cylindrical panels. According to the findings, the natural frequencies go up when the porosity parameter goes down, but they go down when the stiffness of the material goes up. This is because the natural frequencies are more sensitive to changes in the rigidity of the material. In addition, a downward shift in the amplitude-time curve was seen whenever the porosity factor was raised.

Appendix

$$I_{10} = \frac{E_1}{1-v^2}, I_{20} = \frac{vE_1}{1-v^2}, I_{30} = \frac{E_1}{2(1+v)}, I_{11} = \frac{E_2}{1-v^2}, I_{21} = \frac{vE_2}{1-v^2}, I_{31} = \frac{E_2}{2(1+v)},$$

$$I_{12} = \frac{E_3}{1-v^2}, I_{22} = \frac{vE_3}{1-v^2}, I_{32} = \frac{E_3}{2(1+v)},$$

$$A_{11} = \frac{1}{\Delta}I_{10}, A_{22} = \frac{1}{\Delta}I_{10}, A_{12} = \frac{I_{20}}{\Delta}, A_{66} = \frac{1}{I_{30}}, \Delta = I_{10}^2 - I_{20}^2, B_{11} = A_{22}I_{11} - A_{12}I_{21}, B_{22} = A_{11}I_{11} - A_{12}I_{21},$$

$$B_{12} = A_{22}I_{21} - A_{12}I_{11}, B_{21} = A_{11}I_{21} - A_{12}I_{11}, B_{66} = \frac{I_{31}}{I_{30}}, D_{11} = I_{12} - B_{11}B_{12} - I_{21}B_{21}, D_{22} = I_{22} - B_{22}I_{11} - I_{21}B_{12},$$

$D_{12} = I_{22} - B_{12}I_{11} - I_{21}B_{22}, D_{21} = I_{22} - B_{21}I_{11} - I_{21}B_{11}, D_{66} = I_{32} - I_{31}B_{66}.$

$$T_{11}(w) = K_s I_{30} \frac{\partial^2 w}{\partial x^2} + K_s I_{30} \frac{\partial^2 w}{\partial y^2}, T_{12}(\phi_x) = K_s I_{30} \frac{\partial \phi_x}{\partial x},$$

$$T_{13}(\phi_y) = K_s I_{30} \frac{\partial \phi_y}{\partial y}, R_1(w,f) = \frac{\partial^2 f}{\partial x^2} \frac{\partial^2 w}{\partial x^2} - 2 \frac{\partial^2 f}{\partial x \partial y} \frac{\partial^2 w}{\partial x \partial y} + \frac{\partial^2 f}{\partial x^2} \frac{\partial^2 w}{\partial y^2} + \frac{1}{R} \frac{\partial^2 f}{\partial x^2},$$

$$T_{21}(w) = -K_s I_{30} \frac{\partial w}{\partial x}, T_{22}(\phi_x) = D_{11} \frac{\partial^2 \phi_x}{\partial x^2} + D_{66} \frac{\partial^2 \phi_y}{\partial y^2} - K_s I_{30}\phi_x, T_{23}(\phi_y) = (D_{12} + D_{66}) \frac{\partial^2 \phi_y}{\partial x \partial y},$$

$$R_2(f) = B_{21} \frac{\partial^3 f}{\partial x^3} + (B_{11} - B_{66}) \frac{\partial^3 f}{\partial x \partial y^2}, T_{31}(w) = -K_s I_{30} \frac{\partial w}{\partial y}, T_{32}(\phi_x) = (D_{21} + D_{66}) \frac{\partial^2 \phi_x}{\partial x \partial y},$$

$$T_{33}(\phi_y) = D_{22} \frac{\partial^2 \phi_y}{\partial y^2} + D_{66} \frac{\partial^2 \phi_y}{\partial x^2} - K_s I_{30}\phi_y, R_3(f) = B_{12} \frac{\partial^3 f}{\partial y^3} + (B_{22} - B_{66}) \frac{\partial^3 f}{\partial x^2 \partial y}.$$

References

1. Ninh, D.G., Ha, N.H., Long, N.T., Tan, N.C., Tien, N.D., Dao, D.V.: Thermal vibrations of complex-generatrix shells made of sandwich CNTRC sheets on both sides and open/closed cellular functionally graded porous core. Thin-Walled Struct. **182** (2023). https://doi.org/10.1016/j.tws.2022.110161

2. Yang, Z., et al.: Dynamic buckling of rotationally restrained FG porous arches reinforced with graphene nanoplatelets under a uniform step load. Thin-Walled Struct. **166** (2021). https://doi.org/10.1016/j.tws.2021.108103

3. Viet Hoang, V.N., Tien, N.D., Ninh, D.G., Thang, V.T., Truong, D.V.: Nonlinear dynamics of functionally graded graphene nanoplatelet reinforced polymer doubly-curved shallow shells resting on elastic foundation using a micromechanical model. J. Sandw. Struct. Mater. **23**(7), 3250–3279 (2021). https://doi.org/10.1177/1099636220926650

4. Njim, E., Bakhi, S., Al-Waily, M.: Experimental and numerical flexural properties of sandwich structure with functionally graded porous materials. Eng. Technol. J. **40**(1), 137–147 (2022). https://doi.org/10.30684/etj.v40i1.2184

5. Mouthanna, A., Bakhy, S.H., Al-Waily, M.: Frequency of non-linear dynamic response of a porous functionally graded cylindrical panels. J. Teknol. **84**(6), 59–68 (2022). https://doi.org/10.11113/jurnalteknologi.v84.18422

6. Mouthanna, A., Bakhy, S.H., Al-Waily, M.: Analytical investigation of nonlinear free vibration of porous eccentrically stiffened functionally graded sandwich cylindrical shell panels. Iran. J. Sci. Technol. Trans. Mech. Eng. (2022). https://doi.org/10.1007/s40997-022-00555-4

7. Bagheri, H., Eslami, M.R., Kiani, Y.: Geometrically nonlinear response of FGM joined conical–conical shells subjected to thermal shock. Thin-Walled Struct. **182** (2023). https://doi.org/10.1016/j.tws.2022.110171

8. Zhu, C.S., Fang, X.Q., Liu, J.X.: Relationship between nonlinear free vibration behavior and nonlinear forced vibration behavior of viscoelastic plates. Commun. Nonlinear Sci. Numer. Simul. **117** (2023). https://doi.org/10.1016/j.cnsns.2022.106926

9. Li, Y.S., Liu, B.L.: Thermal buckling and free vibration of viscoelastic functionally graded sandwich shells with tunable auxetic honeycomb core. Appl. Math. Model. **108**, 685–700 (2022). https://doi.org/10.1016/j.apm.2022.04.019

10. Singh, A., Naskar, S., Kumari, P., Mukhopadhyay, T.: Viscoelastic free vibration analysis of in-plane functionally graded orthotropic plates integrated with piezoelectric sensors: time-dependent 3D analytical solutions. Mech. Syst. Signal Process. **184** (2023). https://doi.org/10.1016/j.ymssp.2022.109636

11. Sahu, N.K., Biswal, D.K., Joseph, S.V., Mohanty, S.C.: Vibration and damping analysis of doubly curved viscoelastic-FGM sandwich shell structures using FOSDT. Structures **26**, 24–38 (2020). https://doi.org/10.1016/j.istruc.2020.04.007

12. Hung, D.X., Tu, T.M., Van Long, N., Anh, P.H.: Nonlinear buckling and postbuckling of FG porous variable thickness toroidal shell segments surrounded by elastic foundation subjected to compressive loads. Aerosp. Sci. Technol. **107** (2020). https://doi.org/10.1016/j.ast.2020.106253

13. Ninh, D.G., Bich, D.H.: Nonlinear buckling of eccentrically stiffened functionally graded toroidal shell segments under torsional load surrounded by elastic foundation in thermal environment. Mech. Res. Commun. **72**, 1–15 (2016). https://doi.org/10.1016/j.mechrescom.2015.12.002

14. Ghobadi, A., Tadi Beni, Y., Kamil Żur, K.: Porosity distribution effect on stress, electric field and nonlinear vibration of functionally graded nanostructures with direct and inverse flexoelectric phenomenon. Compos. Struct. **259** (2021). https://doi.org/10.1016/j.compstruct.2020.113220

15. Esayas, L.S., Kattimani, S.: Effect of porosity on active damping of geometrically nonlinear vibrations of a functionally graded magneto-electro-elastic plate. Def. Technol. **18**(6), 891–906 (2022). https://doi.org/10.1016/j.dt.2021.04.016

16. Soleimani-Javid, Z., Arshid, E., Amir, S., Bodaghi, M.: On the higher-order thermal vibrations of FG saturated porous cylindrical micro-shells integrated with nanocomposite skins in viscoelastic medium. Def. Technol. **18**(8), 1416–1434 (2022). https://doi.org/10.1016/j.dt.2021.07.007

17. Keleshteri, M.M., Jelovica, J.: Analytical solution for vibration and buckling of cylindrical sandwich panels with improved FG metal foam core. Eng. Struct. **266** (2022). https://doi.org/10.1016/j.engstruct.2022.114580

18. HS, N.K., Kattimani, S., Nguyen-Thoi, T.: Influence of porosity distribution on nonlinear free vibration and transient responses of porous functionally graded skew plates. Def. Technol. **17**(6), 1918–1935 (2021). https://doi.org/10.1016/j.dt.2021.02.003

19. Srikarun, B., Songsuwan, W., Wattanasakulpong, N.: Linear and nonlinear static bending of sandwich beams with functionally graded porous core under different distributed loads. Compos. Struct. **276** (2021). https://doi.org/10.1016/j.compstruct.2021.114538

20. Chan, D.Q., van Thanh, N., Khoa, N.D., Duc, N.D.: Nonlinear dynamic analysis of piezoelectric functionally graded porous truncated conical panel in thermal environments. Thin-Walled Struct. **154** (2020). https://doi.org/10.1016/j.tws.2020.106837

21. Moradi-Dastjerdi, R., Behdinan, K.: Free vibration response of smart sandwich plates with porous CNT-reinforced and piezoelectric layers. Appl. Math. Model. **96**, 66–79 (2021). https://doi.org/10.1016/j.apm.2021.03.013

22. Yadav, A., Amabili, M., Panda, S.K., Dey, T.: Nonlinear analysis of cylindrical sandwich shells with porous core and CNT reinforced face-sheets by higher-order thickness and shear deformation theory. Eur. J. Mech. A/Solids **90** (2021). https://doi.org/10.1016/j.euromechsol.2021.104366

23. Binh, C.T., Van Long, N., Tu, T.M., Minh, P.Q.: Nonlinear vibration of functionally graded porous variable thickness toroidal shell segments surrounded by elastic medium including the thermal effect. Compos. Struct. **255** (2021). https://doi.org/10.1016/j.compstruct.2020.112891

24. Yaghoobi, H., Taheri, F.: Analytical solution and statistical analysis of buckling capacity of sandwich plates with uniform and non-uniform porous core reinforced with graphene nanoplatelets. Compos. Struct. **252** (2020). https://doi.org/10.1016/j.compstruct.2020.112700

25. Heshmati, M., Jalali, S.K.: Effect of radially graded porosity on the free vibration behavior of circular and annular sandwich plates. Eur. J. Mech. A/Solids **74**, 417–430 (2019). https://doi.org/10.1016/j.euromechsol.2018.12.009
26. Babaei, M., Hajmohammad, M.H., Asemi, K.: Natural frequency and dynamic analyses of functionally graded saturated porous annular sector plate and cylindrical panel based on 3D elasticity. Aerosp. Sci. Technol. **96** (2020). https://doi.org/10.1016/j.ast.2019.105524
27. Li, H., Hao, Y.X., Zhang, W., Liu, L.T., Yang, S.W., Wang, D.M.: Vibration analysis of porous metal foam truncated conical shells with general boundary conditions using GDQ. Compos. Struct. **269** (2021). https://doi.org/10.1016/j.compstruct.2021.114036
28. van Vinh, P., Huy, L.Q.: Finite element analysis of functionally graded sandwich plates with porosity via a new hyperbolic shear deformation theory. Def. Technol. **18**(3), 490–508 (2022). https://doi.org/10.1016/j.dt.2021.03.006
29. Amir, M., Talha, M.: Nonlinear vibration characteristics of shear deformable functionally graded curved panels with porosity including temperature effects. Int. J. Press. Vessel. Pip. **172**, 28–41 (2019)
30. Keleshteri, M.M., Jelovica, J.: Analytical assessment of nonlinear forced vibration of functionally graded porous higher order hinged beams. Compos. Struct. **298** (2022). https://doi.org/10.1016/j.compstruct.2022.115994
31. Liu, Y., Qin, Z., Chu, F.: Nonlinear forced vibrations of functionally graded piezoelectric cylindrical shells under electric-thermo-mechanical loads. Int. J. Mech. Sci. **201** (2021). https://doi.org/10.1016/j.ijmecsci.2021.106474
32. Quan, T.Q., Ha, D.T.T., Duc, N.D.: Analytical solutions for nonlinear vibration of porous functionally graded sandwich plate subjected to blast loading. Thin-Walled Struct. **170** (2022). https://doi.org/10.1016/j.tws.2021.108606
33. Zhou, K., Huang, X., Tian, J., Hua, H.: Vibration and flutter analysis of supersonic porous functionally graded material plates with temperature gradient and resting on elastic foundation. Compos. Struct. **204**, 63–79 (2018). https://doi.org/10.1016/j.compstruct.2018.07.057
34. Meksi, R., Benyoucef, S., Mahmoudi, A., Tounsi, A., Adda Bedia, E.A., Mahmoud, S.R.: An analytical solution for bending, buckling and vibration responses of FGM sandwich plates. J. Sandw. Struct. Mater. **21**(2), 727–757 (2019). https://doi.org/10.1177/1099636217698443
35. Rao, S.S.: The finite element method in engineering. Elsevier (2004)
36. Njim, E.K., Bakhy, S.H., Al-Waily, M.: Analytical and numerical investigation of buckling load of functionally graded materials with porous metal of sandwich plate. Mater Today Proc (2021). https://doi.org/10.1016/j.matpr.2021.03.557

Turkish Sign Language Recognition Using a Fine-Tuned Pretrained Model

Gizem Ozgul[1], Şeyma Derdiyok[1](✉), and Fatma Patlar Akbulut[2]

[1] Department of Computer Engineering, Istanbul Kültür University,
Istanbul, Turkey
`s.derdiyok@iku.edu.tr`
[2] Department of Software Engineering, Istanbul Kültür University, Istanbul, Turkey
`f.patlar@iku.edu.tr`

Abstract. Many members of society rely on sign language because it provides them with an alternative means of communication. Hand shape, motion profile, and the relative positioning of the hand, face, and other body components all contribute to the uniqueness of each sign throughout sign languages. Therefore, the field of computer vision dedicated to the study of visual sign language identification is a particularly challenging one. In recent years, many models have been suggested by various researchers, with deep learning approaches greatly improving upon them. In this study, we employ a fine-tuned CNN that has been presented for sign language recognition based on visual input, and it was trained using a dataset that included 2062 images. When it comes to sign language recognition, it might be difficult to achieve the levels of high accuracy that are sought when using systems that are based on machine learning. This is due to the fact that there are not enough datasets that have been annotated. Therefore, the goal of the study is to improve the performance of the model by transferring knowledge. In the dataset that was utilized for the research, there are images of 10 different numbers ranging from 0 to 9, and as a result of the testing, the sign was detected with a level of accuracy that was equal to 98% using the VGG16 pre-trained model.

Keywords: Sign language · Convolutional Neural Networks (CNN) · Transfer learning

1 Introduction

Communication is the process of conveying the most fundamental information, such as emotions and thoughts, to the other party using a variety of means. Although communication is a multifaceted process, language is the most effective component. Because of language, humans can execute their daily tasks with relative ease. While language speeds up communication, it is inaccessible to many individuals with hearing impairments. Every country has a sign language that is unique to its language structure. However, the fact that sign language and current grammar are often dissimilar makes it challenging for hearing-impaired

A. Ortis et al. (Eds.): ICAETA 2023, CCIS 1983, pp. 73–82, 2024.
https://doi.org/10.1007/978-3-031-50920-9_6

individuals to become literate. According to the 2018 data of the World Health Organization, there were 34 million hearing-impaired individuals in Europe, and this number is projected to increase by around 12 million by 2050[1]. It has been observed that the fact that people with hearing impairments experience communication challenges has led to a rise in the number of studies aimed at resolving this issue. The advancement of artificial intelligence research, which has gained momentum in recent years, has led to a rise in sign language research [1,2]. Serious research has been undertaken [3,4] (particularly in the fields of machine learning and deep learning). Convolutional neural networks (CNN), one of the deep learning methods, are commonly employed in domains such as image classification, similarity-based grouping, and object recognition.

Communication is essential for the continued existence of humans on earth. There are two main components in any communication: the recipient and the sender [5]. During communication, a channel is formed between the transmitter and the receiver; through this channel, many acts, such as emotions and thoughts, can be transmitted to the other side. Sign language is a visual language, that is a collection of gestures, mimics, and hand and facial movements intended for hearing-impaired people to communicate. According to the Turkish Statistical Institute's (TUIK) 2015 figures[2], there are 406 thousand disabled men and 429 thousand disabled women in Turkey.

Hearing-impaired individuals can communicate effectively with the norms they have established among themselves, but they cannot interact efficiently with other individuals or institutions. This extremely difficult-to-express mechanism generates social dysfunction. They cannot communicate themselves clearly and cannot even comprehend the other party's posts. As a result, individuals with hearing loss tend to withdraw themselves from society [6]. In 2018, there were 34 million hearing-impaired people in Europe alone, according to data released by prestigious health agencies such as the World Health Organization [7]. In 32 years, or in 2050, it is expected that this data would expand by 35.29%. Even in sports, it is quite difficult to interact with hearing-impaired individuals from diverse groups when several studies are analyzed [8]. There are more than 120 sign languages in the world [9], and although they are closely related, there are still communication gaps between them. The statistics indicate that the development of digital solutions to enhance the communication of individuals with impairments is necessary [10,11]. This study proposes a model in which the numbers in Turkish sign language will be developed with the assistance of CNN in order to contribute to the stated challenge.

2   Related Work

The scientific field of sign language recognition is expanding in the field of gesture recognition. Research on the recognition of sign language has been carried out all around the world utilizing a variety of sign languages. These sign languages

---

[1] Available at http://www.who.int/en/data-and-evidence.
[2] Available at https://data.tuik.gov.tr.

include American Sign Language [29], Chinese Sign Language [28], Japanese Sign Language [27], Turkish Sign Language [26], etc. Numerous systems for sign language recognition employ machine learning due of its capacity to train useful models using limited and sometimes noisy sensor input. There are a variety of sensor options, including data gloves and other tracker systems, computer vision approaches employing a single camera, numerous cameras, and motion capture systems, and handcrafted sensor networks.

Approaches to representing basic units of signed languages vary significantly across researchers. The simultaneous nature of significant left-hand, right-hand, and head gestures in sign languages presents a barrier for many sequential approaches [12]. Others attempt to develop models with a structure resembling phonemes, whereas the majority of studies opt to use the sign as their modeling unit of origin. Utilizing technology means allows for the possibility of locating a solution to the problem that will remove the bottlenecks that are now in place.

Examining the studies in the scientific literature reveals that image processing [13] technologies are commonly utilized to detect human limb motions. Numerous models have been developed in this direction with the contribution of deep learning models [14,15], which have recently acquired prominence in this field. Attractiveness has been drawn to the success of deep learning systems in image processing and classification. In the study of Kemalolu and Sevli [16], for instance, convolutional neural networks (CNN), one of the deep learning techniques, are utilized to train and process an image set including Turkish sign language numbers. In the process of classifying sign languages, a considerable amount of strategies and methodologies have been presented. Pigou et al. [15] completed a deep learning investigation to describe 20 Italian sign language hand movements. In the study, the results of an artificial neural network were mixed with the CNN model. As a result of this combination structure, they attained a 91.7% success rate. Bheda et al. [17] completed another investigation utilizing the deep learning model on American sign language. They utilized a small-scale dataset that they previously developed. Consequently, while expanding the datasets and utilizing them with the CNN model, a 97% success rate has developed. Kalam et al. [18] generated a total of 7000 images by rotating 700 numerals (images) in American sign language from ten different angles, yielding a total of 7000 images. By training the dataset they created using the CNN architecture, they attained a success rate of 97.28%.

3   Material and Method

This section discusses the dataset that was used, the preprocessing techniques that were implemented, as well as the CNN and pretrained models that were developed to train using the dataset.

3.1   Dataset

In this study, Turkish sign language images obtained with the participation of 218 students studying at Ankara Ayrancı Anatolian High School were used as a dataset [19]. The dataset was created in jpeg (rgb) format to represent numbers between 0 and 9 at 100 × 100-pixel resolution. Each student was asked to create 10 different numbers from 0 to 9. In this manner, 2180 image data were obtained. Figure 1 depicts a sub-example of sign images ranging from 0 to 9.

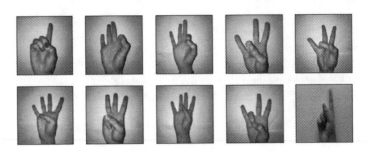

Fig. 1. Samples from Sign Language Dataset

3.2 Data Preprocessing

Red-green-blue (RGB) is the format of the study's data set. RGB (red-green-blue) channels allow for the coloring of images, although working with colored images can be challenging at times. Thus, images will be examined and analyzed in grayscale. The grayscale nature of the images renders them two-dimensional. In this scenario, image colors can have values between 0 and 255 in a single dimension. The images are normalized because the findings of investigations on one-dimensional values between 0 and 255 are often unsuccessful. Normalization identifies the minimum and maximum values of all existing numeric values in a column and reduces these values to 1. As this circumstance falls between 0 to 255 in the present investigation, the image values have been lowered from 0 to 1.

3.3 Methods

In deep learning applications, a learning model may be developed from scratch. However, transfer learning has increased the performance of models. The weights of a previously trained network can be utilized to train the initial model. When comparing these two methods in terms of performance evaluation, it has been discovered that transfer learning is quicker and more efficient. In this investigation, a model was constructed, and transfer learning techniques were utilized to train the sign language visuals. In the study, a 2D-CNN was developed, and it was fine-tuned for different pre-train models, including VGG16, ResNet50V2, EfficientNetB7, InceptionV3, and MobileNetV2, as depicted in Fig. 2. The Adam optimizer, a 0.001 learning rate (lr), and categorical cross-entropy were chosen for the optimization of the specified models. All models employ the same structure since the Adam optimizer and learning rate selections are the metrics that yield the greatest outcomes.

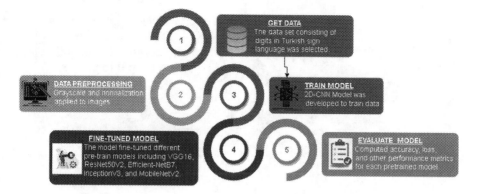

Fig. 2. A graphical representation of the concept of the research methodology.

4   Experimental Results

The aim of this study is to classify sign language gestures consisting of numbers and to further increase the performance of training with transfer learning. In the initial phase of our classification efforts, a two-dimensional CNN model was developed. After the initial 2 convolution layers, the max pooling layer was added, followed by 2 further convolution layers. It was then leveled by going through a layer of maximum pooling. Three dense layers have been traversed to reach the final layer. Ten different classes are predicted by training the last layer using the softmax activation function. Following model training, 86% accuracy was determined. The model's confusion matrix is depicted in Fig. 3. According to the basic model, the distortion rate in the images between the layers was not significant and was found to be normal.

In the second part of our experiment, the model was fine-tuned using the most prominently pretrained models from the literature [25]. We applied the several CNN architectures such as VGG16, ResNet50V2, EfficientNetB7, InceptionV3, and MobileNetV2, each of which offered distinct capabilities. *VGG16* is a convolutional neural network model developed in 2014 by a University of Oxford working group with the same name [20]. As the name suggests, there are sixteen distinct layers. When training our own dataset using the VGG16 architecture's weights, 98% accuracy is attained. In the VGG16 model, the rate of distortion and loss in the images between the layers was quite high without fine-tuning. When the fine-tuning is applied to the model the rate of distortion in images decreased. It is depicted in Fig. 4.

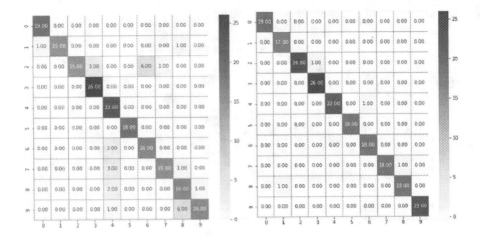

Fig. 3. The Confusion Matrix of a) base and b) fine-tuned VGG16 models

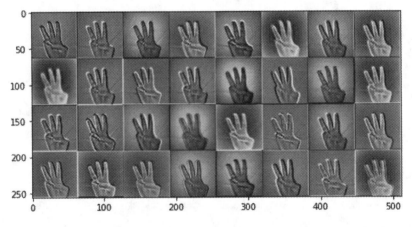

Fig. 4. Visualization of high-level feature map from $conv2d_57$ layer of the fine-tuned VGG16 model using samples from the dataset.

ResNet50V2 is another pretrained model created for imagenet database classification that is designed by Microsoft [21]. There are fifty layers. Its success percentage on the dataset of sign language remained at 89%. *EfficientNetB7*, is also a pretrained model built by Google [22], which can be classified into eight different architectures. It has evolved consistently from B0 to B7. In these cases, the EfficientNetB7 architecture enables more effective training. The proper classification success percentage for the dataset of sign language was determined to be 90%. *InceptionV3* is a model developed by Google with 50 deep layers [23]. It has the ability to classify nearly 1,000 objects using ImageNet weights. The first input size of this network is 299 × 299 pixels. When we pre-train our sign language dataset with the InceptionV3 model, the obtained accuracy value is 97%.

MobileNetV2 is a kind of convolutional neural network developed by Google for mobile display applications [24]. It is a model that uses a limited number of resources. It offers precise validation for tiny datasets. However, it has become clear in our experience that it is not suitable for a sign language dataset. The obtained accuracy value could not exceed 21%.

We adjusted models for our problem by adding a new fully connected layer for each of the 10 classes in our dataset. Backpropagation was then used to fine-tune the original CNN filter weights acquired from natural images such that they more accurately mirrored the modalities in the sign language dataset. It was decided that the VGG16 had the best performance out of all the models. The training and validation error for the 10 epochs of fine-tuned and base models are depicted in Fig. 5. The training error rates for both CNNs follow a consistent trend of a steady decline followed by a plateau. The similarity between the training and validation curves indicates that the proposed fine-tuned model did not overfit the training data.

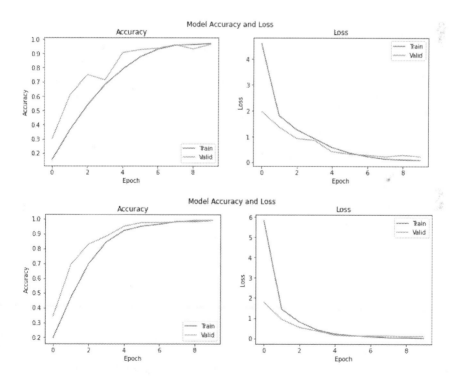

Fig. 5. The accuracy and loss scores of a) base and b) fine-tuned VGG16 models

In Table 1, all results are presented in a comparable manner.

Table 1. Classes-based classification performance of base and fine-tunes models.

Classes	VGG16	ResNet50V2	InceptionV3	MobileNetV2	EfficientNetB7	2DCNN
0	1.0	1.0	1.0	0.0	1.0	1.0
1	1.0	1.0	0.82	0.00	0.94	0.88
2	0.96	0.84	0.92	0.48	0.88	0.60
3	1.0	1.0	1.0	0.00	0.96	1.0
4	0.95	0.95	0.91	0.04	0.86	1.0
5	1.0	0.94	1.0	0.05	0.72	1.0
6	1.0	1.0	0.83	0.72	0.83	0.88
7	0.94	0.73	0.68	0.00	0.63	0.78
8	0.94	0.84	0.84	0.00	0.57	0.84
9	1.0	0.82	0.86	0.00	0.86	0.69
Avg	0.98	0.91	0.89	0.13	0.84	0.86

5   Conclusions

Despite the fact that sign language was developed to aid hearing-impaired individuals in talking with others, it is obvious that they continue to struggle with communication in society. To cover all aspects of sign languages, powerful algorithms that reliably extract characteristic features in uncontrolled contexts were developed. In this research, we present a CNN-based architecture for the classification of sign language gestures. The CNN model has a two-dimensional structure. VGG16, ResNet50V2, EfficientNetB7, InceptionV3, and MobileNetV2 were also trained using a pre-trained model to improve performance and decrease training time. We have observed that transfer learning allows for the creation of more reliable systems. The suggested model outperforms prior state-of-the-art classifiers on average with a recognition rate of 98%. The intriguing results of this study can be used as a starting point for further research into how to recognize complex hand and face movements.

References

1. Sertkaya, M., Ergen, B., Togacar, M.: Diagnosis of eye retinal diseases based on convolutional neural networks using optical coherence images. In: 2019 23rd International Conference Electronics, pp. 1–5. IEEE (2019)
2. Altuntas, Y., Comert, Z., Kocamaz, A.: Identification of haploid and diploid maize seeds using convolutional neural networks and a transfer learning approach. Comput. Electron. Agric. **163**, 104874 (2019)
3. Koller, O., Ney, H., Bowden, R.: Deep learning of mouth shapes for sign language. In: Proceedings of the IEEE International Conference on Computer Vision Workshops, pp. 85–91 (2015)
4. Huang, J., Zhou, W., Li, H., Li, W.: Sign language recognition using 3D convolutional neural networks. In: 2015 IEEE International Conference on Multimedia and Expo (ICME), pp. 1–6. IEEE (2015)

5. Elmas, N.: Örgütsel iletisimin is tatmini üzerindeki etkisi ve bir uygulama. Master Thsesis, Istanbul Ticaret University (2017)
6. Yildiz, Z., Yildiz, S., Bozyer, S.: Isitme Engelli Turizmi(Sessiz Turizm): Dunya ve Turkiye Potansiyeline Yonelik Bir Degerlendirme. Suleyman Demirel University Vizyoner Dergisi 9(20), 103–117 (2018)
7. Campbell, L.: Ethnologue: languages of the world. JSTOR (2008)
8. Togacar, M., Comert, Z., Ergen, B.: Siyam Sinir Aglarini Kullanarak Turk Isaret Dilindeki Rakamlarin Tanimlanmasi. Dokuz Eylul University Muhendislik Fakültesi Fen ve Muhendislik Dergisi 23(68), 349–356 (2021)
9. Haualand, H.: The Two Week Village-The Significance of Sacred Occasions. Disability in Local and Global Worlds. University of California Press, Berkeley (2003)
10. Murray, J.: Coequality and transnational studies: understanding deaf lives. Open Your Eyes Deaf Stud. Talk. 100, 110 (2008)
11. Wang, H., Leu, M., Oz, C.: American sign language recognition using multidimensional hidden Markov models. J. Inf. Sci. Eng. 22(5), 1109–1123 (2006)
12. Patlar, F., Akbulut, A.: Triphone based continuous speech recognition system for Turkish language using hidden Markov model. In: 12th IASTED International Conference in Signal and Image Processing, pp. 13–17 (2010). https://doi.org/10.2316/P.2010.710-059
13. Shanableh, T., Assaleh, K.: User-independent recognition of Arabic sign language for facilitating communication with the deaf community. Digit. Signal Process. 21(4), 535–542 (2011)
14. Cömert, Z., Kocamaz, A.F.: Fetal hypoxia detection based on deep convolutional neural network with transfer learning approach. In: Silhavy, R. (ed.) CSOC2018 2018. AISC, vol. 763, pp. 239–248. Springer, Cham (2019). https://doi.org/10.1007/978-3-319-91186-1_25
15. Pigou, L., Dieleman, S., Kindermans, P.-J., Schrauwen, B.: Sign language recognition using convolutional neural networks. In: Agapito, L., Bronstein, M.M., Rother, C. (eds.) ECCV 2014. LNCS, vol. 8925, pp. 572–578. Springer, Cham (2015). https://doi.org/10.1007/978-3-319-16178-5_40
16. Kemaloglu, N., Sevli, O.: Evrisimsel Sinir Aglari ile Isaret Dili Tanima. In: Proceedings on 2nd International Conference on Technology and Science, pp. 942–948 (2019)
17. Bheda, V. and Radpour, D.: Using deep convolutional networks for gesture recognition in American sign language. arXiv preprint arXiv:1710.06836 (2017)
18. Kalam, M., Mondal, M., Ahmed, B.: Rotation independent digit recognition in sign language. In: 2019 International Conference on Electrical, Computer and Communication Engineering (ECCE), pp. 1–5. IEEE (2019)
19. MAvi, A.: A new dataset and proposed convolutional neural network architecture for classification of American sign language digits. arXiv preprint arXiv:2011.08927 (2020)
20. Simonyan, K., Zisserman, A.: Very deep convolutional networks for large-scale image recognition. arXiv preprint arXiv:1409.1556 (2014)
21. He, K., Zhang, X., Ren, S., Sun, J.: Deep residual learning for image recognition. In: Proceedings of the IEEE Conference on Computer Vision and Pattern Recognition, pp. 770–778 (2016)
22. Tan, M., Le, Q.: Efficientnet: rethinking model scaling for convolutional neural networks. In: International Conference on Machine Learning, pp. 6105–6114. PMLR (2019)
23. Lin, M., Chen, Q., Yan, S.: Network in network. arXiv preprint arXiv:1312.4400 (2013)

24. Sandler, M., Howard, A., Zhu, M., Zhmoginov, A., Chen, L.H.: MobileNetV2: inverted residuals and linear bottlenecks. In: Proceedings of the IEEE Conference on Computer Vision and Pattern Recognition, pp. 4510–4520 (2015)
25. Kocacinar, B., Tas, B., Akbulut, F., Catal, C., Mishra, D.: A real-time CNN-based lightweight mobile masked face recognition system. IEEE Access **10**, 63496–63507 (2022)
26. Yirtici, T., Yurtkan, K.: Regional-CNN-based enhanced Turkish sign language recognition. Signal Image Video Process. 1–7 (2022)
27. Brock, H., Farag, I., Nakadai, K.: Recognition of non-manual content in continuous Japanese sign language. Sensors **20**(19), 5621 (2020)
28. Zhang, J., Zhou, W., Xie, C., Pu, J., Li, H.: Chinese sign language recognition with adaptive HMM. In: 2016 IEEE International Conference on Multimedia and Expo (ICME), pp. 1–6. IEEE (2016)
29. Bantupalli, K., Xie, Y.: American sign language recognition using deep learning and computer vision. In: 2018 IEEE International Conference on Big Data (Big Data), pp. 4896–4899. IEEE (2018)

Efficient Object Detection Model for Edge Devices

Hassan Imani[1]([✉]) [iD], Md Imran Hosen[2] [iD], Vahit Feryad[1] [iD], and Ali Akyol[1] [iD]

[1] Cozum Makina, Yenişehir, Özgür Sk. No: 20, 34779 Ataşehir, Istanbul, Turkey
{himani,vferyad,aakyol}@cozum-makina.com
[2] Department of Computer Engineering, Bahcesehir University, Istanbul, Turkey

Abstract. Deep learning-based object detection methods demonstrated promising results. In reality, most methods suffer while running on edge devices due to their extensive network architecture and low inference speed. Additionally, there is a lack of industrial scenarios in the existing person, helmet, and head detection datasets. This research presents an efficient tiny network (ETN) for object detection that can perform on edge devices with high inference speed. We take the YOLOv5s model as our base model. We compress the YOLOv5s object detection model and minimize the computation redundancy, and propose two lightweight C3 modules (MC3 and SC3). Additionally, we construct two novel datasets: H2 (consists of safety helmet and head) and Person104K (consists of person) that fill the gaps in the earlier datasets with various industrial scenarios. We implemented and tested our method on Person104K and H2 datasets and achieved about 50.6% higher inference speed than the original YOLOv5s without compromising the accuracy. On the Nvidia Jetson AGX edge device, ETN achieves 42% higher FPS compared to the original YOLOv5s. Code is available at https://github.com/mdhosen/ETN.

Keywords: Object Detection · Convolution Neural Network (CNN) · C3 Module · Edge Devices · YOLOv5

1 Introduction

Object detection and classification is identifying the object's location and determining the category of the object in images or video frames. With the increasing computing power capability, deep learning-based approaches have become popular for object detection. Intelligent systems perform detection and classification automatically and more effectively when machine vision is used in place of manual labor. Single target object detection, such as safety helmet, head, and person detection, has got tremendous attention in computer vision (CV) for their divers of applications, including production safety monitoring, surveillance, maritime quick rescue [10,22], and so on. Numerous studies have been reported on object detection while deep learning-based approaches have shown promising

A. Ortis et al. (Eds.): ICAETA 2023, CCIS 1983, pp. 83–94, 2024.
https://doi.org/10.1007/978-3-031-50920-9_7

outcomes. Deep learning-based existing object detection methods can be classified into two major categories: two-stage object detection networks [5–7] and single-stage object detection networks [11, 14, 15, 19].

In order to predict the category and bounding box, two-stage object detection networks establish complete connection layers and obtain proposals via region proposal network (RPN). Girshick et al., [6] proposed R-CNN which is considered as baseline of the two-stage object detection methods. R-CNN performs the object detection task in two steps. In the first step, Region of Interests (ROIs) are extracted. The extracted ROIs are regressed and classified in the second step. Authors utilized selected search to generate region proposal, and the classification network carried out prediction. However, the fixed convolution layer constrained the network's accuracy, and the network was slow for a redundant forward pass. Later, Fast R-CNN [5], and Faster R-CNN [20] were developed based on the R-CNN. In the Fast R-CNN, the authors made the training efficient through a single backward/forward pass in the convolution layer. To the contrary, Faster R-CNN proposed a RPN for a better outcome. Mask R-CNN [7] is a popular object detection technique based on instance segmentation. Mask R-CNN network design was similar to the Faster R-CNN design, and addressed the slicing misalignment to fix the ROIPooling layers observed in [20]. Though these two-stage detection approaches achieved higher accuracy, they demand costly implementation and are inefficient for real-time applications.

The main difference between one-stage detector and two-stage detector is that one-stage detector can directly predict the bounding boxes without using region proposal stage. These types of networks usually have faster inference speed. Single shot MultiBox detector [15] also known as SSD, is one of the pioneering models for one-stage object detection. By encapsulating all processing in a single network, SSD skipped proposal creation and subsequent pixel resampling phases. Though it improved detection significantly, it compromised accuracy. Later, Lin et al., [14] introduced RetinaNet, which addressed the SSD issue, and improved the performance. However, the break-through of one-stage object detection achieved through YOU ONLY LOOK ONCE (YOLO) series [2, 11, 19]. YOLO networks use single neural network and trained on end-to-end manner. They take the input image and directly predict class labels and bounding boxes. One of the efficient recent YOLO frameworks is YOLOv5. In this model, mosaic data augmentation is employed to increase the dataset's variety during the data processing step, and two effective CSPNet structures serve as the framework. YOLOv5 showed great ability in detecting high number of classes such as 80. However, one or two class object detection such as safety helmet, head, and person require minimal architecture. Therefore, compressing YOLOv5 can fit single targets more efficiently and make the model edge device friendly. Additionally, existing safety helmet, head, and person detection datasets need to be improved. We construct two novel and effective datasets for safety helmet, head, and person detection to solve the above mentioned limitations. Then, an efficient tiny network for object detection model based on YOLOv5s is proposed, capable of detecting objects with high inference speed. Overall, the significant contributions and strengths of our work are summarized below:

- We propose an efficient tiny object detection network based on YOLOv5 that can detect objects with higher speed and can be implemented on edge devices.
- We propose two novel object detection datasets: H2 and Person104K. H2 consists of 37K images annotated with safety helmet and head. Additionally, the Person104K dataset contains 104K images. Both datasets include real-world and industry scenarios.
- We have validated our methodology through a considerable experiment and ablation research on H2 and Person104K datasets. Additionally, we have implemented and tested our model on Nvidia Jetson AGX edge devices.

The rest of the paper has been arranged as follows: Sect. 2 describes the Methodology, Sect. 3 presents Dataset Development, Experiments and Discussion has been discussed in Sect. 4, and we concluded our works in Sect. 5.

2   Methodology

The YOLOv5 model has five types, including YOLOv5x (extra large), YOLOv5l (large), YOLOv5m (medium), YOLOv5s (small), and YOLOv5n (Nono). The network design is similar among these types, but they vary in network depth and width. Higher network depth and width provide better precision; however, they suffer high training costs with low inference speed. We take YOLOv5s architecture as our base model. The overall architecture has been shown in Fig. 1. The YOLOv5 model consists of four primary modules: Convolution, Bottleneck, C3, and SPPF modules. We propose three C3 modules: Original C3, Modified C3 (MC3), and simple C3 (SC3). Additionally, two different bottleneck blocks have been used. Finally, we have implemented FReLU [18] activation instead of Sigmoid-weighted Linear Unit (SiLU) [4].

Convolution Module. Convolution module consists of three basic operations including convolution (Conv2d), batch normalization(BN) and activation function (FReLU). First, the input goes through the convolution layer, which extracts features. Then, the BN enables regularization and accelerates the learning rate. By computing the weights and then adding bias to it, the FReLU activation function determines whether or not a neuron should be activated.

Bottleneck Module. Bottleneck module is used to reduce parameters and minimize the calculation. Data training and feature extraction may be carried out more successfully and intuitively after dimensionality reduction. In ETN, we have used the original bottleneck module and modified simple Bottleneck (SBottleneck) shown in Fig. 2. Since the original C3 module used a redundant skip connection, we removed the short connection from the SBottleneck module and used one convolution instead of two.

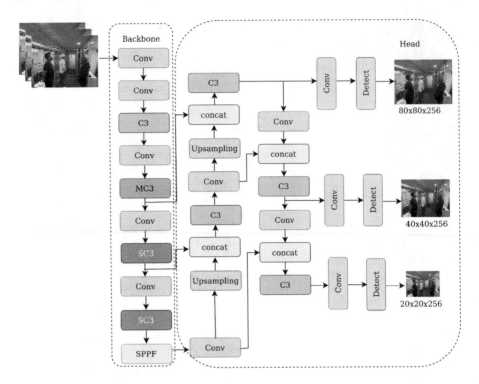

Fig. 1. The architecture of the proposed ETN. The architecture is based on the YOLOv5. C3 module is used along with lightweight MC3 and SC3 module in backbone.

C3 Module. C3 module, a modified version of CSP (cross stage partial layer) bottleneck [21] is the basic building block of YOLOv5 to learn the residual characteristics. It consists of three standard convolution and multiple bottleneck modules, determined by the network depth multiple. C3 uses the same structural design as CSP. However, the correction unit in C3 is different. Additionally, the convolution module is used in the CSP bottleneck after the residual output is removed in C3. Two branches make up the structure. One concatenates the two branches after passing through only a straightforward convolution module, whereas the other employs numerous bottleneck stacks and three regular convolution layers.

The C3 module of YOLOv5 includes n residual blocks and a shortcut structure that joins two neighboring layers of the networks. Multiple residual modules in the C3 module for single targets like a safety helmet, head, and person detection may waste the resources. To compress the model and remove the redundant calculation, we have used two other C3 module (MC3 and SC3) along with the original one. The structure of C3, MC3, and SC3 has been shown in Fig. 2.

MC3 module uses SBottleneck instead of Bottleneck module. It uses one convolution module instead of two and removes the skip connection. On the

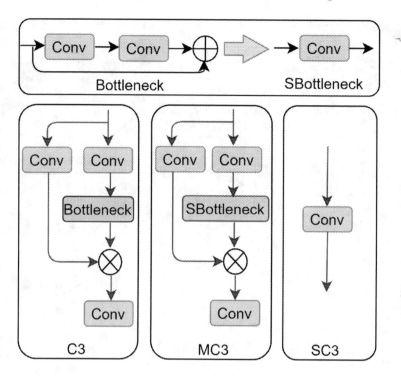

Fig. 2. The structure of C3 and Bottleneck module along with their modified lightweight versions.

other hand, SC3 contains only simple convolution. We placed the SC3 module after C3 and MC3. Since the model implements C3 and MC3 (comparatively large modules) at the beginning of the backbone, the model already learns most of the features. Therefore, we put a simple C3 (SC3) module at the end that reduces the number of layers and model weight.

SPPF Module. SPPF is the optimized version of SPP [8]. The acronym SPP stands for spatial pyramid pooling. The input channel is divided first using a conventional convolution module, and then max-pooling is performed using kernel sizes of 5, 9, and 13. It concatenates the outputs of the three maximum pooling operations with the original data, and the number of channels is doubled after the final merging. This module aims to merge features with different resolutions and provide better feature representations.

Activation Function. The Rectified Liner Unit, mostly known as ReLU [1] is a widely used activation function for its low computational cost. However, it suffers a problem known as dying ReLU for the negative outputs; some neurons die or stop producing anything other than zero throughout training. ReLU can be defined as Eq. 1.

Fig. 3. Examples of samples from H2 and Person104K datasets. First two rows show the sample images from the H2 dataset, while second two rows show samples from the Person104K dataset.

$$\text{ReLU} = \begin{cases} x & if\, x > 0 \\ 0 & if\, x \leq 0 \end{cases} \tag{1}$$

where x is the input.

Flexible Rectified Liner Unit (FReLU) is a variant of ReLU. FReLU can adaptively rectify the ReLU output to collect negative information and offer zero-like behavior. FReLU can be defined as Eq. 2.

$$\text{FReLU} = \begin{cases} x + b_i & if\, x > 0 \\ b_i & if\, x \leq 0 \end{cases} \tag{2}$$

where b_i is the i-th layer learnable parameter. When $b_i = 0$, FReLU acts like ReLU.

We have used FReLU instead of SiLU [4] in the convolution module (most of the modules use a convolution; therefore, it affects overall architecture). While SiLU performs better in reinforcement learning-based systems [12], FReLU performs better for a small network.

3   Dataset Development

To train the proposed model, we need to create a suitable dataset. We have constructed two novel datasets to train and evaluate our model: (i) helmet and head detection dataset, namely "H2," and (ii) a person detection dataset name "Person104K". The developed datasets are constructed from real-world scenarios, and they have the following major characteristics: (a) a sufficient number of data with proper labeling to train and validate the proposed model, (b) different environment condition and visualization, (c) divers data (e.g., different types of helmet). Figure 3 shows some samples from both datasets.

Table 1. Details of our newly developed H2 and Person104K dataset.

Dataset	#Images	#Helmet	#Head	#Person
H2	37K	70K	160K	–
Person104K	104K	–	–	248K

H2 Dataset. Our proposed H2 dataset consists of 37K images, and annotated with two classes of helmet and head. All of the images are collected from real-world and industry scenarios, and annotated manually. While collecting data, we have maintained the major characteristic of the standard dataset such as: different environment conditions (low and high light, captured at day and night, indoor and outdoor), visualization (full and partial), diverse data types. For instance, first row in Fig. 3 contains images with low light, while the second row contains multiple instances with divers samples.

Person104K Dataset. The Person104K dataset contains around 104K images collected from surveillance cameras from various industries to provide real-world prospects. An example of the dataset has been provided in Fig. 3 (second two row). The 3rd row contains samples of partial visualization, while the 4th row shows images which are captured in different lighting conditions. Table 1 shows the details of both datasets.

4   Experiments and Discussion

4.1   Training Details

We use PyTorch (1.10) library with one Nvidia GeForce RTX 2070 GPU and 64GB RAM with Linux OS as our main workstation to implement our proposed

Table 2. Efficiency comparison between YOLOv5s and ETN. The best results are in **bold**.

Method	#Layers	#Param.	GFLOPs	FPS	Model Size
YOLOv5 (default)	157	7.2M	15.8	161	56.8 MB
ETN (Ours)	**112**	**794K**	**3.5**	**250**	**1.9 MB**

experiment for training. We have used the default YOLOv5 hyperparameters setting with image size 640 × 640. The dataset has been divided into training and validation sets with ratio of 7:3. Additionally, we employ data augmentation techniques including random flip, geometric distortion, and lighting distortion. The Adam optimizer has been used in our experiments. We have utilized the YOLOv5 default loss function. The investigation runs for a total of 60 epochs with a batch size of 32.

Evaluation Matrix. Models for object detection predict the category and bounding box of objects in an image. To evaluate the performance of our model and compare it with original YOLOv5s and state-of-the-art method, we have assessed it in terms of precision and mean average precision. Higher values of these metrics indicate better output quality.

Table 3. The quantitative performance of YOLOv5s and ETN. The best results are in **bold**.

Method	Precision	mAP50	Dataset
YOLOv5s	**0.98**	**0.99**	Person104K
ETN (Ours)	0.97	0.98	Person104K
YOLOv5s	**0.96**	**0.97**	H2
ETN (Ours)	0.92	0.93	H2

4.2   Result Analysis

Figure 4 shows the qualitative results of ETN. ETN can detect safety helmets, heads, and persons accurately and confidently, even with narrow descriptions and diverse environmental conditions. For instance, from left-top to right-bottom, in the 2nd, 3rd, 4th, and 11th images, the target object is not clear enough, and only some part is visible; however, our method successfully detects the target object. Additionally, ETN detects objects even in diverse condition such as low light (e.g., 6th, and 7th image), indoor (1st and 5th) and outdoor (8th, 10th and 12th).

Table 2 compares the performance of ETN and the original YOLOv5s model in terms of the number of layers, parameters, GFLOPs, inference time, and model size. Additionally, we provide the results of our proposed approach and

Fig. 4. Visualization of the detection results from ETN. First and second two rows are from the testing set of H2 dataset and Person104K datasets, respectively.

original YOLOv5s in terms of precision (P) and mean average precision (mAP50) in Table 3 for a detailed comparison. While the original YOLOv5 contains 157 layers and has 7.2M parameters, ETN has only 112 layers and 794K parameters. Also, the number of GFLOPs for YOLOv5s is 4.5× more than ETN (YOLOv5: 15.8, ETN:3.5). Due to light architecture, ETN provides efficient performance compared to YOLOv5. For instance, ETN can detect person at 244 FPS while YOLOv5 detect at FPS 161. Additionally, the ETN model weight is 1.9 MB when YOLOv5 weight is 56.8 MB. ETN achieves almost similar accuracy for the Person104K dataset. Although ETN shows slightly lower performance for the H2 dataset, its efficiency compensates for the accuracy. Also, edge device demand trade between the accuracy and lightness of the model.

To compare the state-of-the-art method, we adopt Koksal et al. [13] approach and compare the performance with ETN shown in Table 4. ETN gets almost the same mAP5 as the Koksal method; however, the Koksal method usages 3× more memory, and training time is double that of ETN.

Table 4. Quantitative comparison of state-of-the-art method and ETN on Person104K dataset.

Method	mAP50	Memory Usage	Training Time (H)
Koksal et al., [13]	0.98	3510 MB	22.0
ETN (Ours)	0.97	1240 MB	10.0

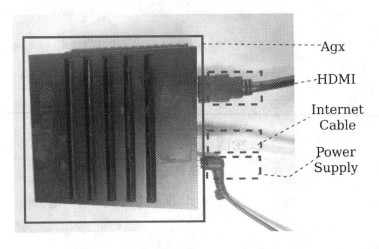

Fig. 5. Visualization of the Nvidia Jetson AGX edge device.

Performance on Edge Devices. Edge devices are lightweight and small that can run efficient computer vision models. Edge computing moves data processing and storage closer to the data's source [17]. Since an edge device can run the processings on the data inside the device, it reduces data transmission costs and vulnerability. Therefore, industries are trending toward the usage of edge devices. However, for being efficient, edge devices demand high accuracy with a small model. We implement the YOLOv5s and ETN on the Jetson AGX shown in Fig. 5. Based on our experiments, ETN achieves significantly better inference for person detection on the edge device. For instance, while keeping the accuracy the same, ENT provides 50 FPS, but YOLOv5 achieves only 35 FPS. The performance on edge devices further proves the efficiency and applicability of ETN. It is important to note that we use the Jetson AGX with low power and Float32 precision.

4.3 Extensive Experiments

We investigated several network designs and depths for ETN. We realized that only reducing the number of parameters can not reduce the inference time, while it decreases the performance. For instance, keeping original structure as YOLOv5s, and reducing the output channel size by half decreases the number of parameters significantly (From 7.2M to 1.7M). However, with this change, the inference time remains the same. C3 has a significant influence on the model parameters. Therefore, we searched for different C3 modules to improve performance and strike a balance between performance and training costs. Adding a single unit of C3 in the backbone increases the number of parameters by 300K while adding MC3 increase the number of parameters by 30K.

We also tested depth-wise convolution. With using depth-wise convolution in our model, though the model parameters decreased by half and GFLOPs decreased around 1.2, we observed that it does not help, and the performance decreased.

The activation function has great influence on the model performance. For instance, using FReLU instead of SiLU improves the inference speed more than 10 FPS. Other activation's such as ELU [3], HardSwish [9], and AconC [16] could not improve the performance, and they are memory inefficient.

For the loss function, we went through different losses such as Varifocal loss [23], loss proposed in [13] and so on. We found YOLOv5 default loss function is an efficient choice for our model's training.

5 Conclusion

This paper introduces a novel efficient tiny network based on YOLOv5 that detects objects with high inference speed while keeping the model accurate. The MC3 and SC3 modules removed the redundant calculations, and significantly reduced model size. We constructed H2 and Person104K datasets which contain sufficient instances and help the model train better. ETN has been evaluated and compared with YOLOv5s and cutting-edge methods on the proposed datasets, showing our model's efficacy. Testing on the Nvidia AGX device showed the model's applicability on edge devices. In future work, we will implement the model for more classes while keeping it efficient.

Acknowledgements. This work is supported by Cozum Makina Corporation.

References

1. Agarap, A.F.: Deep learning using rectified linear units (ReLU). arXiv preprint arXiv:1803.08375 (2018)
2. Bochkovskiy, A., Wang, C.Y., Liao, H.Y.M.: YOLOv4: optimal speed and accuracy of object detection. arXiv preprint arXiv:2004.10934 (2020)
3. Clevert, D.A., Unterthiner, T., Hochreiter, S.: zheng2020distance and accurate deep network learning by exponential linear units (elus). arXiv preprint arXiv:1511.07289 (2015)
4. Elfwing, S., Uchibe, E., Doya, K.: Sigmoid-weighted linear units for neural network function approximation in reinforcement learning. Neural Netw. **107**, 3–11 (2018)
5. Girshick, R.: Fast r-cnn. In: Proceedings of the IEEE International Conference on Computer Vision, pp. 1440–1448 (2015)
6. Girshick, R., Donahue, J., Darrell, T., Malik, J.: Rich feature hierarchies for accurate object detection and semantic segmentation. In: Proceedings of the IEEE Conference on Computer Vision and Pattern Recognition, pp. 580–587 (2014)

7. He, K., Gkioxari, G., Dollár, P., Girshick, R.: Mask r-cnn. In: Proceedings of the IEEE International Conference on Computer Vision, pp. 2961–2969 (2017)
8. He, K., Zhang, X., Ren, S., Sun, J.: Spatial pyramid pooling in deep convolutional networks for visual recognition. IEEE Trans. Pattern Anal. Mach. Intell. **37**(9), 1904–1916 (2015)
9. Howard, A., et al.: Searching for mobilenetv3. In: Proceedings of the IEEE/CVF International Conference on Computer Vision, pp. 1314–1324 (2019)
10. Jha, S., Seo, C., Yang, E., Joshi, G.P.: Real time object detection and tracking system for video surveillance system. Multimed. Tools Appl. **80**(3), 3981–3996 (2021)
11. Jocher, G.: ultralytics/yolov5: v6.0 – YOLOv5n Nano models, Roboflow integration, TensorFlow export, OpenCV DNN support, October 2021. https://doi.org/10.5281/zenodo.5563715
12. Jung, H.K., Choi, G.S.: Improved YOLOv5: efficient object detection using drone images under various conditions. Appl. Sci. **12**(14), 7255 (2022)
13. Köksal, A., Tuzcuoğlu, Ö., İnce, K.G., Ataseven, Y., Alatan, A.A.: Improved hard example mining approach for single shot object detectors. In: 2022 IEEE International Conference on Image Processing (ICIP), pp. 3536–3540. IEEE (2022)
14. Lin, T.Y., Goyal, P., Girshick, R., He, K., Dollár, P.: Focal loss for dense object detection. In: Proceedings of the IEEE International Conference on Computer Vision, pp. 2980–2988 (2017)
15. Liu, W., et al.: SSD: single shot multibox detector. In: Leibe, B., Matas, J., Sebe, N., Welling, M. (eds.) ECCV 2016. LNCS, vol. 9905, pp. 21–37. Springer, Cham (2016). https://doi.org/10.1007/978-3-319-46448-0_2
16. Ma, N., Zhang, X., Liu, M., Sun, J.: Activate or not: learning customized activation. In: Proceedings of the IEEE/CVF Conference on Computer Vision and Pattern Recognition, pp. 8032–8042 (2021)
17. Muniswamaiah, M., Agerwala, T., Tappert, C.C.: A survey on cloudlets, mobile edge, and fog computing. In: 2021 8th IEEE International Conference on Cyber Security and Cloud Computing (CSCloud)/2021 7th IEEE International Conference on Edge Computing and Scalable Cloud (EdgeCom), pp. 139–142. IEEE (2021)
18. Qiu, S., Xu, X., Cai, B.: FReLU: flexible rectified linear units for improving convolutional neural networks. In: 2018 24th International Conference on Pattern Recognition (ICPR), pp. 1223–1228. IEEE (2018)
19. Redmon, J., Farhadi, A.: YOLOv3: an incremental improvement. arXiv preprint arXiv:1804.02767 (2018)
20. Ren, S., He, K., Girshick, R., Sun, J.: Faster r-cnn: towards real-time object detection with region proposal networks. Adv. Neural Inf. Process. Syst. **28** (2015)
21. Wang, C.Y., Liao, H.Y.M., Wu, Y.H., Chen, P.Y., Hsieh, J.W., Yeh, I.H.: CSPNet: a new backbone that can enhance learning capability of CNN. In: Proceedings of the IEEE/CVF Conference on Computer Vision and Pattern Recognition Workshops, pp. 390–391 (2020)
22. Yu, X., Gong, Y., Jiang, N., Ye, Q., Han, Z.: Scale match for tiny person detection. In: Proceedings of the IEEE/CVF Winter Conference on Applications of Computer Vision, pp. 1257–1265 (2020)
23. Zhang, H., Wang, Y., Dayoub, F., Sunderhauf, N.: VarifocalNet: an IoU-aware dense object detector. In: Proceedings of the IEEE/CVF Conference on Computer Vision and Pattern Recognition, pp. 8514–8523 (2021)

Smart Locking System Using AR and IoT

Varun Deshpande, P. Vigneshwaran[✉], and Nama Venkata Vishwak

Department of Networking and Communications, SRM Institute of Science and Technology,
Chennai 603 203, India
vigneshwaran05@gmail.com

Abstract. Modern smart door locks are more prone to damage and attacks, which will reduce robustness. Most smart door locks rely on passcode entry or face recognition outside the door which makes it easy to track and more vulnerable to attacks. Extended (XR) is an emerging term used to denote the amalgamation of multiple immersive technologies such as augmented reality (AR), virtual reality (VR), and mixed reality (MR). Recent research revealed that more than 60% of respondents believed XR will be mainstream in the next five years. Considering the features provided by the XR in terms of visualization and human interaction, XR has a wide range of applications. In this article, we propose the integration of XR with IoT devices to create a Smart door locking system that will be operated using smart glasses or mobile devices. The proposed system aims to implement a smart door-locking system to overcome physical attacks, as no part of the lock is physically intractable. MR is used to take the secret code as input from the user through Smart glasses or mobile devices, which is verified by the mobile application and the corresponding control signal to unlock the door is sent to the NodeMCU if the password is correct. The contactless feature of the lock makes it suitable in hospitals to prevent the spread of diseases and prevent users from touching radioactive components in radioactive areas.

Keywords: Smart Locks · Augmented Reality (AR) · IoT Devices · Virtual Reality (VR) · Home Security · Extended reality (XR)

1 Introduction

Augmented Reality (AR) is the fast-growing technology that is used to superimpose digital data such as images, videos and 3D objects on the Physical world. It is a major part of industry 4.0 as it improves how data is accessed and used. AR using a camera scans the environment and provides the necessary information. It can be used to educate and guide people while they work on complex tasks. Extended Reality (XR) extends the working of AR to interact with virtual environments. Internet of Things (IoT) refers to a network of computing devices (nodes) that includes actuators, sensors, software, servers and communication technologies used for connecting and communicating data with other nodes over the internet.

A. Ortis et al. (Eds.): ICAETA 2023, CCIS 1983, pp. 95–108, 2024.
https://doi.org/10.1007/978-3-031-50920-9_8

Although AR and IoT are two different fields, both depend on the data from the environment and sensors and thus complement each other. So, this combination opens doors to a wide range of interactive and web-based applications. Considering the advantages and the possible wide range of applications, the proposed system focuses on the usage of AR and IoT-based applications to access Smart Lock.

A smart locking system is an electromechanical locking device that authenticates the user and opens wirelessly. It is an extension of the smart home network. It allows users to access their smart home without the traditional key through interconnected devices to make this smart locking system less susceptible to attacks. To increase accessibility, we are integrating IoT and AR into the smart locking system so that the user can use his/her smart glasses/mobile devices to access the lock without any physical contact.

The proposed system was formulated in the recent light of Covid-19 regulations where the emphasis was placed on contactless smart devices. The proposed system has been designed to provide security while being contactless and less prone to physical tampering compared to other smart locks. And the proposed system is developed to support the upcoming Smart devices such as smart glasses, smart lenses and other AR devices.

2   Related Works

Rauschnabel, P. A et al. [1] AR can give precise instructions and feedback for manufacturing to make high-quality products. There are a lot of AR glasses coming into the market. These glasses are used to give training and guidance for the assembly of complex systems. XR is the least developed technology in the world. VR is most developed as it has been used in many games before.

Gong, L., et al. [2] Hardware parameters for XR are Field of View (FOV) and Frames per Sec (FPS). Human eyes have a FOV of 114 horizontally. VR headsets have a FOV of 90–110. Headsets with smaller FOVs have a tunnelling effect. For FPS, it is recommended that 60 frames per second are good, although 90 frames per second should be strived for. The most common platform for AR VR development used is Unity.

Wang, J et al. [3] Collaborative AR environment improves applicability as it can be used. Multi-user AR collaborative system is composed of MUCstudio, MUC view and MUCserver. MUCview is used for AR experience, MUCserver is used for database and collaborative experience and MUCstudio is used for generating 3D content.

Kassem, A et al. [4] proposed that a central control module should be embedded in the door itself which is connected to the local area network (LAN) of the house itself. This provides a robust mechanism for the door and access to the door is limited through the LAN only. He also proposed an offline system to open the door in case of connection loss. The proposed system uses a master key which is stored in both the smart lock system as well as the mobile application to open it.

Kwok, A. O et al. [5] Tourism is highly dependent on location connectivity and destination accessibility. During the covid period, it was affected a lot due to the lockdown. But here XR comes as a saving grace as not only one can have an immersive experience but also stay safe.

Hadis M. S et al. [6] proposed a smart lock system that uses Bluetooth as a medium of communication between the smart Lock system and its mobile application. The smart lock system is made in a manner so that people with disabilities can also access it without contact. The proposed system given is that the smart lock is embedded in the door and the mobile application communicates through Bluetooth. Whenever the mobile is near the door, the smart lock opens. This system is carried out using two different areas concerning the smart lock system 1) Bluetooth area and 2) validation area. For the door to unlock the authorized user has to be in both the Bluetooth area as well as the validation areas. If an unauthorized user is in the validation area, and the authorized person is in the Bluetooth area then access will be denied.

Masood, T et al. [7] AR is an integral part of industry 4.0 because it visualizes and projects digital information over the physical world and supports human interaction making it more accessible. The aggregate market is projected to reach 75 million dollars in 2025. AR allows access to digital information and overlaying of that info over the physical world. Although the efficiency of AR depends on the task at hand, experimentally it is observed that AR-supported tasks are efficient in terms of timely completion and have less error rate. The effectiveness of the AR system also depends on the work experience of the personnel.

Egger, J et al. [8] The vision of Industry 4.0 is to build cyber-physical production systems (CPPS) which connect the physical and the digital world seamlessly. This led to industries moving towards smart factories, which utilize concepts like predictive maintenance or extensive machine-to-machine communication. AR allows us to interact with the digital world of the smart factory. AR makes huge amounts of data generated by CPPS accessible to humans in real-time and hence is a big part of smart manufacturing.

Danielsson, O et al. [9] Implementations of AR can be categorized as head-attached, hand-held and spatial. Head-attached AR, especially smart glasses, are becoming popular due to their lightweight and hand-free operation. These glasses project information on the physical world. AR smart glasses increase the efficiency of the operators in production lines as they don't have to stop working to refer to the instruction books, but those instructions can be displayed using AR.

Saidin, N. F et al. [10] AR enables users to interact with virtual and real-time applications, making the learning process more active and meaningful. This also helps in better understanding and increases the ability to retain the information for a longer time.

Evangelos Anastasioua et al. [11] XR has a huge potential in the primary sector of society. As it can be used in many sectors such as aquaculture, livestock farming and agriculture. There are many ways and devices that allow human–machine interaction in XR environments for the primary sector. These use cases can be achieved by using various XR controllers and viewers that allow interaction with XR environments and help users in achieving higher performance and engagement compared to traditional methods.

Pringle, J. K et al. [12] Extended reality can play a major role in learning and teaching in forensics as it can be accessed irrespective of time and apparatus available, can be used asynchronously and repeatedly interrogated to reinforce learning, and helps users in exploring different strategies with lesser time and can be to simulate any physical fieldwork environment and practical classes. However, the Generation of XR virtual learning environments are time-consuming and the educational eGame needs significant user-experience design expertise, computer programming skills, scientific input and, evaluating and refining the product. Which makes the design and production of applications complex and expensive.

Takanori Sasaki et al. [13] Usage of Mixed reality (MR) has brought a revolution in the field of Oral Surgery. Which is alveolar bone grafting for cheilognathopalatoschisis, temporomandibular joint mobilization for ankylosis, resection of a mandibular calcifying epithelial odontogenic tumour, genioplasty for jaw deformity, open reduction and internal fixation for a mandibular fracture. The usage of MR in complex operations has helped surgeons in making precise and more accurate actions while operating. The visibility of the three-dimensional images was good, and preoperative image information could be observed in real-time.

Alyousify, A. L et al. [14] An AR-based software scans Target Images on a printed book and renders 3D objects and alphabet models and sound and sound. The sound and 3D representation of the letter are played as soon as a target image is detected. The target image depicts 3D objects of the alphabet and the sound of the letter's pronunciation, and an example of a 3D item of a word containing the letter, with the word itself shown on the target image. Augmented reality is in high demand in the educational sector, particularly for educating children.

Pinjala, s et al. [15] proposed a remotely accessible smart lock system using raspberry pi with an HD camera. The proposed system works in the following way: The visitor rings the bell at the door. The visitor's live video is sent to the owner via the raspberry pi to the mobile application. The owner can willingly let the visitor access his house by typing the preset password to the smart lock system. The owner can also leave a voice note to the visitor by typing the message in the app.

Kim, S et al. [16] User-created automation applets, which are used to connect IoT devices and applications, have gained popularity and are widely available. IoT application network with the data of the IFTTT (if this then that) platform is one of the most popular platforms for the self-automation of IoT devices. Triggers and actions are specified using IFTTT applets, which are usually the set of actions to be performed for their corresponding triggering actions.

Zamora-Antuñano, M. A et al. [17] Three phases of AR: first is scanning and recognition of the environment. Second, the virtual information provided is processed and aligned. Finally, the virtual information is augmented onto the physical world. This can be used to integrate with embedded systems for various purposes. Liu and Li presented an AR solution for drone-based building inspection by integrating UAV inspection workflow with the building information model.

Stark, E et al. [18] The following consists of the workflow to be followed while integrating AR and mechatronic devices. The mechatronic system can be recognized by the AR application by scanning the trigger image from the camera. This Application then connects to the server where the mechatronic system's data is stored and its twin is stored in the cloud. The data from the sensors of the mechatronic device is sent to the server, and the device twin in the cloud is synchronized. The application downloads the definition of the user interface and draws a graphical interface for control and monitoring of the system by obtaining information about the mechatronic device. The graphical interface in augmented reality can be used by the user to interact with the mechatronic device. Control commands are sent to the server, which sends them to the connected mechatronic device.

Zhu, Z et al. [19] The implementation of a smart lock based on openCV and Efficient Altitude tracking mechanism made the door lock easily accessible, easy to set up, and easy to operate, less susceptible to errors and damage, and made the lock more secure than the traditional Locks. It also provides the user with the data of the person accessing the door lock.

Croatti, A et al. [20] proposed the combination of Web of Things and Augmented Reality is referred to as Web of Augmented Things (WoAT). WoAT can be considered an extension of WoT. Data generated by different things in WoT is typically collected, stored and managed in the cloud. Thus, in the WoAT it should also reflect the information on augmented entities making a massive impact on people working and collaborating. He also sought the use of augmented or mixed reality in the management and diagnosis of several complex equipments to which only a select number of people have access. This eliminates the requirement for installing physical control panels, to which even a regular employee may gain access.

Park, S et al. [21] Brain-computer interface is an emerging technology that helps with the interaction of devices with brain waves. Integrating BCI (Brain computer interface) with AR and IoT can give hands-free operation of smart devices in the smart home. Experiments conducted showed that it took an average of 2.6 s to switch a device on/off and a false positive rate of the switch operation was 0.015 times/min.

Michael W. Condry et al. [22] proposed the usage of a framework for employing control system gateways to address the issues in security issues, paying special attention to a method of connecting IoT devices directly to control system gateways. Multifactor authentication and authorization over an encrypted communication channel are added to increase the security, and "Real-Time Identity Monitoring" is also included to continuously ensure a legitimate user identity." (Fig. 1).

3   Methodology

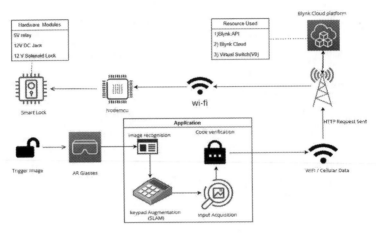

Fig. 1. Architecture Diagram

An AR based application is made using a trained model to detect a specific target image using Harris corner feature points. The smart door has the target image which is recognized by the AR application. The AR application then uses Simultaneous localization and mapping algorithm (SLAM) to augment the keypad on the target image. SLAM also positions the keypad in a real-time scenario to sync the keypad with the movement of the target image. The keys in the keypad are jumbled so no one can guess the password by the pattern. The application uses the camera to capture the input as the user interacts with the augmented virtual keypad using his fingers. The application then calculates the difference in feature points of the actual image and the real-time image to decide the number entered by the user. This Secret code is captured and verified by the AR application. If the Secret code is correct, an HTTP request is sent to the Blynk platform to switch the value of the virtual pin(V0) to 1. The V0 value is then taken by the NodeMCU. The V0 is attached to the D1 GPIO pin of NodeMCU. If the value is of V0 1 the door will be unlocked else it remains locked. If the input pin doesn't match the actual pin it notifies the user regarding the invalid attempt to access (Fig. 2).

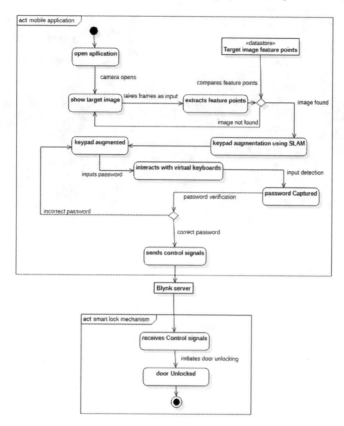

Fig. 2. UML-Activity Diagram

4 Hardware Implementation

The proposed system aims to make a robust locking system. We introduce a door that neither has a door-knob nor a keyhole. This door design makes the door secure from the physical tampering of locks. A smart lock module is attached to the door from the inside consisting of NodeMCU which works in hand with Blynk. Blynk is a platform for IoT applications in which the Virtual pin from the previous chapter exists. A dedicated virtual pin(V0) is constantly synchronized by the NodeMCU so that the given value is reflected during the execution of the program. An Output pin of the NodeMCU is connected to the relay as NodeMCU works in the range of 3 V while the solenoid lock requires a 12 V input signal so this relay is used to handle high-power lock circuits with low-power NodeMCU output. The VCC terminal of the relay is connected to 3V3 NodeMCU, the Common terminal of the relay is connected to the positive terminal of the 12 V DC jack the solenoid lock is connected to the NO (Normally Open) terminal of the relay, the NO relay contact is the pin that is open when the input to relay is LOW. That is If the virtual pin value is 0, the output of the NodeMCU pin is LOW thus unlocking the door and if it's 1 then HIGH is produced by the NodeMCU thus locking the door (Fig. 3).

Fig. 3. Smart Lock Circuit (Model Door)

5 Algorithms

5.1 Harris Corner Detection

Harris corner algorithm is used to detect all the corner points that are in the image; These points are used as feature points to detect the target image. These points are also used in the image difference algorithm to detect interactions with virtual buttons.

Algorithm:

Step 1: Take the input Image.

Step 2: Convert the input image from colour to greyscale.

Step 3: This greyscale is converted into a matrix holding the intensity in terms of numbers (0–255) of each pixel

Step 4: A corner in an image is detected when the differentiation of image intensity with respect to the x and y directions is different. I.e rapid change of image intensity in two different directions.

Step 5: Visualize each corner using a marker (Fig. 4).

Fig. 4. Harris corner detection

5.2   Simultaneous Localization and Mapping (SLAM)

Algorithm to track feature points and Augment digital information. This algorithm is used to Map Augmented data onto a set of feature points so that the information is localized to that set and does not randomly appear on the image. This algorithm also helps in tracking the image dynamically and augments the data in an efficient manner when there is a change in camera distance, angle etc.

Algorithm:

Step 1: Take real-time input from the camera.

Step 2: Detect the image from the real-time input using harris corner.

Step 3: Map and augment the virtual buttons to a set of distinct localized and predetermined feature points.

Step 4: Update the augmented virtual buttons in each frame with respect to their corresponding feature points (Fig. 5).

Fig. 5. The following figure shows the input captured by the application through AR interactions.

5.3 Image Difference Using Feature Points

In this module the differences in the actual image and the real-time image when the finger interacts with a virtual button. The idea behind this module is that while the finger is kept between the camera and the image, it hides some of the feature points. These hidden feature points are mapped to a particular virtual button. From the difference, the hidden feature points can be highlighted and the corresponding virtual button is entered.

Algorithm:

Step 1: Take real-time input from the camera.

Step 2: Extract the Feature points of the image using the Harris corner algorithm.

Step 3: When the user uses his finger, the finger covers a set of feature points.

Step 4: This new image with the finger is captured by the camera and compared with the original image. The comparison results in feature points that are hidden. The results are bordered with rectangular areas.

Step 5: Each set of feature points in the image is mapped to a number. The application then uses this result to decide the corresponding number pressed. Each frame is taken as an input.

Step 6: For example, if the area covered has the number 7 then the application takes the number 7 as input

6   Result and Discussions

6.1   Flaws in Existing Smart Locking Systems

Existing smart lock systems use RFID, Bluetooth, keypad, biometric system etc. These technologies can be eavesdropped on, intercepted or tampered with to create a Denial of Service attack (DOS). Due to intractable input devices and interceptable communication methods, the existing smart locking devices pose multiple security threats.

6.2   Advantages of the Proposed System Over the Existing Systems

A single application to access multiple doors with a smart lock. Unique images on the door are used to identify each door. The intractable AR application is to achieve contactless door access. Applications can be deployed on various platforms such as android, iOS, windows, smart glasses, smart lenses etc. Doesn't require the user to carry a key or an access card. Usage of the Internet to access the smart lock minimises the threats from eavesdrops and interception of communication. Silent and less access time. Safe from external weather and physical threats. Implementation of Two-factor authentication to provide better security (*something you have* - user dedicated Mobile application, *something you know* - passcode).

Given below are images of working prototype of the proposed system (Figs. 6 and 7).

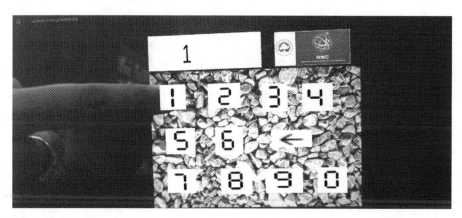

Fig. 6. The following figure shows the input captured by the application through AR interactions.

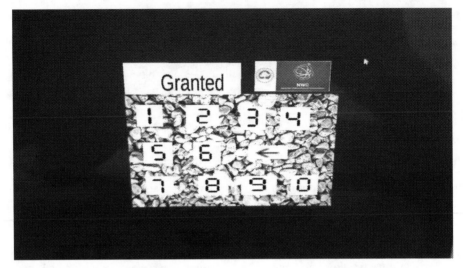

Fig. 7. The following figures show the accepted and rejected status in the text field when correct and wrong codes are entered respectively.

7   Conclusion

The incorporation of AR with IoT devices has made the smart locking mechanism more accessible, interactive and less prone to physical attacks. The authorization to access the lock is limited to the authorized users, and the physical interaction has been eliminated to make "A lock with no holes".The usage of AR made the smart lock contactless, making the lock hidden and limiting exposure to the outer environment This decreases any possible ways of physical intervention hence decreasing the chances of intrusion. The smart lock works on two-factor authentication requiring a mobile application and password for access. Implementation of access using AR and IoT has made the lock easily accessible without a key.

AR is a niche field with huge potential for innovation and improvement. Smart devices like AR smart glasses and AR contact lenses are in development, which gives a boost to our project as wearing these devices the virtual button can only be seen by the authorized user and no one else. The interactiveness of the virtual buttons will also improve in the coming future owing to the technological advances being made. Following the user requirements, many more user-friendly features can be added to the project just by making minor changes such as Remote unlocking, Track of lock usage, and Intrusion detection. Security can be increased by adding advanced authentication methods authentication.

References

1. Rauschnabel, P.A., Felix, R., Hinsch, C., Shahab, H., Alt, F.: What is XR? Towards a framework for augmented and virtual reality. Comput. Hum. Behav. **133**, 107289 (2022)
2. Gong, L., Fast-Berglund, Å., Johansson, B.: A framework for extended reality system development in manufacturing. IEEE Access **9**, 24796–24813 (2021)
3. Wang, J., Qi, Y.: A multi-user collaborative AR system for industrial applications. Sensors **22**(4), 1319 (2022)
4. Kassem, A., Murr, S.E., Jamous, G., Saad, E., Geagea, M.: A smart lock system using Wi-Fi security. In: 2016 3rd International Conference on Advances in Computational Tools for Engineering Applications (ACTEA) (2016). https://doi.org/10.1109/actea.2016.7560143
5. Kwok, A.O., Koh, S.G.: COVID-19 and extended reality (XR). Current Issues Tour. **24**(14), 1935–1940 (2021)
6. Hadis, M.S., Palantei, E., Ilham, A.A., Hendra, A.: Design of smart lock system for doors with special features using Bluetooth technology. In: 2018 International Conference on Information and Communications Technology (ICOIACT) (2018). https://doi.org/10.1109/icoiact.2018.8350767
7. Masood, T., Egger, J.: Augmented reality in support of industry 4.0—implementation challenges and success factors. Robot. Comput.-Integr. Manuf. **58**, 181–195 (2019)
8. Egger, J., Masood, T.: Augmented reality in support of intelligent manufacturing–a systematic literature review. Comput. Ind. Eng. **140**, 106195 (2020)
9. Danielsson, O., Syberfeldt, A., Holm, M., Wang, L.: Operators perspective on augmented reality as a support tool in engine assembly. Procedia Cirp **72**, 45–50 (2018)
10. Saidin, N.F., Halim, N.D.A., Yahaya, N.: A review of research on augmented reality in education: advantages and applications. Int. Educ. Stud. **8**(13), 1–8 (2015)
11. Anastasiou, E., Balafoutis, A.T., Fountas, S.: Applications of extended reality (XR) in agriculture, livestock farming, and aquaculture: a review. Smart Agric. Technol. **3**, 100105 (2022)
12. Pringle, J.K., et al.: eXtended Reality (XR) virtual practical and educational eGaming to provide effective immersive environments for learning and teaching in forensic science. Sci. Justice (2022)
13. Sasaki, T., Dehari, H., Ogi, K., Miyazaki, A.: Application of a mixed reality device to oral surgery. Adv. Oral Maxillofac. Surg. **8**, 100331 (2022)
14. Alyousify, A.L., Mstafa, R.J.: AR-assisted children book for smart teaching and learning of Turkish alphabets. Virtual Reality Intell. Hardw. **4**(3), 263–277 (2022)
15. Pinjala, S.R., Gupta, S.: Remotely accessible smart lock security system with essential features. In: 2019 International Conference on Wireless Communications Signal Processing and Networking (WiSPNET) (2019). https://doi.org/10.1109/wispnet45539.2019.903
16. Kim, S., Suh, Y., Lee, H.: What IoT devices and applications should be connected? Predicting user behaviours of IoT services with node2vec embedding. Inf. Process. Manag. **59**(2), 102869 (2022)
17. Zamora-Antuñano, M.A., et al.: Methodology for the development of augmented reality applications: MeDARA. Drone flight case study. Sensors **22**(15), 5664 (2022)
18. Stark, E., Kučera, E., Haffner, O., Drahoš, P., Leskovský, R.: Using augmented reality and internet of things for control and monitoring of mechatronic devices. Electronics **9**(8), 1272 (2020)
19. Zhu, Z., Cheng, Y.: Application of attitude tracking algorithm for face recognition based on OpenCV in the intelligent door lock. Comput. Commun. **154**, 390–397 (2020)
20. Croatti, A., Ricci, A.: Towards the web of augmented things. In: 2017 IEEE International Conference on Software Architecture Workshops (ICSAW), pp. 80–87. IEEE, April 2017

108   V. Deshpande et al.

21. Park, S., Cha, H.S., Kwon, J., Kim, H., Im, C.H.: Development of an online home appliance control system using augmented reality and an SSVEP-based brain-computer interface. In: 2020 8th International Winter Conference on Brain-Computer Interface (BCI), pp. 1–2. IEEE, February 2020
22. Condry, M.W., Nelson, C.B.: Using smart edge IoT devices for safer, rapid response with industry IoT control operations. Proc. IEEE **104**(5), 938–946 (2016). https://doi.org/10.1109/JPROC.2015.2513672

Simulation of a Wheelchair Control System Based on Computer Vision Through Head Movements for Quadriplegic People

Jirón Jhon[1], Álvarez Robin[1(✉)], Vega David[1], Felipe Grijalva[2], Lupera Pablo[1], and Flores Antonio[1]

[1] Escuela Politécnica Nacional, Av. Ladrón de Guevara E11-253, Quito 170525, Ecuador
{jhon.jiron,robin.alvarez,jose.vega01,pablo.lupera,
luis.flores04}@epn.edu.ec
[2] Colegio de Ciencias e Ingenierías "El Politécnico", Universidad San Francisco de Quito USFQ, Quito, Ecuador
fgrijalva@usfq.edu.ec

Abstract. People with quadriplegia rely on someone else to get around using a manual wheelchair. Using motorized wheelchairs gives them some independence and, simultaneously, the need of methods to control them. In the state-of-the-art, there is a great variety of these minimally invasive methods that use external devices, for example, accelerometers, which could generate discomfort. The present work proposes a non-invasive system based on artificial vision is proposed that does not require placing any device on the user. The proposed system consists of an image acquisition stage, one for face detection using Viola-Jones algorithm and another for tracking using Kanade-Lucas-Tomasi (KLT) algorithm. Additionally, a way to enter or exit the commands mode is proposed so that the user can activate/deactivate the system as required. For this, a specific movement is presented, as long as this movement is not performed, the user can move his head freely without activating the motors in the angle of focus of the camera 20° to 45°, the system correctly interprets 100% of the head moving towards or away. The novel detail is that the system allows entry and exit of the command mode by means of a special movement of the head, if the user enters the commands mode, the following options are available: tilt his head to the left, right, forward or backward.

Keywords: Quadriplegia · Wheelchair commanded by head movements · Computer vision · Viola-Jones algorithm · Kanade-Lucas-Tomasi algorithm

1 Introduction

A deep inspection of the current choices to treat quadriplegia, [1] reveals that these are not enough to improve the quality of life of these people and it is

© The Author(s), under exclusive license to Springer Nature Switzerland AG 2024
A. Ortis et al. (Eds.): ICAETA 2023, CCIS 1983, pp. 109–123, 2024.
https://doi.org/10.1007/978-3-031-50920-9_9

necessary to resort to the use of devices. Such as a motorized wheelchair, which allows moving independently and even being productive, positively influencing mental health. In the last 20 years, several ways have been proposed to control wheelchairs depending on the degree of disability of the person, starting with those with some mobility in the hands. For instance, [2,3] allow control of navigation through a joystick. In [4] the user selects the destination through an application on the cell phone or through a screen built into the wheelchair as [5] and the authors in [6] uses an RGB-D sensor so that the wheelchair is automatically navigated, avoiding any obstacles. In cases where the person cannot manipulate a joystick for wheelchair navigation, a possible solution is [7] that works based on hand movement, which would be monitored by an algorithm based on artificial vision. Systems have also been developed where hand-wheelchair interaction is not necessary using other parts of the body, such as [8], where the researchers use a tongue piercing to control a wheelchair according to its movement. In [9], the authors use encephalogram signals that can be acquired by a EPOC Neuroheadset or [10] that works through electrodes evaluating alpha waves. As for entirely non-invasive systems, some works have been proposed designs based on gaze movement, such as those implemented in [11], while others are based on the detection of voice commands [12].

Concerning systems more similar to our approach, [13] stops the movement of the wheelchair using mouth gestures. However, it only uses face detection through the AdaBoost algorithm [14], and it does not make the tracking which leads to detecting the face in each frame. Therefore, the processing will increase significantly. Another similar work is [15], which is based on recognizing gestures of the face and tracking the face with the Camshift algorithm. According to the authors [16], this algorithm is more likely to fail in light changes, including tracking other objects, and part of the face is excluded. Similarly, [17], showed the disadvantage that the user does not have full and independent control of the wheelchair as soon as the system starts up. Although there are many studies or face detection algorithms, tracking algorithms based on modern algorithms such as machine learning have not been found.

The aforementioned works do not specify a way to enter/exit the commands mode for wheelchair control. This work solves this issue by a movement that would hardly be performed involuntarily. In our case, it is a movement of the head from left to right in a specific time interval. As long as this movement is not performed, the user is free to move the head without affecting the activation of the motors of the wheelchair. In addition, among the similar works mentioned above, only [17] consider possible low-light conditions. Hence, our work deals with this issue by using an infrared camera.

On the other hand, in the literature there are various algorithms to perform the detection of faces, such as convolutional networks [18], Active Shape Model [19], Viola-Jones [14], etc. This work proposes using the Viola-Jones algorithm due to its remarkable accuracy, low consumption of computing resources, and ease of implementation. Furthermore, our system also uses a robust object tracking KLT algorithm explained in [20].

In this article, it is shown that our system can not only have hardware applications, but also within software applications, which is demonstrated by moving an image of a wheelchair on the computer screen based on the head movements detected by our artificial vision algorithm. The system has five commands: 'Forward', 'Back', 'Left', 'Right' and 'Stop'.

The paper is organized as follows: The wheelchair control system based on the detection of the movement of the user's head and nose is described in Sect. 2. Experimental results are presented in Sect. 3 and Sect. 4 concludes the paper.

2   Materials and Methods

2.1   Materials

Our system is implemented with the following resources: night vision camera with RCA connectors, RCA adapter, EasyCap USB 2.0 video capturer, Laptop with Intel Core i7 microprocessor and 16 GB of RAM, Matlab version R2021A, Matlab support Package for USB Webcams, Image Processing Toolbox and Computer Vision System Toolbox.

2.2   Methods

Our system has seven interconnected stages as indicated in Fig. 1.

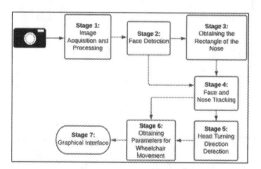

Fig. 1. The wheelchair control system based on the detection of the movement of the user's head and nose. Therefore, it involves catching actions to enter and exit the commands mode.

Stage 1. Images of a person's face are captured in real-time through an infrared vision camera that allows working in environments with poor lighting conditions. The image captured by the camera is in RGB format which is converted to grayscale. Finally, the contrast of the images is improved by equalizing the histogram as explained in [22].

Stage 2. A single person's face is detected using Viola Jones's algorithm, whose operation is detailed in [14,23], with the Local Binary Pattern (LBP) image encoding model. The LBP method takes neighboring pixels to form textures that are not prone to rotation or grayscale variation by creating a histogram using Eq. (1) and is given by:

$$LBP_{P,R}^{riu2} = \begin{cases} \sum_{p=0}^{p-1} s(g_p - g_c)2^p; & U(LBP^{P,R}) \le 2 \\ P+1; & U(LBP^{P,R}) > 2 \end{cases}, \quad (1)$$

where P is the number of pixels of the local pattern (sub-window), R is the radius of the pixel pattern, g_p (p = 0, ..., P − 1) is the intensity of P pixels, g_c is the intensity of the central pixel, s is the sign of the differences, $U(LBP_{P,R})$ is a measure of uniformity that is defined as less than or equal to 2 according to [24] and the $riu2$ superscript reflects the use of rotation-invariant "uniform" patterns that have a U-value of at most 2. The central pixel is located at (0, 0) and the numbering of the p pixels is done counterclockwise by taking the pixel p = 0 at the position (0, R) as shown in Fig. 2.

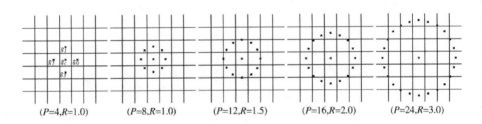

(P=4,R=1.0)   (P=8,R=1.0)   (P=12,R=1.5)   (P=16,R=2.0)   (P=24,R=3.0)

Fig. 2. Circularly symmetrical sets of neighbors for different (P; R) [24].

At the end, all the histograms of the sub-windows are concatenated to form the encoded image as shown in Fig. 3. Subsequently, the characteristics are classified with the AdaBoost algorithm, and the classifiers resulting from the training are organized in the form of a waterfall. Thus, the existence of the desired object in that region is determined only if the final stage classifies it as positive. As illustrated stages focus on regions labeled as positive and discard negative samples as quickly as possible.

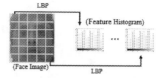

Fig. 3. Concatenation of histograms of LBP characteristics.

A detection threshold is specified for object creation. This threshold is an integer scalar and allows you to define criteria for a final detection in a rectangle where several detections exist around the object. If the threshold value is increased, it improves the detection of the object because it reduces the number of false detections, but if the value is too high, it is too restrictive, and the time to detect a face increases. The image captured and processed in stage 1 serves as input for the created detector object. In this case, there is a possibility that no face has been detected, so the existence of at least one face is validated. In addition, a time of three seconds is available until the head is placed in the corresponding position when a new detection is needed. The other possibility is that one or more faces are detected, so the largest detected face rectangle is chosen, that is, the closer one to the camera.

Stage 3. In this stage, the nose is located is obtained by dividing the rectangle of the face into smaller boxes and positioning the nose according to the standard aesthetic proportions of the face. The horizontal axis of the face is divided into five equal parts equivalents to the width of the nose, whereas facial height is divided into three equal parts. The dimensions of the face are obtained from the vector face, $[x_0 \; y_0 \; w_0 \; h_0]$, and the vector nose, $[x_1 \; y_1 \; w_1 \; h_1]$, is the box obtained from the nose as shown in Fig. 4.

$$x_1 = x_0 + \frac{2 * w_0}{5} \tag{2}$$

$$y_1 = y_0 + \frac{h_0}{3} \tag{3}$$

$$w_1 = \frac{w_0}{5} \tag{4}$$

$$h_1 = \frac{h_0}{3}, \tag{5}$$

where x_0 and y_0 are the 'x' and 'y' coordinates at which they initiate the face, w_0 is the face width and h_0 its height, x_1 and y_1 are the horizontal and vertical coordinates respectively of the beginning of the nose, w_1 is its widht and h_1 is its height.

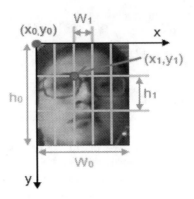

Fig. 4. Obtaining the rectangle of the nose.

Stage 4. In this stage, the face and nose are tracked using the Kanade-Lucas-Tomasi (KLT) algorithm, which has a set of features determined by the Minimum Eigenvalues algorithm. The extraction of features that this function follows is typical of the KLT algorithm to track these characteristics with the aforementioned follower object. The work [20] emphasizes that this tracker object works particularly for objects that do not change shape and [21] mentions as the operating principle of the algorithm the minimum residual error when comparing the intensity functions of two displaced consecutive images:

$$\varepsilon = \iint_w (h(x) - g \cdot \delta)^2 w \, dw, \qquad (6)$$

where ε is the residual error, w is the sub window, g is the gradient of the image intensity function expressed as $\left(\frac{\partial I}{\partial x}, \frac{\partial I}{\partial y}\right)$, δ is the displacement in two dimensions and $h(x) = I(x) - J(x)$ is the difference between the intensity functions of the previous image $I(x)$ shifted by an amount δ and the current image $J(x)$. To obtain the minimum value of the residual error, Eq. (6) is differentiated with respect to δ and the result is equalized to zero, considering that $(g\delta)g = (ggT)$, resulting in a system of two equations with two unknowns:

$$G\delta = e, \qquad (7)$$

where G denotes the matrix of coefficients and is calculated from a frame, e defines the two-dimensional vector and can be obtained from the difference between two frames along with the gradient calculated above. Therefore, the displacement vector δ is the solution of the system. In practice, it takes very few iterations to converge. Also, in the eigenvalues λ_1 and λ_2 characteristic of the G matrix, three cases can be presented: both eigenvalues are small, which means that the intensity profile is constant within a sub-window; one eigenvalue is large and the other small, indicating a one-way intensity pattern; and two large eigenvalues can represent corners, salt-and-pepper textures, or any other pattern that can be tracked reliably. In practice, the smallest eigenvalue is above the noise level, and the maximum pixel value is limited. Then, the window is accepted if:

$$min\,(\lambda_1, \lambda_2) > \lambda, \tag{8}$$

where λ is the predefined threshold by the noise level.

The tracking algorithms need an input of a specific image region to perform the respective process delivered by the detection algorithm. Once this region of interest is defined, face detection is avoided in all frames and processing is reduced. Therefore, geometric perspectives such as straight lines and planes encompassing all features should be considered according to [25]. All the parameters mentioned allow to identify the inliers (matching features) that describe the surface obtaining a transformation matrix H based on the MSAC (M-estimator Sample Consensus) algorithm whose operation is detailed in [26]. For this work, the Transformation Nonreflective Similarity explained in [20] was used, which supports translation, rotation and isotropic scale, that is, equal size change in all directions.

Stage 5. It is very important to mention that to enter/exit of the commands mode, where the movement of the wheelchair is controlled, it is necessary to define a specific movement of the head. This movement does not have to be significantly affected by the tilt of the head, so the nose's movement is chosen. Although the features located in the nose area are within those that identify the face, these have a more significant average displacement when rotating the head because the nose is closer to the camera as shown in Fig. 5. According to the example shown in Fig. 5, the vertices of the nose rectangle moved on average 17,9974 pixels. On the other hand, the vertices of the face rectangle moved on average 1,6138 pixels. This gives about 11 times more displacement of the nose than of the face in general when turning the head, which provides a motion alternative to configure the enter/exit of the commands mode independent of those used to control the movement of the wheelchair.

In this phase, the threshold determines whether the head has been turned to the left or to the right is defined, because the threshold is defined based on the width of the nose, the maximum displacement of the vertices is determined when the head is tilted to move the wheelchair left or right. Based on this displacement, a higher value is determined so as not to misinterpret a rotation towards one of the sides. In this way, a rotation is defined when there is a positive (right) or negative (left) displacement greater than 1,5 times the width of the nose according to Eq. (9). It is entered or exited from commands mode if the turn is first to the left and then immediately turned to the right. For this case, a flag has been defined and the time within which the subsequent right turn can be made is restricted, sets to three seconds.

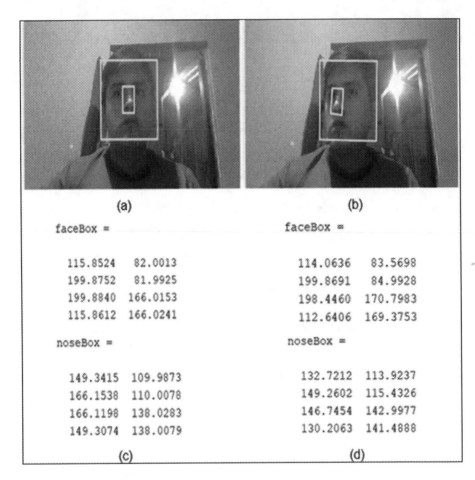

Fig. 5. Movement of the rectangles of the face and nose as the head is turned to the left: (a) rectangle of the initial nose, (b) rectangle of the final nose when rotating the head, (c) initial positions of the vertices of the rectangles and (d) positions of the vertices of the rectangles after rotating the head.

$$\begin{cases} Turn\ left; & Pmov <= Pref - 1,5(w_1) \\ Turn\ right; & Pmov >= Pref + 1,5(w_1), \\ \quad No\ turning; & others\ cases \end{cases} \tag{9}$$

where $Pmov$ are the vertices of the rectangle on the left side when the face moves, $Pref$ are the vertices of the rectangle on the left side of the face at initial detection and w_1 is the width of the nose.

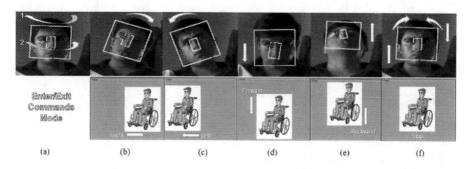

Fig. 6. Available movements to control the wheelchair: (a) enter/exit commands mode, (b) to the right, (c) to the left, (d) forward, (e) backward and (f) stop according to Eqs. (10) and (11).

Stage 6. In this stage, the features extracted from the face's rectangle are defined to adapt them to the movement of the wheelchair. According to the variation of these characteristics, there is greater speed in terms of movement. The characteristics extracted from the rectangle are the slope in degrees θ on the left side and the ratio of the current height of the rectangle of the face h concerning the initial h_o. For both vertical and horizontal axis, there is a range in which some movement of the wheelchair is not considered, and it remains motionless. This range is used for the Stop or Rest command mode. On the y-axis, the interval used is related to height, where the resting range from a 10% decrease in the initial height to a 10% increase in the initial height, i.e., the range covers [0,9 * h0, 1,1 * h0]. On the other hand, for the x-axis, the space considered involves the degree of inclination of the head, which contains 10° on the left and 10° on the right, that is, a movement between -80° and 80°. The movements involved in the movement of the wheelchair are shown in Fig. 6. In general, the vertical motion is given by Eq. (10).

$$\begin{cases} Y = Y - (h - h_{max}); & h > 1,1(h_o) \\ Y = Y + (h_{min} - h); & h < 0,9(h_o) \\ Y = Y; & others\ cases \end{cases} \tag{10}$$

where Y is the position of the wheelchair on the vertical axis, h_{max} is the maximum height of the resting state defined as $h_{max} = 1,1 * h_o$, h_{min} is the minimum height of the resting state defined as $h_{min} = 0,9 * h_o$, h is the current height of the face rectangle and h_o is the initial height of the face rectangle.

On the other hand, the horizontal movement of the wheelchair is defined by:

$$\begin{cases} X = X - (\theta_{max} - \theta); & 0 < \theta < 80 \\ X = X + (\theta - \theta_{min}); & 0 > \theta > -80 \\ X = X; & others\ cases \end{cases} \tag{11}$$

where X is the position of the wheelchair on the horizontal axis, θ_{max} is the maximum leftward inclination of the resting state defined as $\theta_{max} = 80^\circ$, θ_{min}

is the minimum rightward inclination of the resting state defined as $\theta_{min} = -80^{\circ}$ and θ is the current inclination of the face rectangle.

Stage 7. In this stage all the characteristics obtained from the face and nose, the corresponding rectangles, the image of the wheelchair and its movement according to the position parameters obtained in the previous stage of Eqs. (10) and (11) are presented. The movements of the head and their respective impact on the displacement of the wheelchair are shown in Fig. 6.

3   Results

Here, we present the results of the tests carried out offline and online. Offline testing is performed for stages (1, 2, 3, and 4) to vary the parameters of Matlab objects under the same conditions. Whereas online tests are performed for stages (5 and 6) to test the performance of the selected parameters. These parameters are evidenced in the Table 1.

Table 1. Parameters influencing system performance

Parameter	Value	Object
MaxDistance	15	estimateGeometricTransform2D
MergeThreshold	11	vision.CascadeObjectDetector
MaxBidirectionalError	4	vision.PointTracker

MaxDistance is the distance between a point and its projection in the next Frame, MergeThreshold is the number of minimum detections to determinate the existence of the object and MaxBidirectionalError is the distance between the original position of a feature and its final location after tracking it.

Table 2 presents the results of the tests performed for the number of minimum detections in a sub-window. The threshold range 7 to 11 correctly detects the face eleven times, whereas this value is not achieved below and above, and incorrect detections are made. A high threshold value requires more time to detect an object, as there must be more matches in a certain image region to decree it as restrictive existing. On the other hand, a low value requires less time, but when there are low coincidences, more mistakes are made, and there are false positives. By considering the above results, the threshold value is defined in 11 since, within the range of correct detections, there is not much difference in terms of detection time and needing more matches is more robust.

Table 3 indicates that in the first frame of second video 2999 features were found here, as the video goes on, these are lost due to noise. Noise is a change in pixel intensity due to variation in lighting, object rotation or object movement. The more the maximum error is increased, the more permissive the movement

Table 2. Face detection time and number of successful/unsuccessful detections

MergeThreshold	Time to detect face (s)	Number of Successful Detections	Number of Incorrect Detections
40	0,092	5	0
30	0,105	7	0
25	0,095	7	0
15	0,081	10	0
11	0,079	11	0
10	0,081	11	0
9	0,083	11	0
7	0,085	11	0
4	0,078	4	6

of the object, but by allowing more error, the reliability of that characteristic decreases. If the reliability is low, the estimator object cannot accurately represent the object's geometry. Therefore, you must have a balance in the maximum error. The value of 4 pixels of error is the only one that allows you to get four features in frame 150, so you choose that value.

Table 3. Number of features along the video according to the Maximum Bidirectional Error

2 * MaxBidirectionalError	Number of Detected Features			
	Frame 1	Frame 50	Frame 100	Frame 150
0	2 999	0	0	0
2	2 999	58	9	0
3	2 999	69	10	0
4	2 999	58	10	4
5	2 999	41	19	0

The distance between the position of a feature and its projection in the next frame directly influences the number of features as indicated in Table 4. Thus, with a greater distance, more mobility is allowed, but the features may not adequately represent the geometry of the object. The Table 4 shows that with low values of maximum distance there are many features that do not meet this criterion and as the video progresses they are lost. Therefore, the value 15 pixels is chosen as the maximum distance since it contains several features and above this value the difference is not significant in terms of the shape of the rectangle of the face.

Table 4. Number of features according to the Maximum Distance among projected features

2 * MaxDistance	Number of Detected Features			
	Frame 1	Frame 50	Frame 100	Frame 150
2	2 999	254	38	3
4	2 999	659	340	119
6	2 999	1 230	956	407
10	2 999	1557	1 263	694
15	2 999	2 123	1 388	828
20	2 999	2 307	1 817	995
25	2 999	2 319	2 075	1 411

The focus angles of the camera with the most accurate interpretation of movements are from 20º to 45º. The available resolutions of the EasyCap capturer are 480×320, 640×480 and 720×480 observing that when the resolution of the camera is uploaded, fewer images are processed per second. Although the number of features increases with lighting, the difference is insignificant, even decreasing when the lighting is high, as in outdoor environments with the presence of sun. In outdoor environments with more than 40000 lx of lighting, it is possible to detect 1,6 times more characteristics than in indoor environments with less than 100 lx. However, these are lost 1,2 times faster as the face moves from frame 200. For environments with low lighting benefits the use of the night vision camera because even with 0 lx of lighting, many features are found. Also, there is a relationship between the side of the face more illuminated and the interpretation of the direction of rotation of the head. This is how the turn towards the most illuminated side is always interpreted correctly. Whereas the turn of the head to the less illuminated side is detected between 66,67% and 88,89% of the time.

Table 5 shows the results of the tests carried out. The angle of focus of the camera is specified. For this test, 20 forward and 20 backward head movements were made. From these movements it was determined whether the image of the chair actually moved in the corresponding direction. Before 20º and after 45º the system does not correctly interpret the movement of the head. In the range of 20º to 45º, the system correctly interprets all 40 movements of the head moving towards or away.

In the following link: https://youtu.be/OAQRQaRi-2A, we present a video showing implementation of a minimally invasive system using accelerometers and then our non-invasive system. It is very important to mention that the prototype could be implemented in hardware.

Table 5. Correct interpretations of vertical head movement based on camera positioning angle

Camera position (degree)	Movements to the front interpreted correctly	Backward movements performed correctly	Percentage of correct interpretations
20°	18	15	82.5%
25°	20	18	95%
30°	20	20	100%
35°	20	20	100%
45°	20	20	100%
50°	10	16	65%

4 Conclusions

This paper solved the main drawback in the previous works that do not consider a safe way to enter or exit the commands mode. It was revelated that performance in outdoor environments is critical, as the rectangles of the face and nose distort 1,46 times faster than in indoor environments. Furthermore, detecting the face in environments with little or no lighting takes approximately ten times longer than in conditions of good lighting without lenses. For the above reason, a person can use the system without problems if the user is in an indoor environment with good lighting, greater than 35 lx. With the use of lenses, the face was detected only 10% of the time and taking up to 20 s for detection so this system works quite well when the user does not wear lenses.

The results showed that the movement of the chair on the vertical axis depends on the angle at which the camera focuses on the face. This is because, within the established operating range (25º to 45º) a 98,75 % correct interpretation of the movement of the head on that axis was obtained. However, only a 46,79% success rate was achieved outside this range. In indoor environments with good lighting, the user enters and exits the command mode 97,6% of the time, and in environments with low lighting it reduces to 89,42%. The movement of the head is interpreted correctly in the interface 98,75% times and in low light 89,65% of the time.

References

1. Scholten, E.W., et al.: Self-efficacy predicts personal and family adjustment among persons with spinal cord injury or acquired brain injury and their significant others: a dyadic approach. Arch. Phys. Med. Rehabil. **101**(11), 1937–1945 (2020)
2. Fattouh, A., Sahnoun, M., Bourhis, G.: Force feedback joystick control of a powered wheelchair: preliminary study. In: IEEE International Conference on Systems, Man and Cybernetics, vol. 3, pp. 2640–2645 (2004). https://doi.org/10.1109/ICSMC.2004.1400729

3. Nguyen, V.T., Sentouh, C., Pudlo, P., Popieul, J.-C.: Joystick haptic force feedback for powered wheelchair - a model-based shared control approach. In: IEEE International Conference on Systems, Man, and Cybernetics (SMC), pp. 4453–4459 (2020). https://doi.org/10.1109/SMC42975.2020.9283235

4. Scudellari, M.: Self-driving wheelchairs debut in hospitals and airports. IEEE Spectrum **54**(10), 14 (2022). https://spectrum.ieee.org/selfdriving-wheelchairs-debut-in-hospitals-and-airports

5. Sato, F., Koshizen, T., Matsumoto, T., Kawase, H., Miyamoto, S., Torimoto, Y.: Self-driving system for electric wheelchair using smartphone to estimate travelable areas. In: IEEE International Conference on Systems, Man, and Cybernetics (SMC), pp. 298–304 (2018). https://doi.org/10.1109/SMC.2018.00061

6. Burhanpurkar, M., Labbe, M., Guan, C., Michaud, F., Kelly, J.: Cheap or robust? The practical realization of self-driving wheelchair technology. In: International Conference on Rehabilitation Robotics (ICORR), pp. 1079–1086 (2017). https://doi.org/10.1109/ICORR.2017.8009393

7. Rabhi, Y., Mrabet, M., Fnaiech, F.: Intelligent control wheelchair using a new visual joystick. J. Healthc. Eng. **2018**, 1–20 (2018). https://doi.org/10.1155/2018/6083565

8. Kim, J., et al.: Qualitative assessment of Tongue Drive System by people with high-level spinal cord injury. J. Rehabil. Res. Dev. **51**(3), 451–466 (2014). https://doi.org/10.1682/JRRD.2013.08.0178

9. Banach, K., Malecki, M., Rosól, M., Broniec, A.: Brain-computer interface for electric wheelchair based on alpha waves of EEG signal. Bio-Algorithms Med-Syst. **17**(3), 165–172 (2021). https://doi.org/10.1515/bams-2021-0095

10. Carrera, F.F., Chadrina, O., Andrango, E.M., Drozdov, V., et al.: System design to control a wheelchair using brain electric signals. Medisur **17**(5), 650–663 (2019)

11. Al-Fahaidy, F., Al-Doais, H., Al-Fu-Haidy, F., Al-Oqabi, A., Qasabah, E.: Design and implementation of an eye-controlled self-driving wheelchair. Am. J. Sci. Res. Essays **6**, 1–15 (2021)

12. Echefu, S., Lauzon, J., Bag, S., Kangutkar, R., Bhatt, A., Ptucha, R.: Milpet - the self-driving wheelchair. Electron. Imaging Auton. Veh. Mach. **2017**(19), 41–49 (2017). https://doi.org/10.2352/ISSN.2470-1173.2017.19.AVM-019

13. Ju, J., Shin, Y., Kim, E.: Vision based interface system for hands free control of an intelligent wheelchair. J. NeuroEngineering Rehabil. **6**, 33 (2009). https://doi.org/10.1186/1743-0003-6-33

14. Viola, P., Jones, M.: Rapid object detection using a boosted cascade of simple features. In: IEEE Computer Society Conference on Computer Vision and Pattern Recognition (CVPR), vol. 1, pp. I-511–I-518 (2001). https://doi.org/10.1109/CVPR.2001.990517

15. Jia, P., Hu, H.H., Lu, T., Yuan, K.: Head gesture recognition for hands-free control of an intelligent wheelchair. Ind. Robot. **34**(1), 60–68 (2007). https://doi.org/10.1108/01439910710718469

16. Chatterjee, D., Chandran, S.: Comparative study of camshift and KLT algorithms for real time face detection and tracking applications. In: Second International Conference on Research in Computational Intelligence and Communication Networks (ICRCICN), pp. 62–65 (2016). https://doi.org/10.1109/ICRCICN.2016.7813552

17. Manta, F., Cojocaru, D., Vladu, I., Dragomir, A., Mariniuc, A.: Wheelchair control by head motion using a noncontact method in relation to the pacient. In: 20th International Carpathian Control Conference (ICCC), pp. 1–6 (2019). https://doi.org/10.1109/CarpathianCC.2019.8765982

18. Zhang, K., Zhang, Z., Li, Z., Qiao, Y.: Joint face detection and alignment using multitask cascaded convolutional networks. IEEE Sig. Process. Lett. **23**(10), 1499–1503 (2016). https://doi.org/10.1109/LSP.2016.2603342

19. Kumar, A., Kaur, A., Kumar, M.: Face detection techniques: a review. Artif. Intell. Rev. **52**(2), 927–948 (2019). https://doi.org/10.1007/s10462-018-9650-2

20. Shi, J., Tomasi: Good features to track. In: IEEE Conference on Computer Vision and Pattern Recognition, CVPR 1994, pp. 593–600 (1994). https://doi.org/10.1109/CVPR.1994.323794

21. Lucas, B., Kanade, T.: An iterative image registration technique with an application to stereo vision. In: 7th International Joint Conference on Artificial Intelligence (IJCAI 1981), pp. 121–130 (1981). Retrieved from the ResearchGate database

22. Furht, B., Akar, E., Andrews, W.: Digital Image Processing: Practical Approach. Springer, Cham (2018). https://doi.org/10.1007/978-3-319-96634-2

23. Kadir, K., Kamaruddin, M., Nasir, H., Safie, S., Bakti, Z.: A comparative study between LBP and Haar-like features for Face Detection using OpenCV. In: 4th International Conference on Engineering Technology and Technopreneuship (ICE2T), pp. 335–339 (2014). https://doi.org/10.1109/ICE2T.2014.7006273

24. Ojala, T., Pietikainen, M., Maenpaa, T.: Multiresolution gray-scale and rotation invariant texture classification with local binary patterns. IEEE Trans. Pattern Anal. Mach. Intell. **24**(7), 971–987 (2002). https://doi.org/10.1109/TPAMI.2002.1017623

25. Gotardo, P., Bellon, O., Boyer, K., Silva, L.: Range image segmentation into planar and quadric surfaces using an improved robust estimator and genetic algorithm. IEEE Trans. Syst. Man Cybern. Part B Cybern. **34**(6), 2303–2316 (2004). https://doi.org/10.1109/TSMCB.2004.835082

26. Torr, P., Zisserman, A.: MLESAC: a new robust estimator with application to estimating image geometry. Comput. Vis. Image Underst. **78**(1), 138–156 (2000). https://doi.org/10.1006/cviu.1999.0832

Image Encryption Using Quadrant Level Permutation and Chaotic Double Diffusion

Renjith V. Ravi[1] , S. B. Goyal[2(✉)] , and Chawki Djeddi[3]

[1] Department of Electronics and Communication Engineering, M.E.A Engineering College, Malappuram, Kerala, India
[2] City University, Petaling Jaya, Malaysia
drsbgoyal@gmail.com
[3] Laboratoire de Vison et d'intelligence Artificielle, Université Larbi Tebessi, Tébessa, Algérie

Abstract. Information security is becoming more and more important as information technology advances. In the fields of telecommunication networks, diagnostic imaging, and multimedia applications, information security is crucial. The use of images as a medium for the conveyance of information is integral to the information age. However, image data differs from traditional text-based information in that it is dense and has strong pixel correlation. As a consequence, image encryption no longer works with traditional encryption techniques. Consequently, a safe and difficult-to-crack image encryption technique is required. In this research, a new symmetric image encryption technique based on improved Henon chaotic system with byte sequences and a unique method of shuffling a picture's pixels is suggested. This algorithm produces effective and efficient encryption of images. The suggested image encryption technique created a new dimension for safe image transfer in the realm of digital transmission by causing more confusion and diffusion, according to statistical analysis and experimental key sensitivity analysis.

Keywords: Image Encryption · Chaotic Image Encryption · two level permutation · diffusion using henon map

1 Introduction

The study of multimedia involves integrating several types of information, such as texts, images, music, and video files [1]. Digital images are commonly employed in communication nowadays. Any information exchanged online requires strong security against attackers [2]. The art of preserving information by converting readable information (plain data) into an unintelligible format (cipher) with the use of organized encryption algorithms and secret keys is known as cryptography [3]. There are two different forms of cryptography. One is symmetric key cryptography [4], which encrypts data at the sender end and decrypts it at the receiver's end using a single secret key. The alternative method, called the "public key cryptography approach," is focused on asymmetric key cryptography, in which each host has a pair of keys: a private key that is kept secret from other hosts and a public key that is known to every other host in contact. Chaotic

© The Author(s), under exclusive license to Springer Nature Switzerland AG 2024
A. Ortis et al. (Eds.): ICAETA 2023, CCIS 1983, pp. 124–134, 2024.
https://doi.org/10.1007/978-3-031-50920-9_10

patterns [5] and [6] have been employed extensively in encryption systems in recent years. The Greek term for chaos, which denotes unpredictability, is used to study non-linear dynamic systems. Chaotic systems are popular due to their unpredictability and unpredictable nature.

A double permutation and chaotic diffusion-based image encryption technique is used in this study. The related works are covered in Sect. 2, and the materials and techniques are covered in Sect. 3. Section 4 presents the outcomes and a discussion of the outcomes. Section 5 brings the article to a conclusion.

2 Related Works

Numerous articles have been written on the topic of using chaos theory for secure image storage. However, several of the aforementioned surveys, like the one reported in [7], are too outdated to be useful. Furthermore, despite the fact that some researchers created a roadmap for the future, Deepa and Sivamangai [8] asserted that, from a deliberately altered medical image, it is more challenging to identify the real order. The secrecy of clinical images is now more important than ever. On the other side, the encryption procedure might place a significant burden on the systems used for medical transmission and processing. They argued that the best ways to handle this trade-off are via chaos and DNA cryptography. In order to demonstrate how the aforementioned technologies, address the trade-off, their evaluation presented certain qualitative and quantitative measures that were taken from previously published, relevant research papers [9]. Additionally, they developed a few criteria for future study in this field.

According to Yadav and Chaware [10], there are flaws in present encryption and information-hiding systems that make it possible for data to be stolen and copyrights to be violated. They started out by reviewing current image encryption techniques. They paid particular attention to cooperative encoding (error-correcting) encryption techniques. Then, with the assistance of the AES and substitution boxes, they suggested a revolutionary technique based on LDPC code and chaotic maps (S-boxes).

A comparative perspective is used in certain existing assessments. For instance, [11] examined the benefits and drawbacks of the various chaotic image encryption techniques. In [12], the authors evaluated and contrasted several single-dimensional chaotic maps with certain hyper-dimensional ones for their applicability in cryptography. This study is another pertinent one. As another scenario, the authors of [5] cited chaotic encryption as a viable method for securing images and movies in which the correlation between adjacently placed pixels is strong. To find the best chaotic map, they looked at the currently used chaotic techniques for image encryption. They examined maps of tents, logistics, sine, etc. The most effective chaotic map for this objective was proposed as being Arnold's cat map. Furthermore, the authors compared and contrasted the performance of image encryption techniques based on the five traditional algorithms, DES, Blowfish, AES and RSA, and with certain chaos-based techniques in [13].

3 Materials and Methods

3.1 Henon Map

The branch of mathematics known as chaos theory was developed by Edward Lopez. In 1989 [14], a chaotic system based on ambiguity and dispersal was created. Because chaotic systems are sensitive, nonlinear, predictable, and filled in, it is simple to reconstruct an image from them. The Henon map is a kind of chaotic map used to generate the really random sequences needed for secure encryption. The symmetric key stream cypher cryptographic technique Henon chaotic map, discovered in 1978, is employed. That's just math. Nonlinear dynamical system in two dimensions with a discrete time. The following is a description of the Henon chaotic map, which produces a seemingly random binary sequence.

$$\begin{cases} x(n+1) = 1 + y(n) - ax(n)^2 \\ y(n+1) = bx(n) \end{cases} \tag{1}$$

Here the values of $x(n)$ and $y(n)$ are known as the state values and the variable a and b are the control parameters.

The 2D-CHM exhibits a very significant chaotic behavior at the maximal Lyapunov exponent (LE) of $a = 1.4$ and $b = 0.3$. The 2D-CHM does, however, have significant drawbacks, such as straightforward chaotic behaviour and discretized chaotic periods. In [14], the authors enhanced the 2D-CHM to 2D-ICHM, which is stated as follows, in order to address the aforementioned shortcomings:

$$\begin{cases} x(n+1) = \cos(1 - ax(n)^2) + e^{by(n)^2}, \\ y(n+1) = \sin(x(n)^2), \end{cases} \tag{2}$$

Here a and b are the control parameters.

3.2 Pixel Shuffling

The association between the neighboring pixels may be disturbed by shuffling. Depending on how many rows and columns there are, the image will shuffle. The pixel is shuffled twice in this instance.

Step 1: A quadrant is partitioned into sub-quadrants after each iteration.
Step 2: If the kth iteration is odd, the quadrant is rotated in a clockwise manner; else, it is rotated anti-clockwise.

Figure 2 depicts two iterations as an illustration.

First Iteration
Start by dividing the image into four quadrants, Upper Left (UL), Upper Right (UR), Lower Left (LL), and Lower Right (LR), and rotate each one clockwise [15]. UL moves to the UR position, UR shifts to the LR position, and LL shifts to the UL position. In the final quadrant of the image, LR switches to the LL position, as seen in Fig. 1.

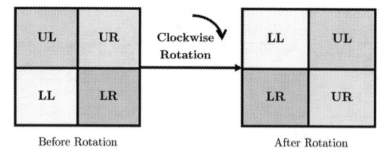

Fig. 1. Shuffling of Quadrants after first Rotation

Second Iteration
Following the first iteration, each quadrant of the randomized image is further separated into sub-quadrants using the same process shown in Fig. 1 but in the other direction as shown in Fig. 2.

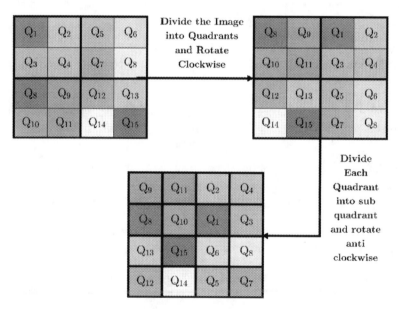

Fig. 2. Image Rearranged After Two Iterations

3.3 Proposed Algorithm

The image to be protected and the default values of the Henon map are fed into the chaotic Henon model and used as a key. We designate a $m \times n$ size image in this document.

In the proposed algorithm (Fig. 3), the plaintext image in RGB color format will be separated into its three-color components, R, G and B, for processing parallelly. Hence,

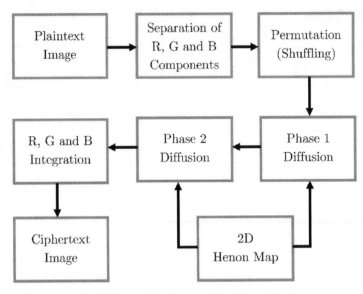

Fig. 3. Proposed Encryption Approach

there will be three parallel processes to encrypt the three-color planes. As mentioned above, the image will be permuted or shuffled in the first encryption stage. As the shuffling alone will not provide perfect encryption, a two-stage diffusion will be carried out according to the chaotic sequences generated by a 2D henon map. Finally, these diffused color planes will be integrated to form the ciphertext image. All of these operations in reverse order will be carried out at the decryption.

Step 1: Set the starting value of the Henon map to (X_1, Y_1). For the Henon map, this value serves as the first secret symmetric key.
Step 2: For the cryptosystem, Henon maps serve as a key stream generator. The length of the sequences is determined by the dimensions of the image. The number of henon segments will be $8 \times m \times n$ according to the equation if the image size is $m \times n$.
Step 3: Separate the color channels in the plaintext image
Step 4: Employ block wise permutation
Step 5: Carry out two-stage diffusion with the sequences from 2D henon map.
Step 6: Combine the diffused R, G and B channels to form the ciphertext image.

3.4 Decryption of Encrypted Image

Since the behavior of a chaotic system is predictable, reconstructing an image using the same key (X_1, Y_1) at the conclusion of the decryption process results in a scrambled image. Further, the sequence of this shuffled image is reversed from how it was done during encryption.

4   Results and Discussions

Experimental findings of the suggested image encryption technique are presented in this part to demonstrate its effectiveness. Figure 4 displays a test image that is 256×256 in size (a). In order to generate a chaotic system, the initial Henon map parameters are set at $a = 1.4$ and $b = 0.3$. The private symmetric key encryption key is made up of the values $X_1 = 0.01$ and $Y_1 = 0.02$.

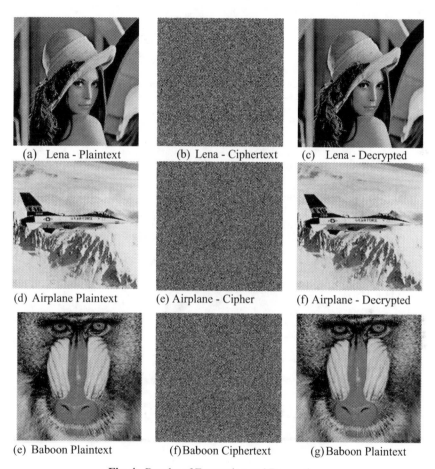

(a)   Lena - Plaintext   (b) Lena - Ciphertext   (c)   Lena - Decrypted

(d) Airplane Plaintext   (e) Airplane - Cipher   (f) Airplane - Decrypted

(e) Baboon Plaintext   (f) Baboon Ciphertext   (g) Baboon Plaintext

Fig. 4. Results of Encryption and Decryption

4.1   Histogram Analysis

A visual depiction of pixel intensity levels is the histogram of an image. A grey image may have 256 distinct intensities; therefore, the visual representation of the histogram will show all 256 intensities and the allocation of pixels among them. Figure 5 makes clear that, in contrast to the histograms of the input image, the distribution of grayscale values in the encrypted image is uniform. Some grayscale values between 0 and 255 do not exist in the input image [16]. On the other hand, the grayscale values in the ciphertext are all evenly between 0 and 255. It is therefore shown that the resultant image does not enable hackers to use a statistical attack against the encryption method.

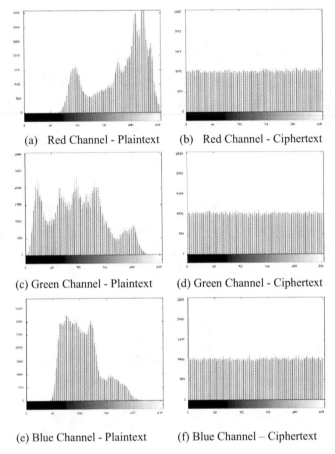

(a) Red Channel - Plaintext   (b) Red Channel - Ciphertext

(c) Green Channel - Plaintext   (d) Green Channel - Ciphertext

(e) Blue Channel - Plaintext   (f) Blue Channel – Ciphertext

Fig. 5. Histograms of plaintext and ciphertext of red, green and blue channels of *Lena* Image

4.2 Information Entropy Analysis

The level of uncertainty in the encryption scheme is used to determine information entropy [16]. It is used to determine the algorithmic efficiency of image encryption. Entropy, a statistical measure of unpredictability, is used to describe the texture of the input image. It is computed in accordance with the Eq. (3).

$$E(m) = -\sum_{i=0}^{255} P(m_i) \log_2(P(m_i)) \qquad (3)$$

where m is the ith information source and $P(m_i)$ is the probability of occurrence of m_i.

An ideal encrypted image should have an entropy of 8, which indicates an unexpected source. Ideal information entropy cannot be attained in practice. It never reaches the desired value. The Table 1 shows the values of entropies obtained for various test images and Table 2 shows its comparison with the literature.

Table 1. Entropies obtained for three channels

Image	Color Channel	Plaintext	Ciphertext
Lena	Red	7.2641	7.9987
	Green	7.6630	7.9982
	Blue	6.9778	7.9954
Airplane	Red	6.7287	7.9976
	Green	6.7981	7.9964
	Blue	6.2241	7.9958
Baboon	Red	7.6642	7.9971
	Green	7.3868	7.9969
	Blue	7.6639	7.9958

Table 2. Comparison of Entropy Values with the literature

Reference	Ciphertext
Proposed Work	7.9982
Alghamdi et al. [17]	7.9997
Alanezi et al. [18]	7.9991
Arif et al. [19]	7.9992
Lu et al. [20]	7.9971

4.3 Key Sensitivity Test

The key must be sensitive and have a large key size to withstand all types of brute force attacks in order to provide safe encryption. The primary idea behind the Henon map is randomness.

For conducting the key sensitivity test, the original secret key was modified slightly as listed in Table 2 in order to evaluate the sensitivity of the key [15, 16]. As a consequence, it was impossible for the receiver to retrieve the original image in the case of wrong keys (Table 3 and Fig. 6).

Table 3. Modified keys

Original Key	Modified key 1	Modified key 2	Modified key 3
$x(1) = 0.01$	$\dot{x}(1) = 0.010001$	$\ddot{x}(1) = 0.010002$	$\dddot{x}(1) = 0.010001$
$y(1) = 0.02$	$\dot{y}(1) = 0.020001$	$\ddot{y}(1) = 0.020011$	$\dddot{y}(1) = 0.020015$

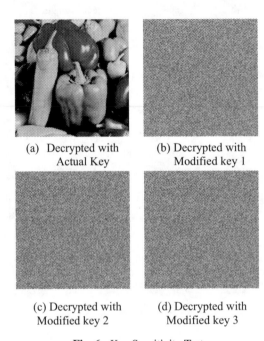

(a) Decrypted with
Actual Key

(b) Decrypted with
Modified key 1

(c) Decrypted with
Modified key 2

(d) Decrypted with
Modified key 3

Fig. 6. Key Sensitivity Test

4.4 Analysis of Ability to Withstand Differential Attack

The ability of an encryption algorithm to withstand differential attack is analysed using two metrics NPCR and UACI. The method of calculating NPCR and UACI are clearly described in [17] by Alghamdi et al. This measure adds the difference between two cypher images that are created by encrypting the identical plaintext image but changing one bit. This number is useful for figuring out how resistant an encryption method is to different kinds of attacks. In image encryption, UACI is used to measure the change in the average intensity between two images encrypted from the same plaintext image with a change of 1 bit.

The NPCR and UACI are determined as in Eq. 4:

$$\left.\begin{array}{l} NPCR = \frac{1}{MN} \sum_i^M \sum_j^N D(i,j) \cdot 100\% \\ D(i,j) = \begin{cases} 1, I_1(i,j) \neq I_2(i,j) \\ 0, I_1(i,j) = I_2(i,j) \end{cases} \\ UACI = \frac{1}{MN} \sum_i^M \sum_j^N \frac{|I_1(i,j) - I_2(i,j)|}{255} \cdot 100\% \end{array}\right\} \quad (4)$$

The values obtained for NPCR and UACI are shown in Table 4 and its comparison with the literature are shown in Table 5.

Table 4. NPCR and UACI values obtained for the test images

Image	NPCR	UACI
Lena	99.6126	33.3426
Airplane	99.5874	33.2545
Baboon	99.3587	33.3256

Table 5. Comparison of NPCR and UACI with the Literature

Reference	NPCR	UACI
Proposed	99.3587	33.3256
Alghamdi et al. [17]	99.6153	33.4718
Alanezi et al. [18]	99.6239	33.5615
Arif et al. [19]	99.6059	33.4375
Lu et al. [20]	99.5743	33.3941

5 Conclusion

The suggested approach was used on a test image, and the outcomes demonstrated a greater degree of image security. The encryption image cannot be decrypted by an eavesdropper. Here, a secret key and an image encryption method are used for security. Chaos is very secure since it is recognized for its unpredictability. Diffusion is again accomplished via the byte sequence produced by the Henon map, and confusion has been accomplished through pixel displacement from the current point to a new one. Therefore, the security of the cryptographic algorithm was strengthened by both the processes of greater confusion and diffusion.

References

1. Dhane, H., Agarwal, P., Manikandan, V.M.: A novel high capacity reversible data hiding through encryption scheme by permuting encryption key and entropy analysis. In: 2022 4th International Conference on Energy, Power and Environment (ICEPE) (2022)
2. Manikandan, V.M., Zhang, Y.-D.: An adaptive pixel mapping based approach for reversible data hiding in encrypted images. Sig. Process. Image Commun. **105**, 116690 (2022)
3. Chaudhary, N., Shahi, T.B., Neupane, A.: Secure image encryption using chaotic, hybrid chaotic and block cipher approach. J. Imaging **8**, 167 (2022)

4. Gupta, M., Singh, V.P., Gupta, K.K., Shukla, P.K.: An efficient image encryption technique based on two-level security for Internet of Things. Multimed. Tools Appl. **82**, 5091–5111 (2023). https://doi.org/10.1007/s11042-022-12169-8

5. Ayad, J., et al.: Image encryption using chaotic techniques: a survey study. In: 2021 International Conference in Advances in Power, Signal, and Information Technology (APSIT) (2021)

6. Dhall, S., Pal, S.K., Sharma, K.: A chaos-based probabilistic block cipher for image encryption. J. King Saud Univ. Comput. Inf. Sci. **34**, 1533–1543 (2022)

7. Sankpal, P.R., Vijaya, P.A.: Image encryption using chaotic maps: a survey. In: 2014 Fifth International Conference on Signal and Image Processing (2014)

8. Deepa, N.R., Sivamangai, N.M.: A state-of-art model of encrypting medical image using DNA cryptography and hybrid chaos map-2D Zaslavaski map. In: 2022 6th International Conference on Devices, Circuits and Systems (ICDCS) (2022)

9. Zolfaghari, B., Koshiba, T.: Chaotic image encryption: state-of-the-art, ecosystem, and future roadmap. Appl. Syst. Innov. **5**, 57 (2022)

10. Yadav, K., Chaware, T.: Review of joint encoding and encryption for image transmission using chaotic map, LDPC and AES encryption. In: 2021 6th International Conference on Signal Processing, Computing and Control (ISPCC) (2021)

11. Suneja, K., Dua, S., Dua, M.: A review of chaos based image encryption. In: 2019 3rd International Conference on Computing Methodologies and Communication (ICCMC) (2019)

12. Bu, Y.: Overview of image encryption based on chaotic system. In: 2021 2nd International Conference on Computing and Data Science (CDS) (2021)

13. Thein, N., Nugroho, H.A., Adji, T.B., Mustika, I.W.: Comparative performance study on ordinary and chaos image encryption schemes. In: 2017 International Conference on Advanced Computing and Applications (ACOMP) (2017)

14. Chen, Y., Xie, S., Zhang, J.: A hybrid domain image encryption algorithm based on improved henon map. Entropy **24**, 287 (2022)

15. Raghava, N.S., Kumar, A.: Image encryption using henon chaotic map with byte sequence. Int. J. Comput. Sci. Eng. Inf. Technol. Res. (IJCSEITR) **3**, 11–18 (2013)

16. Liu, L., Zhang, L., Jiang, D., Guan, Y., Zhang, Z.: A simultaneous scrambling and diffusion color image encryption algorithm based on hopfield chaotic neural network. IEEE Access **7**, 185796–185810 (2019)

17. Alghamdi, Y., Munir, A., Ahmad, J.: A lightweight image encryption algorithm based on chaotic map and random substitution. Entropy **24**, 1344 (2022)

18. Alanezi, A., et al.: Securing digital images through simple permutation-substitution mechanism in cloud-based smart city environment. Secur. Commun. Netw. **2021**, 1–17 (2021)

19. Arif, J., et al.: A novel chaotic permutation-substitution image encryption scheme based on logistic map and random substitution. IEEE Access **10**, 12966–12982 (2022)

20. Lu, Q., Zhu, C., Deng, X.: An efficient image encryption scheme based on the LSS chaotic map and single S-box. IEEE Access **8**, 25664–25678 (2020)

Leveraging Graph Neural Networks for Botnet Detection

Ahmed Mohamed Saad Emam Saad$^{(\boxtimes)}$ (iD)

Texas A&M University-Corpus Christi, Corpus Christi, TX 78412, USA
asaad1@islander.tamucc.edu

Abstract. Guarding the cyberinfrastructure is critical to ensure the proper transmission and availability of computer network services, information, and data. The proliferation in the number of cyber attacks launched on the cyberinfrastructure by making data unprocurable and network services inaccessible is on the rise. Botnets are considered one of the most sophisticated cybersecurity threats to the cyberinfrastructure and are becoming more daunting with time. Developing an efficient and robust botnet detection technique is a priority to ensure the security and reachability of the cyberinfrastructure. In this research, we introduce a solution and explore the use of a novel neural network architecture leveraging a graph-based learning approach, namely Graph Neural Network (GNN) for botnet detection. GNN was used to benefit from the unique architecture of botnets and to omit the feature engineering step of the machine learning pipeline as it is a costly and cumbersome process. Additionally, we report the effectiveness of different GNN variations in terms of detecting botnets to get an insight into the performance of each model. The ISCX-Bot-2014 dataset was used to create a graph data object for the training and testing of our proposed approach. The results show our proposed GNN solution's ability to generalize to unseen botnets and perform better compared to other relevant work from the literature with an accuracy that exceeds 94%.

Keywords: Graph neural network · Representation learning · Cyberinfrastructure · Cybersecurity · Machine learning · Botnet detection system

1 Introduction

Cyberinfrastructure is a terminology that refers to every computing system including data warehouses, Internet of things (IoT) systems, and computer network services all linked together to improve research productivity and facilitate breakthroughs [1]. IoT systems and network services have the capacity for cyber vulnerabilities and are prone to cyber-attacks [2]. Distributed denial-of-service (DDoS) attacks pose a damaging threat to the cyberinfrastructure [3]. Botnets are one of the primary sources of DDoS attacks [4].

A. Ortis et al. (Eds.): ICAETA 2023, CCIS 1983, pp. 135–147, 2024.
https://doi.org/10.1007/978-3-031-50920-9_11

A botnet is a terminology that is used to refer to a collection of infected computers or bots, that are controlled by an attacker by using various command and control (C&C) channels. These channels can use various botnet topologies: centralized, distributed (e.g., P2P) or randomized [5]. It can perform a wide variety of malicious activities like DDoS attacks and phishing. In other words, for the protection and guarding of the cyberinfrastructure, there is an immanent need for an effective solution for detecting botnets. Numerous botnet detection approaches were introduced and categorized under four different categories: DNS-based, signature-based, mining-based, and anomaly-based [6].

Each of the previously mentioned approaches has a downside. The DNS-based approach is used to monitor and detect anomalies in DNS query information initiated by bots to receive commands from C&C channels. Similarly, anomaly-based detection techniques inspect network traffic for abnormalities such as high traffic volume and activities on unusual ports. Both approaches face the issue of producing high false positive rate detection results when detecting botnets since they consider any deviation as an anomaly [6]. The signature-based approach works similarly to rule-based systems such as DNS-BD [7], and it lacks the ability to detect unknown botnets since it is limited by the rules within its database. The mining-based technique originally was proposed to use data-driven solutions such as machine learning to detect anomalies in the botnet C&C communication traffic since it is difficult to identify using the other approaches. The downside of the traditional machine learning solution is that it only uses the network traffic feature to identify botnets, ignoring its architecture. In addition, the feature engineering process to set up a machine learning solution with high accuracy and a low false positive rate is a laborious task.

Therefore, creating a novel solution that is able to identify botnets from both architectures and network traffic features contributing to a better performance with high accuracy and within an acceptable false positive rate is crucial. Moreover, the solution should be scalable and able to identify unknown botnets and avoid going through the cumbersome feature engineering process. To this end, we leverage Graph Neural Network (GNN), a novel neural network architecture with a supervised representation learning approach.

In this research, we propose an efficient solution for the botnet detection problem using representation learning combined with a novel neural network architecture, namely Graph Neural Network for the following reasons. First, it can understand and learn the structure or topology of data in form of a graph in addition to its features. Second, it automatically learns features, meaning it does not require feature engineering for the input data as it is a time-consuming and difficult process. Third, it can scale and generalize to previously unseen botnet types. The introduced approach utilizes different variations of the GNN to get an insight into the performance of each model.

Representation learning utilizes the inherent topological structure and information of the graph data object to learn its architecture and features. The problem is formalized as a node-level binary classification problem on a graph data object. We use the ISCX-bot-2014 botnet dataset [8] to build a graph data object for the learning process. The graph data object is represented as nodes and edges. The nodes resemble each existing IP address in the dataset and the edges resemble the connection between the nodes. Each node is assigned a label depending on its IP address nature. The learning process starts by creating an initial node embedding which will later be updated as the learning process continues. Each node starts automatically learning its features by collecting information from its neighboring nodes in a process called *aggregate*. After that, each node updates its own embedding in a process called *update*. Moreover, each node embedding is trained against the label assigned to the node. Multiple aggregators were utilized to create different instances of the GNN and to get an insight into the performance of each model for botnet detection. The ISCX-Bot-2014 dataset is one of the largest and most diverse in terms of existing botnet types [8]. It is used as a benchmark to assess the performance of botnet detection solutions as it overcomes the three challenges for most of the exiting botnet datasets (i.e., generality, realism, and representatives) [8]. It is the most appropriate choice and will best serve the aim of this research.

The key contributions of this study are as follows:

1. Introducing a novel graph-based botnet detection solution utilizing the botnet's communication network traffic in addition to its architecture. The developed approach excludes the tedious feature engineering process from the machine learning pipeline.
2. Providing a graph data object for the ISCX-Bot-2014 dataset as it adds to the benchmarks used for evaluating GNN approaches for botnet detection solutions.
3. Gaining insight into the performance of multiple GNN models for botnet detection under different variations compared to other solutions from the literature.

The remaining of this paper is organized as follows. Section 2 explores the most relevant related work. A brief background is provided in Sect. 3. The followed methodology is explained in Sect. 4. In Sect. 5, evaluation and benchmarking are discussed. Finally, we conclude in Sect. 6.

2   Related Work

Several botnet detection solutions using machine learning approaches and techniques were developed.

The first attempt to use GNN for botnet detection was proposed by Zhou et al. [9], they formalized their approach as on graph node multi-classification problem. They based their botnet detection solution on the idea of recognizing the topologies of botnets (i.e., centralized and distributed). Their approach ignored the random architecture of botnets. In addition, it is computationally expensive requiring a massive dataset to train on and a large number of neighborhood hops, 12 to be exact. The dataset used for evaluation was a combination of both real and synthetic topologies of background traffic network and botnet traffic network. The authors claimed that their approach was able to identify previously unseen botnet structures compared to other approaches.

Nguyen et al. [10] proposed an IoT botnet detection solution using a deep learning approach based on printable string information (PSI) graph in addition to the linkage between them and Convolution Neural Network (CNN) as a classifier. They demonstrated the potential of CNN as a malware classifier when trained on PSI graph links and information such as IP address, domain name, etc. Their approach was tested on the IoTPOT dataset achieving an accuracy of 92%.

Chowdhury et al. [11] suggested a botnet detection approach based on recognizing multiple topological features of nodes on a graph. Their detection approach used a clustering method known as a self-organizing map. The authors claim that the proposed approach can limit the search time of botnet nodes by its ability to cluster and isolate botnet nodes in sub-clusters of smaller sizes. They tested their approach on the CTU-13 dataset and reported their results against other classification-based detection solutions.

The work that introduced the ISCX-Bot-2014 dataset used in this research was presented by Beigi et al. [8]. In addition, they developed their botnet detection approach based on classifying network traffic features as malicious or benign. They address the issue of modeling the behavior of a botnet depending solely on its flow-based features. They used a shallow learning algorithm, namely a decision tree as a classifier. The authors used a feature selection algorithm that consisted of 2 key parts: group exclusion and feature inclusion. In other words, they experimented with multiple flow-based feature groups to find the optimal feature groups that yield the highest accuracy and detection rate. Then, they eliminated feature groups that weakly contribute to the accuracy of the model. Furthermore, the features in the least performing groups were analyzed and the features that strongly enhance the accuracy of the model were selected. The authors tested their approach in seven different settings and top using the introduced dataset. The top-performing model achieved an accuracy of 75% and a detection rate of 69%.

Hossain et al. [12] proposed a feature selection algorithm combined with a supervised deep learning classifier, namely an artificial neural network (ANN). Their proposed approach integrated the inclusion-exclusion feature selection process to determine the optimal subset of features for a more accurate flow-based botnet detection solution. They introduced a heuristic algorithm that initially tests the accuracy with all existing features, and the yielded accuracy is the baseline. Moreover, an exclusion iterative process takes place where some features are excluded at each iteration step. Furthermore, the features contributing to a model with a higher botnet detection accuracy rate remain, otherwise excluded. The iterations continue till the accuracy of the remaining features model drops below a certain threshold (i.e., the previous iterations' accuracies). Moreover, their highest-performing model included training on only ten features excluding MissingBytes and AvgPBS features. They tested their approach on the ISCX-Bot-2014 dataset, similar to the work in [8]. The experiments show that their model for botnet network traffic achieved a detection rate of 91% and an accuracy of 86.41%.

The most relevant works from the literature are [8,12] for the following reasons. First, they used a feature selection algorithm and ignored botnet structures, considering only network traffic features which we challenge with our proposed GNN approach. Second, they trained and tested their approach on the same dataset used in this research. Therefore, these research papers' results will be used later to be compared with our proposed approach.

3   Graph Neural Network

GNN is a novel neural network architecture introduced to deal with data represented in the form of graphs in graph domains [13]. It is considered the natural evolution of the existing neural network architectures. The GNN architecture is designed to work on different types of graphs such as directed, undirected, etc. Graphs G can be represented as nodes or vertices x and connections between these nodes can be represented as edges l. In GNNs, we have three prediction problem levels we can solve: node level, edge level, and graph level [14]. For the node level prediction, we can predict the label or type of each node based on the information provided by its surrounding or neighbor nodes. Moreover, the edge or link prediction level is concerned with predicting the label of the edge based on the information gathered from the two linked nodes. Furthermore, graph level prediction is a problem in which we predict a label for an entire graph, where multiple graphs exist, and each graph represents a class or a label.

Each GNN approach differs in 2 ways: the way we aggregate f_w the messages (i.e., node features in a vector format) from the neighbors and the way we update g_w the self-node embedding. This process is known as message passing, during this process each node gets information from other nodes in its neighborhood as shown in Fig. 1. The output of the message passing process for every single node is called *embedding*. The depth from which the information is collected is called a hop k, meaning if k is set to value '1', the information will be gathered from only the direct neighbors. Moreover, if k is set to value '2' the information gathered is not limited to the direct neighborhood, but it includes the neighbors' neighborhoods. Equations 1 and 2 further explain the aggregation process of node x_1 and the updating process of its self-node embedding o_n to be combined with information received from its direct neighbors (i.e. $k = 1$), respectively. Both equations are notated as follows: f_w represents the aggregation function, g_w represents the update function, l_n is the label of node n, l_{co} is the label of n edges, x_{ne} is the current embeddings of neighborhood nodes n, and l_{ne} is the label of each node n in the neighborhood.

$$x_n = f_w \left(l_n, l_{\mathrm{co}[n]}, x_{\mathrm{ne}[n]}, l_{\mathrm{ne}[n]} \right) \tag{1}$$

$$o_n = g_w \left(x_n, l_n \right) \tag{2}$$

In addition, different approaches can also include utilizing only node embeddings, only edge embeddings, or both for the message-passing process.

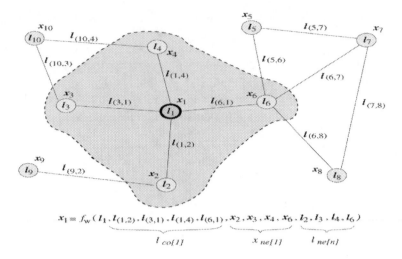

Fig. 1. Graph and the neighborhood of a node [13]

Furthermore, depending on the data, we first structure the data into either a single graph for node and edge level predictions or multiple graphs for graph-level prediction. Moreover, for graph-level prediction, each graph will include its nodes and edges if multiple data objects exist and will be used for classification or clustering.

4   Methodology

This study aims to propose a solution for the botnet detection problem, which utilizes a representation learning algorithm in addition to the topological structure and network traffic presented in the ISCX-Bot-2014 dataset. The problem will be considered a binary classification node-level problem since we attempt to classify whether some nodes (i.e., bots on a graph or network) belong to a botnet. In addition, the omission of the feature engineering process from the machine learning pipeline is considered to provide an efficient end-to-end fashion botnet detection solution. In the developed approach, GNN is used under different variations to enrich the node embeddings in the message-passing process to make the botnet node more recognizable to the classifier from normal nodes. The ISCX-Bot-2014 botnet dataset was selected as it contains many feature groups that make feature engineering a difficult, time-consuming, and computationally expensive process. Therefore, it is used to prove that our approach can perform well without the feature engineering step, hence overcoming the feature engineering process.

4.1   Dataset

The dataset ISCX-Bot-2014 was provided by the Canadian Institute for Cybersecurity, University of New Brunswick [8]. The dataset was obtained from its official sources and is divided into training and test datasets that include 7 and 16 types of botnets, respectively. The data has unlabeled training and testing sets with a record count of 331,851 and 345,746, respectively. Moreover, each record has 82 features of network traffic.

4.2   Dataset Preprocessing

The data comes in PCAP format, hence there is a need to convert it into a convenient format to be manipulated. CICFlowMeter [15] was used to convert the network traffic from PCAP into a CSV file format. The training and testing dataset were combined in one CSV file to follow the transductive learning approach [16] followed by GNN (i.e., train and test data must be batched as one graph). Each network traffic was labeled based on its IP address status (i.e., malicious or benign) using a labeling function since a list of all malicious IP addresses was provided by [8]. The malicious traffic label was given a value of "1", and the benign traffic was given a value of "0". The dataset contains some features that will not be utilized in the training process such as ('Src IP', 'Src Port', 'Dst IP', 'Dst Port', and 'Label') and therefore dropped.

4.3   Graph Data Object Construction

Creating a graph data object was a challenging process given that the dataset is not represented as a graph and that we need to batch the training and testing dataset as one graph. Moreover, tailoring a GNN model to work on both network traffic features and structure for botnet detection was another challenge. The construction of the graph data object will follow the convention of PyTorch Geometric [17] since it is the framework used to develop and test the GNN approach. In order to construct a graph data object, the following steps were followed.

We start by constructing two different matrices: a connectivity matrix and nodes feature matrix combined with a labels list to train against. Moreover, we construct other components necessary for the representation learning process (i.e., training and testing masks, data batch, and class count), which all are discussed below.

Connectivity Matrix. The connectivity or adjacency matrix was constructed by using two features that determine the existence of a connection between two nodes, namely source IP and destination IP. After running some analysis on the data, some nodes were observed to be disconnected from the rest of the nodes, which may decrease the training efficiency. Networkx Python library [18] was used to visualize the entire graph to confirm the analysis results of disconnected nodes. The presence of disconnected nodes was confirmed and then removed from the dataset, and some tests were conducted before and after the removal of those nodes. The accuracy of the GNN model increased after removing the disconnected nodes, and the number of records was reduced to 330,133 for training and 337,907 for testing.

Labels List. Labeling the dataset was done in the data preprocessing step, and all the labels were converted into a list for further usage during training.

Node Features. Assigning initial features to nodes was a cumbersome process since the number of nodes is double the number of features provided (i.e., network traffic records or edge features). In addition, several cases were observed in the data. For example:

* Source node is benign, and the destination node is benign.
* Source node is malicious, and the destination node is benign.
* Source node is benign, and the destination node is malicious.
* Source node is malicious, and the destination node is malicious.

The problem was assigning the network traffic features to which node, since assigning the existing features to source nodes will leave the destination nodes featureless. In addition, if the all ones feature vector $x_n = \{1, ..., 1\}$ method [19] was used to initialize the initial nodes' embeddings for the featureless nodes, it may result in low accuracy. This is because the initial features will be the same for benign and malicious destination nodes.

To address the aforementioned problem, four solutions were proposed and tested:

1. Assigning all one constant vectors to all destination nodes and assigning network traffic features to all source nodes (to confirm our theory).
2. Assigning all one constant vectors to all source nodes and assigning network traffic features to all destination nodes.
3. Duplicating source nodes features and assigning them to all destination nodes.
4. * Duplicating the features for both the source and destination node if both source and destination nodes are benign or malicious.
   * Assigning the network traffic features to the source node, copying a random benign node's traffic features, and assigning them to the destination node if the source is malicious and the destination is benign.
   * Assigning the network traffic features to the source node, copying a random malicious node's traffic features, and assigning them to the destination node if the source is benign and the destination is malicious.

Moreover, the dimension of all one constant vectors is set to be equal to the number of network traffic features. Experiments show that the fourth solution resulted in higher accuracy than the others.

Training and Testing Masks. Graph-based learning follows the convention of transductive learning [16] (i.e., the entire dataset including training and testing sets is fed during the training and testing processes as one graph). To separate the training nodes from the testing nodes, we use another concept special to graph-based learning called masking. A mask is a Boolean array indicating which nodes to be used during training, and those are assigned a Boolean value *'True'*.

Otherwise, it is assigned a Boolean value *'False'*. Similarly, during testing, the test nodes are assigned a Boolean value of *'True'*, and other nodes are assigned a Boolean value of *'False'*. During training, we use a training mask in which the nodes in the training dataset are masked in and test nodes are masked out. Furthermore, during testing, we mask in the nodes in the testing dataset, and we mask out the other nodes.

Data Batch. In GNN, a batch indicates which graph each node belongs to. For example, if we have multiple graphs for a graph classification level problem, then multiple nodes can belong to a different graph. To address this, a batch array is constructed by assigning a unique number to all the nodes belonging to the same graph. In our case, we only have one graph for node classification, and the batch array will contain the same number indicating that all the nodes belong to the same graph.

Class Count. The class count is an integer assigned to indicate the number of classes that exist in the data. In the context of the presented problem, the botnet detection problem is a binary node classification problem in which we have only to classify whether a node belongs to a botnet. Moreover, the class count will be an integer of value "2", indicating the two classes (i.e., malicious and benign).

The final representation of the graph data object was visualized using Networkx [18] and can be viewed in Fig. 2.

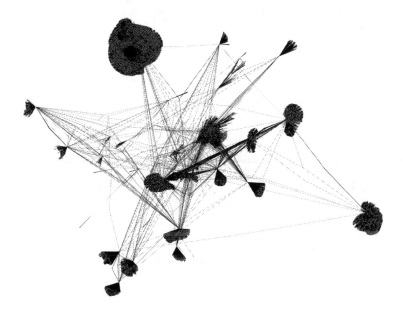

Fig. 2. Graph Representation

5   Evaluation

The method used to evaluate, measure, and describe the performance of the proposed approach is the confusion matrix [20]. It is considered the optimal evaluation method for classifiers. In addition, it shows the actual classification or labels against the predicated ones. Since the developed approach in this study is a binary classifier, the confusion matrix is the selected method for evaluation. The metrics inferred from the confusion matrix include F1 score, accuracy, precision, and recall (detection rate). Those metrics are represented in terms of prediction evaluation categories such as True Positive (TP), False Negative (FN), False Positive (FP), and True Negative (TN). Given the nature of the problem, the evaluation will be heavily reliant on the recall (detection rate) in addition to the accuracy.

5.1   Graph Neural Network Experimental Results

The performance of the proposed GNN approach was examined under four different variations. Each variation utilizes a different operator (i.e., aggregation and update function). Further experiments were conducted, but only top-performing ones are reported. The other experiments were conducted to explore the effect of different operators and neighborhood hop values k on the performance of the proposed approach. In all reported experiments, we used 150 training iterations, a unit size of 136, and a learning rate of 0.001. The top performing experiment is "GNN - Experiment 4" as it scored the highest in every performance metric. As observed, the best results were achieved by using a k value of "2". Moreover, there was no further improvement shown when the k value was increased to "3". Although the training data has only 7 types of botnets as opposed to 16 in the testing set, the results show the GNN model's ability to generalize to unseen botnet types. In addition, the high precision indicates the model's ability to perform with a low false positive rate. An aggregation of the performance metrics of the reported experiments is shown in Table 1.

Table 1. Aggregated Results for the Reported Experiments

Experiment	Operator	k-value	F1 (%)	Recall (%)	Precision (%)	Accuracy (%)
GNN - Experiment 1	GCNConv	2	80.69	80.26	81.12	86.48
GNN - Experiment 2	SAGEConv	2	77.92	73.39	83.05	85.36
GNN - Experiment 3	SGConv	2	81.73	79.17	84.48	87.55
GNN - Experiment 4	TAGConv	2	95.73	93.79	97.75	94.38
GNN - Experiment 5	GCNConv	3	80.75	74.23	88.52	87.54
GNN - Experiment 6	SAGEConv	3	80.43	73.82	88.34	87.36
GNN - Experiment 7	SGConv	3	82.02	76.67	88.18	88.17

5.2 Comparison with Relevant Work

Table 2 is an aggregation of the top-performing GNN model in addition to the performance results of relevant work from the literature. The relevant studies [8,12] have been carried out to find the optimal feature set to improve botnet detection based solely on network traffic features. They evaluated their approach using the same ISCX-Bot-2014 dataset. Both relevant papers suggest a feature selection approach using an inclusion-exclusion algorithm for optimal results. On the other hand, our approach provides an alternative solution based on the topological structure in addition to network traffic features and omits the feature engineering step. A comparison with the relevant studies is shown in Table 2 to emphasize that our approach exceeds relevant solutions in terms of accuracy and detection rate and to show the contribution added by the proposed approach.

Table 2. Comparison with Others' Work

Approach	Detection rate (%)	Accuracy (%)
Our approach (Developed in this research)	93.79	94.38
Group-based Feature selection, decision tree [8]	69	75
Inclusion & exclusion for Feature Selection, feed forward ANN [12]	91	86.41

6   Conclusion

The growth and expansion of the cyberinfrastructure starting from data centers to IoT systems uncovered numerous vulnerabilities in our cyber defense capabilities. Those vulnerabilities can pose a threat to the entire cyber domain. One of the most brutal cyber attacks that threaten the cyberinfrastructure is DDoS. Botnets are considered the origin of most DDoS attacks. The aforementioned problem produced the necessity for a reliable and efficient botnet detection solution. Different botnet detection solutions were developed using mining-based approaches such as machine learning algorithms. However, most of the developed approaches are reliant on detecting botnets using network traffic features and ignoring their topological structure in addition to the tedious task of feature engineering for optimal results.

In this paper, we presented a botnet detection solution by utilizing a representation learning approach, namely graph neural network. We developed a novel approach for botnet detection that utilized the inherent structure of botnets alongside their network traffic features. In addition, we introduced the first graph data object based on the ISCX-Bot-2014 dataset. The aforementioned dataset was used as it is the most diverse in terms of botnet types. The graph object was used to train, evaluate, and test the effectiveness of the developed GNN approach. In addition, we tested multiple GNN variations to get an insight into the performance of different models for botnet detection. The performance results obtained reflect the dominance of the proposed approach when compared to other related solutions from the literature. The proposed approach outperformed other solutions, achieving an accuracy of more than 94%. Multiple experiments were conducted using different GNN operators and hyper-parameters to obtain the optimal model. The results and settings of the experiments conducted were reported, compared, and discussed.

In future work, we can consider the temporal aspect of the network traffic behavior for botnet detection and integrate it with the proposed approach to add further improvements in the performance.

References

1. Stewart, C.A., Simms, S., Plale, B., Link, M., Hancock, D.Y., Fox, G.C.: What is cyberinfrastructure. In: Proceedings of the 38th Annual ACM SIGUCCS Fall Conference: Navigation and Discovery, pp. 37–44 (2010)
2. Djenna, A., Harous, S., Saidouni, D.E.: Internet of Things meet internet of threats: new concern cyber security issues of critical cyber infrastructure. Appl. Sci. 11(10), 4580 (2021)
3. Kaur Chahal, J., Bhandari, A., Behal, S.: Distributed denial of service attacks: a threat or challenge. New Rev. Inf. Netw. 24(1), 31–103 (2019)
4. Hoque, N., Bhattacharyya, D.K., Kalita, J.K.: Botnet in DDoS attacks: trends and challenges. IEEE Commun. Surv. Tutor. 17(4), 2242–2270 (2015)
5. Abu Rajab, M., Zarfoss, J., Monrose, F., Terzis, A.: A multifaceted approach to understanding the botnet phenomenon. In: Proceedings of the 6th ACM SIG-COMM Conference on Internet Measurement, pp. 41–52 (2006)
6. Feily, M., Shahrestani, A., Ramadass, S.: A survey of botnet and botnet detection. In: 2009 Third International Conference on Emerging Security Information, Systems and Technologies, pp. 268–273. IEEE (2009)
7. Alieyan, K., Almomani, A., Anbar, M., Alauthman, M., Abdullah, R., Gupta, B.B.: DNS rule-based schema to botnet detection. Enterp. Inf. Syst. 15(4), 545–564 (2021)
8. Beigi, E.B., Jazi, H.H., Stakhanova, N., Ghorbani, A.A.: Towards effective feature selection in machine learning-based botnet detection approaches. In: 2014 IEEE Conference on Communications and Network Security, pp. 247–255. IEEE (2014)
9. Zhou, J., Xu, Z., Rush, A.M., Yu, M.: Automating botnet detection with graph neural networks. arXiv preprint arXiv:2003.06344 (2020)
10. Nguyen, H.T., Ngo, Q.D., Le, V.H.: IoT botnet detection approach based on PSI graph and DGCNN classifier. In: 2018 IEEE International Conference on Information Communication and Signal Processing (ICICSP), pp. 118–122. IEEE (2018)
11. Chowdhury, S., et al.: Botnet detection using graph-based feature clustering. J. Big Data 4(1), 1–23 (2017). https://doi.org/10.1186/s40537-017-0074-7
12. Hossain, M.I., Eshrak, S., Auvik, M.J., Nasim, S.F., Rab, R., Rahman, A.: Efficient feature selection for detecting botnets based on network traffic and behavior analysis. In: 7th International Conference on Networking, Systems and Security, pp. 56–62 (2020)

13. Scarselli, F., Gori, M., Tsoi, A.C., Hagenbuchner, M., Monfardini, G.: The graph neural network model. IEEE Trans. Neural Netw. **20**(1), 61–80 (2008)
14. Zhou, J., et al.: Graph neural networks: a review of methods and applications. AI Open **1**, 57–81 (2020)
15. Draper-Gil, G., Lashkari, A.H., Mamun, M.S.I., Ghorbani, A.A.: Characterization of encrypted and VPN traffic using time-related. In: Proceedings of the 2nd International Conference on Information Systems Security and Privacy (ICISSP), pp. 407–414 (2016)
16. Rossi, A., Tiezzi, M., Dimitri, G.M., Bianchini, M., Maggini, M., Scarselli, F.: Inductive–transductive learning with graph neural networks. In: Pancioni, L., Schwenker, F., Trentin, E. (eds.) ANNPR 2018. LNCS (LNAI), vol. 11081, pp. 201–212. Springer, Cham (2018). https://doi.org/10.1007/978-3-319-99978-4_16
17. Fey, M., Lenssen, J.E.: Fast graph representation learning with PyTorch Geometric. In: ICLR Workshop on Representation Learning on Graphs and Manifolds (2019)
18. Hagberg, A., Swart, P., Chult, D.S.: Exploring network structure, dynamics, and function using NetworkX. Technical report, Los Alamos National Lab. (LANL), Los Alamos, NM, United States (2008)
19. Lo, W.W., Layeghy, S., Sarhan, M., Gallagher, M., Portmann, M.: E-GraphSAGE: a graph neural network based intrusion detection system for IoT. In: NOMS 2022–2022 IEEE/IFIP Network Operations and Management Symposium, pp. 1–9. IEEE (2022)
20. Vihinen, M.: How to evaluate performance of prediction methods? Measures and their interpretation in variation effect analysis. BMC Genomics **13**, 1–10 (2012)

Voice Commands with Virtual Assistant in Mixed Reality Telepresence

Shafina Abd Karim Ishigaki[✉], Ajune Wanis Ismail, Nur Ameerah Abdul Halim, and Norhaida Mohd Suaib

Mixed and Virtual Reality Research Lab, ViCubeLab, Faculty of Computing, Universiti Teknologi Malaysia, 81310 Johor, Malaysia

{shafina,nurameerah}@graduate.utm.my, {ajune,haida}@utm.my

Abstract. Mixed Reality (MR) telepresence is an expanding field in marketing research and practice especially in computer vision due to its ability to connect people in a remote location. As voice command and speech recognition technology have also been significant milestones in the twenty-first century, this technology has the advantage to be used in MR telepresence for interaction. Meanwhile, the virtual assistant avatar's implementation in MR telepresence can provide support for the interaction by reducing the workload and improving the performance of the interaction in collaborative MR telepresence. Therefore, in this paper, we propose to explore the voice commands interaction and describe the implementation with the virtual assistant avatar in MR telepresence. We implement speech-to-text (STT) conversion and utilize the semantics of Natural Language Processing (NLP) for the voice command. Meanwhile, for the avatar to respond, the speech service is used to convert the text-to-speech (TTS) input. This paper ends with a conclusion, limitations and suggestions for future works.

Keywords: Natural language processing · Telepresence · Mixed Reality · Virtual Assistant · Voice Commands · Speech recognition

1 Introduction

Mixed Reality (MR) is a new form of experience in the actual world that is augmented by a layer of computer graphics-based interaction tied to a specific activity [1]. MR seamlessly overlays two-dimensional (2D) and three-dimensional (3D) objects onto the actual world, including the audio files, films, and textual material. The user of an MR interface sees the actual world through a portable or head-mounted device (HMD) that coats graphics on the surroundings. User interaction in MR has to be intuitive as possible and the robust interaction where user can directly interact to the object [2, 3]. MR system has been described as the mixed between real and virtual items in a real environment, runs interactively in real-time, as in [4] they brought the idea of telepresence where the real user reconstruction into MR space. During combined with augmented data, MR allows users to experience the actual world as a more integrated and better environment. Collaborative interface enables the user to cooperate with other users in a different type of interfaces in simultaneous [5, 6]. As shown in Fig. 1, the related works to this study.

A. Ortis et al. (Eds.): ICAETA 2023, CCIS 1983, pp. 148–158, 2024.
https://doi.org/10.1007/978-3-031-50920-9_12

(a) User interaction in MR [2]   (b) 3D telepresence in MR [3]   (c) Collaborative AR/VR [4]

Fig. 1. Related works

Meanwhile, telepresence is almost similar to Virtual Reality (VR). VR targets to obtain the illusion of presence in the virtual world. While telepresence targets to obtain the illusion of presence at the remote location [7]. Telepresence is a sense of being existing in the virtual environment [8]. The "telepresence" term is used as the experience of presence in an environment by means of a communication medium. By incorporating MR technology with telepresence systems, virtual environments can be made to look more lifelike, and users can experience real-world conversation as if it were happening in the same room as them even though they are not physically present.

The integration of a dedicated telepresence system with MR also has attracted a lot of interest from the industrial and research communities as paved the way for real-time human-to-human interaction over long distances [9]. As the technology has the potential to minimize the traveling cost and has made it possible to volumetrically capture the human body and performance in real-time [10]. The interest in 3D remote collaboration systems also has been increasing both by academic researchers and companies because these systems overcome some of the limitations of traditional 2D video-based teleconferencing systems. By experiencing their 3D, virtual representations as part of their immediate surroundings, users can communicate with remote individuals or control remote devices [11].

As voice command and voice recognition technology have also been significant milestones in the twenty-first century [12], this technology has the advantage to be used in MR telepresence for interaction. It offers interface alternatives that do not require users to physically touch or press buttons in order to complete a certain function. Nowadays, speech technologies are widely accessible for a limited but intriguing variety of applications [12]. Amazon is one such corporation, it released Amazon Echo [13]. In order to play music from a specific streaming service or turn on a smart light bulb connected to the same home internet service, users can simply speak into their Amazon Echo devices, and their voice commands are translated into text and executed by Amazon's intelligent cloud-based personal assistant, Alexa. Other researcher also has implemented the voice command on mobile AR platform [14], VR [15] and MR applications [16, 17].

Besides that, voice recognition also can converts English speech into Indian sign language avatar animation using Natural Language Processing (NLP) [18]. NLP enables immersive experience through speech recognition. This technology enables virtual assistant to accurately and dependably respond to human voices and deliver important and helpful services. Therefore, this paper aim to explore the implementation of voice commands in MR telepresence and enable the virtual assistant avatar. The voice command deployed by the local user will be process to the text output, performing the action and convert the text into a speech for the virtual assistant in the MR telepresence to be respond. The proposed system, test application and conclusion has discussed in the remaining sections.

2 Proposed System

This section introduces the proposed system that allows the HMD user to give commands to the virtual assistant avatar in MR telepresence using a voice. The virtual assistant avatar will be able to receive the voice input information, process the information, execute the action and project audio as feedback to the commands. The process will involve the conversion of Speech-to-Text (STT), Text-to-Speech (TTS), and set up for MR telepresence. Hence, the proposed method will be divided into three processes. The first process is voice commands, following the avatar virtual assistant and MR telepresence application. More details will be explained in each subsection accordingly.

2.1 Voice Commands

NLP in this context is one of the effective technologies that can be integrated with advanced technologies, such as machine learning, AR, VR and MR to improve the process of understanding and processing the natural language [19]. This can enable human-computer interaction in a more effective way as well as allow for the analysis and formatting of large volumes of unusable and unstructured data/text in various industries. This will deliver meaningful outcomes that can enhance decision-making and thus improve operational efficiency.

Besides that, NLP also gives more meaningful interactions and relationships with avatars and other digital characters or assets [20]. With NLP, users can use their voice to navigate in MR world. Voice command and voice recognition technology has also been a significant milestone in the twenty-first century, allowing human connection with devices and applications through speech recognition. Voice input is a natural approach that allows the user to convey their purpose [21]. This eliminates the requirement for the user to use gestures in order to command a virtual item directly. The user has just to concentrate on the virtual item and then verbally speak his or her instruction. Using voice for navigating complicated manual guides is also very helpful as it allows users to navigate layered menus with a single command. Users are able to save time and speed up the interaction using voice commands. Hence, in this study, we make use of a simple voice command to pull the item closer to the user, adjust the texture and colour of the object, and play the video that is stored inside the MR telepresence room. Figure 2

Fig. 2. Speech interaction module [17]

shows the speech interaction module [17], user wears HMD and deliver the commands and language to MR application.

To implement the voice commands, STT conversion is required to identify the command. After the command has been identified, the command will produce the text. The instruction will be executed according to the text output. Figure 3 shows the process of the STT conversion. According to Fig. 3, the voice is required to be analysed before sending it to the speech recognition decoder. Based on the acoustic model and language model, the speech recognition decoder looks through all possible alignments of all possible pronunciations of all possible word sequences to identify the most probable one.

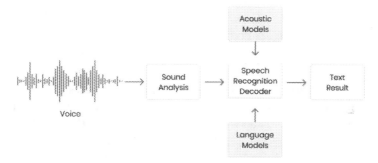

Fig. 3. The process of the STT conversion

2.2 Virtual Assistant

Next, after the STT conversion, the virtual assistant will execute the command according to the text output data from the speech. For example, If the user commands to play a video inside the MR telepresence, the virtual assistant will play the video accordingly. In this study, the voice commands are constrained based on the database provided.

To provide information to the user after performing the voice commands in MR telepresence, the virtual assistant avatar will be able to project audio as feedback to the voice commands. Microsoft Azure Cognitive is a speech service utilized in this study to convert text into humanlike synthesized speech. Speech synthesis is also known as TTS. The process of the TTS conversion using Microsoft Azure Cognitive service is shown in Fig. 4. Based on Fig. 4, the text file is executed using C# script in the unity platform to produce the audio. The process is much easier as Microsoft Azure Cognitive has provided a speech system development kit (SDK) for the unity platform. As long as the internet connection is active, the script can be accessed by the SDK to utilize the services. The script will play the audio from the SDK.

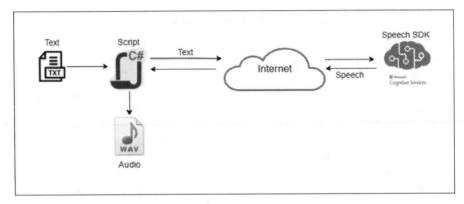

Fig. 4. The process of TTS conversion using a speech SDK

2.3   MR Telepresence

The setup for MR telepresence is based on [22] study. According to the study, the setup will require two workspaces for the local user and the remote user. The setup for the remote user will require two depth sensors to capture the remote user from the front and behind. Meanwhile, the setup for the local user only required an HMD device connected to a personal computer (PC) with spacious space for the play area. Both of these setups are shown in Fig. 5.

Referring to Fig. 5, the local user and the remote user are connected using the local network connection. The Kinect sensors at the remote workspace, capture and transmit the input through the network and reconstructed it at the local user workspace. In our system we use two Kinect sensors to track the remote user in 3D views. At the local user workspace area, the local user will be able to see the 3D reconstruction of the remote user using the HMD in the immersive MR environment. In the virtual environment, the virtual assistant avatar appears as the third person. A table and a few primitive 3D objects are prepared in the environment for the local user to perform the interaction with the remote user. The integration between the virtual 3D environment with the real-time 3D reconstruction has created the MR environment. A stable network was required to perform MR telepresence.

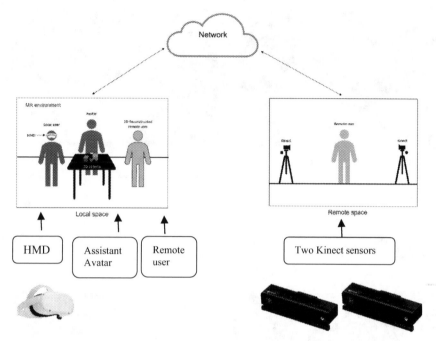

Fig. 5. MR telepresence for the local and remote user.

3   Test Application

This section continues to discuss the results of the proposed method for voice commands and virtual assistant in MR telepresence. Figure 6(a) shows the results of the MR telepresence from the local user's point of view. It can be seen the 3D reconstruction of the remote user appears annotation as Ms Ameerah, while the virtual assistant appears annotation as Alex. In the MR environment, Ms Ameerah will communicate with the local user. However, Ms Ameerah is unable to hear and communicate with Alex. The virtual assistant has been set to assist local user in MR. Alex only can be heard by the local user, which is the HMD user as Alex's role is to assist and interact with the local user. Alex will receive the voice commands from the local user and perform the commands accordingly.

For the first task, the local user instructed Alex to play a video and song in the MR telepresence. As in Fig. 6(b) Alex received the commands and play the video and song in the display panel. Both the local user and remote user will be able to listen to the song.

Another interaction happens in our proposed system is object manipulation. In our MR application primitive objects have been used such as cube, cylinder, triangle and rectangle. With Alex assistant, we able to speed up the process of the local user interaction inside MR telepresence. For example, by using single commands, Alex able to produces 3D objects on the workbench tabletop according to the local user preferences. Figure 7(a) shows the 3D objects created on the table based on the voice commands by the local user. Red colour is the default colour for the 3D objects. As we can see in Fig. 7(a), the

(a) Video panel is hidden   (b) Video panel visible after the voice command

Fig. 6. The environment of the MR telepresence

3D object is far from the local user. Hence, the local user is able to bring the objects closer to them by giving voice commands to the virtual assistant. Figure 7(b) shows the 3D objects become closer to the local user after the voice commands.

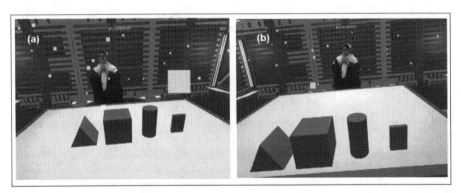

Fig. 7. The 3D object appears (a) at the centre of the table (b) closer to the local user (Color figure online)

Moreover, we also added commands for the local user to change the texture of the object into a different texture. As presented in Fig. 8(a), the voice command for changing the texture of the object is applied to the cube and cylinder objects in the scene. The texture of both cube and cylinder is changed from brick texture to grass texture as shown in Fig. 8(b).

This result is achieved as the virtual assistant responds to the voice command instructed by the local user. The voice command between local user and Alex virtual assistant as listed in Table 1. To begin interacting with Alex, "Hello Alex" as a greeting. Once Alex response, the local user can start to give commands. Ten commands have been tested in this experiment. The data involves such as video, textures in 2D images, 3D object including primitive objects and virtual environment with workbench in the middle space area. Based on the proposed interaction, virtual assistant has planned to speed up the process when in conventional method, user need to bring UI panel to select the

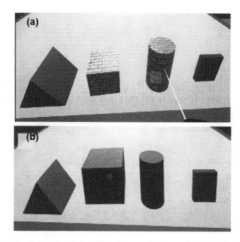

Fig. 8. The texture of the 3D object change to (a) brick (b) grass

texture and the different panel shows the object listing where user can click to instantiate the object.

Table 1. Voice command with response from the virtual assistant.

Voice Commands	Virtual Assistant response
Hello Alex	Hello there
Alex, play the UTM song	Okay, UTM song is play
Alex, pause the song	The song is paused
Alex, load the primitive's object onto the table	We have cube, cuboid, cylinder and prism. Do you want to load it all?
Yes, please	Okay
Alex, what texture do we have?	We have sand, grass, brick wood and metal
Okay apply brick texture to the cube and cylinder	Brick texture applied
Alex, apply grass texture to both of them	Okay, grass texture is applied
Apply wood texture to the cuboid	Wood texture applied
Alex create another cube	A cube is created
Change the cube colour to yellow	The colour is changed to yellow

4 Conclusion

The goal of this study is to explore and implements the voice command and virtual assistant features to enhance the interaction in our MR Telepresence system. The voice command is deployed by the local user using TTS conversation and NLP language to execute the command. Then, the command will be replied by the virtual assistant in the MR telepresence using STT conversion. The setup for the MR telepresence has been discussed. The interaction using voice command is much more reliable and faster without the requirement of heavy physical movement during the interaction. This will allow the user to be more immersed and focused on the scene with reduced distraction from other gestures or movements. In [23], they have performed real hand gesture in MR with speech to speed up the interaction. Our system local user uses virtual assistant to spawn the object and change the texture of the object through speech voice commands. Further improvement in our prototype is to add more six degrees of freedom (6DOF) object manipulation including translation, scaling and rotation.

However, there are several constrained that are required to be highlighted in future works. One of the constrain is the network latency or connection. If the network is lagging, it will increase the chances of the local user being distracted and break the immersive interaction due to the delayed command. Other than that, it is suggested to enhance the virtual assistant avatar by training the avatar for artificial intelligent (AI) to broaden the database of commands and answers. It can be envisioned that the interaction for business for example smart house of the future using a voice command approach with the help of a virtual assistant can be implemented in MR telepresence. In this research, the voice commands have been fixed into several instructions, as this is the initial result for voice commands and virtual assistant in MR telepresence. For future work, this study can be extend by performing the evaluation on the voice command accuracy by referring to [17] for the implementation of voice commands in MR telepresence. Despite that, another suggestion as future research is to enhance the voice commands by implement AI in the virtual assistant, to make the instruction more varied.

Acknowledgment. We would like to express our gratitude to Universiti Teknologi Malaysia (UTM) for funding under UTM Encouragement Research (UTMER) grant number Q.J130000.3851.19J10 for the support towards this research.

References

1. Azuma, R.T.: A survey of augmented reality. Presence Teleoperators Virtual Environ. **6**, 355–385 (1997)
2. Harborth, D.: Augmented reality in information systems research: a systematic literature review (2017)
3. Halim, N.A.A., Ismail, A.W.: Designing ray-pointing using real hand and touch-based in handheld augmented reality for object selection. Int. J. Innov. Comput. **11**, 95–102 (2021). https://doi.org/10.11113/IJIC.V11N2.316
4. Fadzli, F.E., Kamson, M.S., Ismail, A.W., Aladin, M.Y.F.: 3D telepresence for remote collaboration in extended reality (xR) application. In: IOP Conference Series: Materials Science and Engineering, p. 012005. IOP Publishing (2020). https://doi.org/10.1088/1757-899X/979/1/012005

5. Fadzli, F.E., Ismail, A.W., Ishigaki, S.A.K., Nor'a, M.N.A., Aladin, M.Y.F.: Real-time 3D reconstruction method for holographic telepresence. Appl. Sci. **12**, 4009 (2022). https://doi.org/10.3390/APP12084009
6. Nor'a, M.N.A., Ismail, A.W.: Integrating virtual reality and augmented reality in a collaborative user interface. Int. J. Innov. Comput. **9**, 59–64 (2019). https://doi.org/10.11113/IJIC.V9N2.242
7. Silva, R., Oliveira, J.C., Giraldi, G.A.: Introduction to augmented reality (2003)
8. Zahorik, P., Jenison, R.L.: Presence as being-in-the-world. Presence Teleoperators Virtual Environ. **7**, 78–89 (1998)
9. Tonchev, K., Bozhilov, I., Petkova, R., Poulkov, V., Manolova, A., Lindgren, P.: Implementation requirements and system architecture for mixed reality telepresence application scenario. In: International Symposium on Wireless Personal Multimedia Communications, WPMC. IEEE Computer Society (2021). https://doi.org/10.1109/WPMC52694.2021.9700439
10. Cho, S., Kim, S., Lee, J., Ahn, J., Han, J.: Effects of volumetric capture avatars on social presence in immersive virtual environments, pp. 26–34 (2020). https://doi.org/10.1109/VR46266.2020.00020
11. Joachimczak, M., Liu, J.: Real-time mixed-reality telepresence via 3D reconstruction with HoloLens and commodity depth sensors (2017). https://doi.org/10.1145/3136755.3143031
12. Das, P., Acharjee, K., Das, P., Prasad, V.: Voice recognition system: speech-to-text. J. Appl. Fundam. Sci. **1**, 191 (2015)
13. Guamán, S., Tapia, F., Yoo, S.G., Calvopiña, A., Orta, P.: Device control system for a smart home using voice commands: a practical case. In: ACM International Conference Proceeding Series, pp. 86–89 (2018). https://doi.org/10.1145/3285957.3285977
14. Priya, S.S., Rachana, P., Chellani, D.: Augmented reality and speech control from automobile showcasing, pp. 1703–1708 (2022). https://doi.org/10.1109/ICSSIT53264.2022.9716534
15. Fagernäs, S., Hamilton, W., Espinoza, N., Miloff, A., Carlbring, P., Lindner, P.: What do users think about Virtual Reality relaxation applications? A mixed methods study of online user reviews using natural language processing. Internet Interv. **24**, 100370 (2021). https://doi.org/10.1016/J.INVENT.2021.100370
16. Hoppenstedt, B., Kammerer, K., Reichert, M., Spiliopoulou, M., Pryss, R.: Convolutional neural networks for image recognition in mixed reality using voice command labeling. In: De Paolis, L.T., Bourdot, P. (eds.) AVR 2019. LNCS, vol. 11614, pp. 63–70. Springer, Cham (2019). https://doi.org/10.1007/978-3-030-25999-0_6/COVER
17. Siyaev, A., Jo, G.S.: Towards aircraft maintenance metaverse using speech interactions with virtual objects in mixed reality. Sensors **21**, 2066 (2021). https://doi.org/10.3390/S21062066
18. Patel, B.D., Patel, H.B., Khanvilkar, M.A., Patel, N.R., Akilan, T.: ES2ISL: an advancement in speech to sign language translation using 3D avatar animator. In: Canadian Conference on Electrical and Computer Engineering (2020). https://doi.org/10.1109/CCECE47787.2020.9255783
19. Bahja, M., Bahja, M.: Natural language processing applications in business. E-Bus. High. Educ. Intell. Appl. (2020). https://doi.org/10.5772/INTECHOPEN.92203
20. Hirschberg, J., Manning, C.D.: Advances in natural language processing. Science **349**, 261–266 (2015). https://doi.org/10.1126/SCIENCE.AAA8685/ASSET/D33AB763-A443-444C-B766-A6B69883BFD7/ASSETS/GRAPHIC/349_261_F5.JPEG
21. Nguyen, T.V., Kamma, S., Adari, V., Lesthaeghe, T., Boehnlein, T., Kramb, V.: Mixed reality system for nondestructive evaluation training. Virtual Real. **25**, 709–718 (2021). https://doi.org/10.1007/S10055-020-00483-1/TABLES/2

22. Ishigaki, S.A.K., Ismail, A.W.: Real-time 3D reconstruction for mixed reality telepresence using multiple depth sensors. In: Shaw, R.N., Paprzycki, M., Ghosh, A. (eds.) ICACIS 2022, vol. 1749, pp. 67–80. Springer, Cham (2023). https://doi.org/10.1007/978-3-031-25088-0_5
23. Wang, M., et al.: Object selection and scaling using multimodal interaction in mixed reality. In: IOP Conference Series: Materials Science and Engineering, vol. 979, p. 012004 (2020). https://doi.org/10.1088/1757-899X/979/1/012004

Photovoltaics Cell Anomaly Detection Using Deep Learning Techniques

Abdullah Ahmed Al-Dulaimi[1], Alaa Ali Hameed[2], Muhammet Tahir Guneser[1], and Akhtar Jamil[3(✉)]

[1] Department of Electrical and Electronics Engineering, Karabuk University, Karabuk, Turkey
[2] Department of Computer Engineering, Istinye University, Istanbul, Turkey
[3] Department of Computer Science, National University of Computer and Engineering Sciences, Islamabad, Pakistan
akhtar.jamil@nu.edu.pk

Abstract. Photovoltaic cells play a crucial role in converting sunlight into electrical energy. However, defects can occur during the manufacturing process, negatively impacting these cells' efficiency and overall performance. Electroluminescence (EL) imaging has emerged as a viable method for defect detection in photovoltaic cells. Developing an accurate and automated detection model capable of identifying and classifying defects in EL images holds significant importance in photovoltaics. This paper introduces a state-of-the-art defect detection model based on the Yolo v.7 architecture designed explicitly for photovoltaic cell electroluminescence images. The model is trained to recognize and categorize five common defect classes, namely black core (Bc), crack (Ck), finger (Fr), star crack (Sc), and thick line (Tl). The proposed model exhibits remarkable performance through experimentation with an average precision of 80%, recall of 87%, and an mAP@.5 score of 86% across all defect classes. Furthermore, a comparative analysis is conducted to evaluate the model's performance against two recently proposed models. The results affirm the excellent performance of the proposed model, highlighting its superiority in defect detection within the context of photovoltaic cell electroluminescence images.

Keywords: Deep learning · Detection · Solar panel · Electroluminescence image detection

1 Introduction

Photovoltaic (PV) cells, also known as solar cells, are a crucial component in the functioning of solar panels and play a vital role in the generation of clean, renewable energy. As with any technology, PV cells are subject to various anomalies and defects that can impact their performance and efficiency. Detecting and addressing these anomalies and defects in a timely manner is essential to ensuring that solar panels operate at optimal capacity. Anomaly and defect detection in PV cells can be performed through a variety of methods, including visual inspection, electrical testing, and computer-based image analysis. These methods aim to identify and classify different types of anomalies and defects, including cracks, hotspots, broken cells, shading, etc. By detecting and addressing these issues, the efficiency and lifespan of solar panels can be improved, reducing the cost of solar energy production and promoting the growth of the renewable energy sector.

Electroluminescence (EL) is a phenomenon in which a material emits light in response to an applied electric field [1]. This light emission occurs due to the movement of electrons within the material and the recombination of electrons with holes. EL is widely used in various applications, including displays, lighting, and imaging. In solar cells, EL refers to the emission of light from a material as a result of the flow of an electric current. This phenomenon is used in PV cells for detecting anomalies and defects. By applying a voltage to the cell and observing the emitted light, one can identify issues such as cracks, hotspots, and other defects that can reduce the efficiency and lifespan of the cell. EL can also be used to determine the quality of the materials used in the cell and monitor the aging process of the cell. The EL technique is a non-destructive and fast method for detecting and analyzing defects, making it an important tool for improving the performance and reliability of photovoltaic systems.

As shown in Fig. 1, a typical setup for EL involves a high voltage source to supply an electric field across a solar cell sample, the sample is illuminated with a low-intensity light source, usually from a near-infrared LED, to excite the photovoltaic materials and generate an electrical current, the resulting light emissions are captured by a CCD camera, which records the pattern of the light emitted by the sample, the EL image is then analyzed to detect any anomalies or defects in the solar cell, such as cracks, open circuits or others [2]. The information gathered from the EL image can be used to optimize the cell design, improve the cell performance, and identify areas for repair or replacement. A near-infrared (NIR) CCD camera is a device used to capture and record EL signals in solar cells, the camera operates in the NIR spectrum and is sensitive to the NIR EL signals emitted by the solar cell, by using a NIR CCD camera, it is possible to detect and analyze various types of anomalies and defects in PV cells that cannot be seen with the naked eye, this technology helps to improve the quality control process of PV cells and ensure the efficiency and longevity of PV systems [3].

The classification process for PV cell EL typically involves using machine learning algorithms to analyze the images and classify them into different categories based on the presence or absence of defects, first, the images are preprocessed to remove noise and enhance the relevant features. Then, features are extracted from the images using techniques such as convolutional neural networks (CNNs), which are designed to identify patterns in the image data, once the features are extracted, they are used to train a classification model, such as a support vector machine (SVM), to predict the class labels of new images. The model is trained on a labeled dataset, where each image is assigned to a specific category based on the type of defect present, such as cracks or broken cells, the trained model can then be used to classify new images into the different defect categories with a high degree of accuracy. This process allows for automatically detecting and classifying defects in PV cells, which is crucial for maintaining their performance and ensuring their longevity [4–6].

The detection of anomalies in photovoltaic panels refers to the process of identifying any faults, defects, or damages in these panels. Photovoltaic panels convert light into electricity and it is crucial to ensure their proper functioning for the efficient production of electricity. Over time, the method used for detecting anomalies in photovoltaic panels has improved and evolved. Initially, optical images were used to detect anomalies in photovoltaic panels. However, this method had certain limitations, as optical images only provided a limited view of the panels functioning. As technology advanced, more specific images, such as multi-spectral, thermal, optical, etc., were adopted. These images provide a more detailed and accurate view of the panels functioning, enabling better detection of anomalies. Multi-spectral images, for example, capture images in different wavelength ranges, which helps to detect anomalies related to the performance of the panel. Thermal images capture the heat generated by the panel and can be used to detect faults in the electrical connections.

In recent years, a growing number of researchers have focused their efforts on utilizing PV cell EL images to identify and diagnose anomalies in photovoltaic panels. The task of identifying defects in multiscale electroluminescence images from photovoltaic cells is difficult because the features tend to disappear as the network becomes deeper [7]. Bidirectional Attention Feature Pyramid Network (BAFPN) has been proposed in [8], it combines the feature pyramid network (FPN) with a bidirectional attention mechanism, which enables the network to have a better ability to capture and propagate context information throughout different levels of the feature pyramid, BAF-Detector is a modification of Faster RCNN + FPN, which incorporates a type of feature pyramid network (FPN) called BAFPN into the region proposal network, by doing so, it aims to improve the accuracy and efficiency of object detection by utilizing the advantages of both Faster RCNN and BAFPN, the aim of the BAFPN is to extract features from multiple scales and provide high-resolution features for objects of different scales in an image, the bidirectional attention mechanism also allows the network to attend to different regions in the image and make decisions based on the most relevant features, thereby improving its accuracy and robustness. The complementary attention network (CAN) has been proposed in [9], CAN is a novel architecture designed to address the challenging problem of automatic defect detection in solar cell EL images, CAN connects a channel-wise attention subnetwork with a spatial attention subnetwork sequentially to suppress background noise features and highlight defect features simultaneously, the channel-wise attention subnetwork integrates output features extracted by global average pooling layer and global max pooling layer using convolution operations; additionally, CAN is embedded into a region proposal network in a faster R-CNN to extract refined defective region proposals, and construct an end-to-end faster RPANCNN framework for detecting defects in raw EL images.

The PV cell EL imaging detection still has many challenges. The background in EL images can be complex and noisy, making it difficult to identify defects accurately. Additionally, there may be similarities between different types of defects, making classifying and differentiating them challenging. Researchers are developing advanced techniques such as deep learning-based approaches, attention networks, and other machine learning algorithms to overcome these challenges. These approaches aim to improve the accuracy and efficiency of defect detection in EL images and are constantly being improved to address new challenges that arise.

This paper proposes a state-of-the-art detection model called Yolov7 for Photovoltaic cell Electroluminescence. The model is designed to detect five different classes, namely black core (Bc), crack (Ck), finger (Fr), star crack (Sc), and thick line (Tl). The paper presents in-depth details about the model's architecture. Finally, we compare our results with those of two other papers that have used different models.

2 Dataset

A dataset has been created for detecting anomalies in photovoltaic cells on a large scale in [10], this dataset consists of 10 categories, several detection models were investigated based on this dataset, the best model Yolov5-s achieved 65.74 mAP@.5. The provided Table 1 shows the models and their corresponding characteristics for detecting defects in PV cell EL images, the models listed are BAF-Detector, Faster RPAN-CNN, and our model proposed Yolov7, with PVCEL being the images type, the number of training and testing images vary among the models, BAF-Detector was trained on 847 images and tested on 1282, while Faster RPAN-CNN was trained on 847 images and tested on 2782, Yolov7 was trained on 4700 images and tested on 430, the defects that each model detected were listed as Bc, Ck, Fr, Sc, and Tl. Figure 2 shows a comparison of the ground truth samples used in this paper and those used in references [8] and [9].

Table 1. Distribution of PV cell EL images datasets in the training/testing.

Ref	Model	Images type					Training	Testing
[8]	BAF-Detector	PVCEL					847	1282
		Bc	Ck	Fr				
[9]	Faster RPAN-CNN	PVCEL					847	2782
		Bc	Ck	Fr				
Proposed	Yolov7	PVCEL					4700	430
		Bc	Ck	Fr	Sc	Tl		

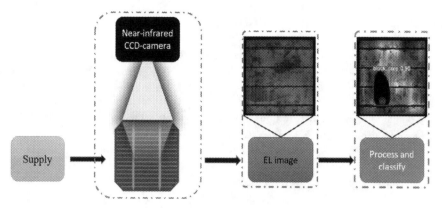

Fig. 1. A typical setup for Electroluminescence (EL)

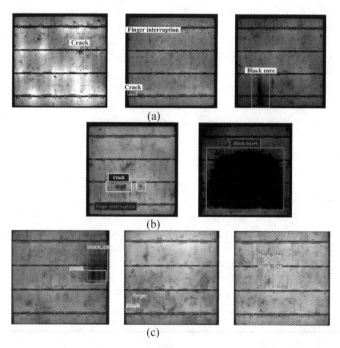

Fig. 2. Compares the ground truth samples prepared to identify the PV cells. a) shows the ground truth prepared by [8], b) shows the ground truth prepared by [9], and c) shows our ground truth.

3 Methodology

3.1 CNN-Based Detection

RCNN [11], Fast-RCNN [12], Faster-RCNN [13], YOLOv3 [14], YOLOv5 [15], and YOLOv7 [16] are all state-of-the-art object detection algorithms. Each algorithm uses different architectures and techniques to detect objects in images or videos. RCNN

(Region-based Convolutional Neural Network) is a two-stage object detection algorithm that uses selective search to extract object proposals and then uses a convolutional neural network to classify each proposal. RCNN was introduced in 2014 and improved with Fast-RCNN and Faster-RCNN. Fast-RCNN was introduced in 2015 and is a faster and more accurate version of RCNN. Instead of using selective search to extract proposals, Fast-RCNN uses a single convolutional neural network to generate feature maps for the entire image, and then selects regions of interest (RoI) from those feature maps. Faster-RCNN was introduced in 2015 and built upon the improvements made in Fast-RCNN. It uses a Region Proposal Network (RPN) to generate proposals instead of using the selective search or a separate network. This makes Faster-RCNN faster and more accurate than previous versions of RCNN. Faster-RCNN achieves state-of-the-art performance on various object detection benchmarks, including COCO and VOC. Yolov3 is a real-time object detection algorithm that identifies specific objects in videos, live feeds, or images. Yolov3 uses a few tricks to improve training and increase performance, including multi-scale predictions, a better backbone classifier, and more. It achieves state-of-the-art performance on the COCO dataset. Yolov5 is a family of compound-scaled object detection models trained on the COCO dataset. It includes simple functionality for Test Time Augmentation (TTA), model ensembling, hyperparameter evolution, and export to ONNX, CoreML and TFLite. Yolov5 achieves state-of-the-art performance in both speed and accuracy. Yolov7 is the fastest and most accurate real-time object detection model for computer vision tasks. It was introduced in 2022 and used a trainable bag of freebies to set a new state-of-the-art for real-time object detectors. Yolov7 achieves state-of-the-art performance on the COCO dataset and outperforms its competitors in terms of speed and accuracy.

3.2 Yolov7 Architecture

The overall architecture of the Yolov7 model is shown in Fig. 3. The stem refers to the initial layers that process the input image before it is passed through the backbone layers. The stem layer is typically composed of a sequence of convolutional layers, max-pooling layers, and batch normalization layers that help extract low-level features from the input image. The output of the stem layer is then passed to the backbone layers for further processing and feature extraction.

The term backbone refers to the feature extractor part of the network, which is responsible for extracting high-level features from the input image. The backbone typically consists of several convolutional layers and is usually pre-trained on a large dataset, E-ELAN (Extended efficient layer aggregation network) is a computational block used in the Yolov7 backbone. It has been designed to improve the speed and accuracy of the Yolov7 model by analyzing the following factors that impact both speed and accuracy: memory access cost, I/O channel ratio, element-wise operation, activations, and gradient path.

Spatial Pyramid Pooling Cross Stage Partial Connection (SPPCSPC) is a technique used in object detection Yolov7, which is applied as a third step after the stem and backbone. SPPCSPC combines two techniques: Spatial Pyramid Pooling (SPP) and Cross Stage Partial Connection (CSPC). SPP is a technique that allows a CNN to receive input images of various sizes and scales by pooling the features over various regions of the image. CSPC, on the other hand, connects the input and output of a convolutional layer across multiple stages, allowing the model to learn more complex features. SPPCSPC uses SPP to generate fixed-length feature maps from input images of various sizes, which are then fed to the CSPC module. The CSPC module connects the output of one stage to another stage's input, allowing the model to learn features of different scales and complexities. By using SPPCSPC, object detection models are able to process images of various sizes and scales while also learning complex features, resulting in better accuracy and performance.

Furthermore, Yolov7 applied CSP-OSA is an improved version of the Cross Stage Partial network (CSPNet), a backbone network used for object detection in computer vision. CSP-OSA stands for Cross Stage Partial network with Orthogonal Spatial Attention, which combines the CSPNet architecture with an attention mechanism called Orthogonal Spatial Attention (OSA). This attention mechanism is designed to improve the ability of the network to attend to important features while suppressing irrelevant features. The combination of the CSPNet and OSA in CSP-OSA has shown improved performance in object detection tasks compared to the original CSPNet.

In object detection models, we have two heads, the lead head and auxiliary head refer to two separate paths in the network used to predict the final output. The lead head is responsible for predicting the bounding box coordinates, objectness score, and class probabilities for the detected objects. The auxiliary head is a secondary path that provides additional information to the network to help improve its predictions. It may take intermediate features from the backbone network and use them to predict additional outputs that can help refine the lead head's predictions. During training, the loss is calculated for both the lead head and auxiliary head based on the same soft labels that are generated. Ultimately, both heads get trained using the same soft labels. Finally, the loss function is calculated for both the lead head and the auxiliary head based on the same generated soft labels. However, the difference between the loss functions for the two sub-networks lies in their weighting. The lead head's loss is given more weight than the auxiliary head's loss, as the lead head is responsible for making the final predictions.

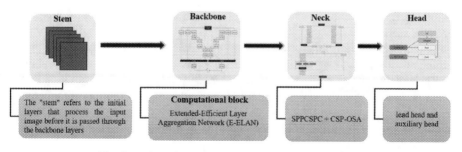

Fig. 3. Yolov7 detection steps: the four key components

3.3   Mathematical Model

In model pipelines, one of the initial steps is to resize the input image to a fixed size to ensure consistency across all input images, the input image is resized to 640 x 640. This fixed size can also simplify subsequent processing steps in the object detection pipeline. We can express the formula in the following way:

$$I_{resized}(x, y, c) = \frac{255}{\max(I_{original})} I_{original}\left(\left\lfloor \frac{x}{s_x} \right\rfloor, \left\lfloor \frac{y}{s_y} \right\rfloor, c\right) \tag{1}$$

where $I_{original}$ is the original input image, $I_{resized}$ is the resized image, (x, y, c) are the pixel coordinates and color channel of the resized image, s_x and s_y are the scaling factors in the x and y directions, and $\max(I_{original})$ is the maximum pixel value in the original image, in image processing, 255 is often used as the maximum value for pixel intensities in an 8-bit image, note that this formula assumes that the scaling factors s_x and s_y are chosen such that the aspect ratio of the original image is preserved.

The second step in object detection involves feeding the resized image obtained from the previous step into a CNN, the purpose of this step is to extract relevant feature maps from the input image that will help the model detect objects accurately. Convolutional layers are a fundamental building block of CNNs, and they consist of a set of learnable filters that are convolved with the input image to produce a feature map, these filters are designed to detect various features of the input image, such as edges, corners, and other relevant patterns. The output feature maps are then passed through additional layers of the network to further refine the detected objects. The formula can be defined as shown below:

$$\text{conv}(x)ij^k = b_k + \sum l = 1^{C_{in}} \sum_{m=1}^{H_f} \sum_{n=1}^{W_f} w_{lmn}^k x_{(i-1)S_h+l(m-1)+p_1}^{(j-1)S_w+n(q-1)+p_2} \tag{2}$$

where b_k is the bias term for the kth filter, w_{lmn}^k is the weight for the kth filter, S_h and S_w are the horizontal and vertical strides, p_1 and p_2 are the padding values, and C_{in}, H_f, and W_f are the number of input channels, filter height, and filter width, respectively, the output dimensions of the convolution layer are determined by the input size, filter size, stride, and padding. The dilations and groups parameters are not explicitly used in this formula, but they affect the weight and bias values used in the convolution operation.

The sigmoid function ensures that the probability values are in the range of [0, 1], which makes it easier to compare them and to decide whether an object is present in the image or not. Therefore, the sigmoid function plays a critical role in the detection process, as it helps to interpret the output of the convolution layer in a meaningful way. We can define the equation as follows:

$$\sigma(x) = \frac{1}{1+e^{-x}} \tag{3}$$

where σ is the sigmoid function and x is the input.

Max pooling is an operation that is commonly used in neural networks, including those used for object detection. It is a form of down-sampling that reduces the spatial size of the feature maps by taking the maximum value within a certain window, typically a 2×2 window, and moving that window across the entire feature map. The output of the max pooling operation is a new feature map with smaller dimensions, but with the same number of channels as the input feature map. The formula can be formulated as follows:

$$O_{i,j,k} = \max_{m=0}^{1} \max_{n=0}^{1} I_{(i+m\cdot\text{strides}[0]),(j+n\cdot\text{strides}[1]),k} \tag{4}$$

where I is the input tensor, O is the output tensor, i and j are the spatial indices of the output tensor, and k is the channel index.

In object detection models, concatenation refers to the operation of concatenating the output feature maps from multiple previous layers into a single tensor, which is then passed to the subsequent layer. The multiplication operation usually involves element-wise multiplication of two tensors. The addition operation usually involves element-wise addition of two tensors. These operations are used to combine information from multiple layers to improve the detection accuracy. We can express the formulas in the following way:

Concatenation:

$$\text{concat}(x_1, x_2, \ldots, x_n) = \left[x_1^\top, x_2^\top, \ldots, x_n^\top \right]^\top \tag{5}$$

Multiplication:

$$y = x_1 \times x_2 \tag{6}$$

Addition:

$$y = x_1 + x_2 \tag{7}$$

where $x_1, x_2, \ldots, x_n$ are the input tensors and y is the output tensor. Note that the shapes of the input tensors must be compatible for these operations to be valid.

The evaluation in object detection is a crucial step in assessing the performance of a model. The evaluation metrics commonly used in object detection are recall, precision, F-score, and intersection over union (IoU). The recall is a metric that measures the ability of the model to detect all positive samples. It is the ratio of the true positive (TP) samples to the sum of TP and false negative (FN) samples. Precision measures the ability of the model to detect only the positive samples correctly. It is the ratio of TP samples to the sum of TP and false positive (FP) samples. F-score, or the F1 score, is the harmonic mean of recall and precision, providing a balanced measure between these two metrics. IoU measures the degree of overlap between the predicted and ground truth bounding boxes. It is the ratio of the area of intersection to the area of union between the predicted and ground truth bounding boxes. These metrics are typically used in object detection to evaluate the performance of a model on a test dataset. The higher the recall, precision, F-score, and IoU, the better the model's performance.

$$\text{Recall} = \frac{\text{TP}}{\text{TP}+\text{FN}} \tag{8}$$

$$\text{Precision} = \frac{TP}{TP+FP} \tag{9}$$

$$F_1 = 2 \cdot \frac{\text{Precision} \cdot \text{Recall}}{\text{Precision} + \text{Recall}} \tag{10}$$

$$\text{IoU} = \frac{\text{Area of Overlap}}{\text{Area of Union}} = \frac{TP}{TP+FP+FN} \tag{11}$$

In the above formulas, TP represents the number of true positive detections, FN represents the number of false negative detections, and FP represents the number of false positive detections.

4   Results and Discussion

4.1   Training Process

The training process involves training an object detection model to classify input images into five classes: black core, crack, finger, star crack, and thick line. The training set consists of 4700 PV-C-EL images that belong to these classes. The number of epochs used for training is 80, which means that the entire training dataset will be used 80 times to update the model parameters. During each epoch, the model will use a batch of images to compute the gradients of the loss function, which measures the difference between the predicted output and the actual output. The optimizer will use these gradients to update the model parameters to minimize the loss. This process is repeated for each batch of images until the end of the training epochs. The evaluation of the trained model will involve bounding box loss, objectness loss, and classification loss. Furthermore, the training process includes the validation of the model, where a set of 432 validation images will be used to evaluate the model's performance. The evaluation of the model on the validation set will include measuring the losses, such as bounding box loss, objectness loss, and classification loss. Additionally, all models were evaluated in terms of recall, precision, and mAP@.5.

As we see in Fig. 4(a), bounding box loss is used to measure the difference between the predicted bounding boxes and the ground-truth bounding boxes; the values of bounding box loss are usually small and gradually decrease during training in the case of the given data, the values of bounding box loss start from 0.0708 and gradually decrease to 0.02157 after 80 epochs of training. Objectness loss is used to measure the confidence score for the predicted bounding boxes, the values of objectness loss are usually smaller than those of bounding box loss and also gradually decrease during training, in the given data, the values of objectness loss start from 0.008271 and gradually decrease to 0.00357 after 80 epochs of training. Classification loss is used to measure the difference between the predicted class labels and the ground-truth class labels, the values of classification loss are usually the smallest among the three types of losses and also gradually decrease during training, the values of classification loss start from 0.01963 and gradually decrease to 0.0008734 after 80 epochs of training. The gradual decrease of loss values during training indicates that the model is learning and becoming more accurate in detecting objects. The values of loss in the given data suggest that the training process is progressing well and the model is achieving good results. As shown in Fig. 4(b), the validation results measure the losses, including bounding box loss, objectness loss, and classification loss, using the same analysis criteria as the training set.

As shown in Fig. 5, the mAP@.5 values provided show the model's performance at different epochs during validation, the mAP@.5 values range from 0.18 to 0.87, with a general increasing trend in performance as the model is trained. A higher mAP@.5 indicates better performance in detecting objects. Similarly, the recall values provided show the fraction of true positives detected by the model, the recall values range from 0.178 to 0.864, with a general increasing trend in performance over epochs. Lastly, the precision values provided show the fraction of true positives to the total number of predicted positives. The precision values range from 0.39 to 0.998, with a general decreasing trend in performance over epochs. In general, a good object detection model should have high precision, recall, and mAP@.5 values.

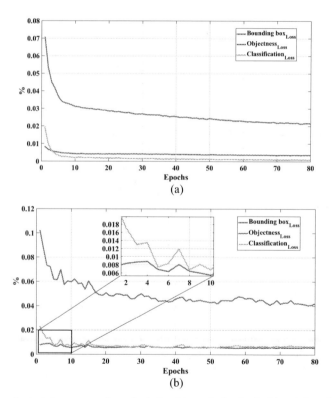

Fig. 4. Shows the training process for calculating losses during both: (a) training, (b) validation.

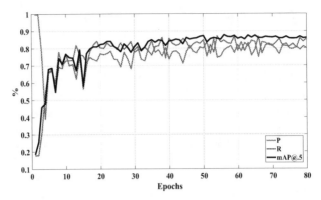

Fig. 5. Shows the validation performance based on recall, precision, and mAP@.5.

4.2   Testing Models

In the testing process, there are a total of 430 images, with five different classes: black core (Bc), crack (Ck), finger (Fr), star crack (Sc), and thick line (Tl), the number of labels for these classes is 99, 165, 319, 20, and 152 respectively, the evaluation metrics used to analyze the results of the testing process include recall, precision, and F-score, with the average of all classes calculated to determine the mAP at the IoU threshold of 0.5. Figure 6 shows the results of all classes based on the metrics mentioned above. In Table 2, we can see that the model achieved high precision and recall for the Bc class, indicating that it is able to identify these objects accurately. The Ck class has lower precision and recall, suggesting that the model has more difficulty with these objects. The Fr class has relatively high precision and recall, indicating that the model is performing well on these objects. The Sc class has lower precision but high recall, suggesting that the model identifies most of the positive examples but also produces many false positives. The Tl class has high precision and recall, indicating that the model is accurately identifying these objects. Overall, the mAP at a threshold of 0.5 for all classes is relatively high, suggesting that the model is performing well across all classes.

Table 3 compares the performance of three different object detection models in detecting anomalies in photovoltaic cell electroluminescence images. The first model is the BAF-Detector, a model proposed in a previous paper that can detect three categories of anomalies. The second model is the RPAN-CNN (GAP&GMP), another previously proposed model that can also detect three categories of anomalies. The third model is the one proposed in this paper, which uses Yolov7 and can detect five categories of anomalies. The table shows the mAP of each model at a confidence threshold of 0.5. The mAP@.5 metric measures how well the model performs in terms of precision and recall. The results show that while the BAF-Detector and RPAN-CNN models perform well in detecting three categories of anomalies, the model proposed in this paper has a slightly lower mAP but can detect five categories of anomalies, making it more versatile for detecting defects in photovoltaic cells.

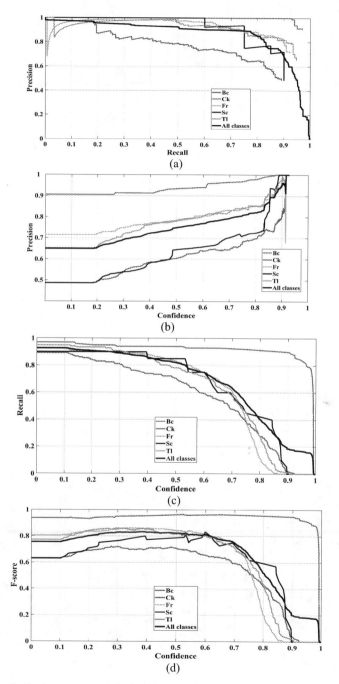

Fig. 6. Testing process results based on recall, precision and F-score metrics.

Figure 7 displays eight images that the proposed detection model for photovoltaic cell electroluminescence has detected. The image set has been chosen to demonstrate the effectiveness of the detection model for identifying different types of defects in photovoltaic cells. The figure shows how the model has successfully detected and localized the different types of defects in the images, including Bc, Ck, Fr, Sc, and Tl. The model accurately identifies the location and type of each defect and produces a bounding box around it. Overall, the visualization of the final detection results in this figure indicates the effectiveness of the proposed detection model for identifying and localizing various types of defects in photovoltaic cell electroluminescence images.

Table 2. Precision, recall, and mAP scores for each class.

Classes	No. of images	No. of labels	P	R	mAP@.5
Bc	430	99	0.96	0.93	0.97
Ck	430	165	0.65	0.74	0.72
Fr	430	319	0.83	0.88	0.88
Sc	430	20	0.71	0.90	0.85
Tl	430	152	0.84	0.87	0.89

Table 3. mAP@.5 for all classes

Detector	Number of categories	mAP@.5
BAF-Detector [8]	3	0.88
RPAN-CNN (GAP&GMP) [9]	3	0.873
Our (Yolov7)	5	0.864

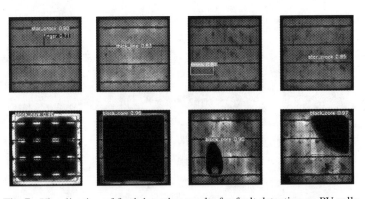

Fig. 7. Visualization of final detection results for fault detection on PV cells.

5 Conclusion

This research presented cutting-edge models for detecting defects in photovoltaic cells using electroluminescence images. The proposed model leverages advanced deep learning techniques, showcasing remarkable performance with an impressive overall mAP@.5 score of 86%. It also exhibits high precision and recall rates across all five classes. These outstanding results serve as evidence for the effectiveness of the proposed model in accurately identifying defects, thereby contributing to the enhancement of the quality and reliability of photovoltaic cells and the overall efficiency of solar energy conversion systems.

Future endeavors could focus on further refining the model's performance for specific types of defects, delving into the intricacies of their detection. Additionally, exploring real-world applications and evaluating the practical implementation of the proposed model would be valuable areas of investigation for future research.

References

1. Otamendi, U., Martinez, I., Quartulli, M., Olaizola, I.G., Viles, E., Cambarau, W.: Segmentation of cell-level anomalies in electroluminescence images of photovoltaic modules. Sol. Energy **220**, 914–926 (2021)
2. Demirci, M.Y., Beşli, N., Gümüşçü, A.: Efficient deep feature extraction and classification for identifying defective photovoltaic module cells in Electroluminescence images. Expert Syst. Appl. **175**, 114810 (2021)
3. Meribout, M., Tiwari, V.K., Herrera, J.P.P., Baobaid, A.N.M.A.: (2023). Solar panel inspection techniques and prospects. Measurement **209**, 112466
4. Akram, M.W., et al.: CNN based automatic detection of photovoltaic cell defects in electroluminescence images. Energy **189**, 116319 (2019)
5. Tang, W., Yang, Q., Hu, X., Yan, W.: Convolution neural network based polycrystalline silicon photovoltaic cell linear defect diagnosis using electroluminescence images. Expert Syst. Appl. **202**, 117087 (2022)
6. Et-taleby, A., Chaibi, Y., Allouhi, A., Boussetta, M., Benslimane, M.: A combined convolutional neural network model and support vector machine technique for fault detection and classification based on electroluminescence images of photovoltaic modules. Sustain. Energy Grids Netw. **32**, 100946 (2022)
7. Al-Dulaimi, A.A., Guneser, M.T., Hameed, A.A., Márquez, F.P.G., Fitriyani, N.L., Syafrudin, M.: Performance analysis of classification and detection for PV panel motion blur images based on deblurring and deep learning techniques. Sustainability **15**(2), 1150 (2023)
8. Su, B., Chen, H., Zhou, Z.: BAF-detector: an efficient CNN-based detector for photovoltaic cell defect detection. IEEE Trans. Industr. Electron. **69**(3), 3161–3171 (2021)
9. Su, B., Chen, H., Chen, P., Bian, G., Liu, K., Liu, W.: Deep learning-based solar-cell manufacturing defect detection with complementary attention network. IEEE Trans. Industr. Inf. **17**(6), 4084–4095 (2020)
10. PVEL-AD: A Large-Scale Open-World Dataset for Photovoltaic Cell Anomaly Detection
11. Girshick, R., Donahue, J., Darrell, T., Malik, J.: Rich feature hierarchies for accurate object detection and semantic segmentation. In: Proceedings of the IEEE Conference on Computer Vision and Pattern Recognition, pp. 580–587 (2014)
12. Girshick, R.: Fast R-CNN. In: Proceedings of the IEEE International Conference on Computer Vision, pp. 1440–1448 (2015)

13. Faster R-CNN: Towards Real-Time Object Detection with Region Proposal Networks
14. Redmon, J., Farhadi, A: YOLOv3: an incremental improvement. arXiv preprint arXiv:1804.02767 (2018)
15. TPH-YOLOv5: Improved YOLOv5 Based on Transformer Prediction Head for Object Detection on Drone-captured Scenarios
16. Wang, C.Y., Bochkovskiy, A., Liao, H.Y.M. YOLOv7: trainable bag-of-freebies sets new state-of-the-art for real-time object detectors. arXiv preprint arXiv:2207.02696 (2022)

Anomaly Detection Algorithm with Blockchain to Detect Potential Security Attacks in the IIoT Model of Industry 5.0

Piyush Pant[1], S. B. Goyal[2(✉)], Anand Singh Rajawat[1], Amol Potgantwar[3], Pradeep Bedi[4], and Chawki Djeddi[5]

[1] School of Computer Sciences and Engineering, Sandip University, Nashik, India
[2] Faculty of Information Technology, City University, 46100 Petaling Jaya, Malaysia
drsbgoyal@gmail.com
[3] Sandip Institute of Technology and Research Center, Nashik, India
[4] Galgotias University, Greater Noida, India
[5] Laboratoire de Vison et d'intelligence Artificielle, Université Larbi Tebessi, Tébessa, Algérie

Abstract. The research presents a model for detecting potential security attacks in the Industry 5.0's Internet of Things (IIoT) model using an Anomaly Detection Algorithm, with Blockchain technology to further enhance security. One-class Support Vector Machines (SVM) is used as the Anomaly Detection Algorithm, to identify any unusual behavior in the IIoT system. The proposed model ensures the integrity of data by implementing the decentralized features of Blockchain technology. This paper aims to address the current security challenges faced by Industry 5.0 and enhance the reliability of the IIoT model. Since Industry 5.0 is not here yet, hypothetical data is used to train the model which is generated after seeding using Numpy. The Blockchain technology enhanced the overall security of the Industrial Internet of Things (IIoT) model whereas, to secure it even further by detecting anomalous activities, the machine learning algorithm is proposed. Anomaly detection algorithm with Gaussian distribution is proposed through One-class SVM. The threshold for an activity to be classified as unusual or anomalous is discussed in the paper along with the difference between classification algorithm and anomaly detection algorithm. The research implemented One-class SVM algorithm to train the model by randomly seeding data using Numpy with an average accuracy of 92.8% after 5 different runs with different datasets. The algorithm also focused on other applications of the model like detection of faulty driver, device, or equipment.

Keywords: Artificial Intelligence · Machine learning · Blockchain · IIoT · Industry 5 · Anomaly Detection Algorithm · Gaussian distribution

1 Introduction

The existing technologies and models are being upgraded to new levels because of the improvement in some of the technologies like Artificial Intelligence, Blockchain, IoT, etc. Most of the domains are grateful to these technologies as their flow is dependent on them. The world is stepping into a new world where the interaction between humans and machine will increase drastically. Even a lot of places like restaurants, stadiums, clubs, etc. are using varieties of machines to ease their work and labour. In the research, the trending, powerful and futuristic technologies are studied along with implementation of the integration. The aim of the research is to provide the security and is useful for the next industry and also for any major model that deals with sensitive data and requires an advanced security model.

Internet of Things (IoT) is everywhere these days, however, it is not the best version of its true capability. As the device communication increase, the network needs to be stronger and after that comes the model power of the device, which is, is it able to communicate well and perform the task well for which it is made. Improving the IoT would surely solve these issues. IoT are classified in various category mostly based on their applications like IoNT, IIoT, etc. In this study, we will focus on the Industrial Internet of Things (IIoT). As its name says, it is an extension of the IoT which is primarily developed for the industrial application. Since we are on the verge from going to Industry 5.0 from Industry 4.0, we could only reach after some major improvements in the IoT. The device-to-device communication can be made faster by using networks like 5G or 6G [1], but its security cannot be guaranteed. Moreover, as the IoT will replace all traditional tools and will spread in every home all over the world, one weak frame or loophole will let the hacker get access to all the devices and its data, which would be a disaster. Therefore it is the need of the model to be secure. In the further sections, the paper discusses how the security can be implemented in the model along with making it intelligent to detect any kind of bugs, objects, etc. that might harm the system [3].

Blockchain technology is considered as an ideal system for security and it does stand up to its reputation. It is a decentralized system, which means that the authority or power is not central (controlled by one who could modify for selfish reasons), but distributed. All the changes are tracked in the system which makes it one of the most organized too. One of the most important feature it has is its ability to encrypt the data and store it in that form, it cannot be changed back to normal without the correct algorithm and methods. Encryption is a major part of the blockchain but this is not the only ability it has, one of the most impressive and the one from where it got its name is the "chains" of blocks. Each block stores the address of the previous hash and so on, this makes a chain of block and so comes the name Blockchain. The research proposes the blockchain for the IIoT model to improve its security at its best [10]. The research proposes blockchain with 256-bit AES encryption method. This would contribute to a better understanding of AI and Blockchain integration in the real world.

Artificial intelligence, often referred to as AI is one of the most powerful technology on the planet if not the best. It is often seen as the future of humankind. However, the development of AI is not enough for it to be called truly intelligent. Artificial intelligence is a broad term and there are its sub field which are its main core, they are machine learning and deep Learning. Machine learning is implemented by the research to detect the anomaly. Now to dive into the machine learning domain, there are mainly two types – Supervised learning and Unsupervised learning [3]. Just alongside of Unsupervised learning, lies another subtypes like Recommender systems, Dimensionality reduction, etc. and this is the domain where the Anomaly detection algorithm belongs to [11].

Anomaly detection means to detect those objects, entities or things which are not usual or which are different than the rest of their own. For example, in a pride of 20 lions, there is a black lion (a very rare specimen), this is a small and easily detected example, 100000 batteries with full charge but 11 of them are faulty as they do not have any charge. These are examples of anomalies in the data, and the detection of these anomalies is called as Anomaly detection which could be based on Visuals or its features from the rest. One important thing to note here is that the anomalies are different than classification. The paper discusses it in later sections, how they are both different and why do we actually need the anomaly detection algorithm if we already had the classification algorithm. The upcoming sections also gives the detailed implementation of the algorithm along with the theoretical concepts of the anomaly detection algorithm.

The advent of Industry 5.0 has brought about a new wave of innovation in the field of industrial internet of things (IIoT). With increasing digitization of industrial processes, the need for securing these systems has become a top priority. Anomaly detection algorithms play a crucial role in identifying potential security threats in IIoT systems. This paper presents an innovative approach to anomaly detection by combining the robustness of blockchain technology with the power of one-class support vector machines (SVM). The proposed system uses blockchain to store and verify the authenticity of the data, while one-class SVM is used to detect anomalies in the data. The proposed system aims to improve the accuracy and security of anomaly detection in the IIoT model of Industry 5.0, thus preventing potential security attacks.

2   Related Work

Various research that were based on the same technologies as this paper were studied and below is the discussion of some of the references.

P. Pant et al. [1], research is based on Artificial intelligence and blockchain integration with IIoT in the 5G environment of industry 5.0. The research proposed the supervised learning algorithm like multivariate linear regression and Artificial Neural Network for the IIoT. This research presents future work as it proposes the unsupervised learning algorithm for the IIoT as a security purpose.

W. Liu et al. [11], an architecture based on a dual-threaded blockchain was suggested in this paper to identify large-scale abnormalities in intelligent networks. Then, an adaptive encoder is used to implement anomaly detection. The research used blockchain on intelligent network and then the anomaly detection. Our research implements the blockchain for real-time data security and then the anomaly detection is added as a secondary layer of security to detect any kind of potential security attacks.

Z. Il-Agure et al. [13], paper proposed a link mining tool for the blockchain that is based on anomaly detection. This is for the IoT devices which had the blockchain network. Our research implements the optimized anomaly detection algorithm for the IIoT that is using blockchain for real-time data security.

M. Signorini et al. [14], presented a framework called as BAD (Blockchain Anomaly Detection). This framework was focused to reduce the false positive rate of the output. The goal was to detect the anomalies in the blockchain based system with a reduction in the false positive result. Our research fills the research gap and proposes a model that is practical as it is implemented for the IIoT.

The concepts of machine learning especially anomaly detection algorithm which were put forward by Andrew NG were studied by the author and they are grateful for the mathematical concept taught by him in his courses.

3   Proposed Methodology

In this section, the implementation of the Anomaly detection algorithm along with its theoretical concept is discussed. After that the blockchain integration is studied as well.

3.1   Understanding the Security Problem in IIoT

IIoT is definitely the future and the medium for human-machine interaction. The security attacks would be much easier and dangerous as there could be increase in devices, networks and models, and so in the weak spots as well. First we need to understand how the interaction between humans and machine would take place, refer the Fig. 1.

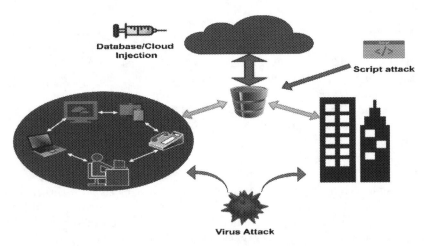

Fig. 1. Security attacks in IIoT

The attackers could inject scripts of malicious code that would alter the original program, they could inject any kind of virus to shut system or even Database injection to get access of the database. The network hijacking is also possible which would allow the hacker to get access to all the devices and communications [20]. All of these attacks could be handled and detected earlier and be dealt with. The data would be the main target of the attacker, so the blockchain would keep it secure and make sure that there are no contacts with the data blocks from unauthorized party. The machine learning, on the other hand would hinder such attacks to happen as it would be different from the rest of the program.

3.2 Why Anomaly Detection Algorithm? Why Not Classification?

To detect any malicious and faulty activity beforehand, the anomaly detection algorithm will be used in the model. To some people, the anomaly detection algorithm might look similar to classification but they are not. Even some may suggest to use classification to classify the anomalies but this would not be optimal.

First thing to understand is that anomalies are quite rare that is, refer Eq. (1)

$$number\ of\ anomalies \ll number\ of\ normal\ objects \tag{1}$$

Second thing is that, the anomalies and normal objects are same and they belong to a same group, however they differ in some property or performance due to their faulty nature, which makes them anomalous. Although the implementation and concept may look similar but the meaning is totally different, hence there is an algorithm for anomaly detection.

For classification, the main idea is to classify various ungrouped objects into a group or a category, example could be, among 1000 animals, classify them as 'cat' or a 'dog'. Now both are different species and hence would belong to a different group. This is classification, however, let's say that in a pack of 1000 dogs, there are 2 different species animals we need to find, this is an anomaly and hence the difference in the concept.

Thirdly, for classification, we already know the number of classes or category and we know all the dataset would go in one of the category. But, for anomaly, we don't know if there is any anomaly, that is why it is unsupervised learning. It is also crucial to understand that the dataset is unlabelled for anomaly detection algorithm whereas the classification requires labelled dataset.

3.3 Requirements for the Model

Before the development of model start, our system needs to be ready to take the load and process the model. The below Table 1 describes the technologies and requirements for the algorithm to be implemented.

Table 1. Requirements for the model

Model Requirements	Solutions/Fulfilments
Programming language	Python (Recommended), R
System RAM	At least 4 GB RAM is recommended, 4+ would be great as larger amount of data could be trained faster
IDE	Anaconda Navigator, Jupyter notebook, VS code, Python IDE – This would be the choice of the developer
Libraries (For Python)	Numpy, Pandas, Matplotlib, Seaborn, Sklearn, etc.
Conceptual knowledge	Probability, Statistics, Linear Algebra, Programming in python, Data Structures, Gaussian Distribution, Threshold, etc.
Data	Data should be cleaned (If not clean, then must be pre-processed), In CSV or excel

3.4 Threshold for an Anomaly

As we have seen above that the anomalies are quite rare, hence the threshold for an example to be an anomaly would be low as well. If the threshold is larger or even half, then almost half of the example would be marked as an anomaly which would create tense condition in the model and it will not work properly. As a probability measure, the anomaly should range below 0.2 or above 0.8. One important thing to understand is that it also depends on the problem statement and the dataset, the value of the threshold may change. Threshold is the most important part of the model as on its basis the anomalies would be marked. If it is too large, even the legal activities would be marked as anomaly and if it is too low, some anomalous activities might escape and cause harm to the system [22].

Let the threshold for anomaly be 'ε'. p(x) is the probability of an example 'x' to be an anomaly. Therefore, the condition for an example to be anomalous is given in Eq. (2).

$$p(x) < \varepsilon \tag{2}$$

3.5 Implementation of the Algorithm

The Gaussian distribution (also referred to as the normal distribution) is required for the implementation of anomaly detection algorithm, so some of the terminologies are –

$$\aleph - Gaussian; \sim - distributed\ as; \mu - Mean; \sigma^2 - Variance \tag{3}$$

For 'x' to be a distributed Gaussian with mean 'μ' and variance 'σ²'

$$x \sim \aleph(\mu, \sigma^2) \tag{4}$$

More the example is in denser region of Gaussian distribution, lesser chances for it to be an anomaly. The Gaussian density would be highest in the part where there are large number of example congested together. It would slowly become less dense as it spreads.

The implementation of the algorithm would require to understand the concept and use of the following equations.

$$\text{Trainingset} - \{x^1, x^2, x^3, x^4, \ldots, x^{m-1}, x^m\} \tag{5}$$

The training set is represented as $x^{\wedge}i$, where $i = 1,2,3\ldots,m$ as shown in Eq. (5). Since this is unsupervised learning so the dataset is not labelled. Each example would be represented as part of a probability equation with mean and variance passed with the input as per Eq. (6). The 'n' represents the feature as 'j' $= 1,2,3,\ldots,n$

$$p(x) = p\left(x_1, \mu, \sigma^2\right) \cdot p\left(x_2, \mu, \sigma^2\right) \cdot p\left(x_3, \mu, \sigma^2\right) \cdots p\left(x_n, \mu, \sigma^2\right) \tag{6}$$

The above Eq. (6) is the conceptual representation to understand the algorithm implementation, this would be simplified as-

$$p(x) = \prod_{j=0}^{n-1} p(x_j^i; \mu_j; \sigma_j^2) \tag{7}$$

This would be the final implementation after the fitting of parameters, which are μ_j and $\sigma\_j^{\wedge}2$, where $j = 0, 1, 2, 3, \ldots, n-1$. After the fitting, the model would be ready for testing and deployment. The Eq. (2) would be used to flag the anomaly that is produced after the Eq. (7).

3.6 Testing of the Model and Deployment

After the successful training of the model with the help of Gaussian distribution, now the testing will take place. First the cross validation set will be used and then the testing set.

The cross validation set - $\{x_{cv}^1, x_{cv}^2, x_{cv}^3, \ldots, x_{cv}^m\}$

The Testing set $- \{x_{test}^1, x_{test}^2, x_{test}^3, \ldots, x_{test}^m\}$

After the successful testing of the model, it would be ready for deployment in the industry 5.0 environment alongside of the IIoT and blockchain.

3.7 Blockchain for the IIoT

Blockchain technology is often considered the ideal technology for security which ensures the integrity of the data and transaction because of its decentralized system. The IIoT deals with a lot of data and communication, hence the data is in real-time which requires a system capable enough to handle it and provide security at the same time. The IIoT consists of network between devices and performs device-to-device or machine-to-machine communication with human interaction. The blockchain would help to store the data and perform secure communication between the devices.

The below Fig. 2 shows the structure of the blockchain and why is it named as it is. It describes the encrypted data and the chains that connects in block in order to ensure the data integrity.

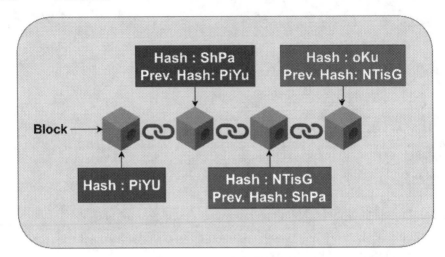

Fig. 2. Structure of Blockchain

As the data would be in real-time so the data storing would also require a strong and secure system. To understand the process of how the data or transaction would take place and added in the blockchain, refer the below Fig. 3 which shows the proposed blockchain for the IIoT and describes the process step by step, how the transaction would be added in the blockchain.

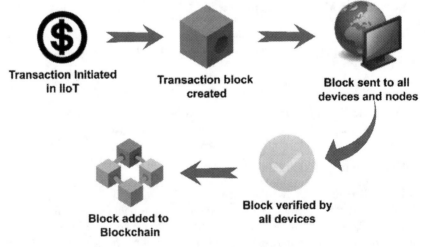

Fig. 3. Process of Block addition

4   Result and Discussion

It is important to understand that the data to train the model does not exist yet because the Industry 5.0 isn't here. Because of this, the research used "Hypothetical data" to train the anomaly detection model. The data is generated using the random seed method by the research. The Fig. 4 is the screenshot of the code that shows the method to randomly seed the data using the Numpy module.

```python
1   import numpy as np
2   import matplotlib.pyplot as plt
3   from sklearn import svm
4
5   # Generate sample data
6   np.random.seed(42)
7   X = 0.3 * np.random.randn(100, 2)
8   X_train = X[:80]
9   X_test = X[80:]
10
```

Fig. 4. Code to randomly seed the data using Numpy

The research used the OC-SVM (One Class Support Vector Machine) for anomaly classification, which is one of the best anomaly detection algorithm. After seeding the

```python
# Plot training data
plt.subplot(121)
plt.title("Training Data")
inlier_idx = y_pred_train > 0
outlier_idx = y_pred_train < 0
plt.scatter(X_train[inlier_idx, 0], X_train[inlier_idx, 1], c='green', marker='o', label='inliers')
plt.scatter(X_train[outlier_idx, 0], X_train[outlier_idx, 1], c='red', marker='x', label='outliers')
plt.legend(loc='best')

# Plot testing data
plt.subplot(122)
plt.title("Testing Data")
inlier_idx = y_pred_test > 0
outlier_idx = y_pred_test < 0
plt.scatter(X_test[inlier_idx, 0], X_test[inlier_idx, 1], c='green', marker='o', label='inliers')
plt.scatter(X_test[outlier_idx, 0], X_test[outlier_idx, 1], c='red', marker='x', label='outliers')
plt.legend(loc='best')
```

Fig. 5. Plotting of Inliers and Outliers for Training and Testing data

data and training the model, the below Fig. 5 shows the plotting of testing and training data which has labels inliers and outliers for non-anomalous and anomalous data respectively.

After training the model, the result is represented by Fig. 6. The Fig. 6 generates two scatter plots that show the training and testing data with the predicted outliers marked in red and the inliers marked in green.

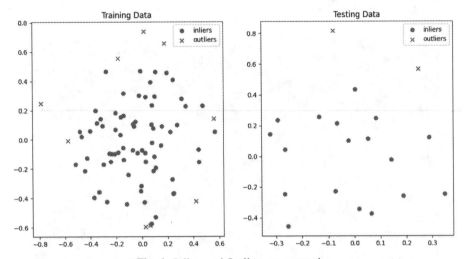

Fig. 6. Inliers and Outliers representation

The above model can now be successfully deployed in the real world after training the model using real world data. The proposed model has an average accuracy of 92.8% after 5 different runs with different datasets to ensure stability of the model. The blockchain is proposed to be added with the 256-bit AES encryption method that would be imported with the encryption modules like Bcrypt which the Industry 5.0 would have inbuilt since it is based on Blockchain.

To have better analysis of the research, the paper have some research questions as RQs. They are based on the research and gives focus on theory of the paper.

RQ1. Can we implement the classification algorithms like Logistic regression for anomaly detection?

Answer: As discussed in the section where difference between classification and anomaly detection is described, the answer would be No. It is because, the dataset would not be labelled like the supervised algorithms need. Along with that, the anomaly detection is not classification of anomaly but to detect the unusual out of the usual. Therefore logistic regression, which is a classification algorithm of supervised learning cannot be used for anomaly detection, even if it is modified to do so, it won't give just as good result as the actual algorithm.

RQ2. Why the threshold matters for the algorithm and why is it so low?

Answer: The threshold is really important as it is the condition to mark the example as an anomaly or not. Anomalies are rare and it would not be advised to mark a large number of example as anomaly as the model may become slow and produce many false positives [14].

RQ3. Pictorial representation of the model.

Answer: The below Fig. 7 shows the architecture of the final model with blockchain and machine learning in IIoT of industry 5.0. The figure shows how the marked anomalies are not allowed to enter the model and the attack is prevented by the proposed model.

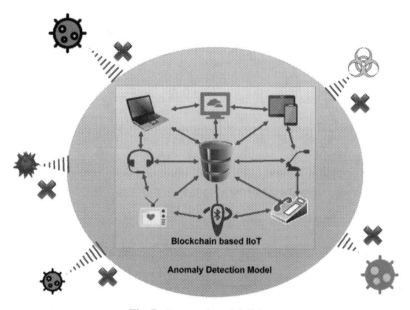

Blockchain based IIoT

Anomaly Detection Model

Fig. 7. Proposed model diagram

RQ5. How is the data divided for the overall development of the model and why?

Answer: The dataset is divided using the 60%-20%-20% rule which means that the training set would be the 60%, cross validation set would be 20% and the testing set would be 20%. This would ensure that the model is efficient and the parameters are fitted correctly.

5 Conclusion

The world is stepping in an era where the Human-Machine and Machine-To-Machine interaction will rise significantly. The drastic rise in these interaction leads to the need for a secure model so that the data could be safe from the hands of attackers. The research integrated the Blockchain technology in the IIoT model, which is an extension of the IoT, to improve its overall security and gave it anti-corruption powers. However, to add an extra layer of security, Machine learning is also proposed and integrated by the research. The anomaly detection algorithm form the domain of machine learning is used by the paper. The anomaly detection algorithm detects any kind of activity that is unusual and marks it as an anomaly. This activity could be any kind of attack from the side of hacker, injection of some virus, unusual data, unwanted scripts, sudden changes in the IIoT, etc. Such activities are usually done by the hackers to get into the model and steal from it. Our secondary layer of security, which is the machine learning algorithm would protect from such attacks whereas the primary layer, the blockchain would prevent any kind of unwanted interaction with the data or the model. This would give the ultimate security to the model for Industry 5.0, hence the research proposes these algorithm and methodologies for the practical approach towards the development of IIoT model for the Industry 5.0. By integrating One-class SVM as an Anomaly Detection Algorithm with blockchain technology in the IIoT model of Industry 5.0, it is possible to improve the overall security of the system and prevent potential security attacks. The use of an Anomaly Detection Algorithm helps to detect any suspicious activity and prevent it from causing damage to the system. By combining this with the decentralized and secure nature of blockchain, the security of the IIoT model can be further strengthened. This integration can play a crucial role in ensuring the safe and secure functioning of Industry 5.0's IIoT model and safeguarding it from potential security attacks. The research encourages other researchers to study and implement some future work of this research like what to do after an anomaly is found, how to deal with it and safeguard the model. Even the Deep Learning can be integrated in the model in place of machine learning to make the system even more advance.

References

1. Pant, P., et al.: Blockchain for AI-enabled industrial IoT with 5G network. In: 2022 14th International Conference on Electronics, Computers and Artificial Intelligence (ECAI), pp. 1–4 (2022). https://doi.org/10.1109/ECAI54874.2022.9847428
2. Lee, C., Kim, J., Kang, S.-J.: Semi-supervised anomaly detection with reinforcement learning. In: 2022 37th International Technical Conference on Circuits/Systems, Computers and Communications (ITC-CSCC), pp. 933–936 (2022). https://doi.org/10.1109/ITC-CSCC55581.2022.9895028
3. Almalawi, A., Tari, Z., Fahad, A., Yi, X.: A Global anomaly threshold to unsupervised detection. In: SCADA Security: Machine Learning Concepts for Intrusion Detection and Prevention, pp. 119–149. Wiley (2021). https://doi.org/10.1002/9781119606383.ch6
4. Babaei, M., Imani, M.: Anomaly detection improvement using sparse representation and morphological profile. In: 2020 6th Iranian Conference on Signal Processing and Intelligent Systems (ICSPIS), pp. 1–5 (2020). https://doi.org/10.1109/ICSPIS51611.2020.9349597
5. Ziemann, A., Simonoko, H., Flynn, E.: Temporal anomaly detection in multispectral imagery. In: IGARSS 2020 - 2020 IEEE International Geoscience and Remote Sensing Symposium, pp. 3975–3978 (2020). https://doi.org/10.1109/IGARSS39084.2020.9324627
6. Potgantwar, A., Aggarwal, S., Pant, P., Rajawat, A.S., Chauhan, C., Waghmare, V.N.: Secure aspect of digital twin for industry 4.0 application improvement using machine learning (2022). https://doi.org/10.2139/ssrn.4187977.SSRN: https://ssrn.com/abstract=4187977
7. Song, J., Nang, J., Jang, J.: Design of anomaly detection and visualization tool for IoT blockchain. In: 2018 International Conference on Computational Science and Computational Intelligence (CSCI), pp. 1464–1465 (2018). https://doi.org/10.1109/CSCI46756.2018.00292
8. Voronov, T., Raz, D., Rottenstreich, O.: Scalable blockchain anomaly detection with sketches. In: 2021 IEEE International Conference on Blockchain (Blockchain), pp. 1–10 (2021). https://doi.org/10.1109/Blockchain53845.2021.00013

9. Rajawat, A.S., Goyal, S.B., Pant, P., Bedi, P.: AI-enabled internet of nano things methodology for healthcare information management. In: Kautish, S., Dhiman, G. (ed.) AI-Enabled Multiple-Criteria Decision-Making Approaches for Healthcare Management, pp. 222–239. IGI Global (2022). https://doi.org/10.4018/978-1-6684-4405-4.ch012

10. Liu, X., Jiang, F., Zhang, R.: A new social user anomaly behavior detection system based on blockchain and smart contract. In: 2020 IEEE International Conference on Networking, Sensing and Control (ICNSC), pp. 1–5 (2020). https://doi.org/10.1109/ICNSC48988.2020.9238118

11. Liu, W., Shen, Y., Yang, H., Bao, B., Yao, Q., Wang, L.: Anomaly detection based on dual-threaded blockchain in large-scale intelligent networks. In: 2022 International Wireless Communications and Mobile Computing (IWCMC), pp. 28–31 (2022). https://doi.org/10.1109/IWCMC55113.2022.9824976

12. Kim, J., et al.: Anomaly detection based on traffic monitoring for secure blockchain networking. In: 2021 IEEE International Conference on Blockchain and Cryptocurrency (ICBC), pp. 1–9 (2021). https://doi.org/10.1109/ICBC51069.2021.9461119

13. Il-Agure, Z., Attallah, B., Chang, Y.-K.: The semantics of anomalies in IoT integrated BlockChain network. In: 2019 Sixth HCT Information Technology Trends (ITT), pp. 144–146 (2019). https://doi.org/10.1109/ITT48889.2019.9075114

14. Signorini, M., Pontecorvi, M., Kanoun, W., Di Pietro, R.: ADvISE: anomaly detection tool for blockchaIn SystEms. In: 2018 IEEE World Congress on Services (SERVICES), pp. 65–66 (2018). https://doi.org/10.1109/SERVICES.2018.00046

15. Morishima, S.: Scalable anomaly detection method for blockchain transactions using GPU. In: 2019 20th International Conference on Parallel and Distributed Computing, Applications and Technologies (PDCAT), pp. 160–165 (2019). https://doi.org/10.1109/PDCAT46702.2019.00039

16. Yu, D., Xie, Y., Long, H., Jin, M., Li, X.: Container anomaly detection system based on rule mining and matching. In: 2022 International Conference on Blockchain Technology and Information Security (ICBCTIS), pp. 102–105 (2022). https://doi.org/10.1109/ICBCTIS55569.2022.00034

17. Iyer, S., Thakur, S., Dixit, M., Katkam, R., Agrawal, A., Kazi, F.: Blockchain and anomaly detection based monitoring system for enforcing wastewater reuse. In: 2019 10th International Conference on Computing, Communication and Networking Technologies (ICCCNT), pp. 1–7 (2019). https://doi.org/10.1109/ICCCNT45670.2019.8944586

18. Kim, J., et al.: A machine learning approach to anomaly detection based on traffic monitoring for secure blockchain networking. IEEE Trans. Netw. Serv. Manage. **19**(3), 3619–3632 (2022). https://doi.org/10.1109/TNSM.2022.3173598

19. Signorini, M., Pontecorvi, M., Kanoun, W., Di Pietro, R.: BAD: a blockchain anomaly detection solution. IEEE Access **8**, 173481–173490 (2020). https://doi.org/10.1109/ACCESS.2020.3025622

20. Pant, P., et al.: Authentication and authorization in modern web apps for data security using Nodejs and role of dark web. Procedia Comput. Sci. **215**, 781–790 (2022). https://doi.org/10.1016/j.procs.2022.12.080

21. Ning, W., Xie, X., Huang, Y., Hu, F.: Data sharing scheme for 5G IoT based on blockchain. In: 2021 International Wireless Communications and Mobile Computing (IWCMC), pp. 327–329 (2021). https://doi.org/10.1109/IWCMC51323.2021.9498712

22. Pant, P., Taghipour, A.: Machine learning and blockchain for 5G-enabled IIoT. In: Taghipour, A. (ed.), Blockchain Applications in Cryptocurrency for Technological Evolution, pp. 196–212. IGI Global (2023). https://doi.org/10.4018/978-1-6684-6247-8.ch012

23. Pant, P., et al.: Using machine learning for Industry 5.0 efficiency prediction based on security and proposing models to enhance efficiency. In: 2022 11th International Conference on System Modeling & Advancement in Research Trends (SMART), Moradabad, India, pp. 909–914 (2022). https://doi.org/10.1109/SMART55829.2022.10047387

24. Pant, P., et al.: AI based technologies for international space station and space data. In: 2022 11th International Conference on System Modeling & Advancement in Research Trends (SMART), Moradabad, India, pp. 19–25 (2022). https://doi.org/10.1109/SMART55829.2022.10046956

COVID-19 Seasonal Effect on Infection Cases and Forecasting Using Deep Learning

Md. Mijanur Rahman(iD), Zohan Noor Hasan(iD), Mukta Roy(iD),
and Mahanaj Zaman Marufa$^{(\boxtimes)}$ (iD)

Southeast University, 251/A & 252, Tejgaon Industrial Area, Dhaka 1208, Bangladesh
{mijanur.rahman,2018000000165,2016000000216,
2019000000086}@seu.edu.bd

Abstract. The COVID-19 pandemic infected billions of people worldwide. The government has taken a number of steps to control infection cases, but due to frequent changes in variants, it is difficult to control the infection rate, and taking precautions over a long period of time is infelicitous. Any correlation between the virus's ascendancy and changes in climate or temperature is difficult to detect. In this research, different countries, and seasons have been investigated to assess the relationship between temperature and the rate of infection. In predicting the infection rate of new cases, popular deep learning (DL) methods, long short-term memory (LSTM), and gated recurrent units (GRUs) have been applied here. Infection cases were visualized, including the time period from previous data. Individual countries have particular weather conditions that vary between countries; for this reason, the temperature has taken in a specified range in seasonal-based forecasting. To specify a particular temperature or season is challenging but combining season and temperature from past data generated a pattern. It shows that the highest number of infection cases reached at a certain time of the season and found a seasonal effect on COVID-19.

Keywords: COVID-19 · Seasonal effect · Deep learning · Temperature · LSTM · GRU · Seasonal temperature · Infection Rates · Temperate weather

1 Introduction

The COVID-19 pandemic began in 2019 and spread throughout the entire world; many organizations or individuals claimed that this pandemic originated in China, albeit China is reluctant to concede such an accusation. Since 2019, the population has been targeted by the coronavirus and its various variations, posing the biggest threat to the economy, healthcare, and governance. At the beginning of the epidemic, exposure to infected patients was difficult and drawn out. Deep learning (DL) is one of the possible solutions that will soon be combined and developed with clinical tests to allow the accurate detection of infection cases and take the initial steps automatically [1]. The spread rate of COVID-19 is contingent on time, situation, weather, population and lifestyle, and quite varied region. Considerable strategies have been solicited in locating and assessing the superseding measure of infectious conditions. Whenever an epidemic extends from a region or country from various perspectives over time, especially climate cycle variations or viral transmission through the period, these records are determined by non-linear attributes [2]. Analyzing diverse COVID-19 data sets, it has been established that infection cases evolved rapidly over a certain period of time. Many mathematical model area units are used to predict and evaluate the progression of proven affected cases individually [3]. Area units acquired several modeling, estimation, and statement approaches to deal with this pandemic.

The DL approach Recurrent Neural Network (RNN) has been used to predict the possible infection rates. Compared to supervised learning, it is a challenge in training input areas to find a perfect pattern which is learned by ML mechanism for solving a complex data set and finding relationships in pre-sized output sets [4]. Artificial Neural Network (ANN) layers have single or multiple combinations of layers, and DL neural networks are composed of ANN, so the structure of these neural networks is parallelly established inside RNN algorithms. Using DL, predicting the next pandemic of coronavirus infection rate is problematic choosing the correct algorithm. Deep literacy styles can relate the structure and pattern of similar data to the non-linearity and avoid the complexity of an algorithm LSTM had used in time-series forecasting [5]. Infection rates and climatic changes varied within selected countries. Thus, data sets shrink, and it becomes obvious that GRU will perform well in prediction since RNN's most eccentric redaction is GRU. It is a refined process of data transformation held to the next iteration. It would be more successful in forecasting COVID-19 transmission if the input data had temporal components and was not based on typical regression methods [6].

Many researchers analyzed COVID-19 time-series data; prudent analysis of time-series data aids in making to make new decisions that are very important for public awareness. Gated recurrent units (GRU) and long short-term memory (LSTM) have similarities, with some differences between the computational sections. Both have the highest performance capability, but the fact arises when the data become smaller or larger. Many researchers have shown a process of work where DL gives a satisfactory result on time series forecasting. Some works in the literature part on time series analysis of COVIDCovid-19 data comports similar to their work through the process.

Chowdhury et al. [7] focused on finding a suitable machine learning algorithm that can predict the COVID-19 daily new cases with higher accuracy, they used (ANFIS) and LSTM to see the newly infected cases in Bangladesh in this study LSTM had shown a favorable result on a scenario-based model with MAPE of 4.51, RMSE-6.55 and correlation coefficient -0.75 accuracy was good enough. Liao et al. [8] have reported a COVID-19 prediction model based on a time-dependent + SIRVE. GRU forecasting accuracy was noticeable, and they showed that the single day prediction accuracy rate improves 51% compared to the best existing single deep learning predictions. Shahid et al. [2] proposed forecast models comparison LSTM, GRU, and Bi-LSTM are assessed for time series prediction of confirming cases, death, and recoveries in ten affected countries due to COVID-19, comparing among them Bi-LSTM predicted well, but the accuracy of GRU also showed well result. Arun Kumar et al. [9] in their work proposed state-of-art DL Recurrent Neural Networks (RNN) models, with GRU and LSTM cells to predict the country-wise cumulative confirmed cases, cumulative recovered cases, and cumulative fatalities and showed that individual model show variations in result for each of 10 countries. Some of the country's LSTMs gave satisfactory results, and some of the country GRU gave well accuracy. Engelbrecht and Scholes [10] tested for seasonal climate permittivity in observed COVIDCovid-19 infection data to show that if the complaint does have a substantial seasonal dependence, and herd immunity isn't established during the first peak season of an outbreak, there's likely to be a seasonality-sensitive alternate surge of infections about one time after the original outbreak.

The remaining part of our paper in holds mathematical equations, data visualization graphs, calculated data, and graphical figures to give a clearer understanding of the process of dual application of deep neural network technique which projected a satisfactory outgrowth. Seasonal changes have a significant impact on new cases where a range of temperatures represents weather and season.

2   Related Work

Chowdhury et al. [7] focused on finding a suitable machine learning algorithm that can predict the COVID-19 daily new cases with higher accuracy, they used (ANFIS) and LSTM to see the newly infected cases in Bangladesh in this study LSTM showed a favorable result on a scenario-based model with MAPE of 4.51, RMSE-6.55 and correlation coefficient −0.75 accuracy was good enough. Liao et al. [8] have reported a COVID-19 prediction model based on a time-dependent + SIRVE. This model combines DL technology with the mathematical implementation of infectious diseases and forecasts the parameters in the mathematical model of infectious diseases by fusing DL time series prediction methods in the result section, GRU forecasting accuracy was noticeable, accuracy rate improves 51% compared to the best existing single deep learning predictions. Shahid et al. [2] proposed forecast models comparison LSTM, GRU, and Bi-LSTM are assessed for time series prediction of confirming cases, death, and recoveries in ten affected countries due to COVID-19, comparing among them Bi-LSTM predicted well, but the accuracy of GRU also showed well result, model ranking from good performance to lowest in their scenario was Bi-LSTM, LSTM, GRU, SVR and ARIMA where Bi-LSTM generates lowest MAE and RMSE values of 0.0070 and 0.0077 respectively. Arun Kumar et al. [9] in their work proposed state-of-art DL Recurrent Neural Networks (RNN) models, with Gated Recurrent Units (GRUs) and Long Short-Term Memory (LSTM) cells to predict the country-wise cumulative confirmed cases, cumulative recovered cases, and showed variations in result for each of 10 countries. The GRU and LSTM cells, along with Recurrent Neural Networks (RNN), were developed to predict the future trends of COVID-19, Some of the countryies LSTMs gave satisfactory results, also for some of the countryies GRU gave good accuracy. Engelbrecht and Scholes [10] had test for seasonal climate permittivity in observed COVID-19 infection data to show that if the complaint does have a substantial seasonal dependence and herd immunity isn't established during the first peak season of an outbreak, there's likely to be a seasonality-sensitive alternate surge of infections about one time after the original outbreak.

3   Methodology

The workflow applied in this study is displayed in Fig. 1. Table 1 outlines the computed precision of DL. The mathematical equations utilized are displayed in Eq. 1–7. The graphical representation or assertion of data visualization & predicted results have merged in Figs. 2, 3, 4, 5, 6, 7, 8, 9, 10 and Table 1 successively. To give a clearer understanding of this research, the whole working process has been provided.

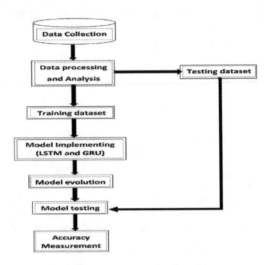

Fig. 1. Methodology diagram and working process.

3.1   COVID-19 Data Set

Data sets have been assembled from unprecedented resources, the final data sets that have been used in this work have been reformed from "OurWorldInData (OWID)" [11], provided publicly accessible daily datasets, and "NASA Prediction of Worldwide Energy Resources" [12] accorded daily cases datasets. Data spanning April 2020 to March 2022, 600 days of data, was ordered in a time series format by date, month & year. Essentially parameters are new cases, new deaths, and new tests from "OWID" [11]. Different temperature parameters were collected in the measurement of latitude and longitude data provided by "NASA". We have utilized the five most populated countries proclaimed on the "WorldOmeters" [13] website. For a clearer observation of data, common parameter called date are present in all separate datasets.

3.1.1   Data Prepossessing

This research concentrated on seasonal COVID-19 affected cases where a range of temperature indicates a season and the data set has a temperature column date. This time series forecasting centered on seasonal effect, a column of affected new cases assembled from the 'OWID' data set. The temperature T2M (temperature in 2 m) was collected from "NASA". In this research, highly populated countries are Bangladesh, the Philippines, Mexico, Vietnam, and Indonesia, as recommended by the "worldometer" population page we cleaned the noisy data and filled up the NaN value linearly, then used the mean value to fill the rest of the process using pandas data frame. In the forecasting, the data set was divided into 5 different sections, because this research is forecasting the seasonal effect of COVID-19 on countries. This research has focused on 3 different seasons summer, winter, and spring.

The main data set has a date, new cases, temperature, season, and location columns. Countries are selected in the Asia region mostly, and countries have a similar range of temperatures. Most populated countries data have accumulated from the "WorldOmeters" website. Temperature is collected by giving latitude and longitude from Google. For Bangladesh 23.6850° N, 90.3563° E, Mexico 23.6345° N, 102.5528° W, Philippines 12.8797° N, 121.7740° E, Vietnam 14.0583° N, 108.2772° E, Indonesia 0.7893° S, 113.9213° E. The data set is separated by season March, April, May, and June; these months are considered summer. Spring is selected as -July, August, September, and October, and the selected month for winter is -November, December, January, and February; these Dates were converted with string to timestamp format. Data normalization is one of the most important steps before training LSTM and GRU models. In this research, MinMaxScaler has been used for normalization. MinMaxScaler turns the training dataset inputs into {0,1} range of data as shown in Eq. (1). Actual values will be turned into minimum 0 and maximum range of 1 for each variable. Normalization avoids scaling problems during training and testing models (Table 1).

$$\text{Scale} = (\text{Input} - \text{minimum of input})/(\text{maximum of input} - \text{minimum of input}) \qquad (1)$$

Exploiting the training set as input and scale are output after scaling, every training set will go through this equation of data scaling for normalization.

3.2 Deep Learning Models Details

DL is one of the significant methods of forecasting. It is a difficult task with traditional programs, hence DL has been shown to significantly improve techniques to predict both structured and unstructured data [14]. The real-time data technique is rather hard to process, beginning with locating statistical data files, transforming them into training and test results, and finally applying RNN to represent the data via visual analysis [15]. In this research, DL RNN models applied time series forecasting.

RNN planned target vectors from the entire history of past information. In this manner, models contrasted with old branches of occurrence data, and RNN are less complex in demonstrating elements of consistent succession data. As a rule, RNN layout associations between units in coordinated circles and recollects past contributions through its internal state. The deeply hidden output feature is beneficial to extract elements of versions into the hidden state, constructing it more simply to expect output summaries of the records of preceding inputs more efficiently [16]. With the help of vanishing gradient descent, the unnecessary data are removed, and the effective data are stored in the memory cell for the next iteration. Stochastic gradients tend to evaporate or expand.

It's hard to keep track of long-term dependencies with such simple RNN to overcome the vanishing or exploding gradient challenges; RNN with LSTM and GRU have been developed [17].

3.2.1 Long-Short-Term-Memory (LSTM)

LSTM works excellently in vanishing and exploding gradients. In the RNN model, problems occur when a large number of data rollovers in this situation memory unit taking spacing with some unnecessary data. To avoid this LSTM was introduced with a memory unit called cell state shown in Eq. (2).

$$\text{Cell state} = (\text{input gate} * \text{new candidate}) + (\text{forget gate} * \text{cell state} - 1) + b \quad (2)$$

The four generalized formulas as input-output and forget gate uses the sigmoid activation function and the tanh activation function is used for new candidates as shown in (3).

$$\sigma/\tanh (W \cdot X + U \cdot h - 1 + b) \quad (3)$$

In LSTM weights are always updating in each layer, to generate new weight automatically from calculated new correction value, in the model new state is introduced as shown in Eq. (4)

$$\text{New state} = \text{output gate} * \text{new candidate} \quad (4)$$

W is weight, b is biased, and (cell state - 1) is the previous output return as input, h-1 is the previously hidden state return as new.

3.2.2 Gated-Recurrent-Unit (GRU)

GRU has two major gates that act as a switch. Either could be 0 or 1. The reset gate considers 0 and the update gate is kept at 1. The reset gate determines how important the information must be discarded [18]. GRU and LSTM had a similarity, only two gating layers reset gates and update gates instead of three gating layers [19]. GRU input gate merges into the reset gate, and the output gate merges into the update gate as shown in Eq. (5) in each hidden state.

$$\sigma (W \cdot X [h - 1, \ X] + b) \quad (5)$$

GRU introduces new memory contents with an adjustable combination; for the fewer gates, the complexity of GRU is much easier.

3.3 Training and Testing

Training data was defined as records between February 24–2020, to December 12–2021, these daily records were considered training data, and this data set was trained for 60 days of a chunk. Testing sets were defined from December 23–2021 to February 24–2022. In the X coordinate, predicted 60 days were added, and in the Y coordinate, new infection cases were added as trained as training and testing datasets. Temperature and new case data these, two features are considered as a perimeter and were used as a feature of the training and testing set. This procedure was applied to all selected countries' datasets.

3.3.1   Prediction Accuracy Measurement

Mean Square Logarithm Error (MSLE) and Root Mean Square Logarithm Error (RMSLE) were used for measuring the loss function of prepared models. These Regression models are used for measuring the forecasting performance and showing the difference between the real value and forecast value, shown in Eq. (6). The specialty of MSLE is matrices that avoid the natural log of possible 0 values for the actual value and forecasting value. MSLE error measurements were used in the validation and testing stage [20]. RMSLE is nothing but root over the MSLE as shown in Eq. (7).

$$\text{MSLE} = \frac{1}{T} \sum\nolimits_{i=1}^{n} ((log(Fi + 1) - log(Ri + 1))2 \qquad (6)$$

$$\text{RMSLE} = \sqrt{\frac{1}{T} \sum\nolimits_{i=1}^{n} ((log(Fi + 1) - log(Ri + 1))2} \qquad (7)$$

where T is the total number of observations, Fi is forecasting a target, Ri is a real target for i, and log(x) is the natural logarithm.

4   Result

In understanding infection rate, data had two parameters; country and seasonal effect. Data visualization of COVID-19 has been performed utilizing several segments, including a bar chart (as shown in Figs. 2 and 9), pie chart (as evident in Fig. 9), and line plotting techniques (can be seen in Figs. 4, 5, 6, 7 and 8). COVID-19 transmission rate As a function of seasonal changes is depicted in Fig. 2. This graph delineates that spring and winter have a paramount number of infections. All countries are visualized separately in 3 axes termed new cases, date, and temperature to give a distinct understanding. First and foremost, Bangladesh's data reveals that in spring 2020-06 to 2020-09 and 2021-06 to 2021-09, the infection rate came to a head at the temperate weather condition of 28 °C demonstrated in Fig. 4. In summer 2020-02 to 2020-05 and 2021-02 to 2021-05 the infection rate was lower at the temperature of 30 °C or above.

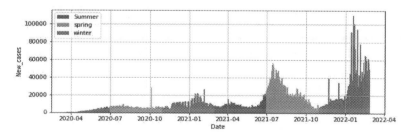

Fig. 2.   Infected country over 3 years

2020-07 to 2020-10 and again in 2021-08 to 2021-09 spring when the temperate weather condition was 28 °C to 29°. In the winter season, the infection rate also rose, but in the summer 2020-02 to 2020-05 and 2021-02 to 2021-05, the infection rate was lower at the temperature of 30 °C or above (shown in Fig. 7). The rest of the countries have a critical situation in the spring season and less infection rate in spring season temperature (as shown in Fig. [5, 6, 8]).

For the Philippines, the infection rate surged in the spring season from the middle of the bar chart (see Fig. 8) indicates that the daily infection rates were high in the winter and spring seasons. It is well established that the beginning of the spring season and middle of the winter season this time period is notable for the vast spread out of the COVID-19 infection rate.

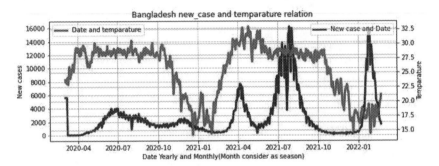

Fig. 3. Bangladesh New case Observation with temperature

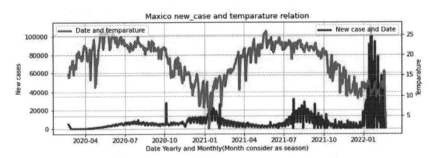

Fig. 4. México New case Observation with temperature.

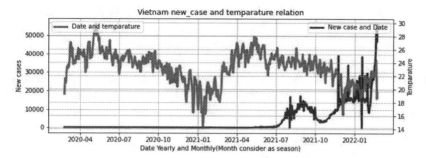

Fig. 5. Vietnam New case Observation with temperature.

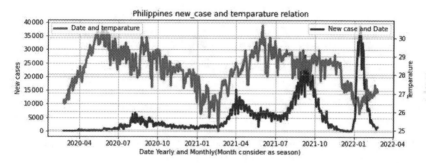

Fig. 6. Philippine New case Observation with temperature.

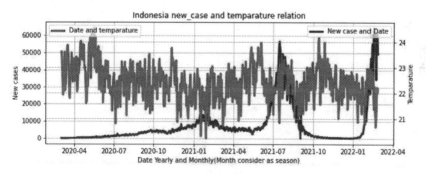

Fig. 7. Indonesia New case Observation with temperature

During the winter season, newly infected cases are 100000 and above in Mexico. In the summer season, the number of infected cases in Mexico is below 20000 Shown in Fig. 8. Here, other countries follow the same pattern of ratio on infection rates (Fig. 8). The positivity rate slowed down during the summer season temperature and evidently increased in winter and spring. Warmer humid climates appear to have less SARS-CoV-2 viral spread, based on the observational process of the research or the inherent potential of distortion, the validity of the data provided was a poor rate of infection [14].

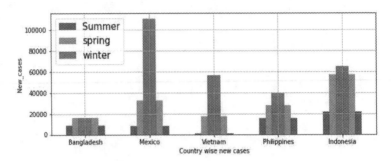

Fig. 8. Seasonal New case Observation temperature.

The parentage of infection transmission pie charts (Fig. 9) had been created by means of new cases that occurred during the time period. The most infected country is Indonesia, where 28.1% of people were infected by a coronavirus, then Vietnam at 15.0% of other countries infection percentage is given (Fig. 9). Here the calculation of the percentage of daily new cases was considered as the mean value of total new cases. Among selected countries again, we can see that a higher number of people were infected in the spring season.

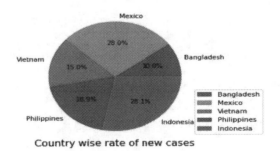

Fig. 9. Country-wise COVID-19 infection percentage.

5 Discussion

After visualizing past data (Figs. 3, 4, 5, 6 and 7) we have uncovered a relationship between season and the spread of COVID-19. A variety of seasonal temperature ranges was selected, and two different RNN techniques LSTM and GRU, with Relu activation function with Adam optimizer (Table 1) have been applied to predict the best result for different countries. In the area of machine learning, Adam was discovered to be strong and well-suited to the optimization problem [21]. Table 1 shows 100 epochs where batch sizes 32 and 64 as these batch sizes are suitable for GRU and LSTM.

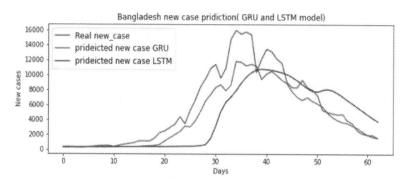

Bangladesh COVID-19 forecasting

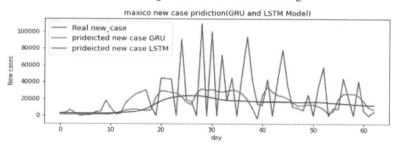

Mexico COVID-19 forecasting.

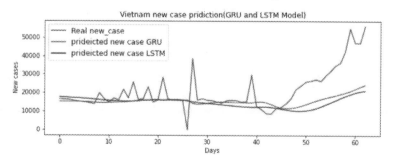

Vietnam COVID-19 forecasting

Fig. 10. Graphical representations of COVID-19 Forecasting.

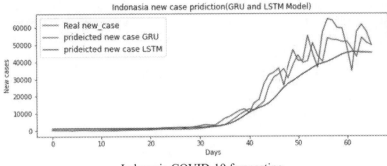

Indonesia COVID-19 forecasting.

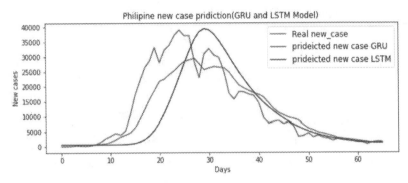

Philippine COVID-19 forecasting

Fig. 10. (*continued*)

According to MSLE and RMSLE evaluation (Table 1), the accuracy for Bangladesh was LSTM (3.903 and 1.975) and GRU (3.470 and 1.862) shown in Fig. 10. Indonesia has MSLE-8.700 and RMSLE-2.949 for LSTM, for GRU MSLE- 11.836 RMSLE-3.440. From these observations, it is clear that GRU performs better than the LSTM model for both MSLE and RMLSE accuracy tests as shown in Fig. 10.

Table 1. COVID-19 LSTM and GRU result

Country	RNN Model	Epochs	Number of Layers	Dropout	Optimizer	MSLE	RMSLE
Bangladesh	LSTM	100	4	0.2	Adam	3.903	1.975
	GRU	100	4	0.2	Adam	3.470	1.862
Mexico	LSTM	100	4	0.2	Adam	18.717	4.326
	GRU	100	4	0.2	Adam	19.429	4.407
Indonesia	LSTM	100	4	0.2	Adam	8.700	2.949
	GRU	100	4	0.2	Adam	11.836	3.440
Philippine	LSTM	100	4	0.1	Adam	4.984	2.232
	GRU	100	4	0.1	Adam	4.590	2.142
Vietnam	LSTM	80	4	0.1	Adam	1.711	1.308
	GRU	80	4	0.1	Adam	1.717	1.310

6   Conclusions and Future Work

People continue to be infected by COVID-19 which continues to be dangerous through prevalence. The purpose of this research is clear visualization of new cases that occur in season and the performance measurement of DL models. Our research demonstrates that COVId-19 has a seasonal effect. Analyzing data demonstrated that the same temperature has different effects on cases in different locations temperatures and newly confirmed cases are very onerous. Where it can be said there is a seasonal effect on new cases, and a particular season has a range in temperature. As a result, we came up with a decision that, the range of temperature during the summer season spread or effectiveness of coronavirus is much slower and becomes inactive. The highest infections happened in temperate weather conditions of spring and the beginning of the winter season. The DL model RNN shows good results on sequential data from different perspectives of model or data deployment, but accuracy varies. LSTM performs well in large data sets, but breaking GRU is used here. As a result, we found that GRU has better accuracy and fast computational abilities. The outcome appears that in temperate weather of the winter and spring seasons, the effect on COVID-19 is considerable and the range of temperatures of these seasons is noticeable, while the temperatures of summer pose less dangeras.

The model's accuracy in this paper could be more significant if there was more data. These models train with two features: a new case and temperature. As per our findings, it is possible to control and maintain summer season temperature or other natural effectiveness artificially in living rooms, offices, organizations institutions, etc. Infection rates could be potentially reduced in other seasons, minimizing infection rates as future work to predict upcoming waves.

References

1. Kumari, M.: An overview on deep learning application in coronavirus (COVID-19): diagnosis, prediction and effects. In: Proceedings - 2nd International Conference on Smart Electronics and Communication, ICOSEC 2021, pp. 1507–1510 (2021). https://doi.org/10.1109/ICOSEC 51865.2021.9591921
2. Shahid, F., Zameer, A., Muneeb, M.: Predictions for COVID-19 with deep learning models of LSTM, GRU and Bi-LSTM. Chaos Solitons Fractals **140**, 110212 (2020). https://doi.org/ 10.1016/j.chaos.2020.110212
3. Zeroual, A., Harrou, F., Dairi, A., Sun, Y.: Deep learning methods for forecasting COVID-19 time-series data: a comparative study. Chaos Solitons Fractals **140**, 110121 (2020). https:// doi.org/10.1016/j.chaos.2020.110121
4. Jamshidi, M., et al.: Artificial intelligence and COVID-19: deep learning approaches for diagnosis and treatment. IEEE Access **8**, 109581–109595 (2020). https://doi.org/10.1109/ ACCESS.2020.3001973
5. Siami-Namini, S., Tavakoli, N., Namin, A.S.: A comparison of ARIMA and LSTM in forecasting time series. In: Proceedings - 17th IEEE International Conference on Machine Learning and Applications, ICMLA 2018, pp. 1394–1401 (2019). https://doi.org/10.1109/ICMLA. 2018.00227
6. Chimmula, V.K.R., Zhang, L.: Time series forecasting of COVID-19 transmission in Canada using LSTM networks. Chaos Solitons Fractals **135**, 109864 (2020). https://doi.org/10.1016/ j.chaos.2020.109864
7. Chowdhury, A.A., Hasan, K.T., Hoque, K.K.S.: Analysis and prediction of COVID-19 pandemic in Bangladesh by using ANFIS and LSTM network. Cognit. Comput. **13**(3), 761–770 (2021). https://doi.org/10.1007/s12559-021-09859-0
8. Liao, Z., Lan, P., Fan, X., Kelly, B., Innes, A., Liao, Z.: SIRVD-DL: a COVID-19 deep learning prediction model based on time-dependent SIRVD. Comput. Biol. Med. **138**, 104868 (2021). https://doi.org/10.1016/j.compbiomed.2021.104868
9. ArunKumar, K.E., Kalaga, D.V., Kumar, C.M.S., Kawaji, M., Brenza, T.M.: Forecasting of COVID-19 using deep layer Recurrent Neural Networks (RNNs) with Gated Recurrent Units (GRUs) and Long Short-Term Memory (LSTM) cells. Chaos Solitons Fractals **146**, 110861 (2021). https://doi.org/10.1016/j.chaos.2021.110861
10. Engelbrecht, F.A., Scholes, R.J.: Test for Covid-19 seasonality and the risk of second waves. One Health **12**, 100202 (2021). https://doi.org/10.1016/j.onehlt.2020.100202
11. covid-19-data/owid-covid-data.csv at master owid/covid-19-data · GitHub. https://github. com/owid/covid-19-data/blob/master/public/data/owid-covid-data.csv. Accessed 09 May 2022
12. POWER I Data Access Viewer. https://power.larc.nasa.gov/data-access-viewer/. Accessed 09 May 2022
13. World Population Clock: 7.9 Billion People (2022) - Worldometer. https://www.worldomet ers.info/world-population/#top20. Accessed 09 May 2022
14. Dash, S., Chakravarty, S., Mohanty, S.N., Pattanaik, C.R., Jain, S.: A deep learning method to forecast COVID-19 outbreak. N. Gener. Comput. **39**(3–4), 515–539 (2021). https://doi.org/ 10.1007/s00354-021-00129-z
15. Mohammad Masum, A.K., Khushbu, S.A., Keya, M., Abujar, S., Hossain, S.A.: COVID-19 in Bangladesh: a deeper outlook into the forecast with prediction of upcoming per day cases using time series. Procedia Comput. Sci. **178**(2019), 291–300 (2020). https://doi.org/10.1016/ j.procs.2020.11.031
16. Pascanu, R., Gulcehre, C., Cho, K., Bengio, Y.: How to construct deep recurrent neural networks. In: 2nd International Conference on Learning Representations, ICLR 2014 - Conference Track Proceedings, pp. 1–13 (2014)

17. Dey, R., Salemt, F.M.: Gate-variants of gated recurrent unit (GRU) neural networks. In: Midwest Symposium on Circuits and Systems, vol. 2017–August, no. 2, pp. 1597–1600 (2017). https://doi.org/10.1109/MWSCAS.2017.8053243

18. Ayoobi, N., et al.: Time series forecasting of new cases and new deaths rate for COVID-19 using deep learning methods. Results Phys. **27**, 104495 (2021). https://doi.org/10.1016/j.rinp.2021.104495

19. Tjandra, A., Sakti, S., Manurung, R., Adriani, M., Nakamura, S.: Gated recurrent neural tensor network. In: Proceedings of International Joint Conference on Neural Networks, vol. 2016–October, pp. 448–455 (2016). https://doi.org/10.1109/IJCNN.2016.7727233

20. Mizoguchi, T., Kiyohara, S.: Machine learning approaches for ELNES/XANES. Microscopy **69**(2), 92–109 (2020). https://doi.org/10.1093/jmicro/dfz109

21. Kingma, D.P., Ba, J.L.: Adam: a method for stochastic optimization. In: 3rd International Conference on Learning Representations, ICLR 2015 - Conference Track Proceedings, pp. 1–15 (2015)

PV Output Power Prediction Using Regression Methods

Abdulhameed Aboumadi[1] and Hilal Arslan[2(✉)]

[1] Department of Electrical and Electronic Engineering, Ankara Yıldırım Beyazıt Univesity, Ankara, Turkey
[2] Department of Software Engineering, Ankara Yıldırım Beyazıt University, Ankara, Turkey
hilalarslan@aybu.edu.tr

Abstract. Over the past few years, the general public has become increasingly aware of climate change and the role of greenhouse gas emissions, especially carbon dioxide, in contributing to it. Therefore, individuals, businesses, and governments around the world have taken steps to reduce their emissions. One of these steps is to increase adoption of renewable energy sources, such as solar power which provides clean energy, in addition to low building and operation costs and minimal maintenance requirements. Accurate estimation of solar energy production is crucial to ensure the stability of electrical networks as the transition to renewable energy sources such as solar power increases. In this study, machine learning regression algorithms including artificial neural networks, support vector regression, regression trees, and k-nearest neighbor are performed to estimate hourly solar energy production of one month using historical production data and various meteorological parameters. The models are optimized using grid search and validated using K-fold cross validation method. The performance of the models is evaluated using the RMSE, MAE, and R2 evaluation metrics. The results showed that the k-nearest neighbor regression model achieves the highest performance with an R^2 score of 0.9715.

Keywords: Power prediction · Machine learning regression · Photovoltaic

1 Introduction

In recent years, renewable energy sources such as solar power have become more popular globally due to advances in photovoltaic (PV) technology. However, there are concerns about the reliability of these energy sources as they are affected by variables such as weather, seasonality, and production patterns. In order to ensure a stable solar power sector, it is important to accurately forecast solar energy production. This is important because as the world transitions to renewable energy sources, such as solar power, the accurate estimation of solar energy production becomes critical for ensuring the stability of electrical networks. Inaccurate predictions of solar energy production can lead to overloading of the electrical grid or, conversely, underutilization of the solar energy that is available, which can result in higher energy costs and grid instability.

© The Author(s), under exclusive license to Springer Nature Switzerland AG 2024
A. Ortis et al. (Eds.): ICAETA 2023, CCIS 1983, pp. 204–213, 2024.
https://doi.org/10.1007/978-3-031-50920-9_16

The use of machine learning regression algorithms provides a reliable and accurate way to estimate solar energy production to increase the adoption of renewable energy sources, which is critical to reducing greenhouse gas emissions and mitigating the effects of climate change. In addition, accurate estimation of solar energy production can help grid operators, distributed energy resource (DER) aggregators, and PV power plant owners make informed decisions about energy storage and distribution, improving the efficiency of the electrical grid.

Li et al. [1] performed Artificial Neural Network (ANN) and Support Vector Regression (SVR) models to predict PV power production for 15 min, 1 h, and 24 h in advance using historical production data from online meteorological services. They used one year of data and converted the historical production data to 15 min and 1-h average values. In their approach, a hierarchical methodology was followed, in which forecasts were made for each inverter separately based on the historical data for that inverter, and a forecast for the entire plant was also made. They found that forecasting production for each inverter individually resulted in more accurate results. Theocharides et al. [2] applied ANN, SVR, and Random Tree (RT) models to predict day-ahead hourly power production for PV systems from historical PV production data, incident global irradiance (GI), and ambient temperature (Tamb) data. Their experimental results showed that the ANN model performed better than the other model. In another study published by Theocharides et al. [3], ANN, K-means clustering, and linear regressive correction models are applied to predict day-ahead hourly power production for PV systems using historical power production data, wind direction (Wa), ambient temperature (Tamb), incident global irradiance (GI), wind speed (Ws), relative humidity (RH), solar azimuth (φs) and elevation (α) angles data. Their method achieved a MAPE of 4.7%. Leone et al. [4] applied to an SVR model to predict day-ahead production at 15 min interval using solar irradiance, ambient temperature, and historical production data. Their model achieved an R2 value exceeding 90%. In a study published by Khandakar et al. [5], historical production data, ambient temperature, dust accumulation, wind speed, solar irradiance, relative humidity, and panel temperature data were applied to an ANN model, linear regression, M5P tree model, and gaussian process regression to predict hourly PV power output. The interval of collected data was not specified. They stated that the ANN model outperformed the other methods and achieved an RMSE of 2.1436. Qu et al. [10] proposed a prediction model called ALSM that uses a combination of CNNs, LSTMs, and an attention mechanism to forecast solar power output over multiple relevant and target variables. This model takes into account a variety of inputs, including historic PV output power, latitude, longitude, array rating, and other geographic data, to capture both short-term and long-term temporal patterns and provide hourly forecasts for the coming day. The model is designed to operate under the multiple relevant and target variables prediction pattern (MRTPP). Visser et al. [11] used historical weather and PV output power data to assess the efficacy of 12 alternative approaches that forecast day-ahead power production based on market circumstances. SVR, deep learning, physical-based techniques, and ensemble learning were among the models used. They also evaluated the effect of aggregating numerous PV systems with different inter-system distances on the forecasting models' efficacy. The models were assessed on their technical and economic performance. Eniola et al. [12] developed a model validation method using more recent input datasets, including temperature, mod temperature, historical production data, wind speed, and solar irradiance, based on an existing prediction model built on a genetic algorithm (GA)-optimized hidden Markov model (HMM). Normalized root mean square error was considered as an evaluation method for the models (nRMSE). Mahmud et al. [13] used various machine learning algorithms to perform short-term and long-term PV output power prediction. They found that random forest regression model outperformed other machine learning algorithms on their dataset that was collected from Alice Springs, which is one of the areas of high PV power generation in Australia. In [14], Mellit et al. predicted short-term PV output power using several kinds of deep learning neural networks. The data used in [14] was gathered from a microgrid in a university in Italy. They found that the case of 1-min with one-step ahead achieved highest accuracy scores, but up to 8 steps ahead gives acceptable results.

In this study, accurate prediction of PV power production predicted for 1 month with 1-h resolution using RT, KNN, SVR, and ANN regression methods. A grid search algorithm was applied to find optimal hyperparameters for the machine learning methods used in this work. In the literature, day-ahead PV production forecast is common (e.g. [3, 10], and [11]). Although day-ahead prediction is important, accurate monthly prediction gives a wider view and better insight for monthly dispatch planning and can be considered a step forward towards managing PV plants as conventional dispatchable power plants. This will be possible because accurate and reliable production values will be used for solving the unit commitment and economic dispatch problems. The significance of this prediction is that it allows long-term (1 month) optimal dispatch of generation and storage assets which helps in maintaining grid resiliency for systems with high PV penetration. Without accurate predictions, it is very hard to optimally dispatch generation units and maintain grid stability in the case of high PV penetration.

This paper is organized as follows: Sect. 2 describes the machine learning methods used in this work, Sect. 3 introduces the dataset and presents the results and discussion, and Sect. 4 concludes the paper.

2 Machine Learning Regression Methods

In this study, support vector regression, k-nearest neighbor, decision tree, and artificial neural networks methods were applied to predict PV output power.

2.1 Support Vector Regression

Support vector machines (SVMs) are statistical learning techniques that are frequently applied to solve regression and classification problems. In SVR, a dataset is first transformed into a high-dimensional space, and a curve is fitted to the data using a "cylinder" that is defined by support vectors, which are the points that determine borders of the cylinder. SVR model estimates the relationships between the inputs and outputs utilizing Eq. 1:

$$f(x) = \omega\varphi(x) + b \qquad (1)$$

where $\varphi(x)$ is the transfer function that maps the input data to high-dimensional feature spaces. The regularized risk function is minimized to estimate the parameters ω and b:

$$\min \frac{1}{2}\omega^T * \omega + C \sum_{i=1}^{n} (\xi_i + \xi_i^*)$$

$$s.t. \tag{2}$$
$$y_i - \omega^T \varphi(x_i) - b \leq \varepsilon + \xi_i$$
$$\omega^T \varphi(x_i) - b - y_j \leq \varepsilon + \xi_i^*$$

In Eq. 2, n is the number of samples used for training, ξ represents the error slacks that ensure the results are within particular tolerances, C denotes a regularization penalty, and ε is the tube's target tolerance range. $\omega^T * \omega$, the first term in the equation, is a regularization term that aids in flattening the curve. The second term is a determined empirical error with ε-insensitive loss function. The loss function being described here measures the difference between expected values and the radius of a cylinder. If the anticipated values are within the cylinder, the loss is 0. If the anticipated values are outside of the cylinder, the loss is equal to the absolute difference between the expected values and the radius of the cylinder ε. This loss function may be used to evaluate the accuracy of predictions made by a machine learning model. The model's goal would be to minimize the loss by making more accurate predictions. The Lagrange multiplier is used to optimize both ε and C, and the corresponding Lagrangian structure of Eq. (2) can be stated as the following equation, where $K(x_i, x_j)$ represents a kernel function:

$$f(x) = \sum_{i=1}^{n} (a_i - a_i^*) K(x_i, x_j) + b \tag{3}$$

Equation 3 defines the Lagrange multipliers a_i and a_i^*, which are obtained by solving the dual version of Eq. (2) in Lagrange structure. The use of a kernel function has the advantage of allowing us to work with feature spaces of any size without having to manually construct the map $\varphi(x)$. Any function that meets Mercer's criteria, such as a polynomial or radial basis function (RBF) kernel [1, 7], can be employed as a kernel function. SVR model with RBF kernel was used in this study.

2.2 Regression Trees

The regression tree approach is a method for constructing a predictive model by dividing a dataset into smaller divisions and fitting a simple prediction model to each partition. The Analysis of Variance (ANOVA) method is used to assess differences or variations between the partitions. To build a numeric prediction regression tree (RT), the dataset is first partitioned at the root node using a decision tree induction algorithm based on the feature that maximizes the gain in homogeneity in the outcome after the split. The tree-growing method assesses homogeneity, which is often measured using statistics such as absolute deviation, variance, and standard deviation from the mean. The standard deviation reduction (SDR) is a common criterion for determining the split, and it is defined as follows:

$$SDR = sd(T) - \sum_i \frac{|T_i|}{|T|} * sd(T_i) \tag{4}$$

The $sd(T)$ function in this equation representing standard deviation of the value of samples in a dataset T, sets of values obtained from a feature split are represented as T_i. The number of observations in the dataset T is represented by the $|T|$ symbol. The splitting criterion is used to measure the decrease in standard deviation from the original value to the weighted standard deviation after the data has been split [2].

2.3 k-Nearest Neighbor

The k-nearest neighbor (KNN) is an approach that uses a predictor variable X to estimate the conditional distribution of a response variable Y and allocates Y to the class with the highest estimated probability. To categorize a new test observation x_0, the KNN method finds the K points in the training data that are closest to x_0 (using the Euclidean distance) and are represented by N_0. The conditional empirical distribution for class j is then calculated as the ratio of the K nearest points categorized as j:

$$\Pr(Y = j | X = x_0) = \frac{1}{K} \sum_{i \in N_0} I(y_i = j) \tag{5}$$

Finally, the class j with the highest estimated probability receives x_0. . It is vital to notice that the value of k influences the KNN classifier significantly. When K is set to one, the decision boundary will overfit the training data, producing a classifier with low bias but large variance. The decision boundary gets more linear as k grows (i.e., low variance but high bias). The bias-variance trade-off is affected by k, which should be considered [9].

KNN can be used to solve regression problems as well. The KNN algorithm in this situation selects the K-nearest neighbors based on some distance metric and assigns the average value of those neighbors as the forecast. The forecast is expressed as follows:

$$\hat{Y} = \frac{1}{K} \sum_{i \in N_0} y_i \tag{6}$$

2.4 Artificial Neural Networks

Artificial neural networks (ANNs) are computer systems that are designed to mimic the way the human brain works. They are composed of interconnected "neurons" that can process and transmit information. ANNs are commonly used in machine learning and artificial intelligence applications, and they can be trained to perform a variety of tasks by being exposed to large amounts of data and adjusting the strengths of the connections between neurons. ANN can be used to represent complex functions and solve real-world issues. To solve challenging nonlinear problems, ANNs employ a network of artificial neurons or nodes. A typical artificial neuron can be represented using a function that processes n input values (also known as "dendrites") to produce a single output value (also known as the "axon"). This function typically combines linear and non-linear operations to weight, sum, and transform the input values in some way. The specific form of the function will depend on the design of the neural network, but it may involve matrix multiplications, activation functions, convolutional filters, and/or pooling operations. A typical artificial neuron with n dendrites can be represented as follows:

$$y(x) = f\left(\sum_{i=1}^{n} w_i x_i\right) \qquad (7)$$

The weights w_i in this equation allow each of the n input variables x to contribute to the total input signals. The activation function f(x) takes the net sum of these input signals and generates the output signal y(x), which is represented by the output axon.

3 Results

In this section, we evaluate results of the machine learning regression methods to predict PV output power. First, we explain our dataset, second, we give performance metrics used in this study and finally we discuss the results.

3.1 Dataset

The dataset was collected from a solar power plant in Konya province in Türkiye and includes hourly measurements of solar production and various meteorological parameters such as incident global irradiance (G_I), ambient temperature (T_{amb}), wind speed (W_s), and mod temperature (T_m) for the period of January 1, 2021, to December 31, 2021. The purpose of the study is to use these features to train machine learning models to accurately forecast solar energy production on an hourly basis, with the goal of improving the reliability of solar power as a renewable energy source. The dataset used in this work is original and has not been used in any previous research.

To evaluate and optimize the performance of the machine learning models, the dataset was split into 11 months of training data and 1 month of test data, and k-fold cross-validation was applied during the hyperparameter optimization process. The value of k was chosen as 12, meaning that the data was split into 12 folds and the model was trained and tested 12 times, each time using a different fold as the test set. The performance of the models was then evaluated using the RMSE, MAE, and R^2 evaluation metrics, and the results showed that the k-nearest neighbor regression model achieved the highest performance with an R^2 score of 0.9715.

3.2 Performance Metrics

In this study, the prediction models' performance was assessed using the following metrics:

- Mean absolute error (MAE) (given in Eq. 8): This is a measurement of the average difference between actual and forecasted data.

$$MAE = \frac{1}{n} \times \sum_{i=1}^{n} |x_i - y_i| \qquad (8)$$

- Root mean square error (RMSE) (given in Eq. 9): The standard deviation of the prediction errors is described by the root mean square error.

$$RMSE = \sqrt{\frac{1}{n} \times \sum_{i=1}^{n} (x_i - y_i)} \tag{9}$$

- Coefficient of determination (R2) (given in Eq. 10): A measure of the fraction of data variability described by the model, ranging from 0 to 1. A number of 0 indicates that the model does not describe the data at all, whereas a value of 1 show that the model explains the data correctly.

$$R^2 = 1 - \frac{RSS}{TSS} \tag{10}$$

3.3 Experimental Setup

The computer used for this work has an Intel Core i5-7200U CPU @ 2.50 GHz 2.70 GHz processor, 8 GB RAM, and 64-bit operating system. Version 5.2.2 of Spyder python development environment is used with Python version 3.9.15. All machine learning models used are from scikit-learn library version 1.0.2.

3.4 Experimental Results

In this work, the results of RT, KNN, SVR, and ANN to predict PV output power were evaluated. We performed grid search method with k-fold cross validation to determine the best hyperparameters of the machine learning methods. In the grid search approach, for the KNN regression model, K values from 1 to 29 were chosen. For the ANN model, 3 options were considered for hidden layers number and sizes. These options are (50,50,50), (50,100,50), and (100), which means 3 layers each containing 50 neurons, 3 layers with 50, 100, and 50 neurons, and a single layer with 100 neurons. Three values were also considered for learning rate, and alpha hyperparameters which are: 0.1, 0.01, and 0.001. Finally, for the SVR model, RBF kernel was used. C and epsilon hyperparameters were chosen as 1, 10, 100, 1000, and 0.01, 0.1, 1, 10 respectively. We note that hyperparameter optimization was applied to all models except RT model, which shows promising results even without optimization.

The dataset was split into 11 and 1 months, 11 months were used for training while 1 month was used to test the model. Results are shown in Table 1. In RT method, minimum samples split is set to 2, minimum samples leaf is set to 1. The RT method achieves an RMSE of 157.13, a MAE of 58.78, and R^2 of 0.96. Prediction results of the RT are shown in Fig. 1. In SVR method, RBF kernel is used. C parameter is set to 100, and epsilon parameter is set 10. The SVR method achieves an RMSE 164.64, MAE of 63.99, and R^2 of 0.95. The prediction results of SVR are shown in Fig. 2. In the KNN method, k is chosen as 29 and Euclidean metric is used. The KNN achieves an RMSE of 137.77, a MAE of 53.95, and an R' of 0.9715. The KNN prediction results are shown in Fig. 3. Finally, in the ANN method, the number of hidden layers is set to 50 and 50 neurons are used in each hidden layer. The learning rate is set to 0.001, and alpha parameter is set to 0.1. The ANN method achieves RMSE of 148.82, MAE of 68.34, and R^2 of 0.96. The prediction results of the ANN are shown in Fig. 4. It is also worth mentioning that hyperparameter optimization was not performed for the RT model because of the high computation complexity which causes very long convergence time. Values of performance metrics shown in Table 1 indicates the high precision of the prediction curves observed in Figs. 1, 2, 3 and 4 quantitatively. It can be clearly seen that reliable predictions are generated, which can be used by grid operators, DER aggregators, and PV power plant owners to optimally dispatch assets, since PV output power is predicted with acceptable tolerance.

Table 1. Results of machine learning methods

Heading level	RMSE	MAE	R^2
RT	157.13	58.78	0.9629
SVR	164.64	63.99	0.9593
KNN	137.77	53.95	0.9715
ANN	148.82	68.34	0.9668

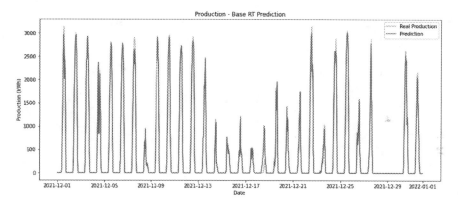

Fig. 1. Prediction results of the RT method

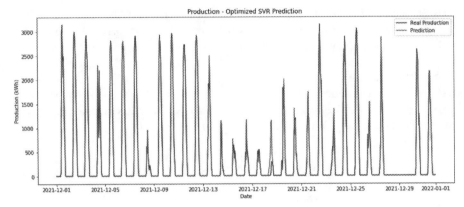

Fig. 2. Prediction results of the SVR method

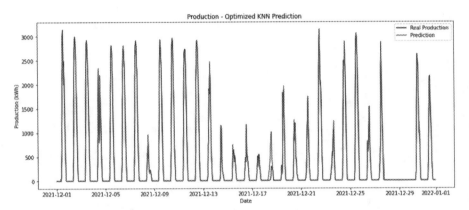

Fig. 3. Prediction results of the KNN method

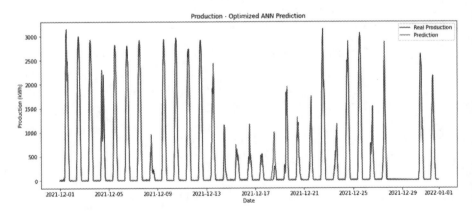

Fig. 4. Prediction results of the ANN method

4 Conclusion

The problem addressed in this study was the need for accurate forecasting of solar power production in order to ensure the stability of electrical networks as the adoption of renewable energy sources such as solar power increases. The solution proposed was the use of machine learning algorithms, including ANN, SVR, RT, and KNN regression, to forecast solar energy production using historical production data and various meteorological parameters. The models were optimized using grid search and validated using the K-fold cross validation method. The results of the study present that the KNN regression model was the most accurate results with an R^2 score of 0.97. These results suggest that machine learning techniques can be effectively used to predict solar energy production and contribute to the stability of electrical networks as the use of renewable energy sources increases. In future studies, deep learning algorithms can also be applied in addition to parallel computing in order to improve the computation speed for the suggested models.

References

1. Li, Z., et al.: A hierarchical approach using machine learning methods in solar photovoltaic energy production forecasting. Energies **9**(1), 55 (2016)
2. Theocharides, S., et al.: Machine learning algorithms for photovoltaic system power output prediction. In: 2018 IEEE International Energy Conference (ENERGYCON). IEEE (2018)
3. Theocharides, S., et al.: Day-ahead photovoltaic power production forecasting methodology based on machine learning and statistical post-processing. Appl. Energy **268**, 115023 (2020)
4. De Leone, R., Pietrini, M., Giovannelli, A.: Photovoltaic energy production forecast using support vector regression. Neural Comput. Appl.. **26**(8), 1955–1962 (2015)
5. Khandakar, A., et al.: Machine learning based photovoltaics (PV) power prediction using different environmental parameters of Qatar. Energies **12**(14), 2782 (2019)
6. Deng, F., et al.: Prediction of solar radiation resources in China using the LS-SVM algorithms. In: 2010 the 2nd International Conference on Computer and Automation Engineering (ICCAE), vol. 5. IEEE (2010)
7. Vapnik, V.: The Nature of Statistical Learning Theory. Springer, Heidelberg (1999)
8. Almeida, M.P., Perpinan, O., Narvarte, L.: PV power forecast using a nonparametric PV model. Sol. Energy **115**, 354–368 (2015)
9. Isaksson, E., Karpe Conde, M.: Solar power forecasting with machine learning techniques (2018)
10. Qu, J., Qian, Z., Pei, Y.: Day-ahead hourly photovoltaic power forecasting using attention-based CNN-LSTM neural network embedded with multiple relevant and target variables prediction pattern. Energy **232**, 120996 (2021)
11. Visser, L., AlSkaif, T., van Sark, W.: Operational day-ahead solar power forecasting for aggregated PV systems with a varying spatial distribution. Renewable Energy **183**, 267–282 (2022)
12. Eniola, V., et al.: Validation of genetic algorithm optimized hidden Markov model for short-term photovoltaic power prediction. Int. J. Renewable Energy Resour. **11**(2), 796–807 (2021)
13. Karim, A., et al.: Machine Learning Based PV Power Generation Forecasting in Alice Springs
14. Mellit, A., Pavan, A.M., Lughi, V.: Deep learning neural networks for short-term photovoltaic power forecasting. Renewable Energy **172**, 276–288 (2021)

Secure Future Healthcare Applications Through Federated Learning Approaches

Maliha Tabassum[1], Murat Kuzlu[1(✉)], Ferhat Ozgur Catak[2], Salih Sarp[3],
and Kevser Şahinbaş[4]

[1] Old Dominion University, Norfolk, VA, USA
{mtaba006,mkuzlu}@odu.edu
[2] University of Stavanger, Rogaland, Norway
f.ozgur.catak@uis.no
[3] Virginia Commonwealth University, Richmond, VA, USA
sarps@vcu.edu
[4] Istanbul Medipol University, Istanbul, Turkey
ksahinbas@medipol.edu.tr

Abstract. The healthcare field is so sensitive to data privacy and security due to including medical and personal information. Almost all healthcare applications are required to increase data security and privacy, which use traditional machine learning approaches relying on centralized systems, both computing resources and the entirety of the data. Federated learning, a sort of machine learning technique, has been used to exactly address this issue. The training data is disseminated across numerous devices in federated learning, and the learning process is collaborative. There are numerous privacy attacks on Deep Learning (DL) models that attackers can use to obtain sensitive information. As a result, the DL model should be safeguarded from adversarial attacks, particularly in healthcare applications that use sensitive medical data. This paper provides a comprehensive review of federated learning on future healthcare applications. It also discusses the types of federated learning along with its implementation in healthcare applications.

Keywords: Federated Learning · Healthcare · Privacy · Machine Learning · Deep Learning · Artificial Intelligence

1 Introduction

Federated Learning (FL) is a machine learning (ML) technique using multiple decentralized servers to exchange data. It is a privacy-protected technology to overcome data sensibility [1]. This is because the main focus of data is on privacy and security, not on how much data is present. The importance of secure data storage has increased in recent years. The current development in the advancement of Artificial Intelligence (AI), ML, and smart production has experienced a giant leap in the engineering sector [2]. Data security is a matter of concern

A. Ortis et al. (Eds.): ICAETA 2023, CCIS 1983, pp. 214–225, 2024.
https://doi.org/10.1007/978-3-031-50920-9_17

nowadays because people are now paying attention to data security [3,4]. However, it has some drawbacks in terms of the security of data handling. Firstly, in order to run a successful AI or ML project, it needs to process a large number of datasets. Accruing such a number of datasets poses a problem if it is related to humans. In the field of AI, data is the key to successfully training a model that would be able to provide close to accurate results. According to the regulation of the European Parliament, a user holds complete power over their own data [5]. Also, according to China's Cyber Security Law of the People's Republic of China [6], and General principles of the Civil Law of the People's Republic of China [7], it is pointed out that network operators should not execute the destruction, tampering, or disclosure of the collected data. Various rules and regulations need to be specified in the contract for operators to use these data in their models. If a company wants to use this data, it would need permission before using it.

When it comes to healthcare, a person's data becomes highly sensitive as it contains private information. Additionally, with a limited number of datasets, the development of ML models is also limited since more data means better models. For instance, training an AI to detect chronic wounds would require a vast dataset [8]. Obtaining this kind of data is challenging due to its sensitivity, and high regulation [9,10]. For instance, AlphaGo's early version used 160,000 sets of human chess data that could defeat entry-level players. In contrast, AlphaZero used a large mixture of human and machine-generated chess data that could beat professional players [11]. FL has a promising training approach for a neural network that includes image classification [12] and the natural language process [13]. With newer methods, a number of new mobile applications have been developed that are based on FL. Hard et al. [14] improved the word prediction FL through Google keyboard. The FL framework was first introduced to train models without sharing raw data [15,16].

This paper presents the applications of FL in the healthcare industry and how it can improve the overall success of ML models in the medical sector. For instance, a FL model studying a large number of datasets containing different wound images obtained from diabetic patients can make it easier to detect wounds that might result in organ amputation. Early medication will also become more accessible for patients living in remote areas. Implementing FL in healthcare can help develop privacy-enhanced ML applications and ensure that everyone gets the much-needed health support they require [17]. It also provides in detail on federated learning and its impact on the healthcare system. Healthcare datasets are challenging to collect and store because they contain the most sensitive and personal information about patients. In such cases, it is possible to reveal the identities of many individuals using such datasets. But there are also ways to prevent this from happening. This paper also summarizes these ways that can help protect privacy.

The rest of the paper is organized as follows. Section 2 discusses FL and its different types. Section 3 provides the various applications using FL in the healthcare system. Section 4 investigates the complications and obstacles faced using FL. Future opportunities are also discussed in this section. Section 5 concludes the paper.

2   Federated Learning

FL is an emerging approach to distributing resources safely while training ML models. It is an elaborate approach where a number of devices are connected at the same time. The raw data is not kept in a central server, but it is decentralized. Keeping data in a server or cluster means that the data can be breached, which is a severe issue regarding privacy. FL can be differentiated from centralized data usage in three aspects:

1. FL does not have a raw data communication approach.
2. FL utilizes multiple sources from different devices instead of relying on a single centralized server.
3. FL uses the data encryption method not to breach any privacy issues.

FL operates differently from the traditional centralized architecture. As shown in Fig. 1, FL collects data from various sources called "Data Sources", such as mobile phones or computers, and sends them to the local AI models on user devices. The data can be encrypted for privacy before being stored in the "Federated Data Flow". This architecture ensures that user data are kept private, and FL offers advantages, such as smart models, low power consumption, and privacy preservation. The local models compute and send aggregated model updates instead of raw data, which are combined to improve the global model without revealing individual data points.

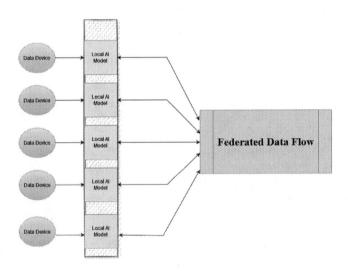

Fig. 1. Federated learning data flow

The critical benefit of FL is that the ML model can be instantly updated by leveraging the personalized experience. If we consider every storing data point as a node, these nodes not only work as data transmission points but can also be used as independent training nodes. FL has the ability to analyze, learn and generate a massive amount of data. According to Lo et al. [18], an FL model has eight phases from creation to completion. However, Kairouz et al. [16] proposed that the life cycle has six phases. Both papers focus on the steps where data is distributed using mobile devices. By combining both models, the steps can be summarized into four stages [19].

- Composition Phase, where the FL model is created with a specific classification, requirements, etc.
- FL Training Phase, where the model is trained with a strategy that has parallelism and aggregation algorithms that updates with parameters. This step improves the accuracy and capacity of the FL model.
- Evaluation Phase, where the trained model is applied in order to observe the performance. If it does not meet the requirements, the model gets modified.
- Deployment Phase, where the FL model is deployed to process real-life data.

The functional architecture of the FL System is layered in four parts, shown in Fig. 2 [20]. A user interacts with the first layer, and from that, the model operates to the client's demand. The four layers are (1) Presentation Layer, (2) User Services, (3) Training, and (4) Infrastructure. Each layer is explained as follows:

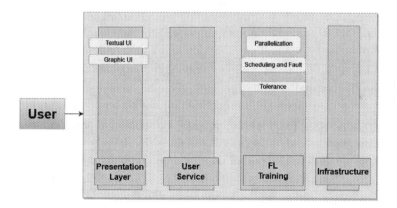

Fig. 2. Layers of functional architecture

1. Presentation Layer: This layer acts as the user interface (UI) for the FL system. It allows users to interact with the FL system and provides them with a graphical or textual user interface. The presentation layer supports the next layer, which is the user service layer.
2. User Service: The user service layer provides monitoring, steering, and logging functionalities to the FL system. It allows users to see real-time data and track the training process. This layer is used to provide clients with information about how the FL system is working and track the execution time.

3. FL Training Layer: This layer is responsible for the distribution process of data and computer resources. It has three modules: parallelization, scheduling, and fault tolerance. The FL training layer generates the execution directives and optimizes the training process of FL models.
4. Infrastructure Layer: This layer provides the interaction between the FL system and the distributed resources, including different types of resources. It has three modules: a data security module, a data transfer module, and a distributed execution module. It also ensures that the data is secure and that the FL system can interact with the distributed resources in a reliable manner.

The distributed training process for FL can be divided into three types in terms of parallelism. The three types of parallelism are data, model, and pipeline parallelism [21]. Additionally, they can be classified into three categories: horizontal, vertical, and hybrid [22,23]. Each type is related to another type, for example, horizontal is related to data parallelism, and vertical exploits model parallelism. Lastly, transfer learning [24] relies on hybrid FL. Parallelization modules can be summarized as follows:

– Data Parallelism: Data parallelism is when data processing is performed in parallel at different computer resources that have the same model but different data paths. It is utilized when the data points are distributed among various computing resources. This means they are distributed horizontally among multiple computing resources. We can compare this with cross-device FL, where a number of edge devices participate in a single model to achieve accuracy.
– Model Parallelism: Different computing resources are assigned to process the data points of specific features with independent data processing nodes. The data processing nodes can be independent or dependent. It is independent when the execution of any node does not depend on the output of the other; on the other hand, there is data dependency when the execution process is dependent.
– Pipeline Parallelism: Dependent data processing nodes are distributed at different computing resources [25]. When the data processing sources are distributed in multiple computing resources, the data processing is parallel. It is not a widely used method in Federated Learning.

Aggregation is being used in the FL models to provide statistical analysis. This is implemented to schedule modules. Aggregation algorithms are summarised as follows:

– Vanilla Aggregation Algorithms: It is used to aggregate the models or gradients generated from each computing resource with forward and backward propagation. This can be centralized, hierarchical, or decentralized.
– Centralized Aggregation: It generally relies on a central server. A single parameter server calculates the average models or gradient sent from multiple computing resources or cellphones. The weights or the gradient are calculated and transferred to the next parameter server in every computing resource. The

complexity of this aggregation is low, but in terms of data owners' trust, it is trusted while it does not have imbalance and high latency.

– Hierarchical Aggregation: Multiple parameter servers are used in hierarchical architecture. It uses a global parameter server and multiple region parameter servers. Each region server is implemented in a cell base station that can connect with computer resources with low latency. In terms of complexity, it is a medium that supports data owners' trust and can address unbalanced data distributed among multiple computing resources [26,27]. Also, it well-clusters the computer resources in groups to address data privacy [28].

FL has the opportunity to provide a secured network that can give client security by keeping their identity anonymous to a more extensive source. This is one of the reasons why this opens a great opportunity in the healthcare field.

3   Healthcare Applications and Datasets

Recent studies on AI/ML/DL have opened up many opportunities in radiology, pathology, and other medical areas [29]. However, this requires a large number of curated datasets to achieve clinical-grade accuracy. For example, training an AI-based detector requires the full spectrum of possible anatomies, pathologies, and input data types. Collecting such a massive amount of data is challenging due to the sensitivity of the data and its regulation. Some models are trained on aggregated training samples obtained from the samples drawn from local clients [30,31]. This model has minimal loss concerning the uniform distribution. In this case, the recent solution is to force the data to adapt to the uniform distribution. Mohri et al. [32] proposed a scheme where the centralized model is optimized by distributing clients. It also optimizes any possible targeted distribution. However, this Agnostic Federated Learning (AFL) has only been applied to small scales. Another method involves sharing data globally, but only a small portion is shared, and the required subset contains a uniform distribution over classes from the central server. Zhang et al. and Wibawa et al. [33,34] proposed a Federated Learning framework to detect COVID-19 infections through X-ray analysis. Moreover, several experiments were conducted to identify COVID-19 through chest X-rays, using various models such as MobileNet, ResNet18, and COVID-Net. Among these, ResNet18 provided the best result [35].

In Fig. 3, the application of Federated Learning in healthcare is illustrated. Different clients provide their data and information via various devices, such as mobile phones and other electronic devices. These data are stored in their respective hospital's server or local server to protect the identity of the patients. Only hospital's own patients' datasets are sent to a local server, and those data that do not contain any personal information can be used to study and predict diseases. Based on the framework, the proposed method performed better than the average FL.

Ahmed et al. [36] proposed an IoT-based framework for detecting COVID-19 infection through X-ray images collected from different sensors. The proposed model achieved 89% accuracy compared to other models. During the coronavirus

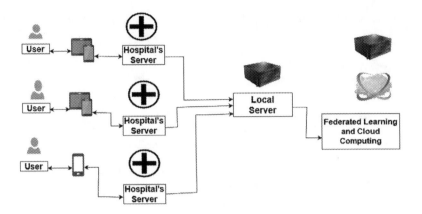

Fig. 3. Application of Federated Learning on Healthcare

epidemic, numerous research projects were developed. Researchers utilized AI and medical image analysis technology, as well as current CT scanning technology, to aid in the process of identifying patients [35]. ML was used to clarify and recognize images generated during CT scan diagnosis, which reduced the workload of healthcare workers. Nowadays, wearable sensors are increasingly popular and play a crucial role in gathering Electronic Health Records (EHRs). However, this comes with the risk of exposing a large amount of data, along with patient information, to disease information and other personal medical records [37,38].

While using the FL model, data owners can be exposed, unless the PFL-IU framework is implemented [39]. PFL-IU is one kind of privacy-preserving FL framework that is efficient and also good at preserving privacy. In this framework first, the irrelevant update is removed and a server-based aggregation protocol is used to communicate with the EHR owner [40]. The PFL-IU system comprises three components: Secret Providers (SPs), Electronic Health Record (EHR) owners, and the server. Each component is explained as follows.

(a) Secret Providers (SPs): It provides each owner with a secret pair of polynomials that generates pairwise secret keys.
(b) EHR owners: The owners send the sign of their local updates to the server. The pairwise secret keys are used to mask and then encrypt the local updates.
(c) Server: It checks relevance, aggregates the updates, and broadcasts the result to each EHR owner.

This architecture is a two-way system that can send and receive data and perform operations simultaneously. It is an efficient and privacy-protected framework that is also robust to Electronic Health Record (EHR) owners, demonstrating practical performance for future work. These are just a few examples of the applications of Federated Learning in healthcare.

4   Opportunities and Challenges

Federated Learning is now a learning paradigm that has the potential to tackle problems encountered in medical data. It provides privacy and security to its clients. The primary source of people's healthcare data for ML is known as Electronic Health Records (EHR) [41]. If ML models use limited data from only one hospital's patients, there will be a big chance of bias while providing predictions. In order to get better predictions, a large number of datasets need to feed into the model.

Furthermore, to train an AI-based model, a large number of datasets are required, including the full spectrum of anatomies, possible medical diagnosis, and data input of patients. These types of data are highly sensitive for each person. Even though data animosity is given, removing some info is not enough to provide complete animosity [42]. Sharing data among organizations will help the model give a proper prediction. However, sharing patient's medical records is considered very sensitive. Utilizing Federated Learning is the best option to overcome these problems.

With the increasing use of different types of sensors in various aspects of life, wearable sensors on humans can help capture changes and record a patient's data. Feeding machine learning models with patient data can enhance the model's ability to provide accurate predictions. However, medical data sharing is not yet a systematic process as collecting and modifying such datasets is time-consuming and expensive. Along with these, many more ways to improve an FL model will open new opportunities in improving the healthcare field. Also, there are many challenges that an FL model might face while dealing with sensitive information. Below some of those are discussed, along with their defense mechanism and ways that might prevent it. The FL model, like other models, has some drawbacks. One of the main concerns is the potential for cyber attacks. FL models are vulnerable to various types of attacks, including those that compromise the central server or local devices within the framework. The following part will discuss some of these attacks and their possible defense mechanisms.

– Membership Inference Attacks: They are a common type of attack on FL models, where raw user data can be inferred from the training data used in FL, even though the raw user data is stored on local devices. One way to defend against these attacks is to use differential privacy, which can provide a privacy guarantee by securing computation in a trusted environment. Secure Multiparty Computation (SMC) and Homomorphic Encryption (HE) are two techniques that can be used for secure computation. In SMC, participants agree to provide inputs to two or more parties and reveal the outputs to a subset of participants without decrypting the computation's first results. Homomorphic encryption allows computation on encrypted input without decrypting it. Differential Privacy adds noise to the clipped model parameter to mask a user's contribution before model aggregation, but this can result in some loss of accuracy. The Trusted Execution Environment (TEE) is a secure platform that can run FL processes with low computational overhead compared to other computation techniques [43], although it is currently only compatible with CPU devices. However, the current version is only compatible with CPU devices. Differential Privacy involves masking the user's contribution by adding noise to the clipped model parameter before model aggregation [44]. However, this approach results in some loss of accuracy.

- Model Poisoning Attacks: These attacks resemble data poisoning attacks. Here the main target is the local models, not the local data. In order to introduce errors in the global model, this poisoning attack is performed. It is done by compromising some devices with modified local mode parameters. This hinders the accuracy of the global market too. The defense mechanism of this model is similar to the previous attack. The common mechanisms are rejection based on error rate and loss function [45]. Here, the models that significantly impact the error rate will get rejected. It is also based on their impact on the loss functions of the global model. The rejection can come from a combination of both error-based and loss-function rejection.

- Backdoor Attacks: Federated Learning allows devices to remain anonymous during the model updating process. However, a device or several devices can introduce a backdoor functionality into the model using the same functionality as the FL model, leading to what is known as a targeted attack [46]. The proportion of compromised devices present is a parameter used to measure the intensity of such attacks along with the model capacity of federated learning [47]. To defend against backdoor attacks, the differential privacy can be weakened. Additionally, participant-level differential privacy can be used as a form of defense against such attacks, but this may come at the cost of the global model's performance [44].

These are some of the challenges the model might face that are related to privacy issues while performing an FL model.

5   Conclusion

This paper discusses the advantages of implementing Federated Learning in healthcare applications. Healthcare information is highly sensitive, and privacy regulations govern its use. However, ML models used for diagnosing and predicting diseases require a vast amount of data. Collecting this data without compromising patients' privacy is difficult. Federated Learning can enable patient data to be shared without violating their privacy. This approach allows the ML model to analyze a large number of real-life data along with some machine-generated data. By combining these data, the model can detect diseases and identify potential solutions. This can save time and potentially save lives. However, DL methods are vulnerable to various attacks that compromise privacy. The paper also discusses ways to prevent privacy invasion attacks while preserving patient privacy and the future opportunities of FL in healthcare systems. Along with this, it also highlights the shortcomings of FL for implementation in the medical field.

References

1. Li, L., Fan, Y., Tse, M., Lin, K.-Y.: A review of applications in federated learning. Comput. Ind. Eng. **149**, 106854 (2020)
2. Li, L., Wang, Y., Lin, K.-Y.: Preventive maintenance scheduling optimization based on opportunistic production-maintenance synchronization. J. Intell. Manuf. **32**(2), 545–558 (2021)
3. Zhang, C., Xiongwei, H., Xie, Yu., Gong, M., Bin, Yu.: A privacy-preserving multi-task learning framework for face detection, landmark localization, pose estimation, and gender recognition. Front. Neurorobot. **13**, 112 (2020)
4. Xie, Yu., Wang, H., Bin, Yu., Zhang, C.: Secure collaborative few-shot learning. Knowl.-Based Syst. **203**, 106157 (2020)
5. Boban, M.: Digital single market and eu data protection reform with regard to the processing of personal data as the challenge of the modern world. In: Economic and Social Development: Book of Proceedings, p. 191 (2016)
6. Chen, Y.-R., Rezapour, A., Tzeng, W.-G.: Privacy-preserving ridge regression on distributed data. Inf. Sci. **451**, 34–49 (2018)
7. Jiang, H., Liu, M., Yang, B., Liu, Q., Li, J., Guo, X.: Customized federated learning for accelerated edge computing with heterogeneous task targets. Comput. Netw. **183**, 107569 (2020)
8. Sarp, S., Kuzlu, M., Wilson, E., Guler, O.: WG2AN: synthetic wound image generation using generative adversarial network. J. Eng. **2021**(5), 286–294 (2021)
9. Van Panhuis, W.G., et al.: A systematic review of barriers to data sharing in public health. BMC Public Health **14**(1), 1–9 (2014)
10. Sarp, S., Zhao, Y., Kuzlu, M.: Artificial intelligence-powered chronic wound management system: towards human digital twins (2022)
11. Holcomb, S.D., Porter, W.K., Ault, S.V., Mao, G., Wang, J.: Overview on deepmind and its alphago zero AI. In: Proceedings of the 2018 International Conference on Big Data and Education, pp. 67–71 (2018)
12. Krizhevsky, A., Sutskever, I., Hinton, G.E.: Imagenet classification with deep convolutional neural networks. Commun. ACM **60**(6), 84–90 (2017)
13. Hinton, G., et al.: Deep neural networks for acoustic modeling in speech recognition: the shared views of four research groups. IEEE Signal Process. Mag. **29**(6), 82–97 (2012)
14. Hard, A., et al.: Federated learning for mobile keyboard prediction. arXiv preprint arXiv:1811.03604 (2018)
15. McMahan, B., Moore, E., Ramage, D., Hampson, S., y Arcas, B.A.: Communication-efficient learning of deep networks from decentralized data. In: Artificial Intelligence and Statistics, pp. 1273–1282. PMLR (2017)
16. Kairouz, P., et al.: Advances and open problems in federated learning. Found. Trends® Mach. Learn. **14**(1–2), 1–210 (2021)
17. Sarp, S., Kuzlu, M., Wilson, E., Cali, U., Guler, O.: The enlightening role of explainable artificial intelligence in chronic wound classification. Electronics **10**(12), 1406 (2021)
18. Lo, S.K., Lu, Q., Zhu, L., Paik, H.Y., Xu, X., Wang, C.: Architectural patterns for the design of federated learning systems. J. Syst. Softw. **191**, 111357 (2022)
19. Liu, J., et al.: From distributed machine learning to federated learning: a survey. Knowl. Inf. Syst. 1–33 (2022)
20. Liu, J., Pacitti, E., Valduriez, P., Mattoso, M.: A survey of data-intensive scientific workflow management. J. Grid Comput. **13**(4), 457–493 (2015)

21. Verbraeken, J., Wolting, M., Katzy, J., Kloppenburg, J., Verbelen, T., Rellermeyer, J.S.: A survey on distributed machine learning. ACM Comput. Surv. (CSUR) **53**(2), 1–33 (2020)
22. Yang, Q., Liu, Y., Chen, T., Tong, Y.: Federated machine learning: concept and applications. ACM Trans. Intell. Syst. Technol. (TIST) **10**(2), 1–19 (2019)
23. Zhu, H., Zhang, H., Jin, Y.: From federated learning to federated neural architecture search: a survey. Complex Intell. Syst. **7**(2), 639–657 (2021)
24. Pan, S.J., Yang, Q.: A survey on transfer learning. IEEE Trans. Knowl. Data Eng. **22**(10), 1345–1359 (2009)
25. Huang, Y., et al.: GPipe: efficient training of giant neural networks using pipeline parallelism. In: Advances in Neural Information Processing Systems, vol. 32 (2019)
26. Briggs, C., Fan, Z., Andras, P.: Federated learning with hierarchical clustering of local updates to improve training on non-IID data. In: 2020 International Joint Conference on Neural Networks (IJCNN), pp. 1–9. IEEE (2020)
27. Mhaisen, N., Abdellatif, A.A., Mohamed, A., Erbad, A., Guizani, M.: Optimal user-edge assignment in hierarchical federated learning based on statistical properties and network topology constraints. IEEE Trans. Netw. Sci. Eng. **9**(1), 55–66 (2021)
28. Wainakh, A., Guinea, A.S., Grube, T., Mühlhäuser, M.: Enhancing privacy via hierarchical federated learning. In: 2020 IEEE European Symposium on Security and Privacy Workshops (EuroS&PW), pp. 344–347. IEEE (2020)
29. Rieke, N., et al.: The future of digital health with federated learning. NPJ Digit. Med. **3**(1), 1–7 (2020)
30. Smith, V., Chiang, C.K., Sanjabi, M., Talwalkar, A.S.: Federated multi-task learning. In: Advances in Neural Information Processing Systems, vol. 30 (2017)
31. Zhao, Y., Li, M., Lai, L., Suda, N., Civin, D., Chandra, V.: Federated learning with non-IID data. arXiv preprint arXiv:1806.00582 (2018)
32. Mohri, M., Sivek, G., Suresh, A.T.: Agnostic federated learning. In: International Conference on Machine Learning, pp. 4615–4625. PMLR (2019)
33. Zhang, W., et al.: Dynamic-fusion-based federated learning for covid-19 detection. IEEE Internet Things J. **8**(21), 15884–15891 (2021)
34. Wibawa, F., Catak, F.O., Kuzlu, M., Sarp, S., Cali, U.: Homomorphic encryption and federated learning based privacy-preserving CNN training: Covid-19 detection use-case. In: Proceedings of the 2022 European Interdisciplinary Cybersecurity Conference, pp. 85–90 (2022)
35. iu, B., Yan, B., Zhou, Y., Yang, Y., Zhang, Y.: Experiments of federated learning for covid-19 chest X-ray images. arXiv preprint arXiv:2007.05592 (2020)
36. Ahmed, I., Ahmad, A., Jeon, G.: An IoT-based deep learning framework for early assessment of covid-19. IEEE Internet Things J. **8**(21), 15855–15862 (2020)
37. Li, Y., Zhou, Y., Jolfaei, A., Dongjin, Yu., Gaochao, X., Zheng, X.: Privacy-preserving federated learning framework based on chained secure multiparty computing. IEEE Internet Things J. **8**(8), 6178–6186 (2020)
38. Guowen, X., Li, H., Dai, Y., Yang, K., Lin, X.: Enabling efficient and geometric range query with access control over encrypted spatial data. IEEE Trans. Inf. Forensics Secur. **14**(4), 870–885 (2018)
39. Chen, H., Li, H., Xu, G., Zhang, Y., Luo, X.: Achieving privacy-preserving federated learning with irrelevant updates over e-health applications. In: ICC 2020-2020 IEEE International Conference on Communications (ICC), pp. 1–6. IEEE (2020)
40. Sannara, E.K., Portet, F., Lalanda, P., German, V.E.G.A.: A federated learning aggregation algorithm for pervasive computing: evaluation and comparison. In: 2021 IEEE International Conference on Pervasive Computing and Communications (PerCom), pp. 1–10. IEEE (2021)

41. Ghassemi, M., Naumann, T., Schulam, P., Beam, A.L., Chen, I.Y., Ranganath, R.: A review of challenges and opportunities in machine learning for health. AMIA Summits on Translational Science Proceedings, vol. 2020, p. 191 (2020)
42. Rocher, L., Hendrickx, J.M., De Montjoye, Y.-A.: Estimating the success of re-identifications in incomplete datasets using generative models. Nat. Commun. **10**(1), 1–9 (2019)
43. Mo, F., Haddadi, H.: Efficient and private federated learning using tee. In: Proceedings of EuroSys Conference, Dresden, Germany (2019)
44. Geyer, R.C., Klein, T., Nabi, M.: Differentially private federated learning: a client level perspective. arXiv preprint arXiv:1712.07557 (2017)
45. Fang, M., Cao, X., Jia, J., Gong, N.: Local model poisoning attacks to {Byzantine-Robust} federated learning. In: 29th USENIX Security Symposium (USENIX Security 2020), pp. 1605–1622 (2020)
46. Bagdasaryan, E., Veit, A., Hua, Y., Estrin, D., Shmatikov, V.: How to backdoor federated learning. In: International Conference on Artificial Intelligence and Statistics, pp. 2938–2948. PMLR (2020)
47. Sun, Z., Kairouz, P., Suresh, A.T., McMahan, H.B.: Can you really backdoor federated learning? arXiv preprint arXiv:1911.07963 (2019)

T-SignSys: An Efficient CNN-Based Turkish Sign Language Recognition System

Sevval Colak[1], Arezoo Sadeghzadeh[1(✉)] [ID], and Md Baharul Islam[1,2] [ID]

[1] Bahcesehir University, 34349 Besiktas, Istanbul, Turkey
arezoo.sadeghzadeh@bahcesehir.edu.tr
[2] Florida Gulf Coast University, Fort Myers, FL 33965, USA

Abstract. Sign language (SL) is a communication tool playing a crucial role in facilitating the daily life of deaf or hearing-impaired people. Large varieties in the existing SLs and lack of interpretation knowledge in the general public lead to a communication barrier between the deaf and hearing communities. This issue has been addressed by automated sign language recognition (SLR) systems, mostly proposed for American Sign Language (ASL) with limited number of research studies on the other SLs. Consequently, this paper focuses on static Turkish Sign Language (TSL) recognition for its alphabets and digits by proposing an efficient novel Convolutional Neural Network (CNN) model. Our proposed CNN model comprises 9 layers, of which 6 layers are employed for feature extraction, and the remaining 3 layers are adopted for classification. The model is prevented from overfitting while dealing with small-scale datasets by benefiting from two regularization techniques: 1) ignoring a specified portion of neurons during training by applying a dropout layer, and 2) applying penalties during loss function optimization by employing L_2 kernel regularizer in the convolution layers. The arrangement of the layers, learning rate, optimization technique, model hyper-parameters, and dropout layers are carefully adjusted so that the proposed CNN model can recognize both TSL alphabets and digits fast and accurately. The feasibility of our proposed *T-SignSys* is investigated through a comprehensive ablation study. Our model is evaluated on two datasets of TSL alphabets and digits with an accuracy of 97.85% and 99.52%, respectively, demonstrating its competitive performance despite straightforward implementation.

Keywords: CNN · Digits and Alphabets · Static Sign language recognition · Turkish Sign Language

This work is supported by the Scientific and Technological Research Council of Turkey (TUBITAK) under the 2232 Outstanding Researchers program, Project No. 118C301 and the 2247-C Trainee Researcher Scholarship Program.

1   Introduction

Sign language (SL) is a form of non-verbal communication means used to convey information through hand gestures and facial expressions. It is utilized by deaf and hard-of-hearing people to interact with others and access services. There are more than 300 SLs worldwide that significantly differ from each other in various terms such as vocabulary and grammatical structures [17]. Due to these wide varieties, learning and interpreting different SLs is time-consuming and infeasible for hearing and hearing-impaired people. On the other hand, accessing human interpreters is costly and not practical in daily life. These challenges form a communication barrier in deaf-to-deaf and deaf-to-hearing people interactions, which not only limits the deaf people's social and professional life (e.g., fewer employment chances, social withdrawal, low academic performance) but also has substantial adverse mental impacts on them, leading to depression, loneliness, and anger. To tackle these issues, researchers have been captivated to develop automated sign language recognition (SLR) models that can accurately identify the signs performed by the signer and assist the public people in interpreting them effortlessly.

Automated SLR systems can significantly improve the life quality of the deaf community by smoothing over their communication and facilitating social service usage. Additionally, they can be used in the form of gesture recognition in many other human-computer interaction applications ranging from virtual/augmented reality (VR/AR) and video games [26] to medical purposes [11]. In these applications, a gesture recognition system tracks the user's hand movements and converts them into actions within a program. The first successful attempts regarding automated SLR have been made using direct measurements through sensor-based devices. These systems are generally more accurate due to their ability to detect the exact position, speed, and other characteristics of the user's hands. However, they require specialized devices [21], which are costly and inconvenient for performing complex signs. These limitations restrict their applicability in real-life scenarios, inspiring researchers to convert their attention to vision-based systems as a viable alternative.

Using images and videos acquired by cameras as input data in the vision-based system makes them appropriate for performing complex signs in different environments. Although the affordability, portability, and high flexibility of the vision-based systems make them superior to the sensor-based systems, they still have their own challenges caused by large varieties that exist either in signers (e.g., hand shapes, skin colors, and way of performing a sign), environment conditions (e.g., complex background, and lighting changes), or both. To address these issues, numerous works have been proposed in the last decade, which are categorized into two main groups considering the input data modality: 1) Static SLR (SSLR), and 2) Dynamic SLR (DSLR).

SSLR is defined as recognizing the digits and alphabets of a sign language from the images [24]. Static sign language (also referred to as fingerspelling) is utilized to perform ages, dates, proper nouns, and technical words with no specific sign, constituting a considerable portion of each SL [29]. This prominent role is important to SSLR attracting researchers' attention to develop highly accurate systems over the years. SSLR approaches can be further divided into two main categories: conventional machine learning (ML)-based and deep learning (DL)-based methods. In ML-based approaches, hand-crafted features are extracted from the input images, then utilized in the ML-based classifiers for final sign recognition. Despite the favorable performance achieved by these methods, their accuracy is highly dependent on the extracted features limiting their applicability to a specified dataset with poor generalization ability [15].

With the advent of DL-based models, highly powerful GPUs, and sign language datasets with large quantities, Convolutional Neural Networks (CNNs) have been widely employed in this domain to improve the performance and the generalization ability [25]. The capability of these models in extracting highly representative features from the complex structures through the backpropagation technique enables them to accurately differentiate the images from one another and achieve high performance. However, most of the approaches in the literature have focused on American Sign Language (ASL) recognition [7]. Although ASL is a globally well-known SL, many other SLs (e.g., Bangla, Arabic, Persian, and Turkish) are also significant for their own communities. However, there are limited studies on these SLs compared to those of ASL. Additionally, even for ASL recognition, alphabet recognition got more attention than digit recognition, while both are significant components of SSLR.

Turkish Sign Language (TSL) is one of the critical SLs used by around 3.5 million deaf people in Turkey. At the same time, its fingerspelling has been studied in only a limited number of research works in the literature. This paper aims to investigate static TSL recognition to ease the communication between deaf people and the general public unfamiliar with it in Turkey. Our proposed *T-SignSys* is based on a novel CNN model for fast and accurate TSL recognition, considering both alphabets and digits. Overall, the main contributions of this paper are listed as follows:

- A novel end-to-end CNN-based model, namely *T-SignSys*, is proposed to efficiently recognize both the alphabets and digits of TSL with a single optimized architecture following a straightforward implementation avoiding the challenging task of hand segmentation for complex backgrounds.
- Carefully tuning the number of layers, activation function, kernel, and filter sizes, and optimizer technique along with taking advantage of dropout layer and L_2 kernel regularizer as regularization techniques, our model obtains high accuracy without overfitting issues even for small-scale datasets.
- A comprehensive ablation study is conducted for the arrangements of the layers, the value of the dropout layer, the learning rate, and the optimization technique to investigate the effectiveness and feasibility of the proposed architecture. As an important factor in achieving high performance and fast convergence during training, Adam and Adamax are selected as optimizer techniques for digits and alphabets, respectively, through our ablation study.

– The performance of the proposed model is evaluated on two benchmark datasets of TSL alphabets and digits and compared with those of existing approaches proving its superiority and capabilities.

2   Related Works

During the last two decades, a significant number of studies have been conducted to advance automated SLR systems, which are categorized into two main groups: conventional machine learning (ML)-based and deep learning (DL)-based approaches. The automated systems in both groups follow three steps to accomplish sign language recognition: pre-processing, feature extraction, and classification. In conventional methods, feature extraction has been performed by extracting hand-crafted features classified using ML-based classifiers. In contrast, feature extraction and classification are carried out in DL-based systems through a single deep network. Some of the recent sign language recognition (SLR) approaches from both categories are briefly discussed in this section.

2.1   Machine Learning-Based Approaches

Machine learning is a subset of artificial intelligence that involves training algorithms to learn patterns and make predictions or decisions based on data. ML techniques have been used for both sensor-based and vision-based SLR. In the sensor-based approaches, the data acquired directly from the sensor-based devices are used in ML-based classifiers (e.g., Support Vector Machine (SVM), K-Nearest Neighbor (KNN)) for sign recognition, while in the vision-based systems, the inputs of the classifiers are the hand-crafted features (e.g., Histogram of Oriented Gradients (HOG), Local Binary Pattern (LBP)) extracted from sign images. Guardino et al. [9] used a Leap Motion sensor-based device and two classifiers of KNN and SVM for recognizing the ASL alphabets. Their acquired data included features from the fingers and palm, such as position, direction, and velocity, which were transmitted to the computer via a USB connection. Instead of employing the raw data, they computed average distance, average spread, and average tri-spread, which were fed into the classifiers for final recognition. In their system, KNN and SVM classifiers obtained an accuracy of 72.78% and 79.83%, respectively, proving the superiority of SVM over KNN. Another sensor-based approach was proposed by Yalçın et al. [30] using a glove equipped with elastic detectors and a gyroscope to detect sign language and translate it into text. Despite its successful performance for SL translation, it was inconvenient and costly for real-life scenarios due to using gloves equipped with detectors and wires connected to an Arduino.

As an affordable and more convenient system, a vision-based approach was proposed in [14] by Kumar et al. using both images and videos as input data. In the pre-processing step, they applied face detection and removal using the Viola and Jones algorithm to prevent the head skin color from interfering with the hand detection. Then, the hand was segmented using HSV thresholding. To improve accuracy, instead of using a global threshold value in the segmentation process, they sampled the skin color of the signer before sign recognition. For static sign language recognition, they utilized the feature vectors extracted through Zernike moments in an SVM classifier achieving an accuracy of 93%. A multi-kernel SVM was trained in [8] using a proper fusion of three different hand-crafted features. The same classifier was also employed in [18], but this time with LBP features.

To tackle the curse of dimensionality as the main limitation of the ML-based approaches, Principal Component Analysis (PCA) was used in [23] to reduce the dimensionality of the feature vector acquired by a combination of different image descriptors. They evaluated their proposed system for user-dependent and -independent scenarios by applying three different classifiers of KNN, Multi-Layer Perceptron (MLP), and Probabilistic Neural Network (PNN). Amrutha and Prabu [4] designed a model capable of recognizing single-handed signs using convex hull features and a KNN classifier. Their training dataset was collected in a controlled environment with a stable and plain background, and image backgrounds were removed in a pre-processing step using the threshold method. Contour-based segmentation was then applied to obtain the contours of the fingers. Despite achieving an accuracy of 65%, the performance of their system was highly dependent on the distance between the camera and the signer. In a recent work by Bansal et al. [5], a system was proposed based on HOG features and SVM classifier whose performance was evaluated on seven different datasets. They utilized Minimum Redundancy and Maximum Relevance (mRMR) and Particle Swarm Optimization (PSO) techniques to remove the feature redundancy while maintaining accuracy. Generally, in ML-based approaches, satisfactory performance is obtained if the extracted features are representative and strong enough. However, they still suffer from poor generalization ability and time-consuming feature extraction steps with high dimensionality.

2.2   Deep Learning-Based Approaches

Deep learning (DL) is another subset of artificial intelligence that involves training artificial neural networks to learn patterns and make predictions or decisions based on data. CNN is a powerful DL model that automatically extracts deeper and more effective low- and high-level features from the input images alleviating the need for manual feature engineering and hand-crafted feature extraction. The superior performance of the CNNs over the conventional ML-based methods and the emergence of the advanced GPUs and large datasets inspired the researchers to employ them in various computer vision domains such as action recognition, object detection, classification, etc. They also have been extensively used in SLR to enhance the systems' accuracy and provide real-time performance.

In DL-based approaches, CNN models can be used in two schemes: 1) only extracting deep features, which are then classified by ML-based classifiers, and 2) conducting both deep feature extraction and classification through a single architecture. Sanchez-Riera [27] conducted a comparative study to analyze and compare the performance of CNN models with the other classifiers using two input modalities of RGB and depth. A CNN-based ASL recognition model was also proposed in [22] to investigate the superiority of the DL models over the ML classifiers of SVM, KNN, and RF. In [20], another novel CNN model was proposed for ASL recognition whose accuracy was enhanced through applying a pre-processing step and hand segmentation. They also developed a new ASL dataset. Das et al. [10] utilized a CNN model to recognize the ASL alphabets, achieving an accuracy of 94.34% for images with plain black backgrounds. Sevli and Kemaloğlu [28] developed a CNN model for recognizing TSL digits, achieving an accuracy of 98.55%. They conducted extensive experiments on four optimizers to investigate their impacts on the model's performance. Another research work regarding TSL recognition was carried out by Öztürk et al. [19] using Faster-R-CNN. They trained their model using the TSL alphabet dataset and tested it using by real-time data, achieving an accuracy of 88%. A new TSL dataset was developed by Aksoy et al. [2], including 10223 images for 29 letters captured at different distances from the camera on plain white background. They performed TSL recognition using different pre-trained models and their proposed model as TSLNet. Among these models, CapsNet and TSLNet models outperformed the others with an accuracy of 99.7% and 99.6%, respectively.

One of the techniques recently attracted the researchers' attention for enhancing the performance of SLR systems is using a combination of several models. Kodandaram et al. [13] used an ensemble of three models (i.e., LeNet-5, MobineNetV2, and a custom CNN) for ASL recognition obtaining an accuracy of 99.89%. Bhaumik et al. [6] created a portable end-to-end CNN model named as ExtriDeNet using two main modules: the intensive feature fusion block (IFFB) and the intensive feature assimilation block (IFAB). The capability of the Extri-DeNet in dealing with challenging environmental conditions (i.e., illumination variations and complex backgrounds) was demonstrated through their experiments. Bousbai et al. [7] enhanced the recognition performance for four ASL datasets by ensembling the features extracted by a custom CNN and a Caps-sNet. Once PCA reduced the dimensionality of the combined feature vector, it was used in an SVM classifier for final recognition. Alnuaim et al. [3] combined ResNet50 and MobileNetV2 architectures for Arabic SLR achieving an accuracy of about 97% after applying various data augmentation techniques. Zakariah et al. [31] used various pre-trained models for Arabic SLR based on transfer learning. Applying different pre-processing and augmentation techniques, the EfficientNetB4 model outperformed the others with an accuracy of 95%. However, it was computationally inefficient due to being a heavy-weight model.

3   Methodology

The overall flowchart of the proposed TSL recognition system is illustrated in Fig. 1. It comprises three main sections: pre-processing (including augmentation and image resizing), feature extraction, and classification. The last two steps are implemented through a single novel CNN model whose layer arrangement and hyper-parameters are optimized to achieve high accuracy for both TSL digit and alphabet recognition. Details of each step are discussed in the following sections.

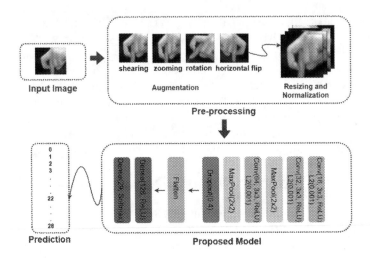

Fig. 1. The main flowchart of the proposed *T-SignSys*.

3.1 Data Preprocessing

Pre-processing is a crucial step in DL-based models to enhance the input data quality and quantity to boost the proposed model's performance. Hence, we first apply data augmentation in our pre-processing step. Data augmentation refers to the techniques employed for modifying and generating new data. It plays a significant role as a regularizer to prevent the model from overfitting and make it robust for various input patterns by increasing the diversity and quantity of the data in the training set (i.e., oversampling). Hence, proper selection of the augmentation techniques is significant to achieve optimized performance. In our model, the quantity of the training samples (RGB images) is increased through four augmentation techniques of shearing (ranges between 0–0.2), zooming (ranges between 0–0.2), rotation (ranges between 0–45), and horizontal flip. All RGB images, including the augmented ones, are resized into the same size (100×100 for digits and 224×224 for alphabets). Then, pixel intensities are divided by 255 for normalization.

3.2 Model Architecture

CNNs are multi-layer architectures trained to extract features from the input images in the form of feature maps that include the corresponding images' main characteristics. A novel efficient CNN architecture, namely *T-SignSys*, is proposed in our paper for accurate and fast TSL digit and alphabet recognition, whose details are presented in Table 1. It is formed by a total of 9 layers. The first 6 layers (i.e., three convolutional layers, two max-pooling layers, and a dropout layer) are selected for feature extraction, and the remaining 3 layers (flatten and fully-connected/dense) are used for classification. Having a hierarchical struc-

Table 1. Summary of the proposed CNN architecture presenting its hyper-parameters.

Layer	Filters	Kernal	Pool	Strides	Activation	Regularizer	Output Shape	Param #
Conv2d	16	3 × 3	–	–	ReLU	L2	(222, 222, 16)	448
Conv2d	32	3 × 3	–	–	ReLU	L2	(220, 220, 32)	4640
Max Pool	–	–	2 × 2	2	–	–	(110, 110, 32)	0
Conv2d	64	3 × 3	–	–	ReLU	L2	(108, 108, 64)	18496
Max Pool	–	–	2 × 2	2	–	–	(54, 54, 64)	0
Dropout(40%)	–	–	–	–	–	–	(54, 54, 64)	0
Flatten	–	–	–	–	–	–	(186624)	0
Dense	–	–	–	–	ReLU	–	(128)	23888000
Dense	–	–	–	–	Softmax	–	(29)	3741

ture, each layer is in charge of non-linearly transforming the input images, and the results are fed into the next layer for another non-linear transformation.

Feature Extraction Module: The inputs of our feature extraction module are RGB images which are resized in the pre-processing step. They are fed into the first convolutional block which is composed of two convolutional layers and a maxpooling layer. These convolutional layers are responsible for learning and extracting the low-level features from the input images, such as edges, curves, and textures. Filters and kernels in these layers are the hyper-parameters responsible for extracting the representative features. Each filter is a small matrix of weights that is used to perform the convolution operation on the input data. The weights in the filter are adjusted during training so that the filter becomes increasingly sensitive to the specific pattern or feature it is trying to detect. The number of filters in a convolutional layer can significantly affect the overall performance of the model. More filters aid in extracting more specific patterns and features by the convolutional layer as the model gets more powerful. However, this filter size increment leads to more parameters and so the risk of overfitting. Considering a good balance between these two issues, we selected the number of filters as 16 and 32 in the first and second layers, respectively. The third convolutional layer which is placed in the second convolutional block is considered the deeper layer for our model. Consequently, we selected a larger filter number in comparison to the first two convolutional layers so that more complex abstract patterns are extracted from the images. Kernel size is another crucial parameter in the convolutional layer which determines the level of abstraction and the number of parameters in the model. A convolutional layer with a larger kernel size can capture more contextual information from the input image and potentially extract more meaningful features. However, similar to the filter size, increasing the kernel size increases the number of parameters leading to overfitting if the model is not properly regularized. On the other hand, a smaller kernel size can capture more fine-grained details from the input image with less number of parameters, but it may not be able to capture the global context as effectively. Making a good trade-off, we select kernel size as 3 × 3 in all convolutional layers achieving the optimal performance with our input images.

In our feature extraction module, two regularization techniques are applied to further improve the performance of the system by increasing the accuracy, improving the convergence, speeding up the training, enhancing the generalization ability, and alleviating the overfitting risk. The first technique is applying L_2 kernel regularizer in all convolutional layers. It is a form of regularization that adds a penalty term to the objective function of the model, which penalizes large weights and encourages the model to learn a simpler, more generalized solution for new unseen data. A dropout layer is used at the end of our feature extraction module as the other regularization technique. Its value is set to 0.4, which means that 40% of the neurons are ignored for weight updates during the training, which prevents the neurons from co-adapting and forces the model to rely on the remaining neurons to make predictions. This dropout layer significantly improves the accuracy of the model and minimizes the gap between the training and the test accuracy (detailed results are discussed in the ablation study section). Two maxpooling layers adopted in our model have identical hyperparameters, i.e., 2×2 kernel size. They are used in both convolutional blocks after convolutional layers to downsample the feature maps by reducing their dimensionality, while still retaining the most important information. It is worth mentioning that the stride value (i.e., the number of pixels that are skipped by the kernel filter in each movement) is selected as one and two in the convolutional and maxpooling layers, respectively, with the "valid" padding scheme. The outputs of the convolutional layers are fed into a non-linear activation function known as Rectified Linear Unit (ReLU). Using this activation function, only the neurons with positive values are activated ($f(x) = max(0, x)$), which leads to fast training and minimizes the possibility of gradient vanishing.

Classification Module: The generated 2D feature maps from the feature extraction module are reshaped into a 1D feature vector using a flattening layer before the fully connected (FC)/dense layers. This feature vector is fed into a FC layer with 128 neurons and a ReLU activation function. To set the number of neurons to 128 in the FC layer as its crucial hyperparameter, we start with a relatively small number of neurons and gradually increase it until reaching the convergence and plateau performance. This layer is followed by the second FC layer, whose number of neurons equals the number of class labels in the corresponding dataset. The last FC layer, known as the classification layer, uses *Softmax* as its activation function. It is a generalization of the logistic function used to convert arbitrary values into a probability distribution, where the sum of the probabilities is 1. It is computed as $Softmax(x_i) = \frac{\exp(x_i)}{\sum_j \exp(x_j)}$, where x_i is a vector of arbitrary values used as its input. The calculated probabilities by this function determine the membership of the input to each class. The highest value probability determines the corresponding input's predicted class. Our model is trained based on the *categorical cross-entropy* loss function as it is dealing with a multi-class classification task. It is defined as $Loss = -\sum_{i=1}^{output\ size} y_i \cdot log\ \hat{y}_i$, where $\hat{y}_i$ and y_i represent the probabilities of the model prediction and the corresponding true target, respectively, and it is demanded to reduce the difference between them during the training process.

4   Experimental Results

4.1   Datasets

Two TSL datasets are selected for our experiments:

1) TSL Digits [1,16]: This dataset includes a total of 2062 RGB images with a
fixed size of 100×100 in 10 classes for the digits from 0 to 9. The signs of
this dataset were performed by 218 different right-handed signers. The images
were all captured on a plain white background. Some samples of this dataset
are illustrated in Fig. 2(a).
2) TSL Alphabets [12]: As the first alphabet dataset for static Turkish Sign
Language, it is composed of 2974 RGB images categorized into 29 classes out
of which 23 categories are for Turkish letters (excluding "Ç, Ğ, İ, Ö, Ş, and
Ü") and the remaining 6 classes are allocated for punctuation marks. The
images in this dataset have different sizes which were captured on cluttered
backgrounds. It contains both single- and double-handed signs performed
by both left- and right-handed signers. Some samples of this dataset are
illustrated in Fig. 2(b).

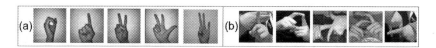

Fig. 2. Illustration of some sample images from (a) TSL Digits, and (b) TSL Alphabets.

4.2   Experimental Setup

All experiments are conducted with Python using Keras and Tensorflow on a PC
with 32 GB RAM, Intel Core i7-9700 CPU, and NVIDIA GeForce RTX 2070
GPU with 8 GB video memory. We use Adam and Adamax optimizers for digit
and alphabet datasets, respectively, with a learning rate of 0.001. Epoch and
batch sizes are selected as 300 and 32, respectively, with data split of 80%–20%.

4.3   Evaluation Metrics

The performance of our proposed model is evaluated in terms of four evaluation
metrics common for classification tasks as follows:

$$Accuracy = \frac{TP + TN}{TP + TN + FP + FN} \tag{1}$$

$$Precision = \frac{TP}{TP + FP} \tag{2}$$

$$Recall/Sensitivity = \frac{TP}{FN + TP} \tag{3}$$

$$F1 - score = \frac{2 \times (Precision \times Recall)}{Precision + Recall} \tag{4}$$

where TP, TN, FP, and FN stand for True Positive, True Negative, False Positive, and False Negative, respectively, obtained from the confusion matrix.

4.4   Performance Assessment

The performance and efficiency of the proposed CNN model are evaluated on two benchmark TSL datasets for digits and alphabets. The experimental results are presented in Table 2 in terms of four evaluation metrics for both training and test sets. Achieving a test accuracy of 99.52% and 97.85% for digit and alphabet datasets, respectively, demonstrates the high capabilities of our proposed model despite its straightforward architecture. As *T-SignSys* is assessed on two diverse datasets with both plain and cluttered backgrounds, single- and double-handed signs performed by left- and right-handed signers, one can draw the inference that it is efficient and robust against hand appearances and environmental conditions. Additionally, our model achieves a great convergence without experiencing any underfitting or overfitting as the difference between the training and test accuracy is minimized in both datasets. Owing to the less number of parameters and benefiting from the regularization techniques in our model, it is fast both in training and test processes making it feasible and practical for real-time application. Training of the digit and alphabet datasets takes 17 and 105 min for 300 epochs, respectively. 418 digit images and 604 alphabet images are recognized in the test phase during 0.47 and 5 s, respectively, proving the real-time performance of our proposed method.

Table 2. Performance evaluation in terms of four evaluation metrics on two datasets.

	Digits Dataset		Alphabet Dataset	
	Training	Test	Training	Test
Accuracy	98.11%	99.52%	99.92%	97.85%
Precision	99.02%	98.56%	99.24%	97.33%
Recall	98.78%	98.33%	98.90%	93.54%
F1 Score	98.50%	98.32%	99.08%	94.96%

Table 3. Performance comparison with state-of-the-art TSL recognition approaches on two TSL datasets.

Dataset	Approaches	Training(%)/Testing(%) Split	Test Accuracy
TSL Digits	Sevli and Kemaloglu [28]	80/20	98.55%
	Bansal et al. [5]	90/10	90.90%
	Proposed Model	**80/20**	**99.52%**
TSL Alphabets	Ozturk et al. [19]	40/-	88.00%
	Proposed Model	**80/20**	**97.85%**

4.5   Comparison with State-of-the-Art

The performance of the proposed model is compared with the available state-of-the-art approaches for TSL recognition on two datasets in Table 3 presenting their test accuracy along with the training and test data split scheme. In the digit dataset, our model outperforms two other approaches with 0.97% and 8.62% accuracy enhancement. For TSL alphabets, it should be mentioned that Ozturk et al. [19] employed the TSL alphabet dataset only for training their model and they tested it by real-time images while we used the TSL alphabet dataset in both training and test phases. Comparing the test accuracy, our model surpasses their approach by achieving 9.85% more accuracy.

4.6   Ablation Study

To deeply investigate the impact of different parameters on the overall performance of our proposed *T-SignSys*, a comprehensive ablation study is carried out over the arrangement of the layers (9 different combinations), dropout layer value (0.25 and 0.40), optimizer (RMSProp, SGD, Adamax, and Adam), and learning rate (for 0.001, and 0.0001). The results for different layer arrangements of our CNN model are presented in Table 4 for two datasets in terms

Table 4. The results of ablation study over 9 different arrangements for CNN layers.

Layer Arrangement	Digits Dataset		Alphabet Dataset	
	Train	Test	Train	Test
1 Conv + 1 FC	93.07%	88.28%	87.55%	80.96%
2 Conv + 1 FC	98.11%	99.52%	97.30%	90.56%
3 Conv + 1 FC	97.81%	97.00%	94.39%	88.08%
3 Conv + 1 MP + 1 FC	96.96%	97.00%	98.82%	92.00%
3 Conv + 2 MP + 1 FC	96.96%	94.00%	98.90%	93.38%
4 Conv + 2 MP + 1 FC	96.53%	96.17%	98.02%	92.38%
5 Conv + 2 MP + 1 FC	94.95%	94.98%	95.02%	91.39%
3 Conv + 2 MP + 2 FC	**99.20%**	97.37%	98.27%	94.04%
3 Conv + 2 MP + 1 DP + 2 FC (ours)	98.11%	**99.52%**	**99.92%**	**97.85%**

Table 5. The results of ablation study over different values for dropout layer.

Dropout Values	Digits Dataset		Alphabet Dataset	
	Training	Test	Training	Test
0.25	**99.09%**	98.33%	99.79%	96.52%
0.40 (ours)	98.11%	**99.52%**	**99.92%**	**97.85%**

Table 6. The results of ablation study over four different optimization techniques.

Optimizers	Digits Dataset		Alphabet Dataset	
	Training	Test	Training	Test
SGD	78.47%	88.04%	83.76%	85.43%
RMSprop	**99.09%**	97.85%	98.23%	95.70%
Adamax	96.00%	94.00%	**99.92%**	**97.85%**
Adam	98.11%	**99.52%**	99.66%	95.53%

Table 7. The results of ablation study over two different learning rates on two datasets.

Learning Rate	Digits Dataset		Alphabet Dataset	
	Training	Test	Training	Test
0.001 (ours)	**98.11%**	**99.52%**	**99.92%**	**97.85%**
0.0001	96.05%	95.21%	98.06%	94.37%

of test accuracy. Employing two convolutional (Conv) layers with the last fully connected (FC) layer leads to high performance for digit dataset while its accuracy is very low for alphabet dataset. To achieve high performance for both digit and alphabet datasets, we gradually increase the number of convolutional layers and add maxpooling (MP) and FC layers to the architecture. Implementing different combinations, the architecture with three Conv layers, two MP layers, and two FC layers obtains a high performance for both datasets. However, there is still overfitting especially for the TSL alphabet. To overcome overfitting, a dropout (DP) layer is added before FC layers, which significantly enhances the test accuracy and minimized the accuracy difference between the training set (98.11% and 99.92% for digits and alphabets) and the new unseen data (99.52% and 97.85% for digits and alphabets). The performance of the dropout layer is investigated for two values of 0.25 and 0.40 in Table 5. The dropout layer with a value of 0.4 outperforms the other with 1.19% and 1.33% higher test accuracy for TSL digit and alphabet datasets, respectively.

Performance of *T-SignSys* for four different optimizers is investigated on two datasets in terms of accuracy for the learning rate of 0.001 in Table 6. The lowest performance is achieved by the SGD optimizer on both datasets as it is highly likely to be stuck in the local minimum and has slow convergence in comparison to the other optimizers for the same number of epochs. The other three optimizers of RMSProp, Adamax, and Adam have significantly higher performance. Achieving the highest test accuracy of 99.52% by Adam for digits and 97.85% by Adamax for alphabet, they are selected as the optimal optimizers for our model. Learning rate is also studied for two values of 0.001 and 0.0001 on both TSL digit and alphabet datasets in Table 7. Adopting a higher value for the learning rate results in faster convergence rather than using a lower value for an equal number of epochs. Comparing the results of two learning rates, the test accuracy of our model is improved by 4.31% and 3.48% for digit and alphabet datasets, respectively, once we set the learning rate as 0.001 without experiencing overfitting.

Fig. 3. Visualization of a sample failure case for our model where the letter "S" is misclassified as "F" when they are performed by two left- and right-handed signers.

4.7 Failure Cases

Although our proposed method can efficiently recognize both TSL digits and alphabets with a straightforward implementation scheme and fast recognition rate, it leads to misclassification when dealing with some specific alphabet classes. One of these classes with a high number of misclassification samples is "S". In most cases, the letter "S" is misclassified as "F". This misclassification occurred mainly due to using both right- and left-handed signers in the dataset. To delve deeper into this issue, one example is illustrated in Fig. 3. Performing letters of "S" and "F" by the same person shows the high inter-class variations between these two classes, which helps the model to distinguish them accurately. On the contrary, when one of them is performed by a left-handed signer and the other is performed by a right-handed signer, the captured images have slight inter-class variations which degrade the performance of our model and lead to misclassification.

5   Conclusion

In this paper, a novel efficient CNN model was proposed for fast and accurate classification of Turkish Sign Language (TSL) digits and alphabets. Despite the great importance of TSL for a large number of deaf people in Turkey, there are only a handful number of studies in the literature conducted on TSL static fingerspelling. To this end, a novel CNN architecture was designed with a total of 9 layers whose number of layers, hyper-parameters, optimizer, and learning rate were carefully adjusted through extensive experiments so that high performance was obtained. To further enhance the performance and prevent the system from overfitting, we used kernel regularizer and dropout layer as two regularization techniques. Conducting a comprehensive ablation study, we investigated the effectiveness of four different optimizer techniques, two learning rates, two dropout values, and 9 different arrangements for layers of the proposed CNN. Achieving an accuracy of 99.52% and 97.85% for TSL digits and alphabets, respectively, with high recognition speed, demonstrated the high capabilities of our model and its feasibility for real-time applications considering cluttered backgrounds. As our future work direction, we plan to enhance the model performance for the signs with slight inter- and high intra-class variations as well as make it robust to environmental variations.

References

1. ASL digits (2017). https://www.kaggle.com/datasets/ardamavi/sign-language-digits-dataset. Accessed 29 Dec 2022
2. Aksoy, B., Salman, O.K.M., Ekrem, Ö.: Detection of Turkish sign language using deep learning and image processing methods. Appl. Artif. Intell. **35**(12), 952–981 (2021)
3. Alnuaim, A., Zakariah, M., Hatamleh, W.A., Tarazi, H., Tripathi, V., Amoatey, E.T.: Human-computer interaction with hand gesture recognition using resnet and mobilenet. Comput. Intell. Neurosci. **2022** (2022)
4. Amrutha, K., Prabu, P.: ML based sign language recognition system. In: 2021 International Conference on Innovative Trends in Information Technology (ICITIIT), pp. 1–6. IEEE (2021)
5. Bansal, S.R., Wadhawan, S., Goel, R.: MRMR-PSO: a hybrid feature selection technique with a multiobjective approach for sign language recognition. Arabian J. Sci. Eng. 1–16 (2022)
6. Bhaumik, G., Verma, M., Govil, M.C., Vipparthi, S.K.: ExtriDenNt: an intensive feature extrication deep network for hand gesture recognition. Vis. Comput. **38**(11), 3853–3866 (2022)
7. Bousbai, K., Morales-Sánchez, J., Merah, M., Sancho-Gómez, J.L.: Improving hand gestures recognition capabilities by ensembling convolutional networks. Expert Syst. e12937 (2022)
8. Cao, J., Yu, S., Liu, H., Li, P.: Hand posture recognition based on heterogeneous features fusion of multiple kernels learning. Multimed. Tools Appl. **75**(19), 11909–11928 (2016)
9. Chuan, C.H., Regina, E., Guardino, C.: American sign language recognition using leap motion sensor. In: 2014 13th International Conference on Machine Learning and Applications, pp. 541–544. IEEE (2014)
10. Das, P., Ahmed, T., Ali, M.F.: Static hand gesture recognition for American sign language using deep convolutional neural network. In: 2020 IEEE Region 10 Symposium (TENSYMP), pp. 1762–1765. IEEE (2020)
11. Ichimura, K., Magatani, K.: Development of the bedridden person support system using hand gesture. In: 2015 37th Annual International Conference of the IEEE Engineering in Medicine and Biology Society (EMBC), pp. 4550–4553. IEEE (2015)
12. Kalkan, S.C.: Turkish sign language (fingerspelling) (2018). https://www.kaggle.com/datasets/feronial/turkish-sign-languagefinger-spelling. Accessed 29 Dec 2022

13. Kodandaram, S.R., Kumar, N.P., et al.: Sign language recognition. Turkish J. Comput. Math. Educ. (TURCOMAT) **12**(14), 994–1009 (2021)

14. Kumar, A., Thankachan, K., Dominic, M.: Sign language recognition, pp. 422–428 (2016). https://doi.org/10.1109/RAIT.2016.7507939

15. Latif, G., Mohammad, N., Alghazo, J., AlKhalaf, R., AlKhalaf, R.: ArASL: Arabic alphabets sign language dataset. Data Brief **23**, 103777 (2019)

16. Mavi, A.: A new dataset and proposed convolutional neural network architecture for classification of American sign language digits. arXiv preprint arXiv:2011.08927 (2020)

17. Murray, J.: World federation of the deaf (2018). https://wfdeaf.org/our-work/. Accessed 29 Dec 2022

18. Muthukumar, K., Poorani, S., Gobhinath, S.: Vision based hand gesture recognition for Indian sign languages using local binary patterns with support vector machine classifier. Adv. Nat. Appl. Sci. **11**(6), 314–322 (2017)

19. Öztürk, A., Karatekin, M., Saylar, İ.A., Bardakci, N.B.: Recognition of sign language letters using image processing and deep learning methods. J. Intell. Syst. Theory Appl. **4**(1), 17–23 (2021)

20. Pinto, R.F., Borges, C.D., Almeida, A., Paula, I.C.: Static hand gesture recognition based on convolutional neural networks. J. Electr. Comput. Eng. **2019** (2019)

21. Qi, J., Jiang, G., Li, G., Sun, Y., Tao, B.: Surface EMG hand gesture recognition system based on PCA and GRNN. Neural Comput. Appl. **32**(10), 6343–6351 (2020)

22. Ranga, V., Yadav, N., Garg, P.: American sign language fingerspelling using hybrid discrete wavelet transform-gabor filter and convolutional neural network. J. Eng. Sci. Technol. **13**(9), 2655–2669 (2018)

23. Sadeddine, K., Chelali, F.Z., Djeradi, R., Djeradi, A., Benabderrahmane, S.: Recognition of user-dependent and independent static hand gestures: application to sign language. J. Vis. Commun. Image Represent. **79**, 103193 (2021)

24. Sadeghzadeh, A., Islam, M.B.: BiSign-Net: fine-grained static sign language recognition based on bilinear CNN. In: 2022 International Symposium on Intelligent Signal Processing and Communication Systems (ISPACS), pp. 1–4. IEEE (2022)

25. Sadeghzadeh, A., Islam, M.B.: Triplet loss-based convolutional neural network for static sign language recognition. In: 2022 Innovations in Intelligent Systems and Applications Conference (ASYU), pp. 1–6. IEEE (2022)

26. Sagayam, K.M., Hemanth, D.J.: Hand posture and gesture recognition techniques for virtual reality applications: a survey. Virtual Reality **21**(2), 91–107 (2017)

27. Sanchez-Riera, J., Hua, K.L., Hsiao, Y.S., Lim, T., Hidayati, S.C., Cheng, W.H.: A comparative study of data fusion for RGB-D based visual recognition. Pattern Recogn. Lett. **73**, 1–6 (2016)

28. Sevli, O., Kemaloğlu, N.: Turkish sign language digits classification with CNN using different optimizers. Int. Adv. Res. Eng. J. **4**(3), 200–207 (2020)

29. Shi, B., Brentari, D., Shakhnarovich, G., Livescu, K.: Fingerspelling detection in American sign language. In: Proceedings of the IEEE/CVF Conference on Computer Vision and Pattern Recognition, pp. 4166–4175 (2021)

30. Yalçin, M., Ilgaz, S., Özkul, G., Yildiz, Ş.K.: Turkish sign language alphabet translator. In: 2018 26th Signal Processing and Communications Applications Conference (SIU), pp. 1–4. IEEE (2018)

31. Zakariah, M., Alotaibi, Y.A., Koundal, D., Guo, Y., Mamun Elahi, M.: Sign language recognition for Arabic alphabets using transfer learning technique. Comput. Intell. Neurosci. **2022** (2022)

Damage Detection on Turbomachinery with Machine Learning Algortihms

Ahmet Devlet Özçelik[1]([✉]) [iD] and Ahmet Sinan Öktem[2] [iD]

[1] Istanbul Gelisim University, Avcılar İstanbul 34310, Turkey
adozcelik@gelisim.edu.tr
[2] Gebze Techincal University, Gebze Kocaeli 41400, Turkey
sinan.oktem@gtu.edu.tr

Abstract. This study uses machine learning methods to find deterioration in turbomachine parts. In turbomachines, damage control procedures are carried out at specific times. Even though these checks take a while, if there is no damage, the components won't be replaced, and it is not anticipated that they will be rechecked until the following control or an unforeseen incident. For this situation, a machine learning algorithm has been developed and 96% accuracy was obtained for overall components.

Keywords: Turbo Machinery · Machine Learning · Predictive Maintenance

1 Introduction

This study aims to detect prior damage in turbo-machine components via machine learning algorithms. The damage control process in turbomachines is performed at certain hours. Although these checks take a quite long time, if there is no damage, the components will not be replaced, and the components are not expected to be checked again until the next control or an unexpected event occurs. The general factors that cause gas turbine damage are as follows:

- **Erosion:** Some particles that come with air passing through the compressor can degrade and damage surfaces inside the compressor instead of binding onto them. Erosion damage caused by solid particles is a frequently occurring problem that can affect the components of aeroengines. Both stationary and rotating airfoils are susceptible to material loss due to the impact of erosive particles. In some cases, this damage can result in negative effects on the hot-section hardware and overall engine performance [1].
- **Abrasion:** Abrasion is a type of wear caused by the mechanical action of one surface rubbing against another. It can be caused by a variety of factors, including the hardness and roughness of the surfaces involved, the presence of foreign particles, and the sliding speed and contact pressure between the surfaces. Abrasion can result in surface damage, material loss, and changes in surface properties such as roughness and hardness [2].

A. Ortis et al. (Eds.): ICAETA 2023, CCIS 1983, pp. 242–253, 2024.
https://doi.org/10.1007/978-3-031-50920-9_19

- **Corrosion:** Sulfur from the fuel and sodium chloride from the air interact during combustion at high temperatures to form sodium sulfate. Following deposition, the sodium sulfate speeds up oxidation (or sulfidation) attacks on hot-section components [3] [4].
- **Foreign Object Damage:** Objects going into the compressor may cause severe damage to industrial gas turbines, which are far more prevalent than turbines used on aircraft with open inlets [5].
- **Fatigue:** The beginning and growth of cracks in a material as a result of cyclic loading is known as fatigue. Fatigue can be caused by dynamic loads, vibrations, impacts, or thermal loads [6]. Fatigue can be extremely dangerous for turbo machinery used in aviation. A sudden loss of power may lead to undesired results. Thus, fatigue detection is essential.
- **Thermomechanical Fatigue:** Hot-section components of gas turbine engines operate in a hostile environment and are constantly vulnerable to failure by the thermal fatigue damage mechanism because of the different heat capacities of the various materials in the component as well as a non-uniform temperature field on the component. A gas turbine engine's starting and stopping can cause temperature redistribution in the parts, which can lead to thermal fatigue damage [7].
- **Creep:** Components of gas turbines working at high temperatures gradually deform under the influence of applied stress. Such deformation eventually builds up and causes a creep rupture mechanism, which causes fracture. The main factor reducing blade life in base-loaded gas turbines is blade creep degradation. Critical component design assessment for high-temperature applications should take these deformation and damage processes into account, and engineering calculations call for knowledge of creep rupture characteristics for the material the structure is made of [8].

Supervised learning and unsupervised learning are two major categories of machine learning algorithms that have been widely studied and applied in various fields in recent years. Supervised learning is a type of machine learning where the algorithm is trained on a labeled dataset, meaning that each data point is associated with a known target value. The algorithm learns to predict the target value for new, unseen data based on the patterns it identifies in the training data. Some popular supervised learning algorithms include linear regression, decision trees, and neural networks [9].

On the other hand, unsupervised learning is a type of machine learning where the algorithm is trained on an unlabeled dataset, meaning that the data points do not have any associated target values. Instead, the algorithm identifies patterns and structure within the data itself, without any prior knowledge of what the data represents. Common unsupervised learning techniques include clustering, anomaly detection, and dimensionality reduction [10].

Both supervised and unsupervised learning have their own strengths and weaknesses, and the choice of which type of learning to use depends on the specific problem and data at hand. In recent years, there have been numerous advancements and innovations in both supervised and unsupervised learning, leading to exciting new applications in fields such as computer vision, natural language processing, and healthcare [11].

- **Regression:** Regression algorithms are categorized as supervised machine learning. They support the explanation or forecast of a numerical value based on a collection of historical facts [12].
- **Classification:** Another sort of supervised machine learning that predicts or explains a class value in classification algorithms. They can help anticipate whether an online buyer would purchase a good, for example. Buyer or non-buyer, the response is either yes or no. Classification systems, on the other hand, are not limited to only two categories [12].
- **Clustering:** Since the goal of clustering algorithms is to group or cluster data with comparable features, they fall within the topic of unsupervised machine learning. Approaches that use clustering don't need output data to train. Instead, this method uses an algorithm to decide the result. Only visualizations can be used by a data scientist to evaluate the quality of a clustering algorithm's answer [12].
- **Dimensionality Reduction:** This technique is used in to remove the least related information from a data set. Since data sets containing a lot of columns are common, it is imperative to lower the overall amount. There are thousands of pixels in a photograph, but not all of them are crucial to research. Similar to this, dozens of measurements and tests may be performed on each chip during the manufacturing process, many of which offer redundant data. To manage the data set in these situations, dimensionality reduction techniques will be needed [12].
- **Ensemble Methods:** Ensemble approaches combine multiple predictive models (supervised machine learning) to make better forecasts than any one model could. For instance, an ensemble method known as random forest techniques combines many decision trees that have been trained using various data sets. As a result, a Random Forest's forecasts are more accurate than a single Decision Tree's [12].
- **Neural Networks and Deep Learning:** Artificial networks aim to capture non-linear patterns in data by incorporating multi-layered parameters into the model, as opposed to logistic and linear regressions. [12].
- **Transfer Learning:** Transfer learning is a method where parts of a pre-trained neural network can be reused and adapted for a new but similar task. Specifically, some of the trained layers from the previous neural network, which was trained on a particular task, can be transferred and combined with a few new layers that are trained on the data from the new task. [12].
- **Reinforcement Learning:** Reinforcement learning is an approach that enables an algorithm to learn from previous experiences in a general sense. By observing actions and using a trial-and-error method in a controlled environment, reinforcement learning can optimize a cumulative reward. [12].
- **Natural Language Processing:** This is a frequently used methodology for preparing text for machine learning. The most widely used text processing package is NLTK (Natural Language ToolKit) [12].

– **Word Embeddings:** TTF-IDF is a numerical representation of text documents that considers only the frequency and weighted frequencies of words. Word embeddings, on the other hand, capture a word's context within a document, enabling us to perform arithmetic with words by measuring the similarity of words based on context. Word2Vec utilizes a neural network to convert words in a corpus into numerical vectors, which can then be employed to identify synonyms, conduct word arithmetic, and represent text documents (by averaging all the word vectors in a document) [12].

Since predictive maintenance with machine learning studies has not been done before, the literature research is mostly focused on the use of machine learning algorithms in mechanical engineering, especially in the energy sector.

Regan et al. combined acoustic with machine learning algorithms, and they detected wind turbine blade damage. In the study, Regan et al. used supervised machine learning to accomplish 98% accuracy [13]. Ghalandari et al. optimized the first row of the compressor blade with an artificial neural network. It has been seen for the aerodynamical view, mass flow increased by 4% and for the structural view, optimized blades met the reduced frequency criteria [14].

In the study "Adaptive Detection and Prediction of Performance Degradation in Off-shore Turbomachinery", Zagorowska et al. tried to detect of degradation in turbomachinery. To accomplish that Zagorowska et al. took the data of weather every day for 2 years and trained and tested the algorithm which showed that it is possible to combine the existing approaches in degradation modeling to improve the accuracy of the prediction, thus making the algorithm useful in industrial performance-based application [15].

In a study Gascon et al. identified the machine learning technique that best estimates the remaining useful life of boiler components using plant operations. The best strategy to anticipate the decline in the life span of the plant with over 90% certainty, according to the authors' testing of five different machine learning algorithms [16].

Chao et al. showed that the capacity to estimate the remaining usufel lifetime (RUL) of its components, is a crucial enabler of intelligent maintenance systems. Datasets with run-to-failure trajectories are required for the creation of data-driven prognostic models. Chao et al. create a new dataset of run-to-failure trajectories for a fleet of aircraft engines under real-world flight conditions to aid the development of prognostics algorithms. The dataset was created using the NASA-developed Commercial Modular Aero-Propulsion System Simulation (CMAPSS) model. The damage propagation model employed in this dataset expands on earlier work's modeling method and adds two new levels of accuracy [17].

As above mentioned, there are no studies on this subject. In this study, various methods were tried to create a database due to the lack of databases or not being shared, and these methods were mentioned in the methodology section. For a quick solution, the classification method mentioned above is used.

2   Methodology

In order to apply Machine Learning technique, the necessary dataset has been obtained from NASA The Prognostics Data Repository [18] which is studied at study of Chao et al. [17]. Figure 1 shows schematic illustration of the engine along with the CMAPSS model's assigned station numbers.

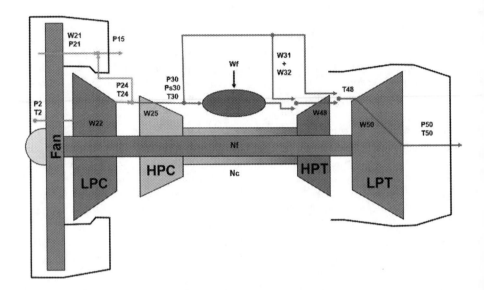

Fig. 1. Schematic representation of the CMAPSS model [17]

The names, descriptions, and units of each input variable in the dataset are can be found in study of Chao et al. [17]. In the CMAPSS model, the variable symbol corresponds to the internal variable name. The model documentation is used to generate the descriptions and units [19].

The output of the data is reaming useful life (RUL). For flight classes 8 and 9 RUL table has been shown in Fig. 2 below. For the study, all flight data has been merged. Due to the enormous number of data (sixty-nine million columns), an algorithm was created for data reduction. This algorithm was used to generate the dataset from 99 rows and 104 columns. The flow diagram of the algorithm is shown in Fig. 3.

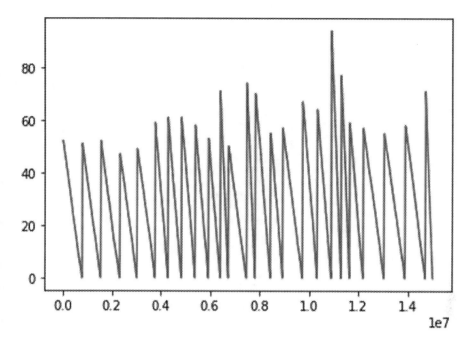

Fig. 2. RUL for Flight Cases 8 and 9. [19]

Due to the lack of maintenance data, failure output data is created with the reduced data. Table 4. shows which damage mechanism affects the sensor or measurement values as follows (Table 1):

Table 1. Sensors that Affect Damage Mechanism

	RPM	Temp	Pre	Flow	phi	Fatigue
Fatigue	+	+	−	−	−	−
Creep	−	+	−	−	−	+
Erosion	−	+	+	+	−	−
Abrasion	−	−	−	+	−	−
Thermomechanical Fatigue	−	+	+	−	+	−
Corrosion	−	+	−	−	+	+

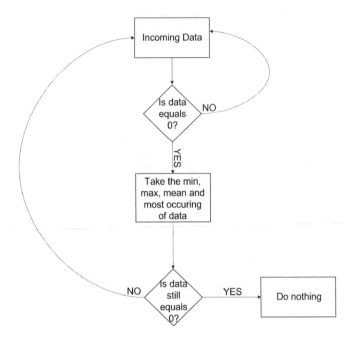

Fig. 3. Flow Chart for Data Reduction Algorithm.

These relations have been gathered from the study of "Taxonomy of Gas Turbine Blade Defects" [20] and other literature surveys. It can be seen in Table 2 that damage mechanisms occurred in components. Due to the lack of output data (which is all zero and because of that A.I. will not learn from it) LPT damage mechanism only consists of fatigue and creep.

Table 2. Damage Types for Components

	Fatigue	Creep	Erosion	Abrasion	Thermomechanical Fatigue	Corrosion
LPC	+	+	+	+	−	−
HPC	+	+	+	+	−	−
Burner	+	+	+	−	+	+
HPT	+	+	+	−	+	+
LPT	+	+	−	−	−	−

According to the relationships in the data set, if the output is one, there is damage, if the output is zero then there is no damage. The classification method was chosen for Machine Learning algorithms because of the probability of damage. All classification techniques have been used and the accuracy scores compared.

3   Results

After data processing is done, the classification technique is chosen for the Machine Learning Algorithm due to quick solution and computer power. Each damage of each component was put into the algorithm separately. 80% of the dataset was reserved for training and the remaining 20% was for testing.

For classification tasks, the mean accuracy can be calculated as the average of the accuracy scores for each class. The formula is:

$$mean\ accuracy = (accuracy\ of\ class\ 1 + ... + accuracy\ of\ class\ n)/n \quad (1)$$

Here, n is the number of classes in the classification problem [21]. Accuracy scores are shown below:

Table 3. Accuracies for LPC

	Fatigue	Erosion	Creep	Abrasion	Mean
LR	.85	.95	1	.95	.9375
KNN	.85	1	1	1	.9625
SVM	.9	.95	1	1	.9625
Kernel SVM	.8	.95	1	1	.9375
Naive Bayes	.8	.75	1	1	.8875
Decision Tree	.8	.9	1	.9	.9
Random Forest	.85	.95	1	1	.95

According to Table 3, SVM performs better than the other models in terms of accuracy, while Naïve Bayes performs worse. This difference can be attributed to the fact that Naïve Bayes assumes that each feature is independent of the others, whereas SVM takes into account how the features interact with one another. Although the results showed that all the features had a value of 1, indicating the possibility of overfitting, the R2 test was repeated to confirm that overfitting was not a problem.

Table 4. Accuracies for HPC

	Fatigue	Erosion	Creep	Abrasion	Mean
LR	.75	1	1	.85	.9
KNN	.95	.95	1	.95	.9625
SVM	.85	1	.95	.95	.9375
Kernel SVM	.8	.9	1	.95	.9125
Naive Bayes	.8	.85	.8	.85	.825
Decision Tree	.95	.9	1	1	.9625
Random Forest	.8	.9	1	.95	.9125

Table 4 shows that KNN has the highest average score, while Naïve Bayes has the lowest. The primary reason for this difference is that KNN is a discriminative classifier, whereas Naïve Bayes is a generative classifier.

Table 5. Accuracies for Burner

	Thermal Fatigue	Fatigue	Erosion	Creep	Corrosion	Mean
LR	.9	.95	.9	1	1	.95
KNN	.9	1	.9	1	.95	.95
SVM	.95	.95	.95	1	1	.97
Kernel SVM	.9	.95	.9	1	1	.95
Naive Bayes	.85	.9	.85	1	1	.92
Decision Tree	.9	.95	.9	1	1	.95
Random Forest	.9	.9	.9	1	1	.94

Table 5 indicates that SVM has the highest average score, while Naïve Bayes has the lowest score, once again. As previously explained, this is due to the different approaches used by these techniques in addressing the problem. While one treats each data point independently, the other considers them as related. It's worth noting that even though the creep scores are all 1, the R2 test has been conducted to confirm that overfitting is not a concern.

Table 6. Accuracies for HPT

	Thermal Fatigue	Fatigue	Erosion	Creep	Corrosion	Mean
LR	.8	1	.95	.9	.8	.89
KNN	.85	1	.95	1	.8	.92
SVM	.85	1	.95	.95	.95	.94
Kernel SVM	.8	1	.95	1	.8	.91
Naive Bayes	.6	1	.95	.75	.75	.81
Decision Tree	.9	1	1	1	.95	.97
Random Forest	.85	1	1	1	.75	.92

Table 6 shows that SVM has the highest average score, while Naïve Bayes has the lowest score, as previously mentioned. This difference can be attributed to the different approaches used by these techniques. While SVM considers how the features interact with each other, Naïve Bayes treats each feature as independent.

Table 7 indicates that the scores for most techniques are similar to each other, except for Naïve Bayes. This is because Naïve Bayes treats each data point as independent, ignoring their relationship to each other. In the LPT section, only fatigue and creep damages are considered due to the possibility of overfitting. Unlike other sections, it is known beforehand that overfitting will occur on damages in the LPT section, prior to the start of the machine learning algorithm.

Table 7. Accuracies for LPT

	Fatigue	Creep	Mean
LR	.95	.95	.95
KNN	.95	1	.975
SVM	.95	1	.975
KernelSVM	.95	1	.975
Naive Bayes	.9	.9	.9
Decision Tree	.95	1	.975
Random Forest	.95	1	.975

In general, SVM has the highest average score of 0.957, or 95.7%. This suggests that the machine learning algorithms were effective in detecting gas turbine damages, achieving high accuracy across all areas. In contrast, Naïve Bayes has the lowest score of 0.8685, or 86.85%. Although this score may be acceptable in other industrial applications, in the energy or aviation sectors, it could lead to catastrophic events.

4   Conclusion

The study showed that machine learning algorithms can accurately detect deterioration in turbomachine parts, with the Support Vector Machine method having the best overall accuracy. This is significant as traditional damage control procedures can be time-consuming and may miss undetected damage, posing risks in the future. Machine learning algorithms offer a way to achieve high accuracy, which can ultimately improve the safety and reliability of turbomachines.

This development is particularly relevant in the aviation industry, where safety is paramount. However, there is a concern that false positive errors could have catastrophic consequences, which may be due to the lack of access to real data. Further research and development could improve the algorithm and reduce the costs and time required for maintenance periods, especially in aviation.

Despite these challenges, the energy sector is expected to benefit from this development as the algorithm could reduce the time needed for damage detection during maintenance work, allowing businesses to resume operations sooner. This could potentially increase their profit margins by reducing downtime and revenue loss.

Overall, the study demonstrates the potential of machine learning algorithms in the field of turbomachines, with further research and development leading to even more effective methods for detecting damage and deterioration. While there are still challenges to overcome, the benefits of this development in the energy and aviation industries are significant.

References

1. Kedir, N., et al.: Erosion in Gas-Turbine grade ceramic matrix composites (CMCs). J. Eng. Gas Turbines Power **141**(1), 1–23 (2019)
2. Hutchings, I.M., Shipway, P.H.: Tribology: Friction and Wear of Engineering Materials, 2nd edn. Butterworth-Heinemann, Oxford (2020)
3. Lai, G. Y.: High-temperature corrosion and materials applications. In: ASM International, pp. 249–257 (2007)
4. Rani, S.: Common failures in gas turbine blade: a critical review. Int. J. Eng. Sci. Res. Technol. **3**, 799–803 (2018)
5. Nowell, D., Duo, P., Stewart, I.F.: Prediction of fatigue performance in gas turbine blades after foreign object damage. Int. J. Fatigue **25**(9–11), 963–969 (2003)
6. Schijve, J.: Fatigue of structures and materials in the 20th century and the state of the art. Int. J. Fatigue **25**(8), 679–702 (2003)
7. Salehnasab, B., Marzbanrad, J., Poursaeidi, E.: Transient thermal fatigue crack propagation prediction in a gas turbine component. Eng. Fail. Anal. **30**, 1–10 (2021)
8. Mazur, Z., Ortega-Quiroz, G. D., Garcia-Illescas, R.: Evaluation of creep damage in a gas turbine first stage blade. In:Proceedings of the ASME 2012 Power Conference (2012)
9. Goodfellow, I., Bengio, Y., Courville, A.: Deep Learning, 1st edn. MIT Press, Boston (2016)
10. Hastie, T., Tibshirani, R., Friedman, J.: The Elements of Statistical Learning: Data Mining, Inference, and Prediction, 2nd edn. Springer, Berlin (2009). https://doi.org/10.1007/978-0-387-21606-5
11. Murty, N.N., Raghava, R.: Unsupervised learning. In: Mukkamala, R., Murty, N. (eds.) Machine Learning for Decision Makers: Cognitive Computing Fundamentals for Better Decision Making. Apress, New York (2019)
12. 10 Machine Learning Methods that Every Data Scientist Should Know. https://towardsdatascience.com/10-machine-learning-methods-that-every-data-scientist-should-know-3cc96e0eeee9. Accessed 8 Apr 2022
13. Regan, T., Beale, C., Inalpolat, M.: Wind turbine blade damage detection using supervised machine learning algorithms. J. Vib. Acoust. **139**(6), 1–14 (2017)
14. Ghalandari, M., Ziamolki, A., Mosavi, A., Shamshirband, S., Chau, K., Bornassi, S.: Aeromechanical optimization of first row compressor test stand blades using a hybridmachine learning model of genetic algorithm, artificial neural networks and design of experiments. Eng. Appl. Comput. Fluid Mech. **13**(1), 892–904 (2019)
15. Zagorowska, M., Spüntrup, F.S., Ditlefsen, A.M., Imsland, L., Lunde, E., Thornhill, N.F.: Adaptive detection and prediction of performance degradation in off-shore turbomachinery. Appl. Energy **268**, 1–17 (2020)
16. Gascon, M., Kumar, N., Ghosh, R.: Predicting power plant equipment life using machine learning. J. Energy Res. Technol. **142**(7), 1–3 (2020)
17. Chao, M.A., Kulkarni, C., Goebel, K., Fink, O.: Aircraft engine run-to-failure dataset under real flight conditions for prognostics and diagnostics. Data **6**(1), 1–14 (2021)

18. PCoE Datasets. https://ti.arc.nasa.gov/tech/dash/groups/pcoe/prognostic-data-repository/. Accessed 8 Apr 2022
19. Frederick, D.K., Decastro, J.A., Litt, J.S.: User's Guide for the Commercial Modular Aero-Propulsion System Simulation (C-MAPSS). Technical Report, NASA, Washington, DC (2007)
20. Aust, J., Pons, D.: Taxonomy of gas turbine blade defects. Aerospace **6**(58), 1–35 (2019)
21. Sai, A.B., Mohankumar, A.K., Khapra, M.M.: A survey of evaluation metrics used for NLG systems. ACM Comput. Surv. **55**(2), 1–39 (2022)

Deep Learning for ECG Signal Classification in Remote Healthcare Applications

Sura Ali Hashim[1]([✉]) and Hasan Huseyin Balik[2]

[1] Electrical and Computer Engineering, Altinbas University, Istanbul, Turkey
sura1464@gmail.com
[2] Computer Engineering, Yildiz Technical University, Istanbul, Turkey

Abstract. Due to several current medical applications, the significance of Electrocardiogram (ECG) classification has increased significantly. To evaluate and classify ECG data, a variety of machine learning methods are now available. Utilizing deep learning architectures, where the top layers operate as feature extractors and the bottom layers are completely coupled, is one of the solutions that has been suggested. In addition to classification results, this work also proposes a learning architecture for ECG classification utilizing 1D convolutional layers and Fully Convolution Network (FCN) layers. We made several changes to get the best result, getting 98% accuracy and 0.2% loss. A comparison has been made and showed that our work is better than other related work. The problem that we found in the rest of the research is the use of less efficient algorithms, so this thing is the reason for the lack of accuracy of the results and an increase in the loss. We used the most efficient algorithm for this work.

Keywords: ECG · CNN · classification · heart arrhythmias

1 Introduction

The main cause of mortality worldwide is heart disease. According to estimates, in 2017, 17.8 million people died from heart disease globally, making up close to 31% of all fatalities. World Health Organization data on human mortality. For effective therapy to work and to lower mortality, early identification of heart disease is essential. Electrocardiography is a low-cost, quick method that helps us understand how the heart works and, in turn, aids in the diagnosis of cardiac disorders. The electrical activity of the heart is recorded by an electrocardiogram (ECG), which a cardiologist uses to detect a variety of illnesses by identifying aberrant cardiac function. However, it takes a lot of time and requires the focus and attention of a qualified specialist to analyze an ECG recording. Using electrodes positioned on the skin. It is important to note that he used his idea for the galvanometer, which measures the strength of electric current. At the time of his drawing, the device was large and difficult to manufacture, but with the passage of time and the introduction of constant improvements, it shrank and became precise in displaying the outcome. Keep a record of your heart's electrical activity. Many cardiac conditions, including arrhythmias, cause changes in the typical ECG pattern. It is also

A. Ortis et al. (Eds.): ICAETA 2023, CCIS 1983, pp. 254–267, 2024.
https://doi.org/10.1007/978-3-031-50920-9_20

possible to monitor how these signals go from the heart to the skin's surface. During an ECG, the fluctuations in electrical signals (or actual voltage) in the different skin layers are measured and graphed. The ECG's resulting graph is known as an electrocardiogram. The ECG measures the depolarization of the heart's muscles, which are all negatively charged. To do this, a collection of positive ions, specifically sodium+ and calcium++, are drawn into the ventricles and cause them to swell. The graphic is either shown on the device screen or created on specialized thermal paper by attaching two electrodes to the sides of the heart. In order to create the layout, more than two electrodes may be used. For instance, one electrode may be placed on the left hand, a second on the right hand, a third on the left leg, and so on. The result is a typical ECG pattern, with the first peak (P wave) illustrating how the electrical impulse (excitation) from the heart travels through the atria. The atria instantly relax after contracting (compressing), pushing blood into the ventricles. The electrical impulses then enter the ventricles. This is demonstrated by the ECG's Q, R, and S waves, or the QRS complex. Ventricles constrict. The ventricles subsequently start to relax again once the electrical impulse ceases propagating, as shown by the T wave [1]. Machine learning, of which deep learning is a subset, only employs neural networks of three layers or more. These neural networks try to function like the human brain, but they fall short, letting the brain "learn" from vast volumes of data. A neural network may be able to approximate predictions with just one layer, but accuracy may be improved by including additional hidden layers. By performing mental and physical activities without requiring human input, deep learning, a technique that serves as the foundation for many artificial intelligence (AI) products and services, encourages automation. Deep learning is used to power both new and old technology, like voice-activated TV remote controls, digital assistants, and credit card fraud detection. Deep learning differs from conventional machine learning in terms of the kind of data it uses and the learning strategies it employs. Machine learning methods employ structured, labeled data to produce predictions, which implies that the model's distinctive properties are established from the input data and organized in tables. This doesn't mean that it doesn't use unstructured data; rather, it only means that, if it does, it usually goes through some pre-processing to organize it. Recent developments have allowed deep learning-based ECG signal classification systems to reach cardiologist-level performance [2]. In terms of precision and memory, the model performed better than the typical cardiologist. Deep convolutional neural networks and sequence-to-sequence models were used by Mousavi et al. [3] to construct an automated heartbeat categorization technique. A lengthy short-term memory and CNN combination was suggested by Murugesan et al. [4]. (LSTM) without any preprocessing, a model-based feature extractor that can be instantly trained. By using an attention mechanism and a recurrent neural network (RNN), Schwab et al. [5] classified a single channel ECG data. For medical professionals to feel confident using computer-assisted electrocardiography, an interpretable model and consistent performance are essential. Although much effort has been done to increase the interpretability of deep learning models in computer vision applications, there Despite recent attention, there hasn't been much progress in the interpretation of ECG classification methods. This paper includes several paragraphs, the first paragraph is the literature review in which we talked about most of the previous research related to our research, then the comparison paragraph, which included a discussion of the research in which we compared our current research, later we talked about the method of work and the algorithms that were used, after that we talked about How to work and the results obtained, and the last paragraph was the conclusion of this work.

2 Related Work

In this part, we have tried to collect the closest works that have a correlation to our proposed work. When Sharda Singh and colleagues applied the method, a sizable amount of standard data, such as ECG time-series data, was used as input to the long-term memory network. Subsets of the data set were created for training and testing. Their method was shown to be effective, accurate, and capable of detecting arrhythmia. Quantitative comparisons with several RNN team models showed an accuracy of 88.1% when they assumed 5 iterations and 3 hidden layers, with 64,256 and 100 neurons for each hidden layer, respectively. This shows that LSTM is superior to RNN and GRU in identifying arrhythmia, whose accuracy is lower than LSTM at 85.4% and 82.5%, respectively. Database with no prior processing carried out. As a result, their model's complexity is substantially lower than that of conventional machine learning techniques. The results of this paper's bilateral categorization of arrhythmias can be enhanced by expanding it to include many categories. The suggested approach produces the same results and leaves room for future research in this area of binary classification (arrhythmia detection), where little significant work has been done. The number of eras can be increased while maintaining classification accuracy. The research illustrates that convolutional neural networks may be used to further classify arrhythmias in the MIT BIH classification dataset, with long-term memory producing the greatest results in binary classification of arrhythmias [13]. The classification of 27 cardiac anomalies using a data collection of 43,101 ECG recordings was the goal of the study that Christian Tronstad and others suggested. A hybrid method that integrates many rule-based deep learning architectures. In this study, the researchers examined two alternative convolutional neural network designs: a completely CNN and an En-coder network, a hybrid of the two that added a second neural network that took into account factors like age and gender. Using derived ECG characteristics, two of these groups were ultimately integrated using a rule-based model. During model development, each of the models was assessed using the validation data [14]. The models are then evaluated on a Challenge validation suite, trained on the supplied development data, and deployed on a Docker image. The best performing models on the challenge validation set were then published and tested on the full challenge test set. A specific challenge score was used to evaluate performance. The best form for their squad, Team UIO, had a complete test score of 0.206 and a challenge validation score of 0.377. We were ranked Based on the outcomes for the whole test group, 20th out of 41 teams made up the official rankings [15]. Minh Huang Nguyen and others an efficient approach for classifying ECGs using 2D convolutional neural networks and ECG images is presented in this study. An ECG recording is transformed into 128×128 grayscale pictures for the MIT-BIH database. Eight different heartbeat types, including a regular pulse and seven abnormal heartbeats, are used to create more than 100,000 ECG pictures. The optimized CNN model was created with key ideas including data augmentation, structure, and K-fold validation in mind. With 0.89 AUC, average accuracy of 96.05%, specificity of 62.57%, sensitivity of 93.85%, and average positive predictive value of 98.55%, this proposed strategy performed well. According to the results of grading arrhythmias on the electrocardiogram, identifying arrhythmias with the use of ECG pictures and a CNN model can be a useful strategy for helping professionals in the diagnosis of cardiovascular illness that can be observed from ECG signals. A medical robot or scanner that can monitor ECG signals and assist medical professionals in more precisely and quickly identifying arrhythmias can also be used to implement the suggested arrhythmia categorization approach. In order to detect the arrhythmia and alert the doctor [16]. In this work, Enbiao Jing suggested a more effective ResNet-18 model for categorizing ECGs. The data was categorized using slicing technology, which aided its pre-processing. The outcomes of the experiment demonstrated that types of arrhythmias may be successfully identified using the enhanced ResNet-18 model. The suggested model also outperformed the most current models that were taken into consideration in terms of classification accuracy, attaining the maximum 96.50%, according to the data. As a result, there are many potential therapeutic applications for the model, which justifies more research and analysis. By changing the loss function and utilizing the weighted loss that arises from batch processing, one way to mitigate the effects of heartbeat class imbalance on model performance is to overweight a small class of losses. Data optimization is a different technique that doubles the data by chopping and dicing the ECG data to enhance training outcomes. Last but not least, to improve the neural network's ability to distinguish small classes of distortions, smaller classes might be given certain features [17]. An effective hybridization method for categorizing electrocardiogram (ECG) samples into key arrhythmia classes to identify irregular heartbeats is presented in Pooja Sharma and colleagues' proposed study. The most frequently used and recognized automated detection technology for keeping track of heart health is the physiological detection utilizing electrocardiogram (ECG) data. Additionally, electrocardiogram (ECG) study focused on elucidating cardiac health state while examining heart rhythm plays a significant role in arrhythmia beat categorization. The authors use discrete wavelet modification to remove the inherent noise of ECG signals during the preprocessing step in order to properly categorize ECG samples into main arrhythmia classifications (DWT). The identification of an ECG signal depends heavily About the QRS complex. Therefore, the position and magnitude of the R peak are calculated to identify the QRS complex. In order to select the collection of the most relevant features, the feature vector is further enhanced using a Cuckoo Search (CS) optimization approach in addition to denoising the signal using DWT. The support vectors trained on the support vector machine (SVM) include the DWT and CS versions by training a feedforward backpropagation neural network (FFBPNN), and the SVM-FFBPNN to classify the signal into five classes. Contains the best training data used for. Various forms of heartbeat are examined using the MIT-BIH arrhythmia database. Heart rate can be calculated with 98.319% accuracy using variant-based classification analysis with feature vectors enhanced using the cuckoo-hunting method and SVM-FFBPNN. In contrast, the FFBPNN variant achieves 97.95 curacy without optimization. Thanks to the improved performance of the new classifier mix, the overall classification accuracy was 98.53%, with precision and recall reaching 98.247% and 95.68%, respectively. The 3600 samples and 1160 heartbeats in the simulation analysis performed better than the existing neural network-based arrhythmia diagnosis. This exemplifies how well the suggested ECG classification model categorizes ECG data in order to categorize arrhythmias [18].

3 Problem Definition

Overserving ECG using a visual method is difficult, time-consuming, costly, and subjective. Due to the complexity of the data amount and clinical content, automatic identification of the arrhythmia in the ECG signal is often a challenging task. Additionally, noise (such as patient movement and disruptions brought on by electrical equipment or infrastructure) often interferes with ECG readings, lowering the quality of the data gathered. The capacity of machine learning (ML) to perform better than conventional classifiers has increased attention in health care systems. In this study, we look into the newest automatic algorithms for identifying aberrant electrocardiograms (ECGs) in a range of cardiac arrhythmias. Choosing which class, the patient's ECG should be allocated is the current ECG classification task. There are four different classes: arrhythmic, highly loud, various types of rhythm, and regular rhythm [19]. The presented data is unbalanced, with 60% of the data falling under the usual sinus rhythm. A quick summary of the data is shown in Table 1. Different processes, such as multiplying an existing ECG for a certain class by moving time values, were utilized to create a balanced dataset. Additionally, there was an attempt to uniformly measure the duration of the ECG using duplicate time-series readings. The input and output layers in the key features of CNN, locally responsive field, shared weights, and pooling are mirrored throughout: The convolutional layer continually learns the entire information from samples while traversing to get many feature maps through weight sharing. It makes use of movable windows that represent sample information pieces (locally acceptable domain). To make its output more straightforward, the pooling layer conducts data compression on the convolutional layer's feature map. The frequently used max-pool method converts the data, removes any values that are not the maximum value in the sample region, and then enhances the method to increase algorithmic robustness. The output of the network is shown in the top complete layer [21].

Table 1. The classification of the data that related to the last works.

part	info	mean	Sd	Maximum	Midd	last
Normal	5154	31.9	10.0	61.0	30	9.0
AF	771	31.6	12.5	60	30	10.0
Other rhythm	2557	34.1	11.8	60.9	30	9.1
Noisy	46	27.1	9.0	60	30	10.2
Total	8528	32.5	10.9	61.0	30	9.0

It is essential to examine patient-provided ECG data before choosing the preprocessing and machine learning technique to be applied, often lasting 30 or 60 s and sampling every 0.003 s. There are several ML that handle time series data and make suitable feature choices. Convolutional neural networks (CNNs) are specific instances of feature extractors, as mentioned in the work's introduction. The developers should no longer be dependent on specialized expertise or custom features thanks to this feature extraction. Rather than eliminating specialist knowledge entirely from the development process, this will speed up the release of the first prototype of ECG classification [23] (Table 2).

Table 2. Literature survey on the ECG classification.

Author	Method	Type of Classification
Sharda Singh et al.	RNN	1 D
Christian Tronstad	CNN	1 D
Huang Minh Nguy	SVM	1 D
Jiaoyang Li1	RNN	1 D
Enbiao Jing	Optimized CNN	1 D

Ten lead wires in a typical ECG provide twelve images of the heart. Your heart rate, rhythm, electrical signal strength, and timing may all be determined via an ECG. Numerous heart-related diseases can be assessed with the test. Because there is no electrical activity and an unequal distribution of ions across the cell membranes while the cardiac muscle cells are at rest, they are said to be depolarized. Ions like sodium $(Na+)$, potassium $(K+)$, and calcium $(Ca2+)$ are at resting potential and have varying concentrations both within and outside the cell. These ions cross the cell membrane in response to an electrical impulse, resulting in a depolarization or action potential. The heart contracts as a result of depolarization. Regular heartbeats and blood flow throughout the body are maintained by the depolarization and repolarization of the heart. The heart has two atria and two ventricles, making up its four chambers. The atria and ventricles' depolarization and repolarization in succession are represented as waves in the ECG. P Wave: Atrial Depolarization Associated Small Deflection Wave The PR interval. The T-wave represents the ventricular repolarization waveform. By blocking atrial repolarization, the QRS complex. The QRS-to-T wave interval (QT interval) is the period of time between these two waves. The cycle from ventricular depolarization to ventricular repolarization is represented by it as shown in Fig. 1 [22].

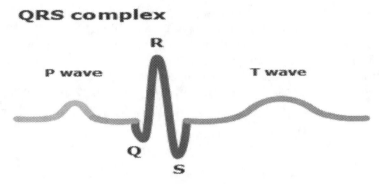

Fig. 1. The typical ECG pattern.

4   The Proposed Solution

This specific situation of ECG is different from the standard picture recognition task for which CNN is used. Contrary to time series, which normally use 1D data, data is always shown in the latter case as 2D data with a few color channels. Whatever the situation, a CNN with the right architecture, which is dependent on the dimensionality and structure of the data, may successfully complete a classification assignment. In this case, CNN 1D is utilized in conjunction with the following buildings: GlovalAveragePooling.

5   The Method of CNN

The neural network in general is designed just like the neurons of the human brain, each in the form of a programmer. Where it contains several cells and nodes, each of which has its function to obtain results that match our brains, more accurate results. These layers are called neural layers. When we talk about deep learning, we are talking about neural networks. We start by talking about the inputs or how the neural network works in general and how to enter data into the network. The inputs are entered on the layers in the neural network and have a certain weight. Each node can be affected by multiple weights because the weights are assigned to the links between these nodes. The neural network takes all the training data in the input layer and then passes this data to the hidden layers, after which it is transformed. These values are based on the weights of each node and finally give us the result in the output layer. Choosing the neural network is very important to determine which features are the most important to use for the model. Also, the process of training data within the neural network can take a long time in order to get better, reliable and coordinated results. A convolutional neural network is an artificial intelligence algorithm, a specific type of neural network that uses a CNN for classification, such as image classification. It has multiple layers. Through these layers, a more accurate and clear result is obtained. Where the CNN is that it helps you get images without pre-processing them. Since the convolutional neural network works accurately and intelligently, thanks to its neural layers, it does not need many parameters to give us the result we are looking for. It also does not take much time to do so. As a result, CNN knows the most important properties of the filters and considers the optimal handling of the filter. The CNN algorithm is very great for dealing with a somewhat huge dataset and working with high-resolution images that contain thousands of pixels because it simply transforms this data into models that are easy to process without losing important features to know what this data represents. One might ask what the difference between a normal neural network and a convolutional neural network is, or what is the importance of having convolutions in a CNN. Scientifically and practically, convolutions deal with math in network programming, but behind the scenes, where convolutions take two functions instead of matrix multiplication in at least one layer of the network. Whereas convolutions return the function instead. What makes CNN so special is that it knows very well how to handle filters and how to set them. Speaking of filters, we will explain what they are in general terms. Filters help us get a better result for our work, as filters are associated with the input. This helps us improve accuracy and reduce loss. This is when convolutions handle these filters well, as they have a different effect on the result, such as opacity, or removing noise from an image. One of the things that can improve the work of CNN is the use of data, as the more training data the better, the better the network will be. Choosing the data to be classified and well formatted will help the network to train better and faster and give us an accurate result as shown in Fig. 2.

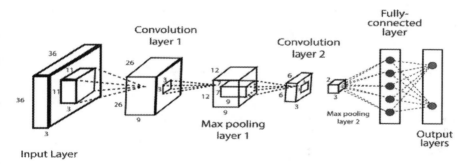

Fig. 2. How maxpooling works.

5.1 A Convolutional Neural Network Works

As mentioned, a convolutional neural network works like a human brain and consists of nodes. It is these nodes that represent the neurons in the network that are like the cells of the human brain. Nodes are specific functions in the network that calculate the weight of the interstitial bits of the network and return the activation map. When the input is an image defined here, these nodes take the pixels from that image and select some visual feature for it, such as colors, and will give you the activation result as a result of this function. Usually the first thing CNN takes from this image is the edges as the definition of the input image, and this definition is passed to the next layer, and the next layer begins to detect other, more subtle parts of the image., such as angles and color combinations, and it is passed to the next layer, and so on until it reaches the last layer, which is the final layer, and this process is called classification. The last layer of CNN that determines the result for us is classification. The meaning of classification in general is the real knowledge of what is inside this image. It is possible for a person, when he looks with his eyes at a certain image, to classify his mind what is the thing he is looking at through the brain cells that he has. He can know if the thing he is looking at is a human, cat, dog, etc. CNN also works in the same way. For example, self-driving vehicles work when they recognize the object, is it a human or another car. For convolutional neural network layers, the first layer is the convolutional layer, which was mentioned in detail, followed by the max pooling layer. CNN has a MaxPooling setting that can be used to improve the performance of networks with high latency. The pooling layer is designed to aggregate features (local image patterns) into a fixed size feature map at each location in the feature space of the input images. These output features are stored in an array that has the same shape as the input images. The goal of the pooling layer is to reduce the spatial dimensionality of the network by extracting the most meaningful information from each part of the input image. As a rule of thumb, the more layers you have, the more computationally expensive your algorithm will be. So, reducing the number of parameters in the network can make it more efficient for large-scale applications. Typically, this is done by selecting fewer weights for each neuron, but this comes at the cost of accuracy. Max Pooling: selecting only the largest activation value at each node in each channel reduces the number of parameters in the network without affecting the accuracy. This reduces the number of parameters in the network without changing the overall structure of the network as shown in Fig. 3.

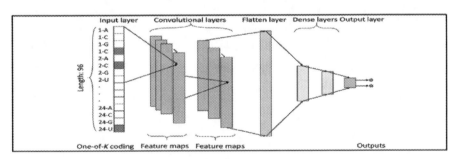

Fig. 3. Convolutional neural network.

The fully connected layer is one of the CNN layers that comes after the pooling layer. The task of this layer is to connect all the inputs in the input vectors with the outputs in the output vector, because not all nodes are connected to each other in the convolutional layer. This means that all the input nodes will connect to the output nodes in the FC layer. This layer consists of weights and biases with neurons and is used to connect neurons with two different networks. The term "fully connected" means that each neuron of the previous layer is connected to the current layer. The number of neurons in a fully connected layer cannot in any way be related to the number of units in the previous layer. You can even place a single fully connected neuron after a layer of 10,000 neurons. The major advantage of fully connected networks is that they are "structure agnostic" i.e., there are no special assumptions needed to be made about the input. While being structure agnostic makes fully connected networks very broadly applicable, such networks do tend to have weaker performance than special-purpose networks tuned to the structure of a problem space [11].

5.2   The Analysis of the Result and Discussion

We utilize a CPU i7 and GPU NVIDIA GeForce GTX for all computational research. Tables 3, 4, and 5 offer a comparison of the results in terms of accuracy scores for the various models. In all of the trials, our model produces the best outcomes. Table 3 demonstrates that, when compared to the other models, our model performed the best, particularly when the parameter no was changed.

Table 3. Dropout effectiveness during the experiments.

Drop out l	Loss	Accuracy
0.8	1.05	0.80
0.75	1.00	0.81
0.7	0.99	0.83
0.63	0.93	0.84
0.19	0.5	0.89

A number of samples of cardiac patients were tested in order to improve the accuracy of the ECG by using the MIT-BIH database by entering it into the layers of the convolutional neural network, where in turn, through neurons, it improves accuracy and reduces loss, and this, in turn, helps cardiologists and recognize as soon as possible on the disease. Table 3 shows the effect of dropout on the results. Dropout is an amazingly popular way to overcome overfitting in a neural network.

Table 4. Maxpoling effectiveness during the experiments.

Maxpooling ld	Loss	Accuracy
16,8	0.55	0.87
14,6	0.5	0.88
13,5	0.5	0.88
13,8	0.5	0.89

One of the most important design aspects of convolutional neural networks is maxpooling (CNN). The main purpose of the pooling layer, a layer of CNN, is to gradually reduce the spatial size of the representation to minimize the number of parameters and computations in the network. Table 4 shows how the pooling layer affected the results. We can see that the smaller value of the pooling layer resulted in higher accuracy and better results.

Table 5. Regpart effectiveness during the experiments.

Regpar 1	Regpar2	Max pooling 1	Max pooling2	Max pooling3	loss	accuracy
0.002	002	31	13	8	0.5	0.88
0.0001	.0001	31	13	8	0.5	0.93
0.00001	.00001	31	13	8	0.3	0.94
0.00001	.00001	32	14	9	0.4	0.92
0.00001	.00001	10	10	7	0.2	0.98

Table 5 shows our latest results. As shown in the table, we made many changes in the maxpooling layer and the regpar. We achieved the best result of accuracy, which is 98%, and the loss is 0.2%. Regpar is an acronym for "Receptive Field Paring" which is what is found in a convolutional neural network. Regpar is the technique you sometimes use when training a CNN on images that have a lot of noise. The idea behind regpar is that you want to filter out all the noise in the input image so that you are left with only the relevant parts of the image.

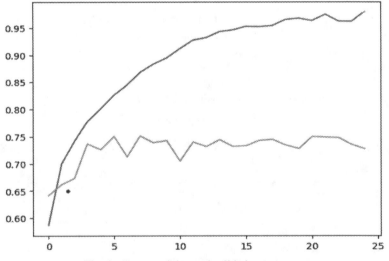

Fig. 4. Shows training and validation accuracy.

Figure 4 shows the training data and validation accuracy, if the training data is unstructured, unbalanced, and can be represented as a 1D time series with typical time lengths for ambulatory monopolar ECG devices, the proposed deep learning solution is It can be used for the task of classifying ECGs according to their results. The classification obtained. The ability of the algorithm to extract features is also very useful when, for various reasons, there are no medical or related experts available for feature engineering. The shortcomings of this study can be characterized as a poor comparison with other DL solutions in terms of processing costs, recommended designs, and optimization methods.

Table 6. A Comparison of different classifications techniques.

Method Signal	Classification Accuracy	Method Signal
ANN & PCA [6] & NCA [29]	ECG	88.5
KNN classifier [30]	HRV	90.4
SVM Classifier	HRV	96
Fuzzy KNN [31]	RNN	92.5
Our proposed method	ECG	98

Table 6 shows a comparison of different classifications techniques. The proposed deep learning solution can be used to classify electrocardiogram (ECG) data when the training data is unstructured and unbalanced and is comparable to standard single-channel portable ECG devices. It can be represented as a 1D time series with time length. The classification obtained. The feature extraction algorithm's ability is also particularly useful when feature engineering is not available to medical or related professionals for a variety of reasons. A drawback of this study is that it does not compare well with other DL solutions in terms of processing costs, design recommendations, and optimization methods.

6   Conclusion and Future Work

The suggested technique is concluded using the preprocessing, feature extraction and classification outcomes. The advantages of the learning algorithm include its infinite ability to make continuous, real-time diagnosis of heart arrhythmias, which helps patients who live in rural areas or in underdeveloped countries and cannot access heart disease care. The person using the circadian rhythm monitor will react immediately and alert emergency personnel to assist the person(s) in diagnosing an occupational arrhythmia when a potentially fatal heart rhythm manifests itself in high-risk groups. Through convolutional neural network layers, ECG devices have been developed, helping doctors in early detection of disease. We used samples from patients with heart disorders in the MIT-BIH database and inserted them into CNN layers. So, we made several changes to get the best result, getting 98% accuracy and 0.2% loss. Our method showed the best performance and great result compared to other related works. Our future work will examine the proposed method using the 2D dataset.

References

1. Le, K.H., et al.: Enhancing deep learning-based 3-lead ECG classification with heartbeat counting and demographic data integration. arXiv preprint arXiv:2208.07088 (2022)
2. Rajkumar, A., Ganesan, M., Lavanya, R.: Arrhythmia classification on ECG using Deep Learning. In: 2019 5th International Conference on Advanced Computing & Communication Systems (ICACCS). IEEE (2019)
3. Gao, X.: Non-invasive detection and compression of fetal electrocardiogram. Interpreting Cardiac Electrograms – From Skin to Endocardium 53–74 (2017). https://doi.org/10.5772/intechopen.69920
4. Gao, Y.: Deep learning based automatic detection of arrhythmia and its applications. Thesis in Shandong University of Science and Technology, pp. 1–61 (2016)
5. Hu, S., Wei, H.X., Chen, Y.D., Tan, J.D.: A real-time cardiac arrhythmia classification system with wearable sensor networks. Sensors **12**, 12844–12869 (2012). https://doi.org/10.3390/s120912844
6. Hu, X., Yu, Z.B.: Diagnosis of mesothelioma with deep learning. Oncol. Lett. 1–8 (2018). https://doi.org/10.3892/ol.2018.9761
7. Hsing, J.M., Hsia, H.H.: Cardiac arrhythmias. In: Criner, G., Barnette, R., D'Alonzo, G. (eds.) Critical Care Study Guide, pp. 341–374. Springer, New York (2010). https://doi.org/10.1007/978-0-387-77452-7_19
8. Isin, A., Ozdalili, S.: Cardiac arrhythmia detection using deep learning. Procedia Computer Science **120**, 268–275 (2017)
9. Jin, L.P., Dong, J.: Deep learning research on clinical electrocardiogram analysis. China Sci. Inf. Sci. **45**(3), 398–416 (2015)
10. Jin, L.P., Dong, J.: Normal versus abnormal ECG classification by the aid of deep learning. In: Artificial Intelligence – Emerging Trends and Applications, pp. 295–315. InTech Open (2018)
11. Lee, Y.N., Kwon, J.M., Lee, Y.H., Park, H.H., Cho, H., Park, J.S.: Deep learning in the medical domain: predicting cardiac arrest using deep learning. Acute Critical Care **33**(3), 117–120 (2018)
12. Kalra, A., Lowe, A., Al-Jumaily, A.: Critical review of electrocardiography measurement systems and technology (2018). https://doi.org/10.1088/1361-6501/aaf2b7
13. Ebrahimi, Z., et al.: A review on deep learning methods for ECG arrhythmia classification. Expert Syst. Appl. **X7**, 100033 (2020)
14. Singh, S., et al.: Classification of ECG arrhythmia using recurrent neural networks. Procedia Comput. Sci. **132**, 1290–1297 (2018)
15. Singstad, B.-J., Tronstad, C.: Convolutional neural network and rule-based algorithms for classifying 12-lead ECGs. In: 2020 Computing in Cardiology. IEEE (2020)
16. Jun, T.J., et al.: ECG arrhythmia classification using a 2-D convolutional neural network. arXiv preprint arXiv:1804.06812 (2018)
17. Jing, E., et al.: ECG heartbeat classification based on an improved ResNet-18 model. Comput. Math. Methods Med. **2021** (2021).
18. Sharma, P., Dinkar, S.K., Gupta, D.V.: A novel hybrid deep learning method with cuckoo search algorithm for classification of arrhythmia disease using ECG signals. Neural Comput. Appl. **33**(19), 13123–13143 (2021)
19. Liu, X., et al.: Deep learning in ECG diagnosis: a review. Knowl.-Based Syst. **227**, 107187 (2021)
20. Obeidat, Y., Alqudah, A.M.: A hybrid lightweight 1D CNN-LSTM architecture for automated ECG beat-wise classification. Traitement du Signal **38**(5) (2021)

21. Nguyen, T., et al.: Detecting COVID-19 from digitized ECG printouts using 1D convolutional neural networks. PLoS ONE **17**(11), e0277081 (2022)
22. Butun, E., et al.: 1D-CADCapsNet: One dimensional deep capsule networks for coronary artery disease detection using ECG signals. Physica Med. **70**, 39–48 (2020)
23. Chen, C.Y., et al.: Automated ECG classification based on 1D deep learning network. Methods **202**, 127–135 (2022)
24. Hasan, Md.A., et al.: Cardiac arrhythmia detection in an ECG beat signal using 1D convolution neural network. In: 2020 IEEE Region 10 Symposium (TENSYMP). IEEE (2020)
25. Giannakakis, G., et al.: A novel multi-kernel 1D convolutional neural network for stress recognition from ECG. In: 2019 8th International Conference on Affective Computing and Intelligent Interaction Workshops and Demos (ACIIW). IEEE (2019)
26. Xiaolin, L., Cardiff, B., John, D.: A 1D convolutional neural network for heartbeat classification from single lead ECG. In: 2020 27th IEEE International Conference on Electronics, Circuits and Systems (ICECS), pp. 1–2. IEEE (2020)
27. Ribeiro, A.H., et al.: Automatic diagnosis of the 12-lead ECG using a deep neural network. Nat. Commun. **11**(1), 1–9 (2020)
28. Venkatesan, C., et al.: ECG signal preprocessing and SVM classifier-based abnormality detection in remote healthcare applications. IEEE Access **6**, 9767–9773 (2018)
29. Rai, H.M., Trivedi, A., Shukla, S.: ECG signal processing for abnormalities detection using multi-resolution wavelet transform and Artificial Neural Network classifier. Measurement **46**(9), 3238–3246 (2013)
30. Saini, I., Singh, D., Khosla, A.: QRS detection using K-nearest neighbor algorithm (KNN) and evaluation on standard ECG databases. J. Adv. Res. **4**(4), 331–344 (2013)
31. Krishnaiah, V., Narsimha, G., Chandra, N.S.: Heart disease prediction system using data mining technique by fuzzy K-NN approach. In: Satapathy, S., Govardhan, A., Raju, K., Mandal, J. (eds.) Emerging ICT for Bridging the Future - Proceedings of the 49th Annual Convention of the Computer Society of India (CSI) Volume 1. AISC, vol. 337, pp. 371–384. Springer, Cham (2015). https://doi.org/10.1007/978-3-319-13728-5_42

Computerized Simulation of a Nonlinear Vibration Sandwich Plate Structure with Porous Functionally Graded Materials Core

Zuhair Alhous[1], Muhannad Al-Waily[1(✉)], Muhsin J. Jweeg[2], and Ahmed Mouthanna[3]

[1] Department of Mechanical Engineering, Faculty of Engineering, University of Kufa, Kufa, Iraq
muhanedl.alwaeli@uokufa.edu.iq
[2] College of Technical Engineering, Al-Farahidi University, Baghdad, Iraq
[3] Department of Medical Instrumentation Engineering Techniques, Al-Maarif University College, Al-Ramadi, Iraq

Abstract. This research presents a novel approximation accurate value of analysis of the nonlinear vibration to evaluate the frequency of sandwich plates that have both functionally graded parts and porosities. Kinematic relations are created and controlled differential equations by making use of the first ordinal differential shear deformation theory. It is assumed that the FGM plates are made of an isotropic material with a porosity distribution consistent over its whole surface. Under the power-law scheme, the only direction in which the qualities of the material fluctuate smoothly is in the direction of thickness. Various variables, including gradient indices, boundary conditions, distribution of porosity and geometrical attributes, are subjected to analyses to determine the effect these alterations have on the parameter of nonlinear vibration of sandwich plates with FG surfaces. Employing the FOSD theory, researchers conduct a thorough numerical examination. The results obtained with a variety of boundary conditions illustrate the distribution of porosity and the nonlinear vibration properties exhibited by FG sandwich plates. The evidence presented here demonstrated that the FOSD theory and the approximation technique had a reasonable agreement with one another. In this work, the FOSD theory was used by deriving the equations for extracting the natural frequency and response of the nonlinear vibration of the functionally graded materials with porosity distribution for one material.

Keywords: Sandwich Plate · Functionally Graded · First-Order Shear Deformation Theory · Nonlinear Dynamic Response · Porous

1 Introduction

Functionally graded materials sometimes referred to as FGMs, are composites that are microscopically graded according to a specific function, inhomogeneous and have mechanical and thermal characteristics that gradually change from one surface to the next. By progressively altering the volume percentage of the component elements, these materials are often fabricated from a combination of metal and ceramics or various metals. The characteristics of FGM vFGM'sous section are expected to change as one moves through the structure's multiple thicknesses. FGMs have different practical uses because of their strong heat resistance. Some examples of these applications are reactor vessels, aeroplanes, space vehicles, military industries, and other technical constructions. Consequently, a great deal of research on the dynamics and nonlinear vibration of the FGM section has been carried out during the last several years.

Qingya Li et al., 2018 [1] Nonlinear vibration and dynamic buckling of a GPL-SFGP plate are studied. Various compressive loading rates at one edge provide axial compressive tension. To determine the plate's dynamic stability, numerical tests are carefully devised. Mirjavadi, 2020 [2], This paper examines nonlinear unconstrained vibrations of porous FGM velar spherical shell slices with flexible circumferential stiffeners. Porous FG material is represented using an advanced power-law function and has uniform and uneven porosities. Ghobadi, et al., 2020 [3] In this study, the influence distribution of porosity on the dynamic with static nonlinear responses of a sandwich.

Structure with thermal, electrical, and elasticity coupling is discussed. The impact of material, geometrical, and boundary conditions on nanostructure mechanical responses was studied. Trinh and Kim, 2021 [4] Study examines deterministic and stochastic sandwich plate dynamics under thermomechanical stresses. The shear deformation theory for 3d-order and Hamilton's rules provide Euler-Lagrange equations. Numerically calculated and visually displayed dynamic plate responses, backbone, and forced response curves. Pradhan and Sarangi, 2021[5], This research examine the nonlinear damping of functionally graded (FG) piezoelectric composite plates. The Golla-Hughes-McTavish approach models the ACLD patch's restricted layer, which is viscoelastic. Liu et al. 2021Attemptmpts to analyze the bifurcation structure and nonlinear vibration response of micro-voided FG piezoelectric shells. The power-law exponent, the porosity volume percentage, the temperature change, and the external stimulation all play significant roles, as do the results of the experiment. Jamalabadi et al., 2021 [7] presented a nonlinear vibration analysis of FGPL-RC cylindrical panels on elastic media.

The two-dimensional differential quadrature technique, arc-length prolongation, and harmonic balancing are used to determine the frequency response. Zaitoun et al., 2022 [8] A sandwich plate consists of three separate layers. Using Navier's idea, we could determine the temperature at which buckling would occur in an axially loaded sandwich plate. Other factors studied include the absorption factor, interpolation method, moisture condition, energy index, and temperature change. Allam et al., 2022 [9] Graphene-on-metal sandwich cylindrical shell with an auxetic honeycomb core. Around its axial axis, the sandwich shell should maintain a constant angular speed while it rotates. Using first-order shear deformation, the displacement field has been characterized. Briquetto and Torre 2022 [10] In this study, the effects of hygrometric stress on multilayered sandwich plates and shells are investigated. The sample is predicated on a generic and perfect three-dimensional shell law. The finding highlights how critical it is to construct the flexible section of the three-dimensions shell sample. Njim, 2021–2022, [11–17] This study introduces a novel analytical model for a sandwich plate's vibration and buckling analysis. The sandwich plate is fabricated from a metal that is porous, while an isotropic metal acts as the outer skin of the plate. It has been determined that porosity coefficients have substantial impacts on vibration and buckling behavior. The study included calculating the analytical model for FGM porosity of vibration and buckling sandwich plate by driving the general equation of motion and buckling equation, from the vibration analysis calculating the natural frequency of plate structure with different porosity parameters and FGM index effect. Also, the authors investigated the vibration and buckling calculating behavior using experimental and numerical.

This work studies the porous functionally graded sandwich plate under nonlinear vibration conditions. Whose mechanical characteristics are varied due to different porosity distributions and changed in the thickness direction based on power-law distributions. The effect of core materials, porosity coefficients, gradient exponents, and geometric parameters on nonlinear vibration characteristics loads is presented and analyzed. A new model of first-order shear deformation theory is formed to discover the nonlinear vibration characteristics according to different FGM parameters.

2 Governing Equations

Considering a thick sandwich FGM plate of a rectangular section composed of both face sheet layers and porous metal core based on the power-law distribution as shown in Fig. 1. The volume percent ceramic (Vc) through-thickness direction of the FG portion can be represented by [18]:

$$V_m + V_c = 1, \quad V_c = V_c(z) = \left(\frac{2z + h}{2h}\right)^N \tag{1}$$

The following are the effective mechanical characteristics of FGMs across the plate's thickness when taking constituent distribution into account [19]:

$$P(z) = P_m - eP_m \left(\frac{2z + h}{2h}\right)^N \tag{2}$$

The plate is described using the Cartesian coordinates x, y, and z, where z is the thickness coordinator and x, y is the midplane of the plate (Fig. 1). The plate is a, b, and h in length, width, and overall thickness, respectively (Fig. 1).

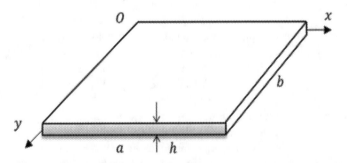

Fig. 1. Cartesian coordinate system

The plate's forces and moments can be expressed as follows [20]:

$$N_x = \int_{-\frac{h}{2}}^{\frac{h}{2}} \sigma_x dz; \ N_y = \int_{-\frac{h}{2}}^{\frac{h}{2}} \sigma_y dz; \ N_{xy} = \int_{-\frac{h}{2}}^{\frac{h}{2}} \sigma_{xy} dz; \ M_x = \int_{-\frac{h}{2}}^{\frac{h}{2}} \sigma_x z dz; \ M_y = \int_{-\frac{h}{2}}^{\frac{h}{2}} \sigma_y z dz$$

$$M_{xy} = \int_{-\frac{h}{2}}^{\frac{h}{2}} \sigma_{xy} z dz; \ (Q_x, Q_y) = \int_{-\frac{h}{2}}^{\frac{h}{2}} \left(\sigma_{xz}, \sigma_{yz}\right) dz$$

$$N_x = \frac{E_1}{1-\upsilon^2}\varepsilon_x^\circ + \frac{\upsilon E_1}{1-\upsilon^2}\varepsilon_y^\circ - \frac{E_2}{1-\upsilon^2}\lambda_x - \frac{\upsilon E_2}{1-\upsilon^2}\lambda_y$$

$$N_y = \frac{\upsilon E_1}{1-\upsilon^2}\varepsilon_x^\circ + \frac{E_1}{1-\upsilon^2}\varepsilon_y^\circ - \frac{\upsilon E_2}{1-\upsilon^2}\lambda_x - \frac{E_2}{1-\upsilon^2}\lambda_y$$

$$N_{xy} = \frac{1}{2(1+\upsilon)}\left(E_1\gamma_{xy}^\circ - 2E_2\lambda_{xy}\right) \qquad (3)$$

$$M_x = \frac{E_2}{1-\upsilon^2}\varepsilon_x^\circ + \frac{\upsilon E_2}{1-\upsilon^2}\varepsilon_y^\circ - \frac{E_3}{1-\upsilon^2}\lambda_x - \frac{\upsilon E_3}{1-\upsilon^2}\lambda_y$$

$$M_y = \frac{\upsilon E_2}{1-\upsilon^2}\varepsilon_x^\circ + \frac{E_2}{1-\upsilon^2}\varepsilon_y^\circ - \frac{\upsilon E_3}{1-\upsilon^2}\lambda_x - \frac{E_3}{1-\upsilon^2}\lambda_y$$

$$N_{xy} = \frac{1}{2(1+\upsilon)}\left(E_2\gamma_{xy}^\circ - 2E_3\lambda_{xy}\right)$$

$$Q_x = K_s I_{30}\gamma_{xz}; \ Q_y = K_s I_{30}\gamma_{yz}$$

The equations calculate and provide the evident analytical equations of E_i ($i = 1, 2, 3$) in the Duc and Tung [20]. The equations of motion are given by the first-order shear deformation plate theory [21]:

$$u: \frac{\partial N_x}{\partial x} + \frac{\partial N_{xy}}{\partial y} = I_0\frac{\partial^2 u}{\partial t^2} + I_1\frac{\partial^2 \phi_x}{\partial t^2},$$

$$dv: \frac{\partial N_{xy}}{\partial x} + \frac{\partial N_y}{\partial y} = I_0\frac{\partial^2 v}{\partial t^2} + I_1\frac{\partial^2 \phi_y}{\partial t^2},$$

$$dw: \frac{\partial Q_x}{\partial x} + \frac{\partial Q_y}{\partial y} + N_x\frac{\partial^2 w}{\partial x^2} + 2N_{xy}\frac{\partial^2 w}{\partial x \partial y} + N_y\frac{\partial^2 w}{\partial y^2} + q = I_0\frac{\partial^2 w}{\partial t^2}, \qquad (4)$$

$$d\phi_x: \frac{\partial M_x}{\partial x} + \frac{\partial M_{xy}}{\partial y} - Q_x = I_2\frac{\partial^2 \phi_x}{\partial t^2} + I_1\frac{\partial^2 u}{\partial t^2},$$

$$d\phi_y: \frac{\partial M_{xy}}{\partial x} + \frac{\partial M_y}{\partial y} - Q_y = I_2\frac{\partial^2 \phi_y}{\partial t^2} + I_1\frac{\partial^2 v}{\partial t^2}$$

where

$$I_i = \left(\int_{-\frac{h_{FG}+h_L}{2}}^{-\frac{h_{FG}}{2}} \rho(z) + \int_{-\frac{h_{FG}}{2}}^{\frac{h_{FG}}{2}} \left(P_m - eP_m\left(\frac{2z+h}{2h}\right)^N\right) + \int_{\frac{h_{FG}}{2}}^{\frac{h_{FG}+h_U}{2}} \rho(z)\right)\left(1, z, z^2\right) dz \qquad (5)$$

The presentation of the function of stress f (x, y, t) looks like this:

$$N_x = \frac{\partial^2 f}{\partial y^2}, \ N_y = \frac{\partial^2 f}{\partial x^2}, \ N_{xy} = -\frac{\partial^2 f}{\partial x \partial y} \qquad (6)$$

Replacing Eq. 6 into 4a and then into Eq. 4 can be rewritten as follows:

$$dw: \frac{\partial Q_x}{\partial x} + \frac{\partial Q_y}{\partial y} + N_x\frac{\partial^2 w}{\partial x^2} + 2N_{xy}\frac{\partial^2 w}{\partial x \partial y} + N_y\frac{\partial^2 w}{\partial y^2} + q = I_0\frac{\partial^2 w}{\partial t^2}$$

$$d\phi_x: \frac{\partial M_x}{\partial x} + \frac{\partial M_{xy}}{\partial y} - Q_x = \left(I_2 - \frac{I_1^2}{I_0}\right)\frac{\partial^2 \phi_x}{\partial t^2} \qquad (7)$$

$$d\phi_y: \frac{\partial M_{xy}}{\partial x} + \frac{\partial M_y}{\partial y} - Q_y = \left(I_2 - \frac{I_1^2}{I_0}\right)\frac{\partial^2 \phi_y}{\partial t^2}$$

The following is a possible representation of the plate's compatibility equation:

$$\frac{\partial^2 \varepsilon_x^\circ}{\partial y^2} + \frac{\partial^2 \varepsilon_y^\circ}{\partial x^2} - \frac{\partial^2 \gamma_{xy}^\circ}{\partial x \partial y} = \frac{\partial^2 w^2}{\partial x \partial y} - \frac{\partial^2 w}{\partial x^2}\frac{\partial^2 w}{\partial y^2} \qquad (8)$$

When Eq. (3) is substituted into Eq. (7), the motion equation is modified in the following manner:

$$R_{11}(w) + R_{12}(\phi_x) + R_{13}(\phi_y) + S_1(w, f) + q = I_0 \frac{\partial^2 w}{\partial t^2},$$

$$R_{21}(w) + R_{22}(\phi_x) + R_{23}(\phi_y) + S_2(f) = \left(I_2 - \frac{I_1^2}{I_0}\right)\frac{\partial^2 \phi_x}{\partial t^2},$$

$$R_{31}(w) + R_{32}(\phi_x) + R_{33}(\phi_y) + S_3(f) = \left(I_2 - \frac{I_1^2}{I_0}\right)\frac{\partial^2 \phi_y}{\partial t^2}$$

$$(9)$$

where

$$R_{11}(w) = K_s I_{30}\frac{\partial^2 w}{\partial x^2} + K_s I_{30}\frac{\partial^2 w}{\partial y^2} - K_1 w + K_2\left(\frac{\partial^2 w}{\partial x^2} + \frac{\partial^2 w}{\partial y^2}\right)$$

$$R_{12}(\phi_x) = K_s I_{30}\frac{\partial \phi_x}{\partial x}; R_{13}(\phi_y) = K_s I_{30}\frac{\partial \phi_y}{\partial y}; S_1(w, f) = \frac{\partial^2 f}{\partial x^2}\frac{\partial^2 w}{\partial x^2} - 2\frac{\partial^2 f}{\partial x \partial y}\frac{\partial^2 w}{\partial x \partial y} + \frac{\partial^2 f}{\partial x^2}\frac{\partial^2 w}{\partial y^2}$$

$$R_{21}(w) = -K_s I_{30}\frac{\partial w}{\partial x}; R_{22}(\phi_x) = D_{11}\frac{\partial^2 \phi_x}{\partial x^2} + D_{66}\frac{\partial^2 \phi_y}{\partial y^2} - K_s I_{30}\phi_x,$$

$$R_{23}(\phi_y) = (D_{12} + D_{66})\frac{\partial^2 \phi_y}{\partial x \partial y}; S_2(f) = B_{21}\frac{\partial^3 f}{\partial x^3} + (B_{11} - B_{66})\frac{\partial^3 f}{\partial x \partial y^2},$$

$$R_{31}(w) = -K_s I_{30}\frac{\partial w}{\partial y}; R_{32}(\phi_x) = (D_{21} + D_{66})\frac{\partial^2 \phi_x}{\partial x \partial y},$$

$$R_{33}(\phi_y) = D_{22}\frac{\partial^2 \phi_y}{\partial y^2} + D_{66}\frac{\partial^2 \phi_y}{\partial x^2} - K_s I_{30}\phi_y; S_3(f) = B_{12}\frac{\partial^3 f}{\partial y^3} + (B_{22} - B_{66})\frac{\partial^3 f}{\partial x^2 \partial y},$$

$$(10)$$

3. In the next part, analysis of nonlinear dynamical of a thick sandwich made of plates with a porous FG core will be discussed. This analysis will make use of a set of Eqs. (8–9), together with boundary conditions.

3 Nonlinear Dynamical Analysis

The behavior of the edge in the plane indicates that the plate must be simply supported on all of its edges. The approximate solutions to the systems of Eqs. (8 and 9) that meet the boundary conditions for the present investigation may be expressed as follows:

$$w(x, y, t) = W(t) \sin \lambda_m x \sin d_n y$$
$$\phi_x(x, y, t) = \phi_x(t) \cos \lambda_m x \sin d_n y$$
$$\phi_y(x, y, t) = \phi_y(t) \sin \lambda_m x \cos d_n y$$
$$\tilde{f}(x, y, t) = \tilde{A}_1(t) \cos 2\lambda_m x + \tilde{A}_2(t) \cos 2d_n y + \tilde{A}_3(t) \sin \lambda_m x \sin d_n y$$

$$(11)$$

The coefficients of the stress function Ai (i = 1, 2, 3) are computed by substituting Eq. (11) into Eq. 8 as follows:

$$
\begin{aligned}
&f(x, y, t) = \tilde{A}_1(t) \cos 2\lambda_m x + \tilde{A}_2(t) \cos 2d_n y + \tilde{A}_3(t) \sin \lambda_m x \sin d_n y \\
&\tilde{A}_1(t) = \frac{d_n^2}{32 A_{11} \lambda_m^2} W^2; \ \tilde{A}_2(t) = \frac{\lambda_m^2}{32 A_{22} d_n^2} W^2 \\
&\tilde{A}_3(t) = \frac{\left(B_{21}\lambda_m^3 + (B_{11}-B_{66})\lambda_m d_n^2\right)\phi_x(t) + \left(d_n^3 B_{12} + (B_{22}-B_{66})\lambda_m^2 d_n\right)\phi_y(t)}{\left(A_{11}\lambda_m^4 + A_{22}d_n^4 + (A_{66}-2A_{12})\lambda_m^2 d_n^2\right)}
\end{aligned}
\tag{12}
$$

Substituting Eq. (11) with motion Eq. (9), then employing Galerkin technique gives,

$$
\begin{aligned}
&t_{11}W + t_{12}\phi_x + t_{13}\phi_y + t_{14}W\#_x + t_{15}W\#_y + t_{16}W + t_{17}W^2 + t_{18}W^3 + L_{32}q = I_0\frac{d^2W}{dt^2} \\
&t_{22}\phi_x + t_{23}\phi_y + n_1 W + n_2 W^2 = \tilde{\rho}_1\, \phi_x, \\
&t_{32}\phi_x + t_{33}\phi_y + n_3 W + n_4 W^2 = \tilde{\rho}_1\, \phi_y
\end{aligned}
\tag{13}
$$

where;

$$
t_{11} = -\left(\frac{\lambda_m^4}{H_{26}R^2}\right); t_{12} = -K_s I_{30}\lambda_m; t_{13} = -K_s I_{30}\delta_n; t_{14} = (L_{12}+L_{13}) - \frac{L_{25}L_{27}}{L_{26}} - \frac{L_{27}L_{29}}{L_{26}}
$$

$$
t_{15} = (L_{14}+L_{15}) - \frac{L_{25}L_{28}}{L_{26}} - \frac{L_{28}L_{29}}{L_{26}}; t_{16} = (L_{23}+L_{24})\phi_1 - K_s I_{30}\lambda_m^2 - K_s I_{30}\delta_n^2;
$$

$$
t_{17} = (L_{16}+L_{17}) - L_{31}; t_{18} = -(L_{18}+L_{19}) - \frac{1}{16}\left(\frac{\lambda_m^2}{A_{22}} + \frac{\delta_n^2}{A_{11}}\right);
$$

$$
t_{22} = -D_{11}\lambda_m^2 - D_{66}\delta_n^2 - K_s I_{30} - B_{21}\frac{\zeta}{S}\lambda_m^3 - (B_{11}-B_{66})\frac{\zeta}{S}\lambda_m\delta_n^2;
$$

$$
t_{23} = -(D_{12}+D_{66})\lambda_m\delta_n - B_{21}\frac{\Psi}{S}\lambda_m^3 - (B_{11}-B_{66})\frac{\Psi}{S}\lambda_m\delta_n^2;
$$

$$
t_{32} = -(D_{21}+D_{66})\lambda_m\delta_n - B_{12}\frac{\zeta}{S}\delta_n^3 - (B_{22}-B_{66})\frac{\zeta}{S}\lambda_m^2\delta_n,
$$

$$
t_{33} = -D_{22}\delta_n^2 - D_{66}\lambda_m^2 - K_s I_{30} - B_{12}\frac{\Psi}{S}\delta_n^3 - (B_{22}-B_{66})\frac{\Psi}{S}\lambda_m^2\delta_n,
$$

$$
\Psi = B_{12}\delta_n^3 + (B_{22}-B_{66})\lambda_m^2\delta_n; S = A_{11}\lambda_m^4 + A_{22}\delta_n^4 + (A_{66}-2A_{12})\lambda_m^2\delta_n^2
$$

$$
n_1 = -K_s I_{30}\lambda_m, n_2 = -\frac{8}{3}\frac{B_{21}}{A_{11}}\frac{\delta_n}{ab}, n_3 = -K_s I_{30}\delta_n, n_4 = -\frac{8}{3}\frac{B_{12}}{A_{22}}\frac{\lambda_m}{ab};
$$

$$
\zeta = B_{21}\lambda_m^3 + (B_{11}-B_{66})\lambda_m\delta_n^2; \tilde{\rho}_1 = \left(I_2 - \frac{I_1^2}{I_0}\right),
$$

$$
a_1 = t_{11} + t_{13}; a_2 = t_{12}\frac{n_3 t_{23} - n_1 t_{33}}{t_{22}t_{33} - t_{32}t_{23}} + t_{13}\frac{n_1 t_{32} - n_3 t_{22}}{t_{22}t_{33} - t_{32}t_{23}} + t_{16}; a_3 = t_{14} + t_{15},
$$

$$a_4 = t_{12}\frac{n_4t_{23} - n_2t_{33}}{t_{22}t_{33} - t_{32}t_{23}} + t_{13}\frac{n_2t_{32} - n_4t_{22}}{t_{22}t_{33} - t_{32}t_{23}}; \, a_5 = t_{14}\frac{n_4t_{23} - n_2t_{33}}{t_{22}t_{33} - t_{32}t_{23}} + t_{15}\frac{n_2t_{32} - n_4t_{22}}{t_{22}t_{33} - t_{32}t_{23}}$$

$$a_6 = t_{14}\frac{n_3t_{23} - n_1t_{33}}{t_{22}t_{33} - t_{32}t_{23}} + t_{15}\frac{n_1t_{32} - n_3t_{22}}{t_{22}t_{33} - t_{32}t_{23}}$$

The following equation can be solved to determine the natural frequencies.

$$\begin{vmatrix} t_{11} + t_{16} + I_0\omega^2 & t_{12} & t_{13} \\ t_{21} + n_1 & t_{22} + \tilde{\rho}_1\omega^2 & t_{23} \\ t_{31} + n_3 & t_{32} & t_{33} + \tilde{\rho}_1\omega^2 \end{vmatrix} = 0 \tag{14}$$

Three angular frequencies are produced by solving Eq. (14), and the lowest one is chosen for analysis. In this work, assume that a uniformly distributed load (q = QsinΩt) is acting on an FG sandwich plate with porosity. As a result of the nonlinear Eq. (12) becomes:

$$\begin{aligned} I_0\frac{d^2W}{dt^2} - t_{11}W - t_{12}\phi_x - t_{13}\phi_y - t_{14}W\phi_x - t_{15}W\#_y - t_{16}W - t_{17}W^2 - t_{18}W^3 &= L_{32}Q\sin\Omega t \\ t_{22}\phi_x + t_{23}\phi_y + n_1W + n_2W^2 &= \tilde{\rho}_1\,\ddot{\phi}_x \\ t_{32}\phi_x + t_{33}\phi_y + n_3W + n_4W^2 &= \tilde{\rho}_1\,\ddot{\phi}_y \end{aligned} \tag{15}$$

When the second and third equations relating to (ϕ_x, ϕ_y) are solved from Eq. (15), the results are then substituted into the first equation to yield:

$$I_0\frac{d^2W}{dt^2} - (a_1 + a_2)W - (a_3 + a_4 + a_6 + r_{17})W^2 - (a_5 + r_{18})W^3 = L_{32}Q\sin\Omega t \tag{16}$$

the fundamental frequencies can be obtained as:

$$\omega_{mn} = \sqrt{\frac{-(a_1 + a_2)}{I_0}} \tag{17}$$

4   Results and Discussion

The sandwich FGM plate is assumed to have an effect under orderly distributed load (q = Q sin Ωt), load frequency illustrated in term Ω, and excited load amplitude demonstrated in term Q. Equation (13) and Eq. (16) are resolved by the fourth-order Runge-Kutta technique. To describe the current design, the FGM core comprises just one material. It is presumed that these elements are dispersed throughout both top with bottom plate components. Table 1 presents the properties used in this study [22]. A comparison of the parameter of the non-dimensional frequency developed in the present work with the findings of Hashemi et al. [23] with Zhao et al. [24] is shown in Table 2. The researchers in these two studies utilized the theory of first-order shear deformation; the element-free KP-Ritz technique was used by Zhao et al. [24], and the displacement functions were used by Hashemi et al. [23]. According to Table 1, there is no statistically significant difference between the values obtained by Hashemi et al. [23] and those obtained by Zhao et al. [24].

Table 1. Mechanical properties of the sandwich FGM plates.

Material Property	FG core	Skins
	Polyethylene	(Aluminum)
Modula's of Elasticity, GPa	110	210
Mass density, Kg/m^3	950	7800
Poisson's ratio	0.42	0.3

Table 2. Comparison of the fundamental nondimensional frequency factor of FGM plates

h/a	References	N				
		0	0.5	1	4	10
0.05	Hashemi et al. [23]	0.0148	0.0128	0.0115	0.0101	0.0096
	Zhao et al. [24]	0.0146	0.0124	0.0112	0.0097	0.0093
	Present study	0.0148	0.0126	0.0113	0.0098	0.0095
0.1	Hashemi et al. [23]	0.0577	0.0492	0.0445	0.0383	0.0363
	Zhao et al. [24]	0.0567	0.0482	0.0435	0.0376	0.0359
	Present study	0.0583	0.0495	0.0447	0.0387	0.0370
0.2	Hashemi et al. [23]	0.2112	0.1806	0.1650	0.1371	0.1304
	Zhao et al. [24]	0.2055	0.1757	0.1587	0.1356	0.1284
	Present study	0.2178	0.1862	0.1685	0.1448	0.1369

Plates' dimensions are supposed to be a = b = 1, porosity parameter (10, 20, and 30%), material gradient (0.5, 1, 5, 10), FG core heights (10, 20, and 25 mm), and face sheet (1, 2, and 3 mm) Figs. 2 and 3 illustrate the impacts of the porosity parameters on the FGM of the plate with a metal core, explicitly concerning the natural frequency and the dynamic response [25–34]. It can be seen from Fig. 2 that the natural frequency rises as a result of a rise in the porosity factor. To put it another way, the sandwich plates' natural frequency is favorably impacted by the distributed porosity. The research shows that as the power-law index rises, the amounts of natural frequencies fall. This is due to a drop in both the elastic modulus and bending stiffness of the FGM plate, which the results may explain. In the transient deflection graphs of the sandwich plates, the impacts of the different pore volume fractions are shown in Fig. 3. The findings demonstrate that the time-deflection of the sandwich FGM plates decreases as the porosity parameter increases due to the improved rigidity of the structure [34–54].

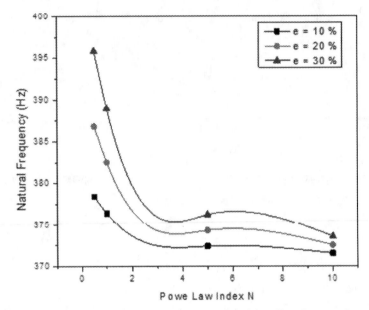

Fig. 2. The natural frequency that arises as a consequence of the effect that porosity parameters have on the FGM plate.

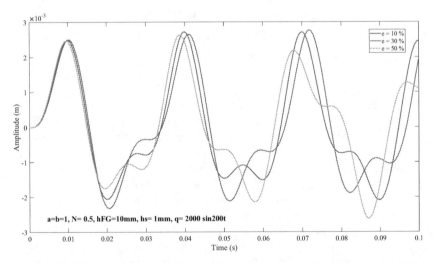

Fig. 3. Result of the porosity parameter on the dynamic response of the FGM sandwich plates.

Figures 4, 5, 6 and 7 evaluate the impact of varying the thickness of the FGM, both core and skin sheet layer, on the natural frequency and the dynamic response since it is one of the study's main objectives. It is evident that in FGM sandwich plates with porous, the natural frequency significantly rises as the thickness of either the FGM core or the face sheet of the plate increases. According to these findings, an increase in either the thickness of the FGM core or the thickness of the face sheet generates an increase in the stiffness of the sandwich of the FGM plate. The nonlinear dynamic performance of the porosity sandwich plate is obtained by employing thicknesses of (10, 20, and 30 mm) illustrated in Fig. 6. The thickness of the face sheet (1, 2, and 3 mm) represent in Fig. 7, together with a material gradient of N = 0.5 and the 20% porosity factor. For the porous FG sandwich plate, it is evident that increasing the FG core thickness reduces the vibration response deflection. Also, in Fig. 7, the findings demonstrate that a sandwich plate's deflection decreases as face sheet thickness increases. This is so that the material's characteristics will improve as the FGM thickens. This implies that the density, elastic modulus, and other features will improve [8, 55–67].

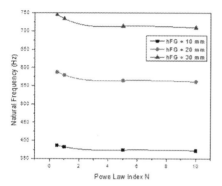

Fig. 4. Natural frequency for different FG core thicknesses of sandwich plates

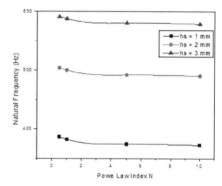

Fig. 5. Natural frequency for different face sheet thicknesses of sandwich plates

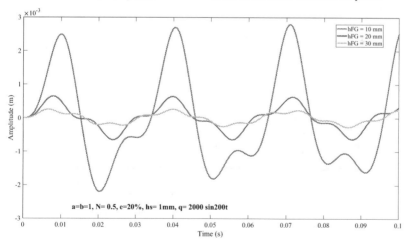

Fig. 6. Nonlinear time-displacement curve and its relationship to FGM core thickness.

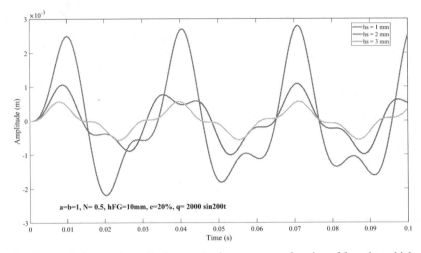

Fig. 7. Changes in the nonlinear displacement - time curve as a function of face sheet thickness.

5   Conclusion

The construction of an FGM sandwich most often consists of FGM cores covered in homogeneous skin. It is possible to consider the donations of almost any FG porous work piece in the design of biomedicine and aircraft structures to achieve superior characteristics such as high-bending solidity. To do this, the kind of core metal and the number of layers must be selected suitably. This research takes into account nonlinear computational simulations of (FGPMs) with a variety of edge circumstances. A porosity metal core when a single-phase is used to construct the sandwich plate, which is then joined to two containers of homogeneous aluminum. The material characteristics, such as the porosity volume concentration, are expected to be changed based on the thicknesses of a power-law distribution. To determine the frequency response of the FG sandwich plate using the various parameters, a straightforward and precise mathematical model that is founded on the FOSD approximation approach has been developed. This study also analyzes the influence of several factors on the nonlinear dynamic properties of FG sandwich plates, including the porous element, gradient indices, and face sheet thickness. Specifically, the porous factor is one of the parameters studied. The results provide the foundation for the next set of observations. It can be seen very clearly from the graphs that the natural frequency goes up as the porosity factor goes up, but it goes down when the power-law index rises. In this study, the nonlinear vibration of the functionally graded plate was calculated using the theoretical technique of the FOSD theory. In this work, it was not possible to calculate the buckling strength, so it is recommending to study the buckling of the porous functionally graded materials by using the same theory.

References

1. Li, Q., Wu, D., Chen, X., Liu, L., Yu, Y., Gao, W.: Nonlinear vibration and dynamic buckling analyses of sandwich functionally graded porous plate with graphene platelet reinforcement resting on Winkler-Pasternak elastic foundation. Int. J. Mech. Sci. **148**, 596–610 (2018)

2. Mirjavadi, S.S., Forsat, M., Barati, M.R., Hamouda, A.M.S.: Geometrically nonlinear vibration analysis of eccentrically stiffened porous functionally graded annular spherical shell segments. Mech. Based Des. Struct. Mach. **50**(1), 1–15 (2020)
3. Ghobadi, A., Beni, Y.T., Żur, K.K.: Porosity distribution effect on stress, electric field and nonlinear vibration of functionally graded nanostructures with the direct and inverse flexoelectric phenomenon. Compos. Struct. **259**, 113220 (2021)
4. Trinh, M.C., Kim, S.E.: Deterministic and stochastic thermomechanical nonlinear dynamic responses of functionally graded sandwich plates. Compos. Struct. **274**, 114359 (2021)
5. Pradhan, N., Sarangi, S.K.: Nonlinear vibration analysis of smart functionally graded plates. Mater. Today: Proc. **44**, 1870–1876 (2021)
6. Liu, Y., Qin, Z., Chu, F.: Nonlinear forced vibrations of functionally graded piezoelectric cylindrical shells under electric-thermo-mechanical loads. Int. J. Mech. Sci. **201**, 106474 (2021)
7. Jamalabadi, M.Y.A., Borji, P., Habibi, M., Pelalak, R.: Nonlinear vibration analysis of functionally graded GPL-RC conical panels resting on elastic medium. Thin-walled-Struct. **160**, 107370 (2021)
8. Jweeg, M.J., Said, S.Z.: Effects of rotational geometric stiffness matrices on dynamic stresses and deformations of rotating blades. J. Inst. Engineers (India) Mech. Eng. Div. **76**, 29–38 (1995)
9. Jweeg, M.J., Ahumdany, A.A., Jawad, A.F.M.: Dynamic stresses and deformations of the below knee prosthesis using ct-scan modelling. Int. J. Mech. Mechatron. Eng. **19**(1), 108–116 (2019)
10. Brischetto, S., Torre, R.: 3D hygro-elastic shell model for the analysis of composite and sandwich structures. Compos. Struct. **285**, 115162 (2022)
11. Njim, E.K., Bakhy, S.H., Al-Waily, M.: Analytical and numerical investigation of buckling behavior of functionally graded sandwich plate with porous core. J. Appl. Sci. Eng. **25**(2), 339–347 (2022)
12. Njim, E.K., Bakhy, S.H., Al-Waily, M.: Free vibration analysis of imperfect functionally graded sandwich plates analytical and experimental investigation. Arch. Mater. Sci. Eng. **111**, 1–17 (2021)
13. Njim, E.K., Bakhy, S.H., Al-Waily, M.: Experimental and numerical flexural analysis of porous functionally graded beams reinforced by (Al/Al2O3) nanoparticles. Int. J. Nanoelectron. Mater. **15**(2), 91–106 (2022)
14. Njim, E.K., Bakhy, S.H., Al-Waily, M.: Analytical and numerical flexural properties of polymeric porous functionally graded (PFGM) sandwich beams. J. Achieve. Mater. Manuf. Eng. **110**(1), 5–15 (2022)
15. Njim, E.K., Bakhy, S.H., Al-Waily, M.: Optimisation design of functionally graded sandwich plate with porous metal core for buckling characterisations. Pertanika J. Sci. Technol. **29**(4), 3113–3141 (2021)
16. Njim, E.K., Bakhy, S.H., Al-Waily, M.: Analytical and numerical investigation of free vibration behavior for sandwich plate with functionally graded porous metal core. Pertanika J. Sci. Technol. **29**(3), 1655–1682 (2021)
17. Njim, E.K., Bakhy, S. H., Al-Waily, M.: Analysis of porous functionally graded materials (FGPMs) sandwich plate using Rayleigh-Ritz method. Academia **110** (2021)
18. Mouthanna, A., Bakhy, S.H., Al-Waily, M.: Frequency of non-linear dynamic response of a porous functionally graded cylindrical panels. Jurnal Teknologi **84**(6), 59–68 (2022)
19. Mouthanna, A., Bakhy, S.H., Al-Waily, M.: Analytical investigation of nonlinear free vibration of porous eccentrically stiffened functionally graded sandwich cylindrical shell panels. Iranian J. Sci. Technol. Trans. Mech. Eng. **47**, 1035–1053 (2022)

20. Duc, N.D., Cong, P.H.: Nonlinear dynamic response of imperfect symmetric thin S-FGM plate with metal–ceramic–metal layers on elastic foundation. J. Vib. Control **21**, 637–646 (2013)
21. Reddy, J.N.: Mechanics of Laminated Composite Plates and Shells, Theory and Analysis. CRC Press, Boca Raton (2004)
22. Liu, Y., Hu, Y., Liu, T., Ding, J.L., Zhong, W.H.: Mechanical behavior of high-density polyethylene and its carbon nanocomposites under quasi-static and dynamic compressive and tensile loadings. Polym. Testing **41**, 106–116 (2015)
23. Hosseini-Hashemi, S., Taher, H.R.D., Akhavan, H., Omidi, M.: Free vibration of functionally graded rectangular plates using first-order shear deformation plate theory. Appl. Math. Model. **34**, 1276–1291 (2010)
24. Zhao, X., Lee, Y.Y., Liew, K.M.: Free vibration analysis of functionally graded plates using the element-free kp-Ritz method. J. Sound Vib. **319**, 918–939 (2009)
25. Al-Shammari, M.A., Al-Waily, M.: Theoretical and numerical vibration investigation study of orthotropic hyper composite plate structure. Int. J. Mech. Mechatron. Eng. IJMME-IJENS **14**(06) (2014)
26. Al-Waily, M., Ali, Z.A.A.A.: Suggested analytical solution of powder reinforcement effect on buckling load for isotropic mat and short hyper composite materials plate. Int. J. Mech. Mechatron. Eng. IJMME-IJENS **15**(04) (2015)
27. Alhumdany, A.A., Al-Waily, M., Al-Jabery, M.H.K.: Theoretical and experimental investigation of using date palm nuts powder into mechanical properties and fundamental natural frequencies of hyper composite plate. Int. J. Mech. Mechatron. Eng. IJMME-IJENS **16**(01) (2016)
28. Al-Waily, M., Deli, A.A., Al-Mawash, A.D., Ali, Z.A.A.A.: Effect of natural sisal fiber reinforcement on the composite plate buckling behavior. Int. J. Mech. Mechatron. Eng. IJMME-IJENS **17**(01) (2017)
29. Al-Waily, M., Resan, K.K., Al-Wazir, A.H., Ali, Z.A.A.A.: Influences of glass and carbon powder reinforcement on the vibration response and characterization of an isotropic hyper composite materials plate structure. Int. J. Mech. Mechatron. Eng. IJMME-IJENS **17**(06) (2017)
30. Ismail, M.R., Ali, Z.A.A.A., Al-Waily, M.: Delamination damage effect on buckling behavior of woven reinforcement composite materials plate. Int. J. Mech. Mechatron. Eng. IJMME-IJENS **18**(05), 83–93 (2018)
31. Abbas, E.N., Jweeg, M.J., Al-Waily, M.: Analytical and numerical investigations for dynamic response of composite plates under various dynamic loading with the influence of carbon multi-wall tube nano materials. Int. J. Mech. Mechatron. Eng. IJMME-IJENS **18**(06), 1–10 (2018)
32. Chiad, J.S., Al-Waily, M., Al-Shammari, M.A.: Buckling investigation of isotropic composite plate reinforced by different types of powders. Int. J. Mech. Eng. Technol. (IJMET) **09**(09), 305–317 (2018)
33. Al-Waily, M., Al-Shammari, M.A., Jweeg, M.J.: An analytical investigation of thermal buckling behavior of composite plates reinforced by carbon nano particles. Eng. J. **24**(3), 11–21 (2020)
34. Abbas, E.N., Jweeg, M.J., Al-Waily, M.: Fatigue characterization of laminated composites used in prosthetic sockets manufacturing. J. Mech. Eng. Res. Dev. **43**(5), 384–399 (2020)
35. Jebur, Q.H., Jweeg, M.J., Al-Waily, M., Ahmad, H.Y., Resan, K.K.: Hyperelastic models for the description and simulation of rubber subjected to large tensile loading. Arch. Mater. Sci. Eng. **108**(2), 75–85 (2021)

36. Mechi, S.A., Al-Waily, M.: Impact and mechanical properties modifying for below knee pros-thesis socket laminations by using natural kenaf fiber. In: 3rd International Scientific Confer-ence of Engineering Sciences and Advances Technologies, Journal of Physics: Conference Series, vol. 1973 (2021)

37. Njim, E.K., Bakhy, S.H., Al-Waily, M.: Optimization design of vibration characterizations for functionally graded porous metal sandwich plate structure. Mater. Today: Proc. (2021)

38. Njim, E.K., Bakhy, S.H., Al-Waily, M.: Analytical and numerical free vibration analysis of porous functionally graded materials (FGPMs) sandwich plate using Rayleigh-Ritz method. Arch. Mater. Sci. Eng. **110**(1), 27–41 (2021)

39. Raad, H., Al-Waily, M., Njim, E.K.: Free vibration analysis of sandwich plate-reinforced foam core adopting micro aluminum powder. Phys. Chem. Solid State **23**(4), 659–668 (2022)

40. AL-Shammari, M.A., Husain, M.A., Al-Waily, M.: Free vibration analysis of rectangular plates with cracked holes. In: 3rd International Scientific Conference of Alkafeel University, AIP Conference Proceedings, vol. 2386 (2022)

41. Jweeg, M.J., Al-Waily, M., Deli, A.A.: Theoretical and numerical investigation of buckling of orthotropic hyper composite plates. Int. J. Mech. Mechatron. Eng. IJMME-IJENS **15**(04) (2015)

42. Kadhim, A.A., Al-Waily, M., Ali, Z.A.A.A., Jweeg, M.J., Resan, K.K.: Improvement fatigue life and strength of isotropic hyper composite materials by reinforcement with different powder materials. Int. J. Mech. Mechatron. Eng. IJMME-IJENS **18**(02) (2018)

43. Al-Shammari, M.A., Al-Waily, M.: Analytical investigation of buckling behavior of honey-combs sandwich combined plate structure. Int. J. Mech. Prod. Eng. Res. Dev. (IJMPERD) **08**(04), 771–786 (2018)

44. Abbas, S.M., Takhakh, A.M., Al-Shammari, M.A., Al-Waily, M.: Manufacturing and analysis of ankle disarticulation prosthetic socket (SYMES). Int. J. Mech. Eng. Technol. (IJMET) **09**(07), 560–569 (2018)

45. Abbas, S.M., Resan, K.K., Muhammad, K.K., Al-Waily, M.: Mechanical and fatigue behaviors of prosthetic for partial foot amputation with various composite materials types effect. Int. J. Mech. Eng. Technol. (IJMET) **09**(09), 383–394 (2018)

46. Abbas, E.N., Al-Waily, M., Hammza, T.M., Jweeg, M.J.: An investigation to the effects of impact strength on laminated notched composites used in prosthetic sockets manufacturing. In: IOP Conference Series: Materials Science and Engineering, 2nd International Scientific Conference of Al-Ayen University, vol. 928 (2020)

47. Abbod, E.A., Al-Waily, M., Al-Hadrayi, Z.M.R., Resan, K.K., Abbas, S.M.: Numerical and experimental analysis to predict life of removable partial denture. In: IOP Conference Series: Materials Science and Engineering, 1st International Conference on Engineering and Advanced Technology, Egypt, vol. 870 (2020)

48. Al-Waily, M., Tolephih, M.H., Jweeg, M.J.: Fatigue characterization for composite materials used in artificial socket prostheses with the adding of nanoparticles. In: IOP Conference Series: Materials Science and Engineering, 2nd International Scientific Conference of Al-Ayen University, vol. 928 (2020)

49. Mechi, S.A., Al-Waily, M., Al-Khatat, A.: The mechanical properties of the lower limb socket material using natural fibers: a review. Mater. Sci. Forum **1039**, 473–492 (2021)

50. Al-Waily, M., Jweeg, M.J., Al-Shammari, M.A., Resan, K.K., Takhakh, A.M.: Improvement of buckling behavior of composite plates reinforced with hybrids nanomaterials additives. Mater. Sci. Forum **1039**, 23–41 (2021)

51. Njim, E.K., Al-Waily, M., Bakhy, S.H.: A review of the recent research on the experimental tests of functionally graded sandwich panels. J. Mech. Eng. Res. Dev. **44**(3), 420–441 (2021)

52. Jebur, Q.H., Jweeg, M.J., Al-Waily, M.: Ogden model for characterizing and simulation of PPHR Rubber under different strain rates. Aust. J. Mech. Eng. **21**, 911–925 (2021)

53. Al-Waily, M., Jweeg, M.J., Jebur, Q.H., Resan, K.K.: Creep characterization of various prosthetic and orthotics composite materials with nanoparticles using an experimental program and an artificial neural network. Mater. Today Proc. (2021)

54. Njim, E.K., Bakhy, S.H., Al-Waily, M.: Analytical and numerical investigation of buckling load of functionally graded materials with porous metal of sandwich plate. Mater. Today Proc. (2021)

55. Bakhy, S.H., Al-Waily, M., Al-Shammari, M.A.: Analytical and numerical investigation of the free vibration of functionally graded materials sandwich beams. Arch. Mater. Sci. Eng. **110**(2), 72–85 (2021)

56. Njim, E.K., Bakhy, S.H., Al-Waily, M.: Free vibration analysis of imperfect functionally graded sandwich plates: analytical and experimental investigation. Arch. Mater. Sci. Eng. **111**(2), 49–65 (2021)

57. Al-Shablle, M., Al-Waily, M., Njim, E.K.: Analytical evaluation of the influence of adding rubber layers on free vibration of sandwich structure with presence of nano-reinforced composite skins. Arch. Mater. Sci. Eng. **116**(2), 57–70 (2022)

58. Hussein, S.G., Al Saffar, I.Q., Al-Shammari, M.A., Al-Waily, M.: Effects of Ni additive on fatigue and mechanical properties of Al-Cu alloy manufactured using powder metallurgy. J. Eng. Sci. Technol. **17**(5), 3310–3325 (2022)

59. Haider, S.M.J., Takhakh, A.M., Al-Waily, M.: Designing a 3D virtual test platform for evaluating prosthetic knee joint performance during the walking cycle. Open Eng. **12**, 590–604 (2022)

60. Jweeg, M.J., Alazawi, D.A., Jebur, Q.H., Al-Waily, M., Yasin, N.J.: Hyperelastic modelling of rubber with multi-walled carbon nanotubes subjected to tensile loading. Arch. Mater. Sci. Eng. **114**(2), 69–85 (2022)

61. Jweeg, M.J., Al-Waily, M., Muhammad, A.K., Resan, K.K.: Effects of temperature on the characterization of a new design for a non-articulated prosthetic foot. In: IOP Conference Series: Materials Science and Engineering, 433, 2nd International Conference on Engineering Sciences, Kerbala, Iraq, 26–27 March 2018 (2018)

62. Abdulridha, M.M., Fahad, N.D., Al-Waily, M., Resan, K.K.: Rubber creep behavior investigation with multi wall tube carbon nano particle material effect. Int. J. Mech. Eng. Technol. (IJMET) **09**(12), 729–746 (2018)

63. Hussein, S.G., Al-Shammari, M.A., Takhakh, A.M., Al-Waily, M.: Effect of heat treatment on mechanical and vibration properties for 6061 and 2024 aluminum alloys. J. Mech. Eng. Res. Dev. **43**(01), 48–66 (2020)

64. Al-Shammari, M.A., Bader, Q.H., Al-Waily, M., Hasson, A.M.: Fatigue behavior of steel beam coated with nanoparticles under high temperature. J. Mech. Eng. Res. Dev. **43**(4), 287–298 (2020)

65. Al-Baghdadi, M., Jweeg, M.J., Al-Waily, M.: Analytical and numerical investigations of mechanical vibration in the vertical direction of a human body in a driving vehicle using biomechanical vibration model. Pertanika J. Sci. Technol. **29**(4), 2791–2810 (2021)

66. Haider, S.M.J., Takhakh, A.M., Al-Waily, M.: A review study on measurement and evaluation of prosthesis testing platform during gait cycle within sagittal plane. In: 14th International Conference on Developments in eSystems Engineering, IEEE Xplore (2021)

67. Ali, Z.A.A.A., Kadhim, A.A., Al-Khayat, R.H., Al-Waily, M.: Review Influence of loads upon delamination buckling in composite structures. J. Mech. Eng. Res. Dev. **44**(3), 392–406 (2021)

The Computer Modelling of the Human Gait Cycle for the Determination of Pressure Distribution and Ground Reaction Force Using a Below Knee Sockets

Sumeia A. Mechi[1], Muhsin J. Jweeg[2], and Muhannad Al-Waily[3($\boxtimes$)]

[1] Department of Materials Engineering, Faculty of Engineering, University of Kufa, Kufa, Iraq
sumiaa.maji@uokufo.edu.iq
[2] College of Technical Engineering, Al-Farahidi University, Baghdad, Iraq
muhsin.jweeg@uoalfarahidi.edu.iq
[3] Department of Mechanical Engineering, Faculty of Engineering, University of Kufa, Kufa, Iraq
muhanedl.alwaeli@uokufa.edu.iq

Abstract. In this research, many composite configurations were presented employing perlon, kevlar, and carbon fibres in addition to the kenaf fibres, which were used to create the original below-knee (B.K.) prosthesis. These modifications were made for convenience and to lengthen the useful life of our prosthesis. The study used a layered experimental design, with some layers containing kenaf and others without. The ideal strength-to-weight ratio (E/ρ) and the desired modulus of elasticity were sought, together with the effective lamination employed in the production of sockets. They prepared the socket produced from natural kenaf fibres and analyzed its performance throughout the gait cycle comprising the experimental phase. A kenaf prosthetic socket was provided to a 58-year-old man who lost his right Leg to diabetes and weighed 80 kg. This socket took group D-f. Tests of the ground reaction force (G.R.F.) showed that the difference between the stance and swing phases did not exceed (-5.3%), or 5.3% improvement over a different instance, signifying a 20.8% improvement.

Keywords: Kenaf Fiber · Below Knee Socket · Natural Fiber · Composite Prosthesis

1 Introduction

Walking is a movement in which the body is supported and propelled by both feet simultaneously. One completes one gait cycle when the heel of the foot makes two successive contacts within a specific time frame. Beginning with a right foot plant, the cycle is complete when the right foot again contacts the ground. The first 62% of the gait cycle occurs when the foot is firmly planted on the ground and is referred to as the stance phase. 38% of each gait cycle is spent in the swing phase. The foot's various parts all advance in tandem during the swing phase of the gait cycle [1].

© The Author(s), under exclusive license to Springer Nature Switzerland AG 2024
A. Ortis et al. (Eds.): ICAETA 2023, CCIS 1983, pp. 283–297, 2024.
https://doi.org/10.1007/978-3-031-50920-9_22

Important parameters are usually a concern at the time an objective analysis is involved: walking speed, steps/min = cadence, step length, step time, stride length, stride width, walking length total per minute Gait analysis results can be used to determine comfort rate of prosthesis during use by the patient, Fig. 1 shows Measurement of walking steps.

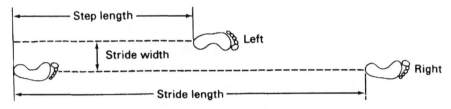

Fig. 1. Measurement of walking steps

Figure 2 shows the eight distinct phases of the gait cycle. The Initial Meeting (I.C.) begins when a baby's feet touch the ground. Pregnancy-related load reactions occur during the gait cycle phase called load response (L.R.). Third, with the body's center of gravity above the foot, step into the middle position (M.S.T.) on the side of the foot opposite the toe. When the body's weight is transferred to the reference foot, the T.S.T. phase of gait starts; this phase concludes when the opposite side foot touches the ground. She lifted her heels off the ground while waiting for the ideal position. When the behaviors start with the lateral toe of the opposite foot and end with that toe, we call it the pre-swing phase. Six, the first swing, or ISW, begins from the toe and continues upward until a reflection occurs over the knee. The upper knee reflex has been achieved by the middle of the golf swing, and it will remain thus until the tibia is perpendicular to the ground. When the shin is parallel to the ground, the terminal swing (TSW) begins, and it ends when the foot makes initial contact [2].

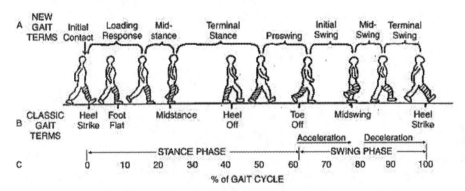

Fig. 2. The basic motions of walking in human [2]

Prosthetics experts have researched the gait cycle for various lower limbs and conducted an extensive study on the most effective techniques for developing prosthetic lower limbs. Using a mathematical approach, Ismail et al. [3] performed a biomechanical investigation of braced and unbraced legs. Then, Al-Waily et al. [4] studied the gait cycle using a variety of artificial hip joints. Abbas et al. [5, 6] studied the fatigue behavior of prosthetic materials subjected to dynamic loads throughout the walking cycle. Five separate categories were selected (person, glass, carbon, Kevlar). Observational data were compared to numerical findings for analysis. As the number of cycles grew, so did the damage factor, and the incidence of failure increased.

Haider et al. [7, 8] examined the walking cycle of a human body using an above-knee prosthesis. Consequently, the research takes both theoretical and practical models into account.

In this paper demonstrates how different laminated composite materials react to natural fiber reinforcement(kenaf) of the gait cycle below the knee of the human body, when comparison with other case. The suggested socket prostheses with natural fiber decrease the butterfly parameters leading to the difference between the injury leg and the intact leg to be minimum. And the gait standards are close to the normal gait of a healthy person, and this is what they set up in the prosthetic limb.

2   Experimental Work

To a lesser extent than composite materials [9–14], natural fibers were employed to manufacture prosthetic limbs. Various basic materials were used to create composite material samples for socket manufacture [15–20]. These materials, discussed below, consist of woven fibers, reinforcing material, and resin [21–26]. Lamination 617H19 includes Carbon Fiber, Kevlar Fiber, Natural Kenaf Fiber, Polyvinyl Alcohol, and Perlon Fiber. Figure 3 depicts a soft socket designed to reduce the stress placed on the patient's remaining Leg by the socket. Before casting, this socket is connected to the positive mould. Using a vacuum chamber and the procedures described below, production was conducted.

In the first stage, the positive mould was created and set on the platform, and the vacuum pressure tube was attached to show the amount of pressure exerted. The inside P.V.A. bag of the positive mould was then sealed at the bottom and top. At ambient temperature, the vacuum apparatus is adjusted to (−0.8 bar) [27–33]. Using kenaf, the following lamination was created: 1 Perlon +2 Carbon fiber +1 Kevlar fiber +1 kenaf fiber +1 Perlon. The exterior P.V.A. bag was sealed from the bottom up, with a tiny hole cut to release trapped air and the opposite side left open for resin supply. Thirdly, 1 L of the layer's resins was combined with 60 grammes of hardener using the typical ratio of 80:20. When combined, they should appear as follows: 100 parts resin for layers and 2–3 parts hardener. The completed matrix was then positioned onto the P.V.A. layer. The substance was thoroughly combined until it cooled by fusing the layers. The socket is complete when the positive mould containing the manufactured socket is separated [6, 34–37]. The foot and pylon are finally linked to the socket.

Fig. 3. The prosthetic socket.

3 Biomechanical Gait Cycle Tests

The ground reaction force dominates everyday actions with vertical components. The patient's ground reaction force (G.R.F.) was experimentally assessed using a treadmill platform during regular daily activities (Zebris, FDM-T). This 58-year-old man patient (80 kg) had his right Leg amputated. Figure 4 displays the results of a two-minute force plate test that yielded a constant reading. The G.R.F. was established to analyze the gait cycle and discriminate between normal and amputee gait. Gait parameters include stride length, stride duration, step length, step width, foot rotation, percentage of time spent in each phase of the gait cycle, velocity (walking speed), and cadence (number of steps taken per minute) [38–42]. Measuring the ground response forces of normal and amputee individuals revealed distinct walking behaviors (G.R.F.). In addition, the distinctive walking patterns of a person without left and right legs. Normal and amputee

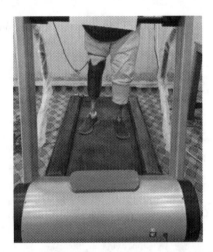

Fig. 4. In-Patient force plate walking test for the artificial limb using Treadmill.

individuals demonstrated a pressure distribution pattern at the ground reaction force (G.R.F.) and center of pressure throughout the gait cycle (C.O.P.).

4 Results and Discussion

4.1 Gait cycle parameters

Amputees and normal persons were used as examples to demonstrate the peculiarities of the gait cycle. This study's typical participant is 32 years old, 179 cm tall, and weighs 125 kg (75 kg). The gait cycle of the amputee differed from that of the ordinary USIA member. A noticeable difference may be seen when comparing a human body's right (damaged) leg to the left (unharmed) Leg. Figure 5 shows typical human behavior, while Fig. 6 shows an amputee wearing a D–f socket. According to the amputee case data, there was a significant difference in the proportion of stance circumstances involving the left Leg (67%) and the right Leg (72,3%). Differences from left to right averaged 65.2%, whereas differences from right to left averaged 64.4% [2, 43–48]. Normal data showed a 1.8% and 8% increase throughout the swing period compared to amputee data. The

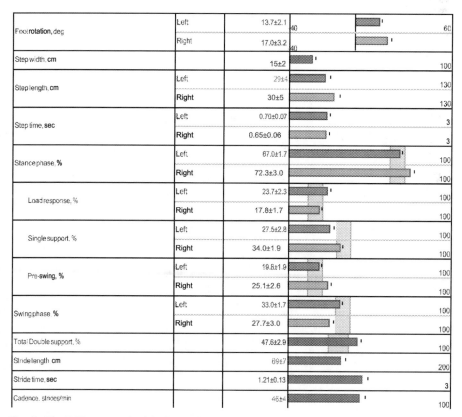

Fig. 5. The B.K. amputee's right-leg socket built from group D–f, and the normative data for his gait cycle parameters is 2–5 km/h.

data also show that the amputee patient's single support case was 27.5% and 34.0% for the left and right legs, respectively, but the findings for the typical person were 35.6% and 34.0% for the left and right legs, respectively. Left pre-swings made up (5.2%) of the total, while right pre-swings made up (10.2%).

In the amputee double-support situation, 18.1% more data was obtained than the typical individual. Assuming Scenario 1, the difference between the stance and swing phases is less than (−5,3%) to (5,3%). The difference between the stance and swing phases in the second scenario was (0.9%), whereas the difference between the swing and stance phases was (-0.9%). The standard case stride length was 29 cm longer than the amputee case stride length, and the standard case cadence was five steps per minute faster. Rapid movement during ambulation created this cadential impact.

Furthermore, step length and time may be differentiated to reveal the subject's skeletal anatomy. The participant's severed ankle did not allow the same degree of plantarflexion when the heel contacted the floor as a normal foot in a neutral position. A patient with a group D–f socket was compared to a patient who underwent a right limb amputation. While the highest difference between the stance and swing phases in the case study was (−6.7%), the differences in a normal occurrence were (0.9%) in the stance phase

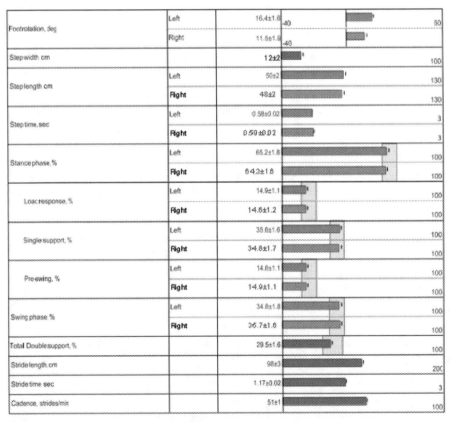

Fig. 6. The typical person's gait cycle characteristics; typical data ranges from 2 to 5 km per hour.

and (-0.9%) in the swing phase, resulting in a (20.8%) improvement. The average person's stride was 33 cm longer than the amputee's, resulting in a 12% improvement. The amputee's performance improved by 28.5% when his cadence was around 7 steps per minute higher than normal. Furthermore, the difference in foot rotation angle between the amputee case and the case study was $3.3°$ in the amputee case but $4.7°$ in the case study, resulting in a 29.7% increase. The patient motion's center of mass (C.O.M.) is similar to that of the natural instance. Their closeness increases the degree of stability and establishes balance [49–55].

In general, the number of steps per min at normal walking: (70–90) steps/min. In this range, the patient walks normal and comfort. Medium speed group with the number of steps: 95 steps/min. High speed group with the number of steps: 120 steps/min [56]. Based on gait analysis results obtained by the average number of steps (46 ± 4) steps/min. These results are within range of normal step group. It can be concluded that patient using prosthesis socket built from group D- f, can walk well in normal condition. In the study of ergonomics, one method to measure level of fatigue by using physiological criteria. If people feel comfortable in walking, then the pulses per minute is low. The comfort level refers to the number of pulse rate of respondent after walking activities is low level (75–100) pulses/min and moderate level (100–125) pulses/min. This method is done by measuring the pulse rate per min before and after gait analysis [54]. Table 1 presents the gait analysis results.

Table 1. Gait analysis results

Gait Parameters	Result
Cadence (steps/min)	46 ± 4
Step time (s)	65 ± 0.06
Right step length cm	30 ± 5
Left step length (cm)	29 ± 4
Step width (cm)	15 ± 2
Stride length (cm)	69 ± 7
Walking length total per minute (m)	36.4 ± 0.57
Walking speed (m/s)	0.6 ± 0.01

4.2 Results of the Ground Reaction Force

The force in the vertical direction is equal to the ground response force detected by a force plate. The typical graphical representation of vertical force is a mountain range with two peaks and a valley. The first impact is what causes the peak to occur. A plantar flexor push-off at the ankle causes the second peak (posterior calf muscles). The G.R.F. for a normal person's right Leg and that of an amputee's right Leg are shown in Figs. 7 and 8, respectively. The findings for first contact and toes off in the event of a good leg amputation were equal to (680 N) and (680 N), respectively, whereas the G.R.F. was found to be equal to (670 N) at first contact and equal to (680 N) at toes off (700 N). Typical G.R.F. (680 N) and amputee G.R.F. (780 N) during initial contact and toe-off of the left Leg are shown in Figs. 9 and 10, respectively (760 N). An amputee patient using a group D-f socket prosthesis exhibited a non-typical walking cycle. These fluctuations are brought on by the gait cycle's effect on the patient's weight on the socket.

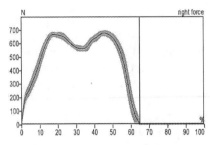

Fig. 7. The normal subject for the G.R.F. of right Leg.

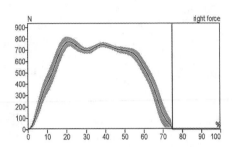

Fig. 8. A subject wearing the socket for the G.R.F. at the right Leg of amputee subject (B.K.).

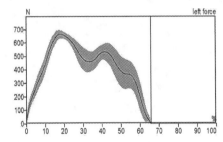

Fig. 9. A typical subject for the G.R.F. on the left Leg.

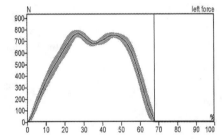

Fig. 10. A subject wearing the socket for the G.R.F. at the left Leg of amputee subject (B.K.)

4.3   Pressure Distribution Through the Gait Cycle

The pressure distributions of a healthy right lower Leg (Fig. 11) and a right lower leg that has been amputated (Fig. 12) are shown. Maximum pressure with a normal stance was (320 kPa) at (46%). (single support phase). During the stance phase, the average maximum pressure felt by the amputee subject (B.K.) wearing the group D- f socket was (410 kPa) at (24%). The initial contact phase of the curve looks different in healthy people because the weight is distributed more evenly over the sole of the shoe. At the same time, the extended valley in amputee patients is attributable to the patient's efforts to keep his center of gravity over his left Leg as he walks. Keeping one's balance when walking might be difficult. Therefore, when the right Leg is weak, the patient will compensate by putting more weight on the left.

Figures 13 and 14 show that since the amputee's left Leg has a smaller initial contact area, it fluctuates more than the right. Maximum pressure for a typical individual was measured at (280 kPa) at 45% of the stance phase (single support phase). On average, the amputee subject (B.K.) wearing the socket felt a maximum pressure of 250 kPa at 20% during the stance phase. Due to the reduced contact area between the shoe and the treadmill's surface, the footprints of both able-bodied and amputee persons are shown in Figs. 15 and 16, respectively. Images presented in Figs. 17 and 18 show a normal person's and an amputee's center of pressure (C.O.P.) for their left and right feet, respectively, in the form of a butterfly to indicate abduction and adduction, respectively, when walking. These graphs show how the C.O.P. beneath the foot varies throughout the gait cycle from heel strike to mid-stance and toe-off. For each state, the maximum pressure is shown. In order to better illustrate the C.O.P. route, the areas of the highest pressure are shown by curved lines. Gait cycle, length, and single point lines in a healthy individual (10 mm) and an amputee (3 mm) are shown in Figs. 17 and 18, respectively (26 mm and 2 mm).

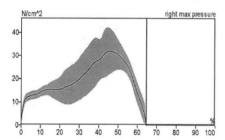

Fig. 11. Average subject for the pressure distribution at the right foot.

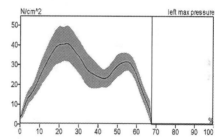

Fig. 12. A subject wearing socket for the pressure distribution at the right foot for amputee subject (B.K.)

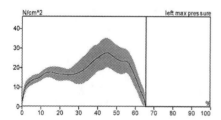

Fig. 13. The normal subject's pressure distribution at the left foot.

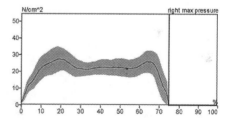

Fig. 14. A subject wearing socket for the pressure distribution at the left foot for amputee subject (B.K.).

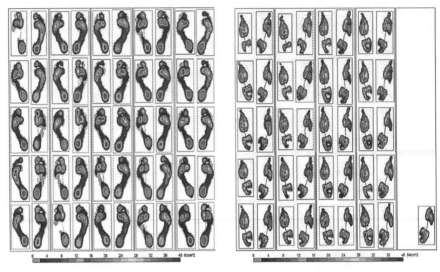

Fig. 15. A typical subject foot print.

Fig. 16. A subject wearing socket foot print for amputee subject (B.K.)

Butterfly Parameters		
	14-06-2011 suhail	
	Left	Right
Gait line length, mm	199±12	209±10
Single support line, mm	117±12	114±9
Ant/post position, mm	126	
Ant/post variability, mm	6	
Lateral symmetry, mm	-23	
Lateral variability, mm	11	
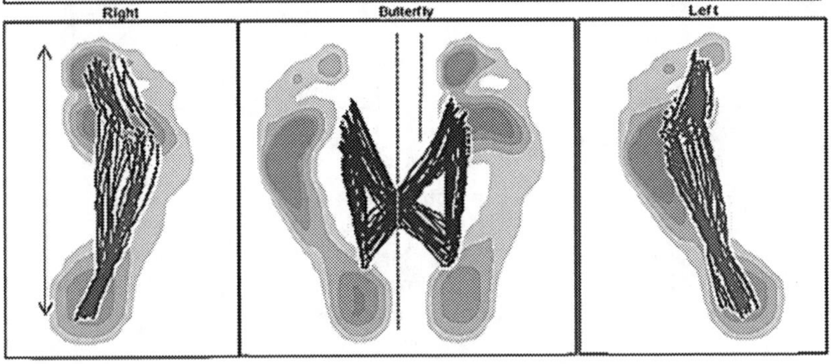 Right	Butterfly	Left

Fig. 17. The gait cycle for the path of the center of pressure (butterfly shape), normal subject, (A) Gait line length. (B) Single support line, anterior/posterior position and lateral.

Parameter, mm	Butterfly Parameters	
	5-4-2021 Gait Analysis	
	Left	Right
Gait line length, mm	243±25	269±18
Single support line, mm	80±10	78±11
Ant/post position, mm	139	
Ant/post variability, mm	7	
Lateral symmetry, mm	-14	
Lateral variability, mm	21	

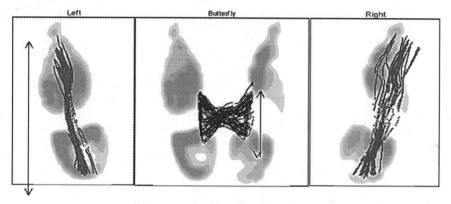

Fig. 18. Amputee subject (B.K.), with socket, demonstrating gait cycle for C.O.P.'s route along (A) gait line length; (B) single support line; (C) anterior/posterior position; and (D) lateral view.

5   Conclusion

A product of Socket prostheses reinforced with natural fibers (Kenaf) was developed with an acceptable strength and comfortable. The gait cycle data are close to the normal gait of a healthy person, and this is what they set up in the prosthetic limb. Because of that the difference between the stance and swing phases is less than (−5.3%) and (5.3), respectively, improvements of 20.8% are indicated as compared with other cases. Such as those presented here, diminish butterfly features and, as a result, the gap between the healthy and injured limb.

It is recommended that the patient wears socks made of layers without natural fibers to show the effectiveness of the natural fibers in strengthening the sockets with a high strength/weight ratio in addition to the comparison study of the walking cycle and the ground reaction forces during the gait cycle using different walking situations such as stairs.

References

1. Bhat, S.G.: Design and Development of a Passive Prosthetic Ankle. M.sc. Thesis, Arizona State University (2017)
2. Al-Waily, M., Jweeg, M.J., Jebur, Q.H., Resan, K.K.: Creep characterization of varous prosthetic and orthotics composite materials with nanoparticles using an experimental program and an artificial neural network. Mater. Today: Proc. (2022)

3. Ismail, M.R., Al-Waily, M., Kadhim, A.A.: Biomechanical analysis and gait assessment for normal and braced legs. Int. J. Mech. Mechatron. Eng. **18**(3), 32–41 (2018)
4. Al-Waily, M., Hussein, E.Q., Al-Roubaiee, N.A.A.: Numerical modeling for mechanical characteristics study of different materials artificial hip joint with inclination and gait cycle angle effect. J. Mech. Eng. Res. Dev. **42**(4), 79–93 (2019)
5. Abbas, E.N., Jweeg, M.J., Al-Waily, M.: Fatigue characterization of laminated composites used in prosthetic sockets manufacturing. J. Mech. Eng. Res. Dev. **43**(5), 384–399 (2020)
6. Abbas, E.N., Al-Waily, M., Hammza, T.M., Jweeg, M.J.: An investigation to the effects of impact strength on laminated notched composites used in prosthetic sockets manufacturing. In: I.O.P. Conference Series: Materials Science and Engineering, 2nd International Scientific Conference of Al-Ayen University, vol. 928 (2020)
7. Haider, S.M.J., Takhakh, A.M., Al-Waily, M.: a review study on measurement and evaluation of prosthesis testing platform during gait cycle within sagittal plane. In: 14th International Conference on Developments in eSystems Engineering. IEEE Xplore (2021)
8. Haider, S.M.J., Takhakh, A.M., Al-Waily, M., Saadi, Y.: simulation of gait cycle in sagittal plane for above-knee prosthesis. In: 3rd International Scientific Conference of Alkafeel University, A.I.P. Conference Proceedings, vol. 2386 (2022)
9. Al-Shammari, M.A., Al-Waily, M.: Theoretical and numerical vibration investigation study of orthotropic hyper composite plate structure. Int. J. Mech. Mechatron. Eng. **14**(6), 1–21 (2014)
10. Kadhim, A.A., Al-Waily, M., Ali, Z.A.A.A., Jweeg, M.J., Resan, K.K.: Improvement fatigue life and strength of isotropic hyper composite materials by reinforcement with different powder materials. Int. J. Mech. Mechatron. Eng. **18**(2), 77–86 (2018)
11. Abbod, E.A., Al-Waily, M., Al-Hadrayi, Z.M.R., Resan, K.K., Abbas, S.M.: Numerical and experimental analysis to predict life of removable partial denture. In: I.O.P. Conference Series: Materials Science and Engineering. 1st International Conference on Engineering and Advanced Technology, vol. 870 (2020)
12. Njim, E.K., Bakhy, S.H., Al-Waily, M.: Analytical and numerical free vibration analysis of porous functionally graded materials (FGPMs) sandwich plate using Rayleigh-Ritz method. Arch. Mater. Sci. Eng. **110**(1), 27–41 (2021)
13. Bakhy, S.H., Al-Waily, M., Al-Shammari, M.A.: Analytical and numerical investigation of the free vibration of functionally graded materials sandwich beams. Arch. Mater. Sci. Eng. **110**(2), 72–85 (2021)
14. Al-Waily, M., Jaafar, A.M.: Energy balance modelling of high-velocity impact effect on composite plate structures. Arch. Mater. Sci. Eng. **111**, 14–33 (2021)
15. Jweeg, M.J., Al-Waily, M., Deli, A.A.: Theoretical and numerical investigation of buckling of orthotropic hyper composite plates. Int. J. Mech. Mechatron. Eng. **15**(4), 1–12 (2015)
16. Jweeg, M.J., Said, S.Z.: Effects of rotational geometric stiffness matrices on dynamic stresses and deformations of rotating blades. J. Inst. Engineers (India) Mech. Eng. Div. **76**, 29–38 (1995)
17. Chiad, J.S., Al-Waily, M., Al-Shammari, M.A.: Buckling investigation of isotropic composite plate reinforced by different types of powders. Int. J. Mech. Eng. Technol. **9**(9), 305–317 (2018)
18. Al-Waily, M., Al-Saffar, I.Q., Hussein, S.G., Al-Shammari, M.A.: Life enhancement of partial removable denture made by biomaterials reinforced by graphene nanoplates and hydroxyapatite with the aid of artificial neural network. J. Mech. Eng. Res. Dev. **43**(6), 269–285 (2020)
19. E.K. Njim, S.H. Bakhy, M. Al-Waily: Optimization design of vibration characterizations for functionally graded porous metal sandwich plate structure. Materials Today: Proceedings (2021)

20. Jebur, Q.H., Jweeg, M.J., Al-Waily, M.: Ogden model for characterizing and simulating PPHR Rubber under different strain rates. Aust. J. Mech. Eng. (2021)
21. Njim, E.K., Bakhy, S.H., Al-Waily, M.: Free vibration analysis of imperfect functionally graded sandwich plates: an analytical and experimental investigation. Arch. Mater. Sci. Eng. **111**(2), 49–65 (2021)
22. Al-Shammari, M.A., Al-Waily, M.: Analytical investigation of buckling behavior of honeycombs sandwich combined plate structure. Int. J. Mech. Prod. Eng. Res. Dev. **8**(4), 771–786 (2018)
23. Al-Waily, M., Al-Shammari, M.A., Jweeg, M.J.: An analytical investigation of thermal buckling behavior of composite plates reinforced by carbon nano particles. Eng. J. **24**(3), 11–21 (2020)
24. Ali, Z.A.A.A., Kadhim, A.A., Al-Khayat, R.H., Al-Waily, M.: Review influence of loads upon delamination buckling in composite structures. J. Mech. Eng. Res. Dev. **44**(3), 392–406 (2021)
25. Al-Waily, M., Jweeg, M.J., Al-Shammari, M.A., Resan, K.K., Takhakh, A.M.: Improvement of buckling behavior of composite plates reinforced with hybrids nanomaterials additives. Mater. Sci. Forum **1039**, 23–41 (2021)
26. Njim, E.K., Bakhy, S.H., Al-Waily, M.: Optimisation design of functionally graded sandwich plate with porous metal core for buckling characterizations. Pertanika J. Sci. Technol. **29**(4), 3113–3314 (2021)
27. Njim, E.K., Bakhy, S.H., Al-Waily, M.: Analytical and numerical investigation of the buckling load of functionally graded materials with the porous metal of sandwich plate. Mater. Today Proc. (2021)
28. Jweeg, M.J., Hammood, A.S., Al-Waily, M.: A suggested analytical solution of isotropic composite plate with crack effect. Int. J. Mech. Mechatron. Eng. **12**(5), 44–59 (2012)
29. Al-Waily, M., Resan, K.K., Al-Wazir, A.H., Ali, Z.A.A.A.: Influences of glass and carbon powder reinforcement on the vibration response and characterization of an isotropic hyper composite materials plate structure. Int. J. Mech. Mechatron. Eng. **17**(6), 74–85 (2017)
30. Abdulridha, M.M., Fahad, N.D., Al-Waily, M., Resan, K.K.: Rubber creep behavior investigation with multi-wall tube carbon nano particle material effect. Int. J. Mech. Eng. Technol. **9**(12), 729–746 (2018)
31. Abbas, S.M., Takhakh, A.M., Al-Shammari, M.A., Al-Waily, M.: Manufacturing and analysis of ankle disarticulation prosthetic socket (SYMES). Int. J. Mech. Eng. Technol. **9**(7), 560–569 (2018)
32. Abbas, E.N., Jweeg, M.J., Al-Waily, M.: Fatigue characterization of laminated composites used in prosthetic sockets manufacturing. J. Mech. Eng. Res. Dev. **43**(5), 384–399 (2020)
33. Al-Waily, M., Tolephih, M.H., Jweeg, M.J.: Fatigue characterization for composite materials used in artificial socket prostheses with nanoparticles. In: I.O.P. Conference Series: Materials Science and Engineering, 2nd International Scientific Conference of Al-Ayen University, vol. 928 (2020)
34. M.A. Al-Shammari, M.A. Husain, M. Al-Waily: Free Vibration Analysis of Rectangular Plates with Cracked Holes. 3rd International Scientific Conference of Alkafeel University, A.I.P. Conference Proceedings 2386 (2022)
35. Al-Waily, M., Deli, A.A., Al-Mawash, A.Z., Ali, Z.A.A.A.: Effect of natural sisal fiber reinforcement on the composite plate buckling behavior. Int. J. Mech. Mechatron. Eng. **17**(1), 30–38 (2017)
36. Jweeg, M.J., Al-Waily, M., Muhammad, A.K., Resan, K.K.: effects of temperature on the characterization of a new design for a non-articulated prosthetic foot. In: I.O.P. Conference Series: Materials Science and Engineering, 2nd International Conference on Engineering Sciences, vol. 433 (2018)

37. Njim, E.K., Al-Waily, M., Bakhy, S.H.: A review of the recent research on the experimental tests of functionally graded sandwich panels. J. Mech. Eng. Res. Dev. **44**(3), 420–441 (2021)
38. Njim, E.K., Bakhy, S.H., Al-Waily, M.: Analytical and numerical flexural properties of polymeric porous functionally graded (PFGM) sandwich beams. J. Achiev. Mater. Manuf. Eng. **110**(1), 5–15 (2022)
39. Abbas, S.M., Resan, K.K., Muhammad, A.K., Al-Waily, M.: Mechanical and fatigue behaviors of prosthetic for partial foot amputation with various composite materials types effect. Int. J. Mech. Eng. Technol. **9**(9), 383–394 (2018)
40. Mechi, S.A., Al-Waily, M.: Impact and mechanical properties modifying for below knee prosthesis socket laminations by using natural kenaf fiber. In: 3rd International Scientific Conference of Engineering Sciences and Advances Technologies, Journal of Physics: Conference Series, vol. 1973 (2021)
41. Mechi, S.A., Al-Waily, M., Al-Khatat, A.: The mechanical properties of the lower limb socket material using natural fibers: a review. Mater. Sci. Forum **1039**, 473–492 (2021)
42. Fahad, N.D., Kadhim, A.A., Al-Khayat, R.H., Al-Waily, M.: Effect of SiO2 and Al2O3 hybrid nano materials on fatigue behavior for laminated composite materials manufacture artificial socket prostheses. Mater. Sci. Forum **1039**, 493–509 (2021)
43. Njim, E.K., Bakhy, S.H., Al-Waily, M.: Experimental and numerical flexural analysis of porous functionally graded beams reinforced by (Al/Al2O3) nanoparticles. Int. J. Nanoelectron. Mater. **152**, 91–106 (2022)
44. Al-Waily, M., Ali, Z.A.A.A.: A suggested analytical solution of powder reinforcement effect on buckling load for isotropic mat and short hyper composite materials plate. Int. J. Mech. Mechatron. Eng. **15**(4), 1–16 (2015)
45. Ismail, M.R., Ali, Z.A.A.A., Al-Waily, M.: Delamination damage effect on buckling behavior of woven reinforcement composite materials plate. Int. J. Mech. Mechatron. Eng. **18**(5), 83–93 (2018)
46. Hussein, S.G., Al-Shammari, M.A., Takhakh, A.M., Al-Waily, M.: Effect of heat treatment on mechanical and vibration properties for 6061 and 2024 aluminum alloys. J. Mech. Eng. Res. Dev. **43**(1), 48–66 (2020)
47. Kadhim, A.A., Abbod, E.A., Muhammad, A.K., Resan, K.K., Al-Waily, M.: Manufacturing and analyzing a new prosthetic shank with adapters by 3D printer. J. Mech. Eng. Res. Dev. **44**(3), 383–391 (2021)
48. Bakhy, S.H., Al-Waily, M.: Development and modeling of a soft finger in robotics based on force distribution. J. Mech. Eng. Res. Dev. **44**(1), 382–395 (2021)
49. Jweeg, M.J., Hammood, A.S., Al-Waily, M.: Experimental and theoretical studies of mechanical properties for reinforcement fiber types of composite materials. Int. J. Mech. Mechatron. Eng. **12**(4) (2012)
50. Alhumdany, A.A., Al-Waily, M., Al-Jabery, M.H.K.: Theoretical and experimental investigation of using date palm nuts powder into mechanical properties and fundamental natural frequencies of hyper composite plate. Int. J. Mech. Mechatron. Eng. **16**(1), 1–11 (2016)
51. Abbas, E.N., Jweeg, M.J., Al-Waily, M.: Analytical and numerical investigations for dynamic response of composite plates under various dynamic loading with the influence of carbon multi-wall tube nano materials. Int. J. Mech. Mechatron. Eng. **18**(6), 1–10 (2018)
52. Njim, E.K., Bakhy, S.H., Al-Waily, M.: Analytical and numerical investigation of free vibration behavior for sandwich plate with functionally graded porous metal core. Pertanika J. Sci. Technol. **29**(3), 1655–1682 (2021)
53. Al-Baghdadi, M., Jweeg, M.J., Al-Waily, M.: Analytical and numerical investigations of mechanical vibration in the vertical direction of a human body in a driving vehicle using biomechanical vibration model. Pertanika J. Sci. Technol. **29**(4), 2791–2810 (2021)

54. Jweeg, M.J., Ahumdany, A.A., Jawad, A.F.M.: Dynamic stresses and deformations of the below knee prosthesis using ct-scan modelling. Int. J. Mech. Mechatron. Eng. **19**(1), 108–116 (2019)
55. Njim, E.K., Bakhy, S.H., Al-Waily, M.: Analytical and numerical investigation of buckling behavior of functionally graded sandwich plate with porous core. J. Appl. Sci. Eng. **25**(2), 339–347 (2022)
56. Irawan, I.A.P., Sukania, W.: Gait analysis of lower limb prosthesis with socket made from rattan fiber reinforced epoxy composites. Asian J. Appl. Sci. **03**(01), 2321–0893 (2015)

Evaluation of the Existing Web Real-Time Signaling Mechanism for Peer-to-Peer Communication: Survey

Sanabil A. Mahmood$^{(\boxtimes)}$ ⓘ and Monji Kherallah ⓘ

University of Mosul, Mosul 41002, Iraq
Sanabil_2000@uomosul.edu.iq, monji.kherallah@fss.usf.tn

Abstract. Web Real-Time Communication (WebRTC) is prepared to allow the co-event of sound, video, and data interconnecting. Also, it is a set of criterions, libraries, and JavaScript APIs. WebRTC had several advantages, including the lack of plug-ins, the ease of usage, no licensing, and the excellent quality of the RTC applications. However, signalizing technicality that setup, establish and end a interconnect amidst peers has not been specified in the WebRTC. This paper reviews general studies and methods that are used and suggested for WebRTC signalling protocols. Moreover, it focuses on the limitations of multi-web crawlers interconnecting, the network topology for exchanging data and multimedia and leveraging public groundwork to manage privacy as a service or information, or public architecture to handle signalling protocols. Therefore, this work focuses on research in related work for the existing WebRTC signalling technicalities/protocols to find out the limitations and the main gap at this time and also for a thorough knowledge of signalling in WebRTC including their advantages and disadvantages.

Keywords: Web Real-time Communication (WebRTC) · Signaling mechanism/Protocols · And Network Topologies

1 Introduction

W3C was used in web crawler API, and IETF, for the wire protocol, developed a new criterion renowned as WebRTC. The WebRTC is prepared to allow the co-event of sound, video, and data interconnecting [31]. In May 2011, the We.b.R.T.C began. At that time, Google announced an open-source initiative to Supply peer-to-peer (P2P), web crawler-based, and real-time interconnecting [30]. WebRTC is a set of criterions, libraries, and JavaScript APIs [46]. By way of explanation [59], declared that WebRTC is a (P2P) protocol as opposed to a customer/server protocol. JavaScript APIs are utilized directly in WebRTC to facilitate interactive connections amidst web crawlers that employ different types of data [16].

Datagram Transport Layer Security (DTLS), Secure Real-Time Protocol (SRTP), and Stream Control Transmission Protocol (SCTP) over DTLS are all methods that can be used to create or operate a secure channel or channels, respectively, for media and data channels. By avoiding intermediary hardware servers, this approach can eliminate security issues like data theft by hackers [14]. Likewise, It can Supply system secrecy and authentication [28]. As a result, [35] confirmed that WebRTC had been used by more than one billion endpoints and devices. Besides, by 2018, it has anticipated that 4.7 billion gadgets will be capable of supporting WebRTC [65].

When WebRTC uses the coequal website page for downloading and enables both web crawlers to use the coequal online application, then it can be one of the cases [48]. Therefore, WebRTC signalling can be performed using any method that enables web crawlers to transmit and receive messages through a server [52]; It merely needs to employ a suitable HTML5 API [60]. The adoption of HTML5 has increased due to the web crawlers' quick development, headed by Firefox, Chrome, and Opera. Server-Sent Events (SSE), a unidirectional protocol, and a full duplex Web Socket are the two ways available in HTML5 for transmitting data from the server to the customer [25]. Accordingly, data eliminates the need for a server and opens up a direct line of interconnecting for web crawlers [65]. WebRTC has three main parts as follows:

- Get. User. Media API (aka, Media. Stream),
- RTC. Peer. Connection API (aka, Peer. Connection)
- RTC. Data. Channel API (aka, Data. Channel).

This paper is coordinated as follows, The WebRTC libraries is given in Sect. 2.
The limitations of WebRTC Signalling Technicality are presented in Sect. 3.
Several proposed solutions and performance for Web.RTC signaling technicality is explained in Sect. 4 has the conclusion.

1.1 UserMedia API

This API allows a web crawler to access native devices like a microphone and camera while representing synchronized streams as audio and video input and output [62]. WebRTC disposes of the requirement for Adobe Blaze to utilize media gadgets and the module prerequisite because of its Programming interface. Additionally, this API makes

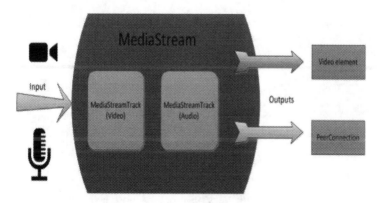

Fig. 1. The structure of WebRTC MediaStream [62].

audio and video available for use as HTML elements in all types of web applications [38]. (See Fig. 1) demonstrates the Media Stream structure.

1.2 RTCPeerConnection API

Direct interconnecting and the foundation of voice or video calls amidst web crawlers are made conceivable by the RTC Peer Connection Programming interface. The getUser-Media () technique accumulates the web crawler all inside a working person, including media streams, and permits a web crawler to get to nearby gadgets like a mouthpiece and camera to transmit information to another party [3]. Also, handling all swap signalling messages requires particular JavaScript methods [34].

1.3 RTCDataChannel API

Web crawlers can transmit data connections thanks to the RTC Data Channel API. Additionally, it offers a two-way data transmission that allows two web crawlers to swap random data. While avoiding the use of servers and intermediaries, each "RTC Data Channel" can lower the cost of service and Supply network utilization like low-latency data swaps in a more modifiable way [20]. (See Fig. 2), depicts the architecture of the WebRTC.

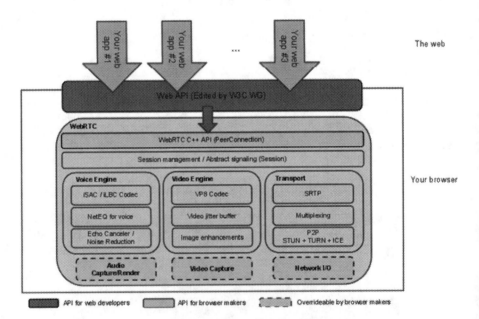

Fig. 2. WebRTC architecture [56].

As mentioned in [22], many devices have one or more layers of Network Address Translation (NAT). NAT is a method for mapping private addresses to public addresses and the variable address data contained within IP packets as they travel over the routing device. Developers of WebRTC technology can use ICE which facilitates the Internet addressing system's complexity. When choosing the finest connectivity option in constrained locations and when it comes to media interconnecting, ICE has proven to be dependable. It is utilized for NAT traversal using Relay NAT and Session Traversal Utilities for NAT (STUN) (TURN) [6]:

1. STUN is used to find the external address of a specific peer [21].
2. TURN is used to support STUN by avoiding the Symmetric NAT determination by connecting to and relaying all information through a TURN server [60].

2 Limitations of WebRTC Signalling Technicality/Protocols

To lay out interconnecting among various clients or gadgets, WebRTC requires a sort of flagging instrument or backing from conventions [51]. However, To assess WebRTC and control the interconnecting design, the IETF and W3C have not yet agreed on a last flagging system or a last Programming interface convention [32]. A signalling technicality is not a component of WebRTC even though it uses the "RTCPeer.Connection" API to transmit data streaming across peers. [57] Chen-Fu, 2014). In addition, [19] revealed that the key component of the application, signalling, has been shown, has not yet been specified on the WebRTC amidst web crawler and server. As a result, the developer must Supply the signalling protocol at the application level [72]; what's more, the two substances should settle on it, either with the focal hub or the other client [38]. As a result, It is undeniable that WebRTC does not criterionize the signalling amidst web crawlers and servers [15].

A signalling technicality decides the interconnecting amidst at least two companions [12] to see as one another, lay out direct P2P streams, and offer contact data [5]. As a result, the core of peer identification, which finds peers and orchestrates interconnecting amidst them, is signalling. It facilitates the development of user-to-user interconnecting through the swap of data across channels [13]. Additionally, it establishes a connection amidst the web crawler and a server and allowance s interconnecting amidst peers via this server, including support for the SDP for mixing network addresses and port numbers for media sharing [50]. Several fixes were proposed, and WebRTC utilized many protocols to obtain the signalling method detailed below:

2.1 Socket.io Signalling Protocol

To enable real-time bidirectional interconnecting amidst web crawlers and a server, a transport protocol written in JavaScript is used [17]. Since its inception, Socket.io has used Node.js as its primary backend [49]. Google Chrome created the open-source Nore.js server platform to Supply a powerful JavaScript engine. It is also a suitable platform for developing web applications because it is based on a non-blocking I/O and event-driven model, which can handle high throughput and performance [33]. As a result, it offers an HTTP server to help create a web application but is neither an application server like Tomcat nor a web server like Apache HTTP [49].

It has been explained in [41], socket.io is a concept for WebSockets using XHR, Flash, and JavaScript Object Notation because it enables developers to use WebSockets and determine various synchronized interconnecting protocols controlled by the customer's web crawler (JSON) [49]. Otherwise, it uses Adobe Flash Socket, AJAX long polling and JSON Polling. Additionally, [17] expressed that socket.io offers a basic server and customer library for supporting constant bi-directional interconnecting; additionally it has the idea of rooms that empowers distributed association, as three WebSocket associations can happen from three different customers [8].

2.2 Jingle Signalling Protocol

This is an open-source innovation arranged to start and deal with a distributed media meeting amidst two substances, so it is a flagging convention [4]. Likewise, it is an extension of XMPP and is a criterion specified by the IETF to carry an instant messaging service [2]. Otherwise, it Supplys a pluggable model that allows the core session management to be applied to a wide diversity of applications (e.g., video chat and file transfer) and with different transport methods: TCP, UDP and ICE [55]. It presented the Specification and Description Language (SDL) model in Web.RTC participates with STUN, TURN and ICE servers to solve network issues. The performation and analysis of this model have confirmed multimedia sessions amidst two peers, but it shows latency in the server response [4].

2.3 Web Socket Protocol

A bi-directional interconnecting session amidst a user's web crawler and a server can be opened with the aid of the WS protocol [23]. Instead of connecting to another peer, it opens a pipe to a server; a customer transmits messages to the server, which then forwards them to the participants [14]. WS Supplys many benefits, including a low rate of data loss and high user concurrency [67]. Additionally, [36] revealed that WS, as long as it employs the web server as a proxy for connections to customers, uses the coequal interconnecting port as HTTP (80/443) and is also an initial connection and handshaking amidst a web customer and a web server [42]. WS leaves both directions of interconnecting in the session open. As a result, users can transmit and receive a lot of messages [62]. In addition, [24] WS is rumoured to Supply a tunnel and leave it open among peers. It performs better than AJAX/Polling in terms of lower latency and increased user data bandwidth [47]. Otherwise, [11] WS's inefficiency with all web crawlers was explained. Additionally, it cannot guarantee scalability and dependability when a high number of peers are using the WebRTC application or when entering and exiting the network. Different developers attempted to use WS to propose or develop a signalling technicality for the WebRTC as addressed below:

An interconnecting service prototype-based RE-presentation State Transfer (REST) API with SIP User Agent over Web Socket has been prepared and performed in [64]. The signalling of this prototype should be supported by a middle component (REST service) to swap messages and establish a media channel. However, the interconnecting is delayed 5 s and done with only two REST web crawlers. By the coequal token, REST relies on an HTTP polling technicality to push notifications from the server to the customer. Moreover, [60] analysed WebRTC video call performance utilizing the Node.js framework and Web Socket protocol (for signalling), TURN servers, etc. Both mesh and star topologies were used for this evaluation. The calls established amidst the three players in each topology, otherwise, use a false video sequence as opposed to a live camera. Additionally, different switches were used for the mesh, and an MCU device was used for the star. Additionally, in this test, all calls are forced to stream the video among participants through the TURN servers. By contrast, [56] stated that TURN servers are being used to relay data amidst endpoints as shown in Fig. (6). Similarly, [14] demonstrated that using the TURN server for simultaneous calls can force more overhead on the bandwidth, and it is not wholly free. It is also necessary to highlight that the WebRTC interconnecting is using a TURN server depending on traffic amidst peers, which suffers regression of media quality and latency.

It has been demonstrated in [61], that the WebRTC chat application using Web Socket as a signalling server based on the node.js platform was prepared and performed. Although this application applied to two peers using adapter.js to support Google Chrome, users should connect a web server sequentially to communicate with each other. In addition, because a server cannot recognize a customer's messages or know where to transmit them, there is a problem amidst customers and a web server. In a similar vein, the author acknowledged that the setup and signalling preformation were flawed. Another trial employing the Firefox and Chrome web crawlers for video conferences used in e-Health over a LAN network proved successful. The signalling was built based on WebSocket using node.js and the socket.io library to establish an interconnecting amidst two peers [14].

2.4 Session Initiation Protocol

It has been illustrated in [53] that no signalling protocol (such as SIP) will be recommended to give developers more flexibility for innovation in web applications. However, SIP is more complex than other protocols like XMPP [37]. In addition, various developers attempted to make video calls using WebRTC and SIP, but SIP still required software installation on such servers [62]. Moreover, [39] emphasized that conventional SIP customers do not yet support the protocols that WebRTC requires. Additionally, employing a new type of integrated interconnecting environment, the coupling of WebRTC features with the SIP platform needs to be improved to support multimedia sessions. Besides, the current real-time interconnecting APIs in an application are more cost-efficient and faster than developing a SIP customer [26].

Several solutions were proposed to combine the SIP protocol with WebRTC to enhance the interconnecting quality or Supply a signalling protocol. As detailed in [37] a theoretical solution was discussed to gather Real-Time Interconnecting Web (RTCWeb) with IP Multimedia Subsystem (IMS) via a web crawler as well as using a signalling gateway with suggested protocols such as SIP and Web Socket. This proposed integration presented some drawbacks: (a) it leads to different directions, (b) it used SIP protocol, which is complicated for some web developers, and (c) used an IMS system with RTCWeb producing little destruction. Moreover, [43] built WebRTC video chat application over LAN network. This application was employed by three peers based on different methods: JSON via XHR (WebRTC-JSON/XHR) and SIP via Web Socket (WebRTC-SIP/WS). Instead, there were no outbound connections, and the WebRTC-SIP/WS overhead can affect the quality of experience, while the Internet connection speed was low. Indeed, WebRTC-SIP/WS has more overhead than WebRTC-JSON/XHR for a completed session.

Table 1. Some differences among SIP, XMPP and WebSocket protocols

SIP	XMPP	WebSocket
Especially utilized for media transmission	Used in the decentralized framework for constant informing	Does not support every browser, web servers and proxies
Applied in additional convoluted frameworks with an intermediary server, area server, recorder and client specialist	Versatility is restricted to XMPP and can't supply adjustment of double information	It suffers from reconnections in services
Messages handling can consume transmission capacity	It has a raised organization above	
The friend should enroll first and afterward can communicate the greeting message, which has enormous data	It has a long interaction to lay out a meeting	

3   Several Proposed Solutions and Perforations for WebRTC Signalling Techncality

[10] created a server less-WebRTC project to decouple the "signalling server" and Supply email and file transfer amidst peers without a web server. However, it often faces failure when transferring a large file. Similarly, [11] proposed a multi-video conference utilizing the XMPP server and lib-jingle to establish a WebRTC P2P connection (plugin). However, without installing lib-jingle, the user was unable to access the system or receive signalling. This test was conducted using two different web crawlers. Another program was developed in [59] for WebRTC video conferencing to allow interconnecting amidst two customers utilizing the Node.js platform. In addition, [9] performed a centralized system for instant messaging, video and audio conferences, and file sharing. Nevertheless, the video preformation is limited to two participants. What is more, [7] created a WebRTC-based video chat system for senior citizens. The system used an email to transmit the chat request rather than signalling. It was also applied to one-to-one video chatting amidst two users. Not only that but also [33] developed a WebRTC video conference to be performed on an e-learning platform using the Node.js platform, and socket.io APIs for signalling and mesh topology. Instead, this test was performed amidst two web crawlers. Besides, the author clarified that it seeks an improvement to solve an echo problem with noise cancellation.

A new approach to online learning called (web-based interactive virtual classroom), which must be accessed through HTTPS, was described. All functions cannot be worked out in a real connection and latency occurred in some cases. This system is specified for video conferencing (unidirectional) as screen sharing amidst 10 users. The video was being shared as a one-to-one strategy. As an example, from A to B, B to C, and C to A [44]. Equally, [1] illustrated that a network architecture named Ufo.js offering a channel that empowers two web crawlers to lay out interconnecting of document moves was carried out. Ufo.js was run through the Node.js stage by utilizing an outside server to hold and deal with an association with each friend; this association is utilized to coordinate all the motioning among peers.

3.1 Using Several Techniques

[66] introduced a decentralised web crawler-based Open Publishing (BOPlish), which consists of a bootstrap server to bear joining peers using the WebRTC offer/answer technicality. On the contrary, this approach was utilised amidst two web crawlers for sharing files, and the signalling depended on using a bootstrap server, which was not described. More suggestions can be found in [45, 58], which proposed a P2P video broadcasting over WebRTC using two servers: one is a streaming server for transmitting captured video and the second is (Splitter + Signalling) server for restoring the video that is coming from the streaming server and broadcasting it to the customers. Indeed, only three customers can be served by this scenario. In addition, the streaming server was playing the essential part that captures and transmits data, instead of the Splitter Signalling server, which was not explained. Moreover, [29] presented different models and architecture of P2PTV using Eclipse to support the conferencing service using WebRTC. Alternatively, these models have no media server for video streaming. Therefore, the video was captured via web camera instead of the media file as planned and was not performed yet. Furthermore, a benchmark for performance evaluation of WebRTC was proposed in [63], so various architectures and platforms were considered. In this proposition, a server acts as both a web server and a signalling server using Node.js. The coequal video sequences should be employed because it is impossible to have a unique input for each test. Besides, the used signalling server was not clarified, and the number of served peers was not stated. Alternatively, when a party idles for 10 s, the connections will be expired, and not all viewers may receive every chunk. Adding to that, different issues occurred such as some nodes being served by the disconnected broadcaster, the children of the node can cause the increase of buffer head due to an inefficient pull policy and the existing performation failing to reach 100% of the nodes in the network. Also, it has a bandwidth bottleneck, and the execution is not satisfactory which did not broadcast the chunks to all nodes. Apart from this, the initial signalling has not kept unwanted nodes out of the network, so it is not sufficient [42].

4 Conclusion

In this work, various WebRTC signalling techniques and protocols were examined and described. Moreover, in-depth research has been done on most signalling platforms and commercial servers with their limitations. In particular, While the IETF and W3C have not yet decided on the precise signalling technicality or a definitive protocol to perform WebRTC and control interconnecting architecture, one is required. Equally, the most current applications were dependent on the public signalling server, signalling service, libraries or APIs to be Supplyd by a signalling protocol. Apart from this, they utilised different topologies such as one-to-one, star (one-to-many) and mesh (many-to-many). However, they claimed that they were not able to achieve interconnecting amidst more than eight peers for audio/video calls and video streaming. According to the aforementioned literature reviews, it was discovered that they primarily focused on the accomplishment of their design, test, or results without clarifying how they created or used the signalling technicalitys/protocols to determine interconnections amidst peers to discover each other, establish direct streams, and swap contact details of the session.

References

1. Bevilacqua, A., Boemio, P.S.P.R.: Introducing ufo. js: a web crawler-oriented p2p network. In: International Conference on Computing, Networking and Interconnecting (ICNC), pp. 353–357. IEEE, Honolulu, HI (2014). https://doi.org/10.1109/ICCNC.2014.6785359
2. Aliwi, H.S., Haj, N.K.A., Alajmi, P., Sumari, K.A.: A comparison amidst inter-asterisk swap protocol and jingle protocol: session time. Eng. Technol. Appl. Sci. Res. **6**(4), 1050–1055 (2016). http://etasr.com/index.php/ETASR/article/viewFile/664/354
3. Aranjo, S., et al.: Review of real-time collaboration frameworks, libraries and products. Int. J. Res. Appl. Sci. Eng. Technol. **10**(4), 653–660 (2022). https://doi.org/10.22214/ijraset.2022.41326
4. El Hamzaoui, A., Bensaid, H., En-Nouaary, A.: A formal model for WebRTC signaling using SDL. In: Abdulla, P., Delporte-Gallet, C. (eds.) Networked Systems. NETYS 2016. LNCS, vol. 9944, pp. 202–208. Springer, Cham (2016). https://doi.org/10.1007/978-3-319-46140-3
5. Bertard, A., Töpfer, M.: How web-based voice interconnecting system customers can change operation in mission control. In: 16th International Conference on Space Operations. SpaceOps, pp. 1–11 (2021). https://spaceops.iafastro.directory/a/proceedings/SpaceOps-2021/SpaceOps-2021/8/manuscripts/SpaceOps-2021,8,x1304.pdf
6. Bhatla, D., et al.: Video/audio conferencing using. Int. J. Eng. Appl. Sci. Technol. **7**(4), 276–280 (2022)
7. Chiang, C.Y., Chen, Y.L., Tsai, P.S.S.M.Y.: A video conferencing system based on WebRTC for seniors. In: International Conference on Trustworthy Systems and their Applications, Taichung, pp. 51–56. IEEE, Taichung, Taiwan (2014). https://doi.org/10.1109/TSA.2014.17
8. Calpe, X.L.: Study, design and performation of Web.RTC for a real-time multimedia messaging application. Universitat Politècnica de Catalunya (2017)
9. Chen, X.: Unified Interconnecting and WebRTC. Norwegian University of Science and technology (2014). https://brage.bibsys.no/xmlui/bitstream/handle/11250/2352722/12293_FULLTEXT.pdf?sequence=1&isAllowed=y
10. Ball, C.: WebRTC without a signaling server, HTML5 Rocks (2013). https://developers.google.com/web/updates/2015/05/high-performance-video-with-hardware-decoding
11. Cola, C., Valean, H.: On multi-user web conference using WebRTC. In: 18th International Conference on System Theory, Control and Computing (ICSTCC), pp. 430–433. IEEE, Sinaia, Romania (2014). https://doi.org/10.1109/ICSTCC.2014.6982454
12. Edan, N. and Mahmood, S.A.: Design and perform a new technicality for audio, video and screen recording based on WebRTC technology. Int. J. Electr. Comput. Eng. **10**(3), 2773–2778 (2020). https://doi.org/10.11591/ijece.v10i3.pp2773-2778
13. Edan, N.M., Abdul, R., Abdul, H.: Design and performation Of WebRTC screen sharing. Webology **19**(3), 1778–1792 (2022)

14. Azom, E., Emmanuel, B.D.D.: A Peer-to-peer architecture for real-time interconnecting using WebRTC. J. Multidiscip. Eng. Sci. Stud. **3**(4), 1671–1683 (2017). http://www.jmess.org/wp-content/uploads/2017/04/JMESSP13420330.pdf
15. Figueira, G., Barradas, D., Santos, N.: Stegozoa: enhancing Web RTC covert channels with video steganography for internet censorship circumvention. In: ASIA CCS 2022 - Proceedings of the 2022 ACM Asia Conference on Computer and Interconnectings Security, pp. 1154–1167 (2022). https://doi.org/10.1145/3488932.3517419
16. Carullo, G., Tambasco, M., Di Mauro, M., Longo, M.: A performance evaluation of WebRTC over LTE. In: Carullo, G., Tambasco, M. (ed.) 12th Annual Conference on Wireless On-demand Network Systems and Services (WONS), pp. 170–175. IEEE, Cortina d'Ampezzo (2016). http://ieeexplore.ieee.org/abstract/document/7429067/
17. Grinberg, M.: Socketio Documentation (2017). https://media.readthedocs.org/pdf/python-soc ketio/latest/python-socketio.pdf
18. Ha, P.J., Hoon, L.D.: Scalable signaling protocol for web real-time interconnecting based on a distributed hash table. Comput. Interconnectings **70**, 28–39 (2015). https://doi.org/10.1016/j.comcom.2015.05.013
19. Hadeed, S.M.: A performance evaluation of peer-to-peer media interconnecting over. Webology **19**(3), 10 (2022)
20. Halder, D., et al.: fybrrChat: A Distributed Chat Application for Secure P2P Messaging (2022). in arXiv preprint arXiv. http://arxiv.org/abs/2207.02487
21. Heikkinen, A.: Performance evaluation of distributed data delivery on mobile devices using WebRTC. In: IEEE, pp. 1036–1042 (2015)
22. Hwang, S.H., Yeh, C.Y.: Session traversal utilities for network address translator (STUN)-based traversal approach using port assignment prediction technicality. Sens. Mater. **34**(5), 1791–1801 (2022). https://doi.org/10.18494/SAM3750
23. Castillo, I.B., Villegas, J.M., Pascual, V.: The WebSocket Protocol as a Transport for the Session Initiation Protocol (SIP). Spain (2014). https://tools.ietf.org/pdf/rfc7118.pdf
24. Fette, I., Melnikov, A.: The WebSocket Protocol. UK (2011). https://tools.ietf.org/pdf/rfc 6455.pdf
25. Grigorik, I.: Server-Sent Events (SSE), O'Reilly Media (2013). https://hpbn.co/server-sent-events-sse/. Accessed 9 Jan 2018
26. Lajtos, I.D., O'Byrne, E.E.: WebRTC to complement IP Interconnecting Services. 1 (2016). https://www.gsma.com/futurenetworks/wp-content/uploads/2016/02/Web.RTC_to_complement_IP_Interconnecting_Services_v1.0.pdf
27. Uberti, J., Jennings, C.E.R.: JavaScript Session Establishment Protocol. USA (2017). https://tools.ietf.org/pdf/draft-ietf-rtcweb-jsep-24.pdf
28. Jaouhari, S.E.L., et al. 'Securing the interconnectings in a WoT/WebRTC-based smart health-care architecture. In: 11th International Conference on Frontier of Computer Science and Technology (FCST). Exeter: (FCST), pp. 1–6 (2017)
29. Edisson, J., Padilla, V.: 김.정호, 이.재오 An performation of P2PTV service based on WebRTC. Netw. Oper. Manag. **16**(2), 7 (2014). http://www.knom.or.kr/knom-review/v16n2/4.pdf
30. Luís, J., Domingos, G.: Website file download acceleration using WebRTC. Uppsala University (2015). https://fenix.tecnico.ulisboa.pt/downloadFile/1970719973966152/thesis.pdf
31. Jang-Jaccard, J., Nepal, S., Celler, B., Yan, B.: WebRTC-based video conferencing service for telehealth. Computing **98**(1–2), 169–193 (2016). https://doi.org/10.1007/s00607-014-0429-2
32. Karam, S.J., Abdulrahman, B.F.: Using socket. io approach for many-to-many bi-directional video conferencing. Al-Rafidain J. Comput. Sci. Math. (RJCM) **16**(1), 81–86 (2022)

33. Bissereth, K., Lim, B.B.L., Shesh, A.: An interactive video conferencing module for e-Learning using WebRTC. In: International Conferences, USA, pp. 1–4 (2014). http://www.cita.my/cita2015/docs/shortpaper/31.pdf

34. Kasetwar, A., et al.: A WebRTC based video conferencing system with screen sharing. Int. J. Res. Appl. Sci. Eng. Technol. **10**(5), 5061–5066 (2022). https://doi.org/10.22214/ijraset.2022.43595

35. López-Fernández, L., et al.: Designing and evaluating the usability of an API for real-time multimedia services in the Internet. Multimed. Tools Appl. **76**(12), 14247–14304 (2016). https://doi.org/10.1007/s11042-016-3729-z

36. Grahl, L.: SaltyRTC Seriously Secure WebRTC. University of Applied Sciences (2015). https://lgrahl.de/pub/ba-thesis-saltyrtc-by-lennart-grahl-revised-v1.pdf

37. Li, L., Zhang, X.: Research on the integration of RTCWeb technology with IP multimedia subsystem. In: 5th International Congress on Image and Signal Processing, (CISP), pp. 1158–1161. IEEE, Chongqing, China (2012). https://doi.org/10.1109/CISP.2012.6469705

38. Lozano, A.A.: Performance analysis of topologies for Web-based Real-Time Interconnecting (WebRTC). Aalto University (2013). https://aaltodoc.aalto.fi/bitstream/handle/123456789/11093/master_Abelló_Lozano_Albert_2013.pdf

39. Deshpande, M., Mohani, S.P.: Integration of WebRTC with SIP – current trends. Int. J. Innov. Eng. Technol. (IJIET) **6**(2), 92–96 (2015). http://ijiet.com/wp-content/uploads/2015/12/14.pdf

40. Nasir, M.S., Saeed, K., Hussain, S., Usama, M., Saeed, M.: A comparison of SIP with IAX an efficient new IP telephony protocol. In: International Conference of Engineering and Emerging Technology, p. 7. IEEE, Lahore (2015). https://www.researchgate.net/publication/274705522_A_Comparison_of_SIP_with_IAX_an_Efficient_new_IP_Telephony_Protocol

41. Mathieu Nebra: Socket.io: let's go to real time, OPENCLASSROOMS (2017). https://openclassrooms.com/courses/ultra-fast-applications-using-node-js/socket-io-let-s-go-to-real-time. Accessed 30 June 2017

42. Melhus, M.K.: P2P Video Streaming with HTML5 and WebRTC. Norwegian University of Science and Technology (2015). https://brage.bibsys.no/xmlui/bitstream/handle/11250/2352761/12874_FULLTEXT.pdf?sequence=1

43. Adeyeye, M., Makitla, I., Fogwill, T.: Determining the signalizing overhead of two common WebRTC methods: JSON via XMLHttpRequest and SIP over WebSocket. In: Africon, Pointe-Aux-PimentsConference, p. 4. IEEE, Pointe-Aux-Piments, Mauritius (2013). https://doi.org/10.1109/AFRCON.2013.6757840

44. Buasri, N., Janpan, T., Yamborisut, U., Wongsawang, D.: Web-based interactive virtual classroom using HTML5-based technology. In: Proceedings of the 3rd ICT International Senior Project Conference, (ICT-ISPC). Nakhon Pathom, IEEE, Thailand, pp. 33–36 (2014). https://doi.org/10.1109/ICT-ISPC.2014.6923212

45. Nursiti, S.: Design and performance of peer-to-peer video and chat interconnecting. Tech. Soc. Sci. J. **17**, 235–243 (2021)

46. Phankokkruad, M., Jaturawat, P.: An evaluation of technical study and performance for real-time face detection using web real-time interconnecting. In: International Conference on Computer, Interconnecting, and Control Technology (I4CT), pp. 162–166. IEEE, Sarawak, Malaysia (2015). http://ieeexplore.ieee.org/abstract/document/7219558/

47. Radogan, G.M.K.: Evaluating WebSocket and WebRTC in the Context of a Mobile Internet of Things Gateway. KTH Royal Institute of Technology (2013). http://kth.diva-portal.org/smash/get/diva2:686624/FULLTEXT01.pdf

48. Rahaman, M.H.: A survey on real-time interconnecting for web. Sci. Res. J. (SCIRJ) **III**(Vii), pp. 39–45 (2015). http://www.scirj.org/papers-0715/scirj-P0715273.pdf

49. Rai, R.: Socket. IO Real-time Web Application Development. Birmingham - Mumbai: PACKT (2013)

50. Rajab, S.: Comparing different network topologies for Web.RTC conferencing. Kthroyal Institute of Technology (2015). https://doi.org/10.1002/ejoc.201200111
51. Raswa, R., Sumarudin, S., Ismantohadi, E.: WebRTC signaling technicality using npRTC topology for online virtual classroom. In: Proceedings of the 5th FIRST T1 T2 2021 International Conference (FIRST-T1-T2 2021), pp. 264–270. Atlantis Press, Negeri Indramayu (2022). https://doi.org/10.2991/ahe.k.220205.047
52. Rob, M.: Getting started with WebRTC. In: Akram Hussain, A.R., Basu, S., Pratik, K., et al. (Eds.). Packt Publishing Ltd., Birmingham (2013). https://doi.org/10.1111/j.2041-210X.2010.00056.x
53. Rodríguez, P., et al.: Advanced videoconferencing services based on WebRTC. In: IADIS International Conferences Web Based Communities and Social Media 2012 and Collaborative Technologies 2012, pp. 180–184. IADIS, Madrid, Spain (2012). http://www.researchg ate.net/publication/235639869_Advanced_Videoconferencing_Services_Based_on_Web RTC/file/9fcfd51233ddc9a053.pdf%5Cnfile:///Users/marcin/Documents/Library.papers3/ Articles/2012/Rodr?guez/2012_Rodr?guez.pdf%5Cnpapers3://publication/uuid/FC
54. Rouse, M.: REST (REpresentational State Transfer), TechTarget (2017). http://searchmicros ervices.techtarget.com/definition/REST-representational-state-transfer. Accessed 2 Jan 2018
55. Ludwig, S., Beda, J., Saint-Andre, P., McQueen, R., Egan, S.J.H.: XEP-0166: Jingle, XMPP Criterions Foundation (2016). https://xmpp.org/extensions/xep-0166.html#schema-errors. Accessed 22 Sep 2016
56. Dutton, S.: WebRTC in the real world: STUN, TURN and signaling, HTML5 (2013). https:// www.html5rocks.com/en/tutorials/WebRTC/groundwork/. Accessed 29 Nov 2016
57. Dutton, S.: Getting Started with WebRTC, HTML5 Rocks (2014). https://www.html5rocks. com/en/tutorials/Web.RTC/basics/#toc-simple. Accessed 1 July 2016
58. Sandholm, T.: SnoW: Serverless n-Party calls over Web.RTC, in arXiv preprint arXiv, pp. 1–17 (2022). http://arxiv.org/abs/2206.12762
59. Shane, H.: Video-to-video using WebRTC. In: JavaScript Creativity: Exploring the Modern Capabilities of JavaScript and HTML5, p. 184. Apress (2014)
60. Singh, V., Lozano, A.A., Ott, J.: Performance analysis of receive-side real-time congestion control for WebRTC. In: 20th International Packet Video Workshop, pp. 1–8. IEEE, San Jose, CA, USA (2013). https://doi.org/10.1109/PV.2013.6691454
61. Sørlie, M.: Congestion Control for WebRTC Services, (June), p. 110 (2017). https://brage. bibsys.no/xmlui/bitstream/handle/11250/2456126/17761_FULLTEXT.pdf?sequence=1
62. Sredojev, B., Samardzija, D., Posarac, D.: WebRTC technology overview and signaling solution design and performation. In: 38th International Convention on Information and Interconnecting Technology, Electronics and Microelectronics, MIPRO - Proceedings, pp. 1006–1009. IEEE, Opatija, Croatia (2015). https://doi.org/10.1109/MIPRO.2015.7160422
63. Taheri, S., et al.: WebRTCbench: a benchmark for performance assessment of WebRTC performations. In: ESTIMedia- 13th Symposium on Embedded Systems for Real-Time Multimedia, p. 8. IEEE, Amsterdam, Netherlands (2015). https://doi.org/10.1109/ESTIMedia.2015. 7351769
64. Ambra, T., Paganelli, F., Fantechi, A., Giuli, D., Mazzi, L.: Resource-oriented design towards the convergence of Web-centric and Telecom-centric services. In: Second International Conference on Future Generation Interconnecting Technologies (FGCT), pp. 120–125. IEEE, London, UK (2013). https://doi.org/10.1109/FGCT.2013.6767203
65. Vashishth, S., Sinha, Y., Babu, K.H.: Addressing challenges in web crawler based P2P content sharing framework using WebRTC. In: 30th International Conference on Advanced Information Networking and Applications (AINA), pp. 850–857. IEEE, Crans-Montana, Switzerland (2016). https://doi.org/10.1109/AINA.2016.143

66. Vogt, C., Werner, M.J., Schmidt, T.C.: Content-centric user networks: WebRTC as a path to name-based publishing. In: Proceedings - International Conference on Network Protocols, ICNP. Goettingen, pp. 1–3. IEEE, Germany (2013). https://doi.org/10.1109/ICNP.2013.673 3652
67. Zhang, L., Shen, X.: Research and development of real-time monitoring system based on WebSocket technology. In: International Conference on Mechatronic Sciences, Electric Engineering and Computer (MEC), pp. 1955–1958. IEEE, Shenyang, China (2013). https://doi.org/10.1109/MEC.2013.6885373

Autonomous Agent Using AI Q-Learning in Augmented Reality Ludo Board Game

Fazliaty Edora Fadzli[✉], Ajune Wanis Ismail, Norhaida Mohd Suaib,
and Lau Yin Yee

Mixed and Virtual Reality Research Lab, ViCubeLab, School of Computing, Universiti
Teknologi Malaysia, 81310 Kuala Lumpur, Johor, Malaysia
{fazliaty.edora,ajune,haida}@utm.my

Abstract. An autonomous agent works with Artificial Intelligence (AI) can
decide its actions to adapt and respond to the changes in a dynamic environment.
The autonomous agent can be developed in games as a Non-Player Character
(NPC) to interact with the changes of state in the game environment. Traditional
board games such as Ludo have had many players since the olden days but slowly
lost attraction to the public, especially the younger generations as digital games
become more popular. Although the Ludo board game can be digitized to fasci-
nate the players through implementing Augmented Reality (AR) technology in
handheld devices, common NPCs found in games have determined actions and
are unable to learn from experience and adapt to the changes of the game environment. Therefore, this research aims to develop an autonomous agent for board
game in handheld AR (HAR). The first step in the three main phases is to examine
the autonomous agent for the HAR board game. The second phase is developing
the AR board game with Q-learning and Minimax algorithms for board game
agents. Finally, the third phase is integrating the AR board game with Q-learning
and Minimax agents in handheld. The novel contribution of this research is the
redesign of Ludo for AR with autonomous agent and generate the training data
using the Q-Learning algorithm to create autonomous agent in AR.

Keywords: Artificial Intelligence · Augmented Reality · Handheld Augmented
Reality · Q-learning · Minimax

1 Introduction

Throughout history, board games have been played by many experienced players [1].
Ludo is a popular example of a traditional board game [2]. Nonetheless, the number of
players of traditional board games has decreased since the introduction of digital games
[3]. Hence, board games such as Ludo can be digitized to maintain existing players and
attract new ones.

Augmented Reality (AR) is a technology that combines the real world with computer-
generated virtual environments or objects [4]. The combination of this technology to a
board game enhances the gaming experience [3]. Nowadays, AR technology is widely

A. Ortis et al. (Eds.): ICAETA 2023, CCIS 1983, pp. 311–323, 2024.
https://doi.org/10.1007/978-3-031-50920-9_24

utilized in handheld devices with advanced features, such as smartphones [5]. Handheld devices are equipped with sensors that can be used to capture the image, movement, and touch [6], as the devices' computational power increases [7]. Therefore, the Ludo board game can be adapted for handheld AR (HAR).

An autonomous agent is an agent that can decide to perform an action by itself in a given environment [8]. The development of an autonomous agent is often closely linked with Artificial Intelligence (AI), particularly in the decision-making process [9]. A Non-Player Character (NPC) can act as an autonomous agent in video games [10]. Common NPCs in games are typically scripted with predetermined actions and unable to adapt to a dynamic environment [11]. Consequently, an autonomous agent can be developed in a game using a machine learning algorithm to adapt to the environment's changes, thereby increasing the game's difficulty and making it more enjoyable [10]. Q-learning is an example of a reinforcement learning algorithm within machine learning that enables the agent to explore the environment and discover the optimal action for a given state [12]. The Q-learning algorithm is frequently employed in a dynamic environment with multiple players [13]. Hence, the autonomous agent works with a Q-learning algorithm that can be developed for the AR Ludo board game.

AR is defined as a technology that combines real and virtual contents in the physical environment [4]. This technology overlays the digital data in the real environment to present various information such as text or images [14]. An AR system focuses on four domains that are sensing, tracking, interaction and display [15]. An example of a tracking technique is feature-based tracking provided by Vuforia Engine to detect, identify and track targets such as images or three-dimensional (3D) objects [16]. According to [17], Vuforia analyses and detects the features of an image uploaded by a user. More features detected in an image make it more suitable to be used as the image target. The analyzed result is represented through a star rating range from 0–5. The highest rating indicates the easiest for Vuforia to track the image target.

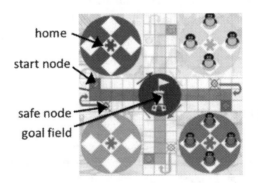

Fig. 1. Ludo board game [19]

Board game is defined as a game that involves a board that allows pieces to move on top of it [18]. According to [2], Ludo board game involves 2 to 4 players taking turns to roll a dice and move the tokens on the Ludo board. Figure 1 shows the Ludo board game.

The rules for the Ludo board game are shown below [19]:

- At the beginning of the game, the players take turns to roll the dice and a player that rolls a six can move a token to the start node. Each player has four tokens at the beginning of the game.
- The players can only move a token according to the number rolled if the token is not at home.
- Once a player rolls a six, that player can take another turn.
- If a player moves a token to the place that is occupied by the opponent's token, the opponent's token is kicked and returned to his home.
- A token is safe from being kicked back home once landing on the safe square. Each player's start node is also considered as a safe square.
- The game ends once all tokens of a player landed on the goal field.

Autonomous agent is an agent that can determine its behavior at runtime based on the current state of its environment [20]. AI contributes to the evolution of autonomous agents, especially in terms of their reasoning ability [21]. Q-learning is a reinforcement learning algorithm within machine learning that enables an agent to independently make decisions in its environment based on its policy, according to [22]. The agent can continue to collect data from the environment and regulate its responses in response to dynamic input from the environment [23].

A handheld device is a computing device that can be held by one's palm. Smartphones and other hand-held devices with various functionalities, such as an integrated camera and internet access, are commonly utilized in modern society [5]. Due to advances in hardware such as processors, cameras, and sensors, augmented reality is also widely used in handheld devices [24]. In many HAR applications, a touch screen interface is used, and a user can interact with the device's screen using only one hand while the other is used to hold the device [25].

This research is essential as a reference for the AR and AI research communities. Although this research is intended for handheld applications, the community can use this research as a point of reference to explore or conduct additional research on AR applications for handheld devices. Moreover, this study introduces the autonomous agent and the non-autonomous agent, both of which utilize different AI algorithms. The autonomous agent uses Q-learning algorithm, whereas the non-autonomous agent uses Minimax algorithms. This paper discusses the methodology, the Ludo Game and the implementation of the AI algorithms in Ludo Game. Then, the paper demonstrates the results and ends with a conclusion.

2 Methodology

This section describes the methodology guiding the creation of the AR Ludo board game. The methodology of the research work is sequentially divided into four phases. In the first phase, the autonomous agent is investigated in order to comprehend its definition and identify the AI algorithms that can be applied to both the autonomous agent and the non-autonomous agent. In addition, the fundamentals of augmented reality technology are investigated in order to select a suitable AR tracking technique for handheld devices.

The data of the board game Ludo is also gathered. All the information gathered in this phase serves as the foundation for the subsequent development phase.

In the second phase, the autonomous Agent A is built using the Q-learning algorithm, while the non-autonomous Agent B is created using the Mini-max algorithm. The board game Ludo is also developed with AR tracking capabilities. The development of autonomous Agent A begins with the design of a Q-learning-based algorithm, as depicted in Fig. 2. Based on Fig. 2, the algorithm begins with the initialization of Q-table values to zero. In accordance with the epsilon-greedy (ε-greedy) policy, Agent A may select a random action through exploration or a greedy action through exploitation after rolling the dice. A greedy action is one that has the highest value in the given state's Q-table. The epsilon value, ε is initialized as 1.0 and decreases continuously with each action taken by Agent A. If the randomly generated value between 0 and 1 is less than, random action is taken; otherwise, greedy action is taken. The reward, r, is then computed according to the state-action pair.

Bellman equation, $Q(s, a) = Q(s, a) + \alpha[r + \gamma max_{a' \in A} Q(s', a') - Q(s, a)]$ used to update the value of the state-action pair taken in Q-table. The learning rate, α and discount factor, γ are set as 0.7 and 0.1 respectively. These steps are repeated until the end of an episode. The training session is set up for Agent A with 20 episodes. An episode is ended once all the four tokens of a player successfully moved into the goal field.

```
Initialize Q(s, a), ∀s ∈ S, ∀a ∈ A
For episode = 1,2,3…do
        Repeat
                Roll dice
                For each token of the player
                        Initialize current state, s
                        Store s in the array
                        Initialize action, a based on the state and dice number
                        Store a in the array
                End For
                Choose action, a using epsilon-greedy policy:
                        If a random value < epsilon, ε
                                Then choose a random action, a
                        Else
                                Choose an action, a with highest Q(s,a) value
                Take action, a
                Observe reward, r and next state, s'
                Update Q-table with Bellman equation
                        Q(s, a) = Q(s, a) + α [r + γ max_{a'∈A} Q(s', a') − Q(s, a)]
                Until s is terminal // 4 tokens of a player reach goal field
End For
```

Fig. 2. Pseudocode of algorithm designed based on Q-learning

The subsequent phase involves the development of non-autonomous Agent B begins by designing a Minimax-based algorithm. Beginning with Agent B attempting to move at MaxNode, the possible actions that Agent B could take are determined. Each action that Agent B could take results in a MinNode. The children nodes of the MinNode are determined based on the opponent's potential actions. Each opponent's possible move results in a tempMin-Node that is also the terminal node. At the terminal node, the evaluationValue is calculated based on the total distance left by the opponent player minus the total distance left by Agent B. As each token requires 57 steps to reach the goal field and each player has four tokens, the total distance for each player is set to 228. Figure 3 illustrates the likelihood of discovering an opponent's actions using the Minimax algorithm.

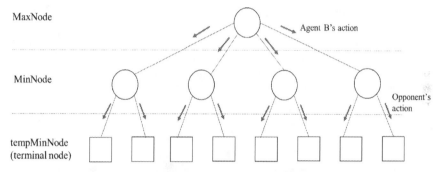

Fig. 3. Possibility of finding opponent's actions by Minimax algorithm

The third phase is the handheld integration of the AR Ludo board game with Agents A and B. The human player of AR Ludo can choose to play against either Agent A or Agent B. The FBX and PNG data sources are imported into the game to create the user interface (UI) and game objects. The AR Ludo board game rewards players with badges as a gamification element. Feature-based tracking is selected for the AR Ludo board game on handheld Android devices. Using one-finger gestures and touch-based interaction on the handheld screen, users can interact with the application. Following this is the testing procedure based on [26].

3   Ludo Game

The Ludo board game consists of dice and four sets of tokens that correspond to the four distinct colors in each corner. However, only two sets of tokens are implemented for the AR Ludo board game, as only two players are permitted. One player is given four blue tokens at the beginning of the game, while the other player receives four green tokens. The four tokens of each player are placed in the home. Then, each player rolls the dice to move a token from the beginning node to the goal field. The mission of the game is to move all four tokens into the goal field in the center of the board to win. According to Fig. 4, there are four purple nodes on the board known as safe nodes, and the dashed line illustrates an example path for blue tokens from the starting node to the goal field.

In order to implement AR tracking on a handheld device, Vuforia is integrated with Unity3D software to track and compare the marker's features with the database of target resources. Vuforia Target Manager receives a PNG image based on the Ludo board's design in RGB format. This image is converted to grayscale, and the extracted features are collected and saved in a database. The virtual AR Ludo board game is displayed once the marker is detected and its features match the data stored in the database with the green color tokens. During training, the remaining three players take random actions to advance in the game. Figure 5 shows a flowchart designed for Agent A to execute a game action using the Q-learning algorithm. Based on the flowchart, the Q-learning algorithm operates when Agent A is required to take a specific action at a given game state. The Q-table values are initialized at the outset, and Agent A can choose between exploration and exploitation after rolling the dice. Exploration permits the selection of a random action, while exploitation permits the selection of the action with the highest value in the Q-table. The execution of the selected action is followed by the observation of either a positive or negative reward based on the action taken. The subsequent step is to update the value of the action performed at the given state in the Q-table. If Agent A rolls a die with a value less than six, the turn passes to the following player. Agent A receives an additional roll if the dice value equals six.

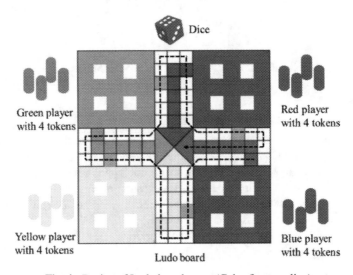

Fig. 4. Design of Ludo board game (Color figure online)

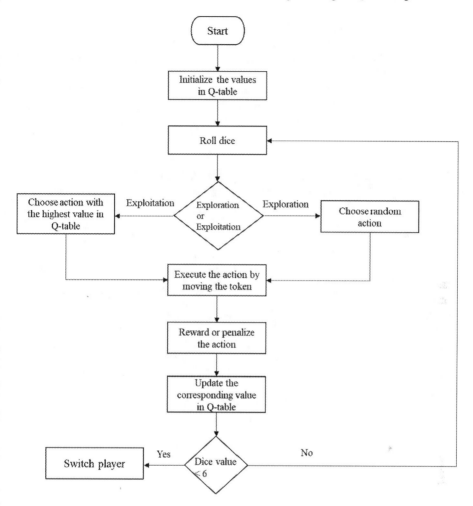

Fig. 5. Agent A's flowchart

At the end of the training, an output file that includes the finalized values of the Q-table is generated as in Fig. 6. Based on Fig. 6, the rows represent the states; meanwhile the columns represent the actions.

	Out Home	Get Into Goal	Move To Last Track	Move To Safe Square	Kill Opponent	Just Move
In Home	0.56337	0	0	0	0	0
On Last Track	0	1.035681	0.671286	0	0	0
On Safe Square	0	0	0.7215118	0.06530653	0.6852753	-0.1390742
On Free Space	0	0.9216789	0.7854184	0.4650407	0.6144565	-1.482576
In Goal	0	0	0	0	0	0

Fig. 6. Q-table file in txt format

Figure 7 illustrates that tokens inside the home have the state "In Home," and the action of moving a token from the home to the start node is "Out Home." The state of a token on a white node that surrounds the board is "On Free Space," and the action of moving from one white node to another without an opponent's token is "Just Move." Agent A may choose to move a token six spaces forward or out of his home based on Fig. 7 when a six is rolled. As the value in the Q-table for this state-action is approximately 0.563, which is greater than the action, Agent A decides to move a token out of the home state.

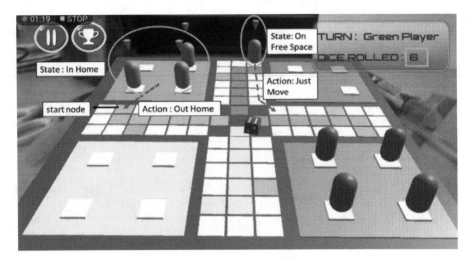

Fig. 7. States and actions in Ludo Game

4 Results

The AR Ludo game workspace is configured as depicted in Fig. 8, with the AR marker printed in color on an A4-sized sheet of paper and placed at a height of approximately 75 cm. The user sits approximately 30 cm in front of the table, holding the handheld device with one hand while interacting with the AR Ludo board game via touchscreen input with the other. The handheld device with camera used to track and display the AR Ludo board game is 45 cm away from the marker.

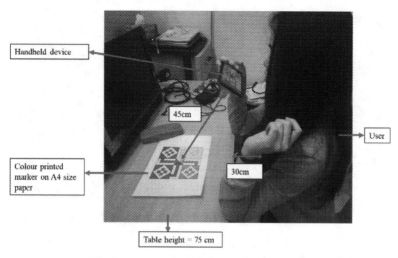

Fig. 8. Experiment setup (Color figure online)

The AR Ludo board game allows the human player to select either Agent A or Agent B as their opponent. If Agent A is selected, the human player is represented by the blue tokens, whereas Agent A is represented by the green tokens. Agent A and the human player compete to win the game. The dice are automatically rolled during Agent A's turn. As Agent A utilizes the Q-learning algorithm, Agent A's actions are referred to the values in the Q-table. After rolling a six on a human player's turn, the player can select the dice to roll or a token to move via touchscreen input as shown in Fig. 9. If the human player faces off against Agent B, Agent B is also represented by green tokens. Agents A and B use different algorithms when performing an action. The human player with blue tokens also rolls the dice according to Fig. 10 or selects a token to move through touchscreen input.

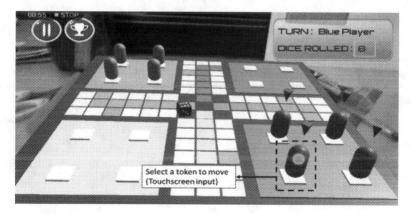

Fig. 9. Human player chooses a token to move (Color figure online)

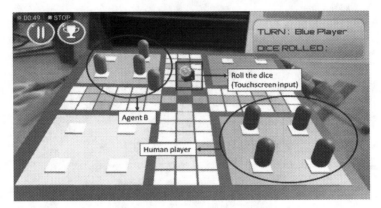

Fig. 10. Human player plays against Agent B (Color figure online)

The Minimax algorithm designed is implemented on Agent B. All of Agent B's movements are referenced by the evaluationValue that was propagated back. Assuming Agent B is a green-colored player and the opponent is a blue-colored player. While Agent B rolls a six, he or she may move an exterior token six spaces forward or an interior token to the start node, as depicted in Fig. 11.

Agent B decides to move a token from home to the start node instead of moving the outside token six spaces forward because he must consider all possible actions that his opponent could take on his next turn to reduce the evaluationValue. In accordance with Fig. 11, if Agent B moves the outside token six spaces forward, passing an opponent's token, the opponent may attempt to move its token four spaces forward on the following turn to send Agent B's token back home if the roll of a four on the dice.

Fig. 11. A token can be moved out from home or an outside token can be moved six steps forward passing an opponent's token (Color figure online)

5 Conclusion

This research required the use of the handheld device's built-in camera in order to detect and track the marker's features. After detecting a marker, the location and orientation of the virtual objects to be displayed on top of the real environment are calculated by matching the marker's features with the database of target resources. This allowed the virtual Ludo board game displayed in the physical world to be viewed on the handheld device's display.

The novel finding in this paper is to redesign the Ludo game for AR with autonomous agent and generate the training data using Q-Learning algorithm to produce autonomous agent in AR. The non-autonomous agent utilizing Minimax is also implemented in the AR Ludo game in order to compare the behavior of the agents in relation to the Ludo gaming. The AR Ludo game application is complete after the AR Ludo board game is integrated with both agents in the handheld. This integration gives the human player the option of playing against Agent A or Agent B. Agent A, unlike Agent B, utilizes the Q-learning algorithm. Agent B is a non-autonomous Minimax agent that, unlike Agent A, does not require retrieval of training data. Touchscreens are a common input method for handheld devices, and the basis of touch-based interaction is tapping the screen of a device with one finger. Therefore, the user interface and virtual game objects, such as dice and tokens, are interacted with by tapping with a single finger. As the ray casting that detects the collision between the ray and a 3D object in the scene is carried out, the virtual dice and tokens are selectable via touchscreen.

Based on the finding, we figured the limitation of Agent A which has been trained with random players that perform random actions in the game environment. Random actions taken the players did not lead to the best action each time. Hence, the training result of Agent A may not be the best as the opponents are not the players with strategies. Agent A only won the game three times out of ten plays. Some future works are suggested to improve this limitation is for Agent A can be trained with one random player, one player with defensive strategy who prioritizes to keep own tokens away from opponents' tokens and one player with aggressive strategy who prioritizes to chase and send opponents' tokens back to home. Minimax algorithm has been implemented to Agent B in AR Ludo board game. However, Agent B did not play well as the Minimax algorithm usually works well on deterministic games with no chance element. Agent B only won the game two times. The results show Agent A performs better than Agent B in the AR Ludo board game due to the aspect of artificial intelligent movement decision. Further evaluation may require to measure the performance and the interaction which trade-off to the tracking accuracy in AR [25]. Improvement can be made to modify the collaborative framework to enhance multi-user interaction since this research only focus on AI autonomous agent [26].

References

1. ChePa, N., Alwi, A., Din, A.M., Mohammad, S.: Digitizing Malaysian traditional game: e-Congkak. In: Knowledge Management International Conference (KMICe), Malaysia (2014)
2. Singh, P.R., Abd Elaziz, M., Xiong, S.: Ludo game-based metaheuristics for global and engineering optimization. Appl. Soft Comput. 1(84), 105723 (2019)

3. Rizov, T., Đokić, J., Tasevski, M.: Design of a board game with augmented reality. FME Trans. **47**(2), 253–257 (2019)
4. Ismail, A.W., Billinghurst, M., Sunar, M.S.: Vision-based technique and issues for multimodal interaction in augmented reality. In: Proceedings of the 8th International Symposium on Visual Information Communication and Interaction, pp. 75–82 (2015)
5. Zsila, Á., et al.: An empirical study on the motivations underlying augmented reality games: the case of pokémon go during and after pokémon fever. Personality Individ. Differ. **15**(133), 56–66 (2018)
6. Grandi, J.G., Debarba, H.G., Bemdt, I., Nedel, L., Maciel, A.: Design and assessment of a collaborative 3D interaction technique for handheld augmented reality. In: 2018 IEEE Conference on Virtual Reality and 3D User Interfaces (VR), pp. 49–56. IEEE (2018)
7. Sanches, S.R., Oizumi, M.A., Oliveira, C., Sementille, A.C., Corrêa, C.G.: The influence of the device on user performance in handheld augmented reality. J. Interact. Syst. **10**(1) (2019)
8. Coutinho, L.R., Galvão, V.M., Jr, A.D.A.B., Moraes, B.R., Fraga, M.R.: Organizational gameplay: the player as designer of character organizations. International J. Comput. Games Technol. **2015**, 6 (2015)
9. Moharir, M., Mahalakshmi, A.S., Kumar, G.P.: Recent Advances in Computational Intelligence. In: Kumar, R., Wiil, U.K (eds.), pp. 255–261. Springer, Cham (2019)
10. Feng, S., Tan, A.H.: Towards autonomous behavior learning of non-player characters in games. Expert Syst. Appl. **56**, 89–99 (2016)
11. Lim, M.Y., Dias, J., Aylett, R., Paiva, A.: Creating adaptive affective autonomous NPCs. Auton. Agent. Multi-Agent Syst. **24**, 287–311 (2012)
12. Lee, J., Kim, T., Kim, H.J.: Autonomous lane keeping based on approximate Q-learning. In: 2017 14th International Conference on Ubiquitous Robots and Ambient Intelligence (URAI), pp. 402–405. IEEE (2017)
13. Selvakumar, J., Bakolas, E.: Min-Max Q-learning for multi-player pursuit-evasion games. Neurocomputing **475**, 1–4 (2022)
14. Ismail, A.W.: User Interaction Technique With 3D Object Manipulation in Augmented Reality Environment (Doctoral dissertation, Universiti Teknologi Malaysia) (Doctoral dissertation) (2011)
15. Rabbi, I., Ullah, S.: A survey on augmented reality challenges and tracking. Acta graphica: znanstveni časopis za tiskarstvo i grafičke komunikacije, **24**(1–2), 29–46 (2013)
16. Borycki, D.: Programming for Mixed Reality with Windows 10, Unity, Vuforia, and UrhoSharp. Microsoft Press (2018)
17. Hameed, Q.A., Hussein, H.A., Ahmed, M.A., Omar, M.B.: Development of augmented reality-based object recognition mobile application with Vuforia. J. Algebraic Stat. **13**(2), 2039–2046 (2022)
18. Dewi, E.: The influence of using board game as a media on the students' speaking ability atthe third semester of English education department of state Institute for Islamic Studies of Metro in the academic year of 2018/2019. IAIN Metro (2019)
19. Chhabra, V., Tomar, K.: Artificial intelligence: game techniques ludo-a case study. ACSIT **2**(6), 549–553 (2015)
20. Hoffmann, R., Ireland, M., Miller, A., et al.: Autonomous agent behaviour modelled in PRISM – a case study. In: Bošnački, D., Wijs, A. (eds.) Model Checking Software. SPIN 2016. LNCS, vol. 9641, pp. 104–110 (2016). https://doi.org/10.1007/978-3-319-32582-8_7
21. Albrecht, S.V., Stone, P.: Autonomous agents modelling other agents: a comprehensive survey and open problems. Artif. Intell. **258**, 66–95 (2018). https://doi.org/10.1016/j.artint.2018.01.002
22. Finnman, P., Winberg, M.: Deep reinforcement learning compared with Q-table learning applied to backgammon (2016)

23. Zamstein, L.M., Smith, B.A., Hodhod, R.: A comparative study of opponent type effects on speed of learning for an adversarial Q-learning agent. In: 2019 SoutheastCon (2019). https://doi.org/10.1109/southeastcon42311.2019.9020449

24. Ong, Y.H., Ismail, A.W., Iahad, N.A., et al.: A mobile game SDK for remote collaborative between two users in augmented and virtual reality. IOP Conf. Ser. Mater. Sci. Eng. **979**, 012003 (2020). https://doi.org/10.1088/1757-899x/979/1/012003

25. Fadzli, F.E., Yusof, M.A., Ismail, A.W., et al.: Argarden: 3D outdoor landscape design using handheld augmented reality with multi-user interaction. IOP Conf. Ser. Mater. Sci. Eng. **979**, 012001 (2020). https://doi.org/10.1088/1757-899x/979/1/012001

26. Nor'a, M.N.A., Ismail, A.W.: Integrating virtual reality and augmented reality in a collaborative user interface. Int. J. Innovative Comput. **9**(2) (2019). https://doi.org/10.11113/ijic.v9n2.242

Modelling and Estimating of VaR Through the GARCH Model

K. Senthamarai Kannan⑩ and V. Parimyndhan$^{(\boxtimes)}$⑩

Department of Statistics, Manonmaniam Sundaranar University, Tirunelveli 627012, India
senkannan2002@gmail.com,senthamaraikannan@msuniv.ac.in, parimyndhanvmp@gmail.com

Abstract. This study focuses on the analysis of fiscal series with time-varying conditional variance utilizing the ARIMA-GARCH with Value at Risk (VaR) model. ARIMA-GARCH can predict risk when stock variance is Heteroscedasticity. The price of the Reliance stock is analyzed for fifty months. This research indicates that the VaR is a useful technique to reduce risk exposure and perhaps avoid losses when investing in the Reliance stock. The findings show that ARIMA (0,0,0)-GARCH (1,1) has the best fit, with an Akaike information criterion (AIC) value of −5915.325, at a confidence level of 95%. The GARCH technique is used to determine the conditional variance of the residuals and contrasts it with the delta-normal method. At a 95% confidence level, the VaR is used to calculate the likelihood of losing an investment by 2.7% or more in a single day.

Keywords: ARIMA · forecasting · GARCH · Reliance · stock · VaR

1 Introduction

One of the most crucial asset groups for any investor is stock. There are various benefits when investing in the stock market. Stocks often offer liquidity and a respectable return over time. Fundamental analysis and technical analysis make up the two primary types of stock price prediction techniques [6]. Fundamental analysis is the procedure used to estimate the fundamental value [15]. Technical analysis is a different kind of analysis and it uses a variety of graphs, charts, and statistical techniques like linear regression, ARMA, ARIMA, GARCH, etc., to eliminate human subjectivity and emotion.

In this work, the return of Reliance stock is considered. Reliance Industries Limited which is found in 1973 and is a large-cap corporation with a market valuation of Rs. 1,735,511 crores working in a diversified area. The primary revenue of Reliance Industries Limited segments includes oil & gas, income from financial services, petrochemicals, various services, income from retail and many

Supported by Manonmaniam Sundaranar University.

others. The value of the stock returns could be used to determine the gains and losses of stock investing.

Stock returns can be classified as homoscedastic or Heteroscedasticity based on variance values [24]. A time series model is a vital data tool for forecasting in the face of future needs [17]. The ARIMA model is used to forecast returns with homoscedastic variance [26]. The ARIMA-GARCH model or ARIMA-GARCH Ensemble can be used to forecast returns with heteroscedasticity variance [16]. In this research article, the ARIMA model is employed as one of the most popular forecasting and econometric analysis tools [12].

The quantiles of a conditional distribution may be easily derived for the computation of VaR by using conditional mean and conditional variance obtained from the estimated GARCH model [25]. The present paper demonstrates the GARCH method which is used to assess the most significant Reliance Stock from the perspective of Value at Risk. By exploiting the GARCH, VaR is estimated at the 95% confidence level.

The rest of the article is structured as follows after this introduction. Section 2 describes the literature study. The fundamental principles of the approaches used are provided in Sect. 3. The reasons for selecting the Modeling approach are briefly discussed in Sect. 4 along with the findings of VaR research on the Reliance stock. Finally, the conclusion of the paper is given in Sect. 5.

2   Literature Survey

Deb et al., (2003) using eight various univariate models, predicted the volatility of both the market indexes of the Indian markets [5]. The GARCH (1,1) model outperforms some other models in this set of models' out-of-sample forecasting results. Additionally, Karmakar (2005) noted that the GARCH(1,1) model offers a respectably accurate forecast using conditional volatility models to estimate the volatility of fifty different equities in the Indian stock market [11].

Belghi et al., (2018) using parametric and non-parametric techniques, examined the possibility of predicting stock prices, and used a variety of parametric time series approaches [3]. Sarah et al., (2018) proposed a hybrid technique that makes use of the distinct strengths of the GARCH + ANN model and the GARCH + SVM model in forecasting stock indexes and demonstrated that the provided hybrid model provides the best forecasting when compared to other models [9].

Sanchia (2020) suggested the ARIMA-GARCH models are used to predict stock values over the following several periods, and the forecast data plot almost matches the actual data pattern, demonstrating the algorithm's ability to provide accurate forecasts [19]. Rahul et al., (2020) In their paper, the S&P BSE stock price is predicted through the Autoregressive Integrated Moving Average (ARIMA) methodology [20].

Financial ratios are simply one way to create a portfolio; other methods include Smart beta, alpha, diversification, and Value at Risk (Salim et al., 2020) [18]. By integrating a variety of ASEAN indices, Waspada et al. (2021) created an ASEAN investment of indices that produced the best earnings with the least amount of risk [27].

Solomanchuk and Shchestyuk (2021) purposed to conduct and contrast the Markowitz technique and the VaR Monte Carlo approach for the Student-like model in terms of investment risk, and to demonstrate that the ideal Markowitz portfolio typically exhibits the minimum Value at Risk when compared to other portfolios [23].

Iswanto and Ramadhan (2022) were carried out to evaluate each individual's risk and profit level using the Value at Risk in January 2022 for the Top 10 Best Stocks edition of Kontan.co.id. [8]. Singh et al., (2022) established if the share market indicators are acceptable decision-aid techniques within the framework of intraday managing risk, the body of research findings on stock index forecasts combined with machine learning approaches for both short- and long-term risk management is used [21].

Weronika (2022) discovered that the LSTM performs similarly to GARCH estimators; however, on actual market data, it is more sensitive to rising or falling volatility and surpasses all other value-at-risk estimators in terms of exception rate and mean quantile score [13]. Salem et al., (2022) explored several statistical techniques to calculate the VaR for stock return in the BRICS nations from 2011 to 2018 [4].

3   Material and Methods

3.1   Material

Data used are the daily observations of the Reliance stock price collected between 01.01.2018 and 31.10.2022. Here is a total of 1193 daily data points. The data was extracted from Yahoo Finance. It has captured data on Open, High, Low, Close, and Volume. Open: The price of the stock when the market opens in the morning, Close: The price of the stock when the market closed in the evening, High: Highest price the stock reached during that day, Low: Lowest price the stock is traded on that day, Volume: The total amount of stocks traded on that day. Here the close price of the Reliance is used.

3.2   Methods

ARIMA. ARIMA is a popular model for analyzing financial time series since it is easy to understand and describes a diverse set of processes [14]. ARIMA, or Auto Regressive Integrated Moving Average, is a well-known forecasting model and is otherwise known as the Box-Jenkins approach [22]. Box and Jenkins state that differencing Y_t can make non-stationary data stationary. Y_t is generally modeled as follows:

$$Y_t = c + \phi_1 y_{d_{t-1}} + ... + \phi_p y_{d_{t-p}} + ... + \theta_1 e_{t-1} + \theta_1 e_{t-q} + e_t \qquad (1)$$

Where Y_t denotes the time series difference, ϕ and θ represent the unknown parameters, and e denotes the error term with zero mean. Y_t is current and past values of error are expressed here. The model employs three fundamental techniques like Auto Regressive (AR), Differencing(I for Integrated), and Moving Average(MA). In AR the model has a "p" value. In differencing if d = 1 the difference between two consecutive time series entries is examined. If d = 2 the differences of the differences acquired at d = 1 are examined, and so on. q value in the MA model represents the number of lagged error values. This model is known as the ARIMA(p, d, q) of Y_t. To construct our model, we will follow the methods outlined below.

Test for Stationarity: A time series must be stationary to be modeled using the Box-Jenkins method. A stationary time series seems to have no trend, a constant mean, and variance throughout time, which makes values easy to predict. Stationarity check - To check the stationarity the ADF unit root test is used. The p-value from the ADF test must be less than 0.05 or 5%. If the p-value indicates that the process is non-stationary if it is greater than 0.05 or 5%.

Identifying p and q: The stock closing is used to validate stationarity in the previous step. The Autocorrelation Function (ACF) and Partial Autocorrelation Function can be used to estimate the p and q order of the ARIMA model (PACF). The AIC is the alternative model for determining the model (AICc) and it calculates the quality of each model. An overfitting issue might develop when more lag parameters were included in the model which reduces the sum of the square of the residuals. Therefore, the model with the lowest AIC is selected [2].

Diagnostics Checking: Examining the residual plot, its ACF and PACF diagrams, as well as the Ljung-Box test, are all included in the approach. If the ACF and PACF of the model residuals show no substantial delays therefore the chosen model is adequate.

GARCH. The residuals in the time series exhibit a few clustering volatility. The volatility model is used by ARCH (Autoregressive Conditional Heteroscedasticity). ARCH represents variance as a function of the magnitude of the error term in the prior period in a time series model ARCH is extended by GARCH, which allows the variance to be affected by its delays as well as the lags of the squared residuals. This model is one of the most commonly used for modelling the volatility of time series, i.e., when the volatility of the series is not constant [7]. GARCH can capture larger changes such as increased or decrease volatility.

$$\sigma_t^2 = \alpha_0 + \alpha_1 u_{t-1}^2 + \alpha_2 u_{t-2}^2 + \ldots + \alpha_p u_{t-p}^2 + \beta_1 \sigma_{t-1}^2 + \ldots + \beta_p \sigma_{t-q}^2 \quad (2)$$

α represents a reaction parameter. When the α is high, the market is spiky or nervous, and when the α is low, the market is stable or calm. β represents a volatility persistence, there is a clustering of volatility when β is high, indicating strong persistence. High β is usually associated with low α and vice versa. There are two parameters for GARCH (p, q). p: number of lag residual errors, q: number of lag variance

Value at Risk: A common risk measurement in finance is Value at Risk (VaR). VaR is referred to as the supreme potential losing value over a specific period with a given level of confidence. VaR measures investment risk by demonstrating the maximum possible loss. A VaR is an amount calculated with a confidence level based on a certain amount and loss. Thus, it measures the loss that may be incurred from a given amount over a given t period with a certain $\propto$ level. Under specific market conditions and at a specific level of risk, VaR is calculated. It is common for VaR estimation to use the standard method which assumes one variable and a normal distribution with μ as its mean and σ as its standard deviation.

Estimating VaR involves finding the percentile of the standard normal distribution z_1-$\propto$ to (1-$\propto$),

$$1-\propto = \int_{-\infty}^{q} f(r)dr = \int_{-\infty}^{z_{1-\propto}} \psi(z)dz = N(z_1-\propto) \tag{3}$$

Where quartiles q= $z_{(1-\propto)}$ σ + μ, $\psi(z)$ is the pdf (probability density function) based on the standard distribution, N(z) is the cumulative normal distribution function, r represents the random variable value, R denotes the return of the stock, and f(r) denotes the density function with mean μ and variance σ. Then,

$$r_t = \mu_t + \sigma_t z_t, \tag{4}$$

$$VaR_\alpha^t(r_t) = -inf(r_t|F(r_t) \geq \propto) \tag{5}$$

4   Results and Discussion

ARIMA Model. The time series must be stationary in order to use the ARIMA model. There will be no significant differences between ordinary returns and logarithmic returns since the return are distributed as a leptokurtic with a mean close to zero. The return of the Reliance stock has been checked for stationary. **Stationary Sequence:**

In left side of the Fig. 1, the time series of return has a zero mean, and the return shows very high volatility on some random days, especially after 2020 which will be visible. The return histogram has been displayed in right side of the Fig. 1. The histogram appears to be symmetrical and is well-centered about zero.

Table 1, shows the results of the ADF test, the stationary time series was suggested by the Null Hypothesis, to confirm the stationary of the return.

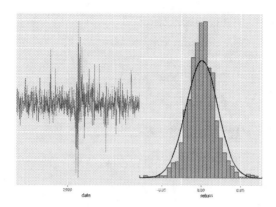

Fig. 1. Reliance stock return plot and Return of the Histogram diagram

Table 1. Result of ADF test.

ADF Value	Significance Level	p-value	Decision
−10.348	5%	0.01	Data is stationary

Identification: In this section, the best ARIMA model is fitted using the auto.arima command in R. The best model according to auto.arima is ARIMA(0,0,0) with a non-zero mean AIC: −5915.325. The lowest AIC of ARIMA(p,d,q) is chosen by the auto.arima function.

Diagnostics Check: ACF and PACF diagrams, as well as Ljung-Box results, are viewed in the residual plot. Autocorrelation appears to be greater than zero in both ACF and PACF plots. Figure 2 shows the histogram of residues compared to a normal distribution $N(0, \sigma^2)$.

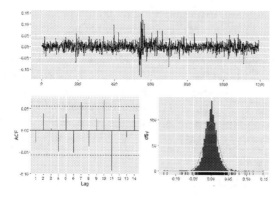

Fig. 2. Residuals form ARIMA(0,0,0) with non-zero mean

The Ljung-Box test is used to verify the idea that the residuals are uncorrelated (Table 2).

$$Q_{LB} = T(T+2) \sum_{s=1}^{m} \frac{\hat{p}_s^2}{(T-s)} \tag{6}$$

Table 2. Result of the Ljung-Box Test.

Ljung-Box Test	Significance Level	p-value
13.999	5%	0.3008

Implementing GARCH (1,1). Even though the residuals of ACF and PACF have no discernible delays, there is considerable cluster volatility in the residual time series plot. The ARCH model is used to model volatility. As a function of the magnitude of earlier error terms, the ARCH statistical model for time series data describes the variance of the present error term. The GARCH process is reliable when the squared residuals are correlated A significant association is easily seen in the ACF and PACF charts.

Table 3. Result of GARCH(1,1).

	Estimate	Std. Error	$Pr(> |t|)$
μ	0.000000000000006252	0.0005145440223	1.0000
ω	0.000000469062831869	0.0000009532932	0.6227
α_1	0.070377641423155821	0.0083417493834	0.0000
β_1	0.917191718194824923	0.0086856871742	0.0000

Table 3, The parameter estimates for α_1 and β_1 are typical of stock return estimation (α_1 smaller, β_1 larger, their total is almost 1). Using α_1 here could determine how well a volatility shock today transfers into the volatility of the next period. 7% of the volatility from the previous period will be carried over to the following day in our model.

A significant device from zero, the β_1 coefficient is 0.9171917. This implies that, in addition to the effects of the squared error from yesterday, about 91% of the return variance from yesterday is passed to the return variance from today. This coefficient is more than 0.7, which indicates that the volatility is clustering. For the process to be stationary for covariance, $\alpha_1 + \beta_1 < 1$. The amount is closer to $\alpha_1 + \beta_1$ to 1, As time goes on the squared residuals of autocorrelation decay more slowly.

The estimates, in this case are 0.9875693, which is almost completely integrated and close to the limit. A high GARCH coefficient indicates persistent and clustering volatility. The presence of a leverage effect in the series is indicated by a positive, statistically significant coefficient for a leverage effect.

Value at Risk. Value at Risk (VaR) is a statistical measure of downside risk based on the current position. It calculates the potential losses of a group of investments under typical market circumstances over a specified time frame. The worst daily loss will not be greater than the VaR calculation at a 95% confidence level that could be predicted. By calculating the value of the 5% quantile, The VaR may be calculated using historical data, and the estimate based on the data is −0.02668343. Figure 3 shows the maximum loss which can be faced by the data at a 95% confidence level, return below the 5% quantile are represented by red bars.

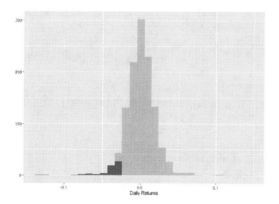

Fig. 3. Illustrates a hypothetical probability density function for profit-and-loss scenarios with a 5% value at risk

Delta-Normal Approach Vs GARCH VaR: In delta-normal analysis, all stocks are assumed to have normal distributions. Calculating the variance of past returns is part of this approach. VaR is characterized as:

$$VaR(a) = \mu + \sigma * N^{-1}(a) \tag{7}$$

where σ is the standard deviation of stock return and μ is the mean stock return, N^{-1} represents the inverse PDF function, and (a) represents the chosen confidence level, constructing a normal distribution's appropriate quantile based on (a). The outcomes of such a straightforward approach are frequently disappointing and hardly applied in modern practice. The return show time-varying volatility. Therefore, the conditional variance is provided by the GARCH(1,1) model to estimate VaR. VaR is represented as follows for this method (Fig. 4):

$$VaR(a) = \mu + \hat{\sigma}_{t|t-1} * t^{-1}(a) \qquad (8)$$

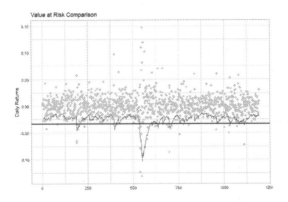

Fig. 4. VaR Comparison

A delta-normal VaR is represented by the green line, while the GARCH model produces a red line.

5   Summary and Conclusion

The data considered in this paper was collected from the Yahoo Finance API for the period between January 1, 2018, and October 31, 2022. The appropriate model is determined using the Box-Jenkins methodology. To choose the optimal model, the AIC test criteria are used in the data provided before. The ARIMA model is unconditional and steady-state. As a result, in the process of Diagnostics Checking, there is no sign of a pattern that could be modeled because the residuals behave like white noise.

This work, here could demonstrate the application of GARCH and ARCH models to model the changing variance in time series with observed periods of volatility. Additionally, we have seen why ARIMA models fall short when it comes to forecast time series with unpredictable patterns. Although the GARCH model used in this work successfully captures the volatility in stock data analysis, it is unable to forecast the volume of stock returns.

VaR is frequently used to estimate market risk. It is excessively straightforward and frequently results in risk quantification. In this paper, VaR with time-varying volatility is used. The possibility of losing investment by 2.7% or more in a single day is 5%. The VaR employed in this article allows for fat tails and leverage effects in volatility in addition to the dynamic aspect. Further work will compare GARCH models with the conditional Extreme Value Theory

(EVT) specifications with a focused on the distribution of heavy tails in order to improve the efficiency of VaR forecasting.

Acknowledgement. The authors would like to express their gratitude to the editor and learned reviewers for their valuable comments and suggestions to improve the earlier version of this manuscript. There is no conflict of interest as declared by the authors.

Financial Support and Funding. This work was financially supported by Bharat Ratna Dr. M. G. Ramachandarn Centenary Research Fellowship by Manonmaniam Sundaranar University, Tirunelveli.

References

1. Artzner, P., Delbaen., Eber, J. M., Heath, D.: Coherent measures of risk. Math. Finance **9**, 203–228 (1999)
2. Ashik, A.M., Kannan, K.S.: Time series model for stock price forecasting in India. In: Logistics, Supply Chain and Financial Predictive Analytics, pp. 221-231 (2019)
3. Belaghi, R.A., Aminnejad, M., Alma, Ö. G.: Stock market prediction using non-parametric fuzzy and parametric GARCH methods. Turkish J. Forecast. **2**(1), 1-8 (2018)
4. Ben Salem, A., Safer, I., Khefacha, I.: Value-at-risk (VaR) estimation methods: empirical analysis based on BRICS markets (2022)
5. Deb, S.S., Vuyyuri, S., Roy, B.: Modeling stock market volatility in India: a comparison of univariate deterministic models. ICFAI J. Appl. Finance, 19–33 (2003)
6. Devadoss, A.V., Ligori, T.A.A.: Forecasting of stock prices using multi layer perceptron. Int. J. Comput. Algorithm **2**(1), 440–449 (2013)
7. Gonzales, M.F., Burgess, N.: Modeling market volatilities: the neural network perspective. Eur. J. Financ. **3**, 137–157 (1994)
8. Iswanto, P., Ramadhan, A.R.: Pengukuran Tingkat Risiko Dan Keuntungan Saham individual Dengan Menggunakan Pendekatan Historis Pada Metode value at risk (VaR)(Studi Kasus Top 10 Saham Terbaik Januari 2022). Jurnal Akuntansi dan Manajemen Bisnis **2**(1), 46–55 (2022)
9. Johari, S.N.M., Farid, F.H.M., Nasrudin, N.A.E.B., Bistamam, N.S.L., Shuhaili, N.S.S.M.: Predicting stock market index using hybrid intelligence model. Int. J. Eng. Technol. (UAE) **7**, 36–39 (2018)
10. Kannan, K.S., SulaigaBeevi, M., Fathima, S.S.A.: A comparison of fuzzy time series and ARIMA model. Int. J. Sci. Technol. Res. **8**(8), 1872–1876 (2019)
11. Karmakar, M.: Modeling conditional volatility of the Indian stock markets. Vikalpa **30**(3), 21–38 (2005)
12. Liu, M.D., Ding, L., Bai, Y.L.: Application of hybrid model based on empirical mode decomposition, novel recurrent neural networks and the ARIMA to wind speed prediction. Energy Convers. Manage. **233**, 113917 (2021)
13. Ormaniec, W., Pitera, M., Safarveisi, S., Schmidt, T.: Estimating value at risk: LSTM vs GARCH. arXiv preprint arXiv:2207.10539 (2022)
14. Tsay, R.S.: Analysis of Financial Time Series. Wiley, Hoboken (2010)
15. Raghunathan, V., Rajib, P.: Stock exchanges, investments and derivatives: straight answers to 250 nagging questions. Tata McGraw-Hill (2007)

16. RiaFaulina, S.: Hybrid ARIMA-ANFIS for rainfall prediction in Indonesia. International Journal of Science and Research (IJSR). 2319-7064
17. Rubio, L., Alba, K.: Forecasting selected Colombian shares using a hybrid ARIMA-SVR model. Mathematics **10**(13), 2181 (2022)
18. Salim, D.F., Rizal, N.A.: Portofolio optimal Beta dan Alpha. Jurnal Riset Akuntansi dan Keuangan **9**(1), 181–192 (2021)
19. Sanchia, N.G.: Implementasi model Arima-Garch menggunakan metode maximum likelihood: Studi kasus harga saham Jakarta Islamic Index (2020)
20. Si, R.K., Padhan, S.K., Bishi, D.B.: Application of box – Jenkins ARIMA (p, d, q) model for stock price forecasting and detect trend of S&P BSE stock index: an evidence from Bombay stock exchange (2020)
21. Singh, B., et al.: ML-based interconnected affecting factors with supporting matrices for assessment of risk in stock market. Wirel. Commun. Mob. Comput. (2022)
22. Situngkir, H.: Value at risk yang memperhatikan sifat statistika distribusi return (2006)
23. Solomanchuk, G., Shchestyuk, N.: Risk modelling approaches for student-like models with fractal activity time (2021)
24. Tarno, T., Di Asih, I.M., Rahmawati, R., Hoyyi, A., Trimono, T., Munawar, M.: ARIMA-GARCH model and ARIMA-GARCH ensemble for value-at-risk prediction on stocks Portfolio (2020)
25. Tsay, R.S.: Analysis of Financial Time Series. Wiley, Hoboken (2005)
26. Wabomba, M.S., Mutwiri, M., Fredrick, M.: Modeling and forecasting Kenyan GDP using autoregressive integrated moving average (ARIMA) models. Sci. J. Appl. Math. Stat. **4**(2), 64–73 (2016)
27. Waspada, I., Salim, D.F.: Smart beta in index country ASEAN. Eur. J. Mol. Clin. Med. **7**(11), 906–918 (2020)

Nuclei Instance Segmentation in Colon Histology Images with YOLOv7

Serdar Yıldız[1,4]([✉]), Abbas Memiş[1,2], and Songül Varlı[1,3]

[1] Department of Computer Engineering, Faculty of Electrical and Electronics Engineering, Yıldız Technical University, İstanbul, Turkey
serdar.yildiz@std.yildiz.edu.tr, abbas.memis@istun.edu.tr,
svarli@yildiz.edu.tr
[2] Department of Software Engineering, Faculty of Engineering and Natural Sciences, İstanbul Health and Technology University, İstanbul, Turkey
[3] Health Institutes of Türkiye, İstanbul, Turkey
[4] BİLGEM, TÜBİTAK, Kocaeli, Turkey

Abstract. In histology image analysis, instance-based nuclei segmentation is one of the challenging tasks within the segmentation-guided studies since it is quite troublesome to detect each distinct nuclei instance of each nuclei type in images in contrast to the semantic segmentation in which all the image pixels of a nuclei type are labelled with the same mask ID although the segmented region may comprise of multiple instances. In this paper, an instance-based medical image segmentation task is addressed, and in this context, instances of multiple types of nuclei in colon histology images are aimed to be delineated distinctly. For the instance-based segmentation of the nuclei in colon histology images, the YOLOv7 algorithm and its built-in instance segmentation module are utilized. In the experimental studies performed on Colon Nuclei Identification and Counting (CoNIC) Challenge 2022 colon histology image dataset by using a 5-fold cross-validation performance evaluation strategy, nuclei instances belonging to 6 classes as the neutrophil, epithelial, lymphocyte, plasma, eosinophil and connective were segmented. To calculate the overall system accuracy, the quantification metrics of mean average precision (mAP) and mean panoptic quality (mPQ) were measured. In performance evaluations, quite promising accuracy values were obtained. The mAP values of 0.2885 and 0.2903, and mPQ values of 0.1659 and 0.1704 were observed by using the YOLOv7 algorithm. To the best of our knowledge, this is the first nuclei instance segmentation study with YOLOv7.

Keywords: Nuclei instance segmentation · Nuclei segmentation · Colon histology images · YOLOv7 · CoNIC 2022 dataset

1 Introduction

In histology, pathologists analyze tissue and cell structures under microscopes to identify various components. Even for most experts, the identification and classification of the tissue components by eyes is a difficult task. It is a time-consuming process due to its difficulty, and the probability of making a mistake is quite high. Moreover, the long duration of tissue analysis causes a delay in the treatment of potentially deadly diseases that can be avoided with an early diagnosis. Histology slides have been digitized, and human-oriented processes have been automated in high-capacity computing systems to solve these kinds of problems in digital pathology.

Histology images stained with Haematoxylin and Eosin (H&E), which is one of the most widely used image staining techniques in digital pathology, or other types of histology slides are digitized using scanning devices to overcome the difficulty of analyzing tissue slides under the microscopes. State-of-the-art advances in artificial intelligence and computational pathology have made it possible to automatically analyze digitized histology slide images and generate supplementary outputs that can be interpreted by the pathologist. In a computer-based intelligent analysis of the digital histology slide images, essential image processing and machine learning tasks such as detection, segmentation, quantification and classification can be utilized depending on the requirements. In such systems, detection models can calculate cell density per square millimeter, whereas the classification model can identify cell type. Furthermore, the semantic segmentation model can also determine cell boundaries.

In histology image analysis, instance-based nuclei segmentation is one of the challenging tasks within the segmentation-guided studies since it is quite troublesome to detect each distinct nuclei instance of each nuclei type in images in contrast to the semantic segmentation in which all the image pixels of a nuclei type are labelled with the same mask ID although the segmented region may comprise of multiple instances. By using an instance-based segmentation model, the tasks of detection, classification and segmentation can be used and unified in a single model. Therefore, the total system complexity may also be reduced. In the current literature, various studies propose a solution to the problem of nuclei instance segmentation. In a related work, Graham et al. [6] proposed HoVer-Net with one encoder and three decoders for instance segmentation. A decoder segments the input image as a nucleus or not and another decoder segments the input image semantically. The final decoder estimates the normalized distance of the pixel from the center of the nucleus in the $[-1,1]$ range. In the post-processing process, instance maps are obtained using the normalized distance information and the number of the nuclei instances is reduced by using the binary nuclei segmentation output. Finally, nuclei class labels are assigned by utilizing the semantic segmentation map. In a recently reported study, Rumberger et al. [8] proposed a HoVer-Net-like model. The decoder module that performs binary segmentation has been changed to predict 3 classes: background, corner, and nucleus. Liu et al. [7] proposed to create an ensemble model by combining HoVer-Net and Cascaded Mask-RCNN [2] models. Moreover, they proposed a non-maximum suppression algorithm that combines the outputs of the two models according to the weights of the classes in the dataset. Weigert et al. [12] proposed to predict a polygon for each nucleus using the Stardist [9] model.

Böhland et al. [1] proposed to train outputs to perform instance segmentation using the L1 loss function with a U-Net structure. An instance segmentation map is obtained by performing a Watershed post-processing to the U-Net model output.

In this study, we also addressed the instance-based nuclei segmentation task, and in this context, instances of multiple types of nuclei in colon histology images are aimed to be delineated distinctly. In other words, it is aimed to predict an instance mask and a nuclei type (class) label for each nucleus in the images. For the instance-based segmentation of the nuclei in colon histology images, the YOLOv7 algorithm and its built-in instance segmentation module are utilized. The experimental studies are performed on Colon Nuclei Identification and Counting (CoNIC) Challenge 2022 colon histology image dataset [4,5]. The nuclei instances belonging to 6 classes as the neutrophil, epithelial, lymphocyte, plasma, eosinophil and connective are segmented. The rest of the paper is organized as follows: The methods used in the proposed study are explained in Sect. 2. In Sect. 3, the experimental dataset is described briefly and the results observed within the scope of the experimental studies are given. Finally, conclusions are stated in Sect. 4.

2   Methods

2.1   Instance Segmentation and YOLOv7

Instance segmentation is a difficult computer vision task used to localize and mask an object. An instance segmentation model, like an object detection concept, predicts a bounding box for each object in the image and masks it by determining the object's border pixels. Instance segmentation is more troublesome than the object detection task in general since occlusion in a predicted bounding box makes the instance segmentation problem difficult to solve [14]. In particular, it is very difficult to determine the boundary lines between the objects belonging to the same class. In this study, we performed a nuclei instance segmentation using the YOLOv7 algorithm [11]. As is known, YOLO (You Only Look Once) single-stage network model is a very popular and widely used object detector [10] and it is used in so many applications such as garbage detection, medical face mask detection, pole detection and counting in distribution network, human detection and ship detection [3]. The latest version of YOLO, version 7, not only provides object detection functionality but also allows for instance segmentation. The instance segmentation feature of YOLO is quite new and the research studies employing the instance segmentation with YOLOv7 are also very limited. In this study, together with YOLOv7, we also used the YOLOv7x distribution (with $\cong$72.3M parameters), which uses much more parameters than pure YOLOv7 including approximately 37.8M parameters.

2.2   Network Training and Instance Segmentation

The YOLOv7 model was trained by using the RGB input image sets comprised of $256 \times 256 \times 3$ histology image patches. The training hyper-parameters of batch size and epoch were set as 32 and 300, respectively. The stochastic gradient descent algorithm was used as the optimizer algorithm and the learning rate parameter was set as 0.01. In the CoNIC dataset, samples are labelled as instance masks by default. However, instance labels were converted to polygon type to perform training using the YOLOv7 algorithm. In this way, the ground-truth bounding boxes that are used when calculating the loss were obtained. In order to achieve higher performance, YOLOv7 default data augmentation methods are used on the input image.

The YOLOv7 model produces a box and a mask for each nucleus as a result of the non-maximum suppression algorithm. Using the masks obtained as a result of the output, the mean average precision value for the instance mask was calculated. However, to calculate the proposed panoptic quality (PQ) metric in the CoNIC challenge, the output masks for all nuclei must be converted to a single mask for the entire input image. Due to model constraints, the intersection of some instance masks occurs during the merge process. To solve this problem, the mask with the highest confidence value is assigned to the intersections of the masks, and a single mask is created for the input image. We performed 5-fold cross-validation on 4,981 samples to obtain more realistic experimental results. The entire dataset is divided into three parts in each round: the training set (60%), the validation set (20%), and the test set (20%). At the end of this process, 5 different models were trained to be used in each test set's instance segmentation performance analysis. This data folding technique was also used in the same way in our previous studies [13].

2.3   Segmentation Evaluation Metrics

To evaluate the accuracy of the proposed nuclei instance segmentation, we used the metrics of mean panoptic quality (mPQ) and mean average precision (mAP). The mathematical notation indicating how to calculate the panoptic quality (PQ) value is presented in Eq. 1 as follows:

$$PQ_t = \underbrace{\frac{|TP_t|}{|TP_t| + \frac{1}{2}|FP_t| + \frac{1}{2}|FN_t|}}_{\text{detection quality}} \times \underbrace{\frac{\sum_{(p_t,g_t)\in TP} IoU(p_t, g_t)}{|TP_t|}}_{\text{segmentation quality}} \qquad (1)$$

where t indicates a nuclei type, p denotes the predicted segmentation, g denotes the ground-truth segmentation and IoU is metric of Intersection over Union. In addition, the symbols of TP, FP and FN in Eq. 1 indicate the true positives, false positives and false negatives, respectively. If $IoU(p, g) > 0.5$ is provided for an instance of a nuclei type t, the predicted segmentation p and ground-truth segmentation g match uniquely. Therefore, all available instances of a nuclei type t within the dataset are divided into matched pairs (TP), unmatched ground-truth instances (FN) and unmatched predicted instances (FP). To calculate the mPQ, the mean of the PQ values observed for all the nuclei types is measured by dividing the sum of PQ values to the total number of nuclei types T as stated in Eq. 2:

$$mPQ = \frac{1}{T} \sum_{t=1}^{T} PQ_t \qquad (2)$$

The second evaluation metric used for the performance evaluation is the mAP as stated previously. Similar to the mPQ, the mean of the AP values (see Eq. 3) of all the nuclei types is measured by dividing the sum of AP values to the total number of nuclei types T to calculate the mAP as shown in Eq. 4.

$$AP_t = \frac{|TP_t|}{|FP_t| + |TP_t|} \qquad (3)$$

$$mAP = \frac{1}{T} \sum_{t=1}^{T} AP_t \qquad (4)$$

3 Experimental Analysis

3.1 Dataset

To perform our YOLOv7-based nuclei instance segmentation study, the Colon Nuclei Identification and Counting (CoNIC) Challenge 2022 dataset [4,5] was used. This dataset is stated to be organized to help drive forward research and innovation for automatic nuclei recognition in computational pathology [5]. The CoNIC Challenge dataset consists of 4,981 histology image patches that are encoded in RGB three-channel color space, and the dimensions of the image patches are 256 × 256 × 3. In image patches, a maximum of 6 different types of nuclei as the neutrophil, epithelial, lymphocyte, plasma, eosinophil and connective appear on a background. Sample image patches from the CoNIC Challenge 2022 dataset are presented in Fig. 1.

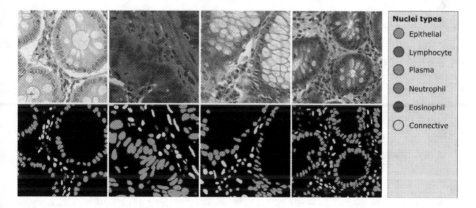

Fig. 1. Sample image patches and corresponding ground-truth segmentation maps from the CoNIC Challenge dataset [4,5].

3.2   Results

As stated in the previous section related to the training of the deep neural network and segmentation of the nuclei instances, a 5-fold cross-validation folding method was employed on the CoNIC colon histology image dataset. Therefore, all 4,981 images in the CoNIC dataset were evaluated in the both training and testing phases. In experimental studies, performances were evaluated for YOLOv7 and YOLOv7x. The accuracy statistics observed in the experimental analyses are given in Table 1 in terms of the AP and PQ metrics.

Table 1. Comparative performance evaluation statistics of YOLOv7 and YOLOv7x.

Nuclei type	Performance statistics for YOLOv7		Performance statistics for YOLOv7x	
	Average precision (AP)	Panoptic quality (PQ)	Average precision (AP)	Panoptic quality (PQ)
Neutrophil	0.2437(±0.0748)	0.0205(±0.0149)	0.2454(±0.0406)	0.0366(±0.0172)
Epithelial	0.5863(±0.0030)	0.3803(±0.0025)	0.5881(±0.0054)	0.3831(±0.0041)
Lymphocyte	0.2465(±0.0059)	0.1806(±0.0027)	0.2444(±0.0066)	0.1808(±0.0023)
Plasma	0.2028(±0.0142)	0.1169(±0.0086)	0.2081(±0.0138)	0.1223(±0.0100)
Eosinophil	0.0000(±0.0000)	0.0000(±0.0000)	0.0000(±0.0000)	0.0000(±0.0000)
Connective	0.4520(±0.0080)	0.2968(±0.0051)	0.4560(±0.0093)	0.2999(±0.0053)
Mean	0.2885(±0.0126)	0.1659(±0.0013)	0.2903(±0.0087)	0.1704(±0.0038)

In the calculation of the AP and PQ values of each class stated in Table 1, the mean of the values obtained in all the cross-validation rounds (folds) was computed. In addition, the standard deviations of the accuracy values in the folds were computed and they are also given in the parentheses next to the mean values presented in Table 1. The last row of Table 1 represents the mean of the statistics in the previous rows that indicate distinct nuclei types. In Table 2, nuclei instance distributions in all the cross-validation folds are presented. Although each test fold contains approximately $\cong$996 images, the distributions of the nuclei instances in the folds are different as can be expected. As can be seen in Table 2, instance distributions are quite different from the uniform distribution. Therefore, very low performances were observed in the classes where the number of samples was quite insufficient, and higher performance values were achieved in the classes with a relatively large number of samples. In the CoNIC dataset, there is a significant imbalance problem in the number of samples between classes, and also the appearances of nuclei labelled with different classes are very similar. For these reasons, the nuclei instance segmentation in the CoNIC dataset is a difficult task to solve.

As clearly seen in Table 1, YOLOv7x outperforms the YOLOv7. However, if class-based statistics indicating the performances of distinct nuclei types are taken into account, it will be seen that there is a significant performance difference between the classes. This case is thought to be related to the distribution of the nuclei types in the entire dataset as stated previously. In Fig. 2, a set of

Table 2. Distributions of the nuclei instances according to the cross-validation folds (img:images, ins:instances).

Nuclei type	Nuclei instance distributions in folds				
	Fold-1 (996 img.) (113,392 ins.)	Fold-2 (996 img.) (115,227 ins.)	Fold-3 (996 img.) (113,815 ins.)	Fold-4 (997 img.) (110,646 ins.)	Fold-5 (996 img.) (113,754 ins.)
Neutrophil	≅ 1% (1,125)	≅ 1% (912)	≅ 1% (927)	≅ 1% (1,083)	≅ 1% (1,012)
Epithelial	≅ 49% (55,848)	≅ 51% (59,238)	≅ 49% (55,273)	≅ 49% (54,417)	≅ 49% (55,991)
Lymphocyte	≅ 22% (24,485)	≅ 20% (23,563)	≅ 22% (24,710)	≅ 21% (22,870)	≅ 22% (24,528)
Plasma	≅ 6% (6,414)	≅ 6% (6,412)	≅ 6% (6,399)	≅ 6% (6,241)	≅ 6% (6,333)
Eosinophil	≅ 1% (778)	≅ 1% (756)	≅ 1% (683)	≅ 1% (851)	≅ 1% (768)
Connective	≅ 22% (24,742)	≅ 21% (24,346)	≅ 23% (25,823)	≅ 23% (25,184)	≅ 22% (25,122)

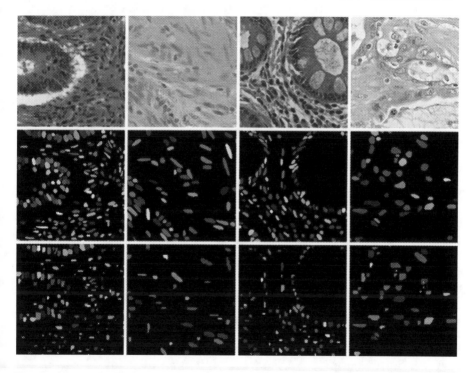

Fig. 2. A set of YOLOv7 instance segmentation output images (3rd row) with the corresponding histology image patches (1st row) and ground-truth segmentation masks (2nd row).

instance segmentation output images obtained with YOLOv7 are presented with their corresponding original histology image patches and ground-truth segmentation masks as follows.

4   Conclusions

In this paper, we proposed to perform YOLOv7 for the task of nuclei instance segmentation in colon histology images. In experimental tests performed on the CoNIC Challenge 2022 dataset, promising nuclei instance segmentation statistics were observed with YOLOv7. Furthermore, the performance of YOLOv7 in nuclei instance segmentation was compared to the YOLOv7x. It was also seen that the YOLOv7x has higher segmentation accuracies with a mean average precision (mAP) value of 0.2903 and a mean panoptic quality (mPQ) value of 0.1704. Additionally, YOLOv7 performed a segmentation with a mean average precision (mAP) value of 0.2885 and a mean panoptic quality (mPQ) value of 0.1659. This paper presents a preliminary study that covers our initial experiments for nuclei instance segmentation with YOLOv7. To the best of our knowledge, this is the first nuclei instance segmentation study with YOLOv7 in the current literature. Within the scope of our future studies, we plan to perform more comprehensive analyzes and improve our performance statistics on segmentation.

References

1. Böhland, M., et al.: Ciscnet-a single-branch cell nucleus instance segmentation and classification network. In: 2022 IEEE International Symposium on Biomedical Imaging Challenges (ISBIC), pp. 1–5. IEEE (2022)
2. Cai, Z., Vasconcelos, N.: Cascade R-CNN: delving into high quality object detection. In: Proceedings of the IEEE Conference on Computer Vision and Pattern Recognition, pp. 6154–6162 (2018)
3. Diwan, T., Anirudh, G., Tembhurne, J.V.: Object detection using yolo: challenges, architectural successors, datasets and applications. Multimedia Tools Appl. 1–33 (2022)
4. Graham, S., et al.: Lizard: a large-scale dataset for colonic nuclear instance segmentation and classification. In: Proceedings of the IEEE/CVF International Conference on Computer Vision, pp. 684–693 (2021)
5. Graham, S., et al.: Conic: colon nuclei identification and counting challenge 2022. arXiv preprint arXiv:2111.14485 (2021)
6. Graham, S., et al.: Hover-net: simultaneous segmentation and classification of nuclei in multi-tissue histology images. Med. Image Anal. 58, 101563 (2019)
7. Liu, L., Hong, C., Aviles-Rivero, A.I., Schönlieb, C.B.: Simultaneous semantic and instance segmentation for colon nuclei identification and counting. arXiv preprint arXiv:2203.00157 (2022)
8. Rumberger, J.L., Baumann, E., Hirsch, P., Janowczyk, A., Zlobec, I., Kainmueller, D.: Panoptic segmentation with highly imbalanced semantic labels. In: 2022 IEEE International Symposium on Biomedical Imaging Challenges (ISBIC), pp. 1–4. IEEE (2022)
9. Schmidt, U., Weigert, M., Broaddus, C., Myers, G.: Cell detection with star-convex polygons. In: Frangi, A.F., Schnabel, J.A., Davatzikos, C., Alberola-López, C., Fichtinger, G. (eds.) MICCAI 2018. LNCS, vol. 11071, pp. 265–273. Springer, Cham (2018). https://doi.org/10.1007/978-3-030-00934-2_30

10. Sultana, F., Sufian, A., Dutta, P.: A review of object detection models based on convolutional neural network. In: Intelligent Computing: Image Processing Based Applications, pp. 1–16 (2020)
11. Wang, C.Y., Bochkovskiy, A., Liao, H.Y.M.: Yolov7: trainable bag-of-freebies sets new state-of-the-art for real-time object detectors. arXiv preprint arXiv:2207.02696 (2022)
12. Weigert, M., Schmidt, U.: Nuclei instance segmentation and classification in histopathology images with stardist. In: 2022 IEEE International Symposium on Biomedical Imaging Challenges (ISBIC), pp. 1–4. IEEE (2022)
13. Yıldız, S., Memiş, A., Varlı, S.: Nuclei segmentation in colon histology images by using the deep CNNs: a U-net based multi-class segmentation analysis. In: 2022 Medical Technologies Congress (TIPTEKNO), pp. 1–4. IEEE (2022)
14. Zheng, Z., et al.: Enhancing geometric factors in model learning and inference for object detection and instance segmentation. IEEE Trans. Cybern. **52**(8), 8574–8586 (2021)

Evolutionary Approach to Feature Elimination in House Price Estimation

Yusuf Şevki Günaydın[1]([✉])[iD] and Ömer Mintemur[2][iD]

[1] Ankara Yıldırım Beyazıt University Computer Engineering, Faculty of Engineering and Natural Sciences, Ankara, Turkey
yusufsevkigunaydin@aybu.edu.tr
[2] Ankara Yıldırım Beyazıt University Software Engineering, Faculty of Engineering and Natural Sciences, Ankara, Turkey
omermintemur@aybu.edu.tr

Abstract. One of the most basic human needs is the need for shelter. Since ancient times, people have been looking for a house that is both safe and affordable. However, in modern times, although safety is no longer an issue, the definition of an affordable house has changed. In the past, affordability did not depend on many parameters as it does today. However, today, this definition depends on different features, such as the location of the house, the year of construction, the number of rooms, etc. These features affect the level of affordability, and consequently the price of the house. Since houses are also used as an investment option, correct estimation of house prices is an important issue. The determination of features that have a significant impact on the price of a house is a subjective notion and, therefore, requires an objective approach. Thanks to technological developments, Artificial Intelligence algorithms remove the human factor in most of the decision-making processes. In this study, a naive approach was proposed to estimate house prices by selecting the most effective features of a house (https://github.com/OmerMintemur/Feature-Elimination-Using-GA.). To select the most effective features, Genetic Algorithm approach was utilized. For estimation, LightGBM was used. The AmesHouse data set was used for the experiments. The results suggested that the proposed method both reduced the features and produced lower estimation errors than other proposed methods that used the same dataset.

Keywords: Prediction · Genetic Algorithm · Feature Reduction · House Prices

1 Introduction

One of the building blocks of all economies are houses. While in ancient times, they were used primarily for basic needs such as a place to sleep and protection from danger, they now serve more complex purposes such as investment and renting. Since people realize the upward trend of house prices, they have an

A. Ortis et al. (Eds.): ICAETA 2023, CCIS 1983, pp. 344–355, 2024.
https://doi.org/10.1007/978-3-031-50920-9_27

inclination to invest more money in the housing industry. Buyers and sellers want to maximize their profits. Also, house prices could be the indicator of the economic wealth of any country. Therefore, this sector has generated many opportunities for people [8,12].

Although there are a lot of opportunities in the housing industry, determining the price of a house for businesses is a time-consuming job since it requires an extensive amount of research. The factors that affect house prices are vast and make the decision-making process a hard and complex task. A person cannot continuously and properly monitor the industry. Therefore, the person who buys or sells a house needs an objective structure to determine the both market value of her/his house and the factors that effect the house prices the most.

Human factor in estimating house prices and leaving people to choose the most important features that affect house prices can produce misleading or biased results. The mechanism of objectivity should be human-free. Thanks to developments in technology, solutions to these problems could be transferred to computers. The problem at hand can be divided into two sub problems. (1) selecting the most important features that affect house prices and (2) estimating house prices with a low error rate. With the rapid rise of Artificial Intelligence (AI) approaches in modern era, different methods which are called Machine Learning (ML) algorithms have been used to solve mentioned problems. The literature in this area is vast, some of them deal with both of the problems (1) and (2), and some of them are interested in them individually.

One of the latest papers that deals with both house price prediction and effect of features that play role in estimating house prices conducted detailed experiments using different ML algorithms [13]. The authors added a new "indicator" that could be effective in house price predictions. The findings of this study indicate that the utilization of the proposed indicator could significantly enhance the accuracy of the house price estimation process.

Another study focused specifically on the factors that affect the parcel prices of multiplex houses in Korea [7]. The results of the experiments suggested that effecting factors in house prices could be reduced to land portion, total parking area, and the jeonse price of neighboring. The house industry does not solely depend on new houses; it is also concerned with second-hand houses. The problem of estimating prices for second-hand houses is addressed in [14]. This research considered attributes such as property type, orientation, floor, construction area, and facility. The study tried different Artificial Neural Networks (ANN) schemes and reported the results of the constructed networks.

A study focused on the Turkey housing market can be found in [11]. The authors proposed a new hybrid ML approach to predict house prices. The methodology was applied to two datasets: one created by the authors collected from the web and the second being the AmesHouse dataset that was also used in this paper. The authors investigated the correlations between features from both datasets and selected the most correlated features.

Previous studies in this area have typically focused on reducing the error rate, with the primary objective being the estimation of house prices. However, the features used for house price estimations should not be overlooked, as not all features contribute equally to the estimation phase. A mechanism that aims to select effective features and keep the error rate low could provide more reliable results for the industry. From this perspective, another crucial step for any ML operation is the feature selection process. Not all features in a given dataset are always useful for a ML method, so selecting the relevant, effective features is an important step. The methods for this step can be divided into two groups. The first group is based on methods that can be called "automatized" methods. Feature elimination[1] using ML methods such as RandomForest [6] or statistical methods such as correlation ranking [11,15] can be considered as automatized methods.

However, these automated methods for feature selection offer limited flexibility in modification. The second group of methods for feature selection processes are those that offer more flexibility. In this regard, a Genetic Algorithm (GA) approach was proposed for ease of modification to select the most relevant features while maintaining low estimation error rates. LigthGBM was selected as an ML method to estimate house prices for given features. AmesHouse [2] dataset was selected to validate the proposed method.

The remainder of the paper is organized as follows: Sect. 2 provides information on the dataset and the methodology employed in this study. Section 3 presents the experiments, results, and general inferences that can be deduced from the results. Finally, Sect. 4 concludes the paper with a general discussion and suggestions for future work.

2   Dataset and Methodology

This section summarizes the dataset and methodology used in our work to select the most important features for the estimation of house prices while keeping the estimation error low. In the housing industry, people can be overwhelmed by the large amount of information available about houses. Apart from the classical features of a house such as age, building area, and building material, there can be many additional features such as roof style, neighborhood, fireplaces, number of bathrooms, etc. To test our methodology, the AmesHouse dataset, which can be obtained from the Kaggle website [1], was utilized given its extensive number of attributes [2], since the proposed methodology requires a dataset that have high number of features.

2.1   Data

The AmesHouse dataset consists of information about houses sold in Iowa between 2006 and 2010. Each house was characterized by a number of features

---

[1] Feature selection and feature elimination have been used interchangeably in this study.

and the dataset was prepared by De Cock Dean [2]. The dataset contains 80 attributes, with 45 of them being categorical and the rest of them are numerical. The main objective of the dataset is to build a regression method to estimate the sale price of a house. The dataset was originally modified for a Kaggle competition and comes in two different sets: one for training and one for testing. These sets were used in our experiments without further division. The training set contains 1460 house records, and the test set contains 1458 house records. Some of the features of a house include the physical location, the type of building and the quality of the external material. More information about the dataset can be found in [1].

With a large number of features, the AmesHouse dataset is suitable for our purposes as there is a high chance that some features are irrelevant for estimating house prices in this dataset.

2.2   Data Preprocessing

Although the AmesHouse dataset contains 80 attributes for each house, some of them have missing values. There are various methods for dealing with missing values, such as filling with the mean or filling them with a specified value etc. But in this study, the attributes with missing values were discarded. This resulted in a reduction of the number of attributes to 46, which is still a relatively high number.

Another preprocessing step that was taken was handling the categorical variables. Since the AmesHouse dataset has a significant number of categorical variables, these were converted from categorical to numerical values. Finally, since each attribute has its own range of values, a normalization step was applied to standardize the range of the features.

2.3   Genetic Algorithm

Genetic Algorithms (GA) are optimization models that are based on evolution. They work by representing data as chromosomes and applying the recombination operation to this data. GAs are commonly viewed as function optimizers, but they have been applied to a wide range of problems. A typical GA begins with a number of chromosomes, referred to as the population, where each chromosome is a randomly generated individual. Then, these individuals are evaluated, and reproduction operations are performed to find better solutions to the problem at hand. In general, individuals with high-quality solutions, referred to as the fitness value of an individual, have a higher chance of reproducing [10]. This process continues repeatedly until the predetermined number of generations. A classical GA process is given in Algorithm 1.

2.4   LightGBM

LightGBM is a gradient-boosting decision tree algorithm (GBDT). It is a new algorithm and was proposed in 2017 [5]. LightGBM has been used to address a

---

Algorithm 1. Conventional Procedures of the Genetic Algorithm

---

$population \leftarrow initialPopulation()$
while $N \neq maxGenerationNumber$ **do**
$\quad fitness \leftarrow getFitnessValues()$
$\quad selectedParents \leftarrow getSelectedParents()$
$\quad childs \leftarrow crossover()$
$\quad childs \leftarrow mutation()$
end while
$saveResults()$

---

variety of problems, including classification and regression. The primary goal of the LightGBM algorithm is to increase the speed of classification or regression tasks when dealing with large amounts of data. Like Decision Tree (DT) algorithms, LightGBM combines weak learners to produce a strong learner. The algorithm is based on GBDT, but it improves upon GBDT in several ways. GBDT can be time-consuming to construct a tree, as finding the optimal splitting point can be challenging. LightGBM reduces the time needed for tree construction by using a histogram algorithm to determine splitting points. Furthermore, the usage of the histogram algorithm could have a regularization effect and prevents the model from overfitting [4]. The parameters of the LightGBM algorithm are the same for any DT algorithm. For the experiments, all parameters of the LightGBM algorithm were left as default.

2.5   Proposed Methodology

In this section, the proposed methodology is briefly introduced and the general structure of the algorithm is shown in Algorithm 2. A vector consisting only of 1 s and 0 s was treated as a chromosome. Since the dataset has 46 features after the elimination of missing values, each chromosome had a length of 46. A set of chromosomes was randomly generated and referred to as the initial population, and each population was referred to as one generation. Later, each chromosome was masked with a constant feature vector. If an attribute was used in training, it was indicated as 1 and otherwise it was indicated as 0 as can be seen in Fig. 1. This process generated a set of features for each individual. Since each individual is a possible solution to the problem, each of them was fed into LightGBM algorithm to obtain a fitness value. Individuals were ordered from best (low error rate) to worst (high error rate) according to their fitness values.

After the calculation of fitness values for each individual, a new generation was produced by using the current generation's individuals. In the proposed algorithm, two parents needed to be chosen to create a single child. The individuals that have better fitness values have a higher chance of being selected as a parent in the population. The proposed method eliminates a certain number of worst solutions in the parent selection progress in each generation. A crossover mechanism to breed a child was applied after the selection of two parents.

MsSubClass	LotArea	Street	...	...	...	RoofStyle	YearBuilt	HouseStyle
0	1	0	...	...	...	1	0	1

Fig. 1. A Chromosome

Algorithm 2. Proposed Method

---

  while *populationSize* ≠ *maxPopulationNumber* **do**
    solution ← *createRandomSolutions*()
    population ← *addSolutionToPopulation*(*solution*)
  end while
  while *generationCount* ≠ *maxGenerationNumber* **do**
    while *newPopulationSize* ≠ *maxPopulationNumber* **do**
      selectedParents ← *getSelectedParents*()
      child ← *produceChild*(*selectedParents*)
      child ← *crossover*()
      child ← *mutation*()
      newPopulation ← *addSolutionToPopulation*(*child*)
    end while
    population ← *newPopulation*
    getFitnessValues(*population*)
    sortPopulation(*population*)
  end while
  bestSolution ← *getBestSolution*(*population*)
  attributes ← *convertSolutionToAttributes*(*bestSolution*)
  LightGBM(*bestSolution*)

Fig. 2. Splitting Point Selection

Generally, one or two splitting points for both parents are used to perform the crossover operation in GA. In this study, a simple crossover mechanism, the one-splitting point approach, was used. We preferred to use a random splitting point for both parents (think parents as the vectors have a length of 46) for each child production in the proposed method. This process is shown in Fig. 2. The red line represents the location of the splitting point and the location changes randomly for each production. After selecting the splitting point, crossover operation is done.

Figure 3 shows the children generated after the crossover operation. The part after the splitting point was crossed between two parents, and two new children

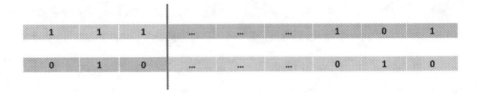

Fig. 3. Crossover Operation

were produced. For each child that was produced, randomly changing 1 to 0 or 0 to 1 in a random position in that child was regarded as a mutation. The process is given in Fig. 4. Therefore, a feature of the selected child was randomly chosen and changed.

Finally, the child was added to the new generation. Creating a new child continues until the maximum number of individuals in the population is reached. Then, population is replaced by the new generation and proposed method starts from the beginning and makes same operations for each generation until the maximum number of generation is reached. Finally, the best individual is accepted as the solution to the problem.

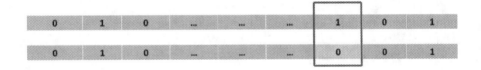

Fig. 4. Mutation

3   Experiments and Results

The experiments were conducted using the Python programming language (Version 3.7). The size of the population was set to 1000 and the GA was run for 10 generations. A mutation operation was applied to each child produced during the genetic operations. The original Kaggle dataset was divided into two separate sets, as previously mentioned. Each individual in the population was trained using the training set and its fitness value was evaluated using the test set. Unless otherwise stated, the performance of each individual was calculated using the test set, and the selection of an individual was based on its performance in the test set. The top 30% of individuals with the lowest error values, were selected as the successful individuals to be used for reproduction. The experiments parameters are given in Table 1.

Table 1. Parameters of the Experiments

Parameters	Values - Method
Number of Generation	10
Number of Individual in Each Generation	1000
Crossover	One Point Split
Mutation Rate	100% for Each Child
Number of Parents for Breeding	30% in Each Generation

Since each individual in the population needs a metric to assess itself, the suggested error metric by the Kaggle competition was selected. The error metric used in the experiments is given in Eq. 1.

$$RMSE = \sqrt{(\frac{1}{n}) \sum_{i=1}^{n} (log(\hat{y}_i) - log(y_i))^2}$$
(1)

where $\hat{y}_i$ is the error produced by an individual that was provided as input to LightGBM, and y_i is the original sale price. n is the total number of records in the test set.

The visual showing the best fitness and the average fitness values per generation is given in Fig. 5. In each generation the fitness values are getting smaller. The improvement seems to stop between the 8th and 10th generation.

The results suggested that a small number of generation was enough to achieve an acceptable error rate. After 10 generations, the lowest error rate, which was 0.0846, was achieved with the proposed method. The average fitness value is 0.1495 after 10 generations. The proposed method not only to maintained a low error rate, but also reduced the number of features. The number of features was reduced while preserving the low error rate. The feature count per generation is given in Fig. 6.

Figure 5 and Fig. 6 show that the proposed method successfully eliminated the features while preserving a low error rate. The lowest error rate which was 0.0846 achieved by using only 4 features. The results also revealed that more features do not necessarily lower the error rate. The proposed method eliminated nearly 91% of the features that were irrelevant for the task at hand.

When the AmesHouse dataset is considered, the proposed method achieved error rates lower than some of the methods proposed to achieve competitive error rates for the task. The method proposed by the authors interpolated the missing values rather than removing them [3]. In addition, they increased the number of features. Based on their results, their proposed methodology produced an error rate of 0.12019 in the test set. The authors' results align with the implications of our experiments. Our results suggested that more features do not contribute to reducing the error rate.

Another study also used a feature expansion approach for the same dataset and reported an error rate of 0.11260 according to their results [9]. Similarly,

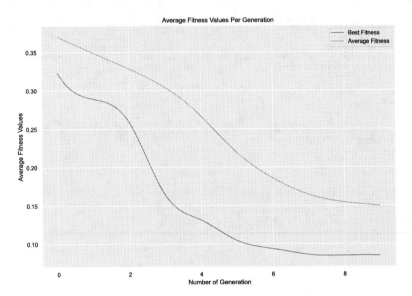

Fig. 5. Best Fitness and Average Fitness Per Generation

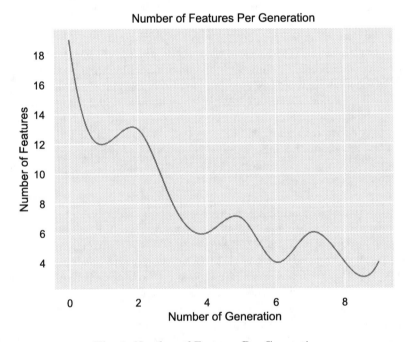

Fig. 6. Number of Features Per Generation

our methodology achieved a lower error rate than the mentioned methodology. The comparison of results is given in Fig. 7.

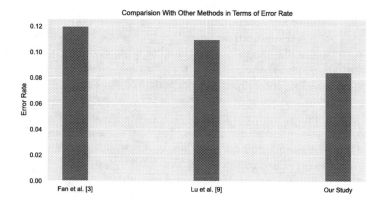

Fig. 7. Comparison of Our Study with Others Studies

During our experiments, we extensively tested different numbers of parameters in GA and according to the results, we limited our study by setting the number of generations to 10 and the population size to 1000. Our experiments revealed that increasing either of these parameters did not necessarily improve the performance of the method. Although GA provides flexibility, it requires additional effort to speed up the process. One potential drawback of our approach is that the GA methodologies are slow and take time to converge. The experiments were carried out on a computer having the specifications of Intel I7-8750H 2.20 GHz, 16 GB RAM, and GeForce Gtx 1050Ti 4 GB. The experiment took 16 min for one trial. However, in general, the proposed method has shown stable performance during our experiments.

The AmesHouse dataset extensively represents the properties of a house; however, it is important to note that feature elimination should not be performed without careful consideration. This study focused only on the AmesHouse dataset and proposed a methodology for feature elimination, with the primary objective of keeping the error rate low. In some cases, in contrast to feature elimination, a high number of features may produce better results.

4   Conclusions

The housing market presents many opportunities, as people's perspectives on housing have changed and now view it as a means of investment. However, investments must be made on objective structures. When it comes to the housing industry, many features must be considered and their impact on the sale price of the house evaluated. AI approaches, such as machine learning algorithms, can help to achieve these objective structures. This study proposed a structure to eliminate irrelevant features in the estimation of house prices while maintaining a low error rate.

For this purpose a classical GA approach was combined with LightGBM regressor. The experiments suggested that the proposed approach has the ability to produce low error rates compared to the error rates found in the literature. Although the proposed method was used only in house price estimation, this method can be applied to any dataset that has a myriad number of features. However, an important takeaway is that the proposed methodology may not be suitable datasets which have low number of features, since children that are bred will not be different from each other.

This methodology proposes a general framework for both feature reduction and flexible selection of fitness functions, making it an open area for further research to apply it to different datasets. However, to maintain the focus of the paper, these investigations have been left for future studies. Moreover, feature reduction part of the method can be compared with the automatized feature reduction algorithms to assess the method more to show the performance on feature reduction. Another avenue for future studies would be to experiment with different mutation and crossover mechanisms to further evaluate the performance of the proposed method.

References

1. Ames housing dataset. https://www.kaggle.com/competitions/house-prices-advanced-regression-techniques/overview. Accessed 09 Dec 2022
2. De Cock, D.: Ames, Iowa: alternative to the Boston housing data as an end of semester regression project. J. Stat. Educ. **19**(3) (2011)
3. Fan, C., Cui, Z., Zhong, X.: House prices prediction with machine learning algorithms. In: Proceedings of the 2018 10th International Conference on Machine Learning and Computing, pp. 6–10 (2018)
4. Ju, Y., Sun, G., Chen, Q., Zhang, M., Zhu, H., Rehman, M.U.: A model combining convolutional neural network and lightGBM algorithm for ultra-short-term wind power forecasting. IEEE Access **7**, 28309–28318 (2019)
5. Ke, G., et al.: LightGBM: a highly efficient gradient boosting decision tree. In: Advances in Neural Information Processing Systems, vol. 30 (2017)
6. Kim, D., Pham, K., Oh, J.Y., Lee, S.J., Choi, H.: Classification of surface settlement levels induced by TBM driving in urban areas using random forest with data-driven feature selection. Autom. Constr. **135**, 104109 (2022)
7. Kim, J.J., Cho, M.J., Lee, M.H.: An analysis of the price determinants of multiplex houses through spatial regression analysis. Sustainability **14**(12), 7116 (2022)
8. Louati, A., Lahyani, R., Aldaej, A., Aldumaykhi, A., Otai, S.: Price forecasting for real estate using machine learning: a case study on Riyadh city. Concurrency Comput. Pract. Experience **34**(6), e6748 (2022)
9. Lu, S., Li, Z., Qin, Z., Yang, X., Goh, R.S.M.: A hybrid regression technique for house prices prediction. In: 2017 IEEE International Conference on Industrial Engineering and Engineering Management (IEEM), pp. 319–323. IEEE (2017)
10. Mathew, T.V.: Genetic algorithm. Report submitted at IIT Bombay (2012)
11. Özöğür Akyüz, S., Eygi Erdogan, B., Yıldız, Ö., Karadayı Ataş, P.: A novel hybrid house price prediction model. Comput. Econ. 62, 1215–1232 (2023)
12. Priya, P., Mahalakshmi, G., Manojkumar, K., Kumararaja, V.: Property price prediction and possibility prediction in real estate using linear regression (2022)

13. Soltani, A., Heydari, M., Aghaei, F., Pettit, C.J.: Housing price prediction incorporating spatio-temporal dependency into machine learning algorithms. Cities **131**, 103941 (2022)
14. Xu, X., Zhang, Y.: Second-hand house price index forecasting with neural networks. J. Prop. Res. **39**(3), 215–236 (2022)
15. Zhou, H., Wang, X., Zhu, R.: Feature selection based on mutual information with correlation coefficient. Appl. Intell. **52**(5), 5457–5474 (2022)

A Hybrid Machine Learning Approach for Brain Tumor Classification Using Artificial Neural Network and Particle Swarm Optimization

Emre Dandıl$^{(\boxtimes)}$ (iD)

Department of Computer Engineering, Bilecik Seyh Edebali University, Bilecik 11230, Turkey
emre.dandil@bilecik.edu.tr

Abstract. Brain tumors have an increasing trend in recent years and are one of the main causes of death. Therefore, computer-assisted secondary tools that can help diagnose brain tumors at an early stage are needed. It is crucial to use machine learning methods which can help brain tumors classification. In this paper, a hybrid machine learning approach is proposed for brain tumor classification using particle swarm optimization and artificial neural network on magnetic resonance (MR) images. The approach is composed of six steps. The first step includes enhancement of MR images and the second step consists of eliminating the skull region. The third step is composed of extracting the region of interest through segmentation of the masses with s-FCM method. In the fourth step, feature extraction of the segmented tumors is undertaken with four different methods and feature selection is applied with relief-f and sequential floating forward selection (SFFS) methods in fifth step. In the last step, benign and malignant tumors are classified using Bayes method, artificial neural networks (ANN), support vector machines (SVM) and particle swarm optimization-based artificial neural networks (PSO-ANN) classification methods and the results are compared with each other. In the experimental studies, the proposed PSO-ANN approach provides high scores such as 96.28% accuracy, 97.58% sensitivity, 93.75% specifity for brain tumor classification. As a result, the proposed approach can facilitate the decision making of radiologists for brain tumor classification.

Keywords: Brain Tumor Classification · Machine Learning · Hybrid Method · Artificial Neural Network · Particle Swarm Optimization

1 Introduction

Brain tumors, which give rise to the majority of central nervous system (CNS) tumors, grow uncontrollably, starting from the brain and surrounding structures. These tumors may adversely affect the CNS and brain tissues, resulting in a reduction in the life expectancy of patients [1]. In addition, malignant brain tumors (brain cancer) are one of the main causes of death in recent years [2]. The brain tumors are classified into two types such as benign or malignant. Normal brain tissue, benign and malignant tumor samples are shown in Fig. 1. Benign brain tumors grow up slowly, have a homogeneous structure and don't spread to neighboring tissues. Malignant brain tumors grow rapidly, have a heterogeneous structure and are aggressive [3]. They may spread to other organs. Therefore, they are known as brain cancer. Brain cancer is spreading with a rising trend. In Global Cancer Statistics 2020 [4], more than 308.000 new brain and CNS cancers were diagnosed and more than 251.000 brain cancer-related deaths were reported worldwide in 2020 alone.

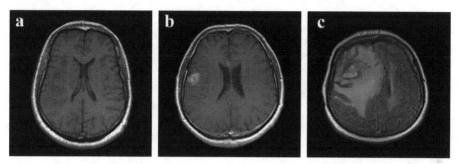

Fig. 1. Classification of brain tumors in MR images (a) normal brain tissue (b) benign brain tumor (c) malignant brain tumor

Nowadays, the most common imaging method used for diagnosis of brain tumors is the magnetic resonance (MR) imaging. Experts diagnose the tumors by analyzing the MR images. There are three types treatment such as surgical operations, chemotherapy, and radiotherapy for brain tumors [5]. The best treatment is applied according to expert's exact detection of the size, location, borders and malignancy of brain tumors [6]. The most reliable distinction of malignancy is carried out with biopsy. Accurate classification of the benign and malignant tumors is extremely difficult for experts in some cases. There is a need to be developed of new methods for classification of brain tumors. Experts can use computer-assisted detection (CAD) systems using machine learning methods which can help classification of brain tumors with high success scores. CADs can be preferred as another opinion and make the final decisions for determination or classification [7].

There are various CAD systems for classification and detection of brain tumors. The proposed systems can be analyzed into two categories such as preliminary methods and holistic methods. Preliminary methods are generally based on segmentation algorithms for brain tumors [8]. For example, some studies [9–11] underline that successful results can be obtained using FCM for brain tumor. In another study, k-means clustering method was proposed by Juang et al. [12] in brain tumor segmentation. Ambrosini et al. [13] developed an automated 3D template matching-based algorithm for brain metastases detection in MR scans. Popuri et al. [14] provided automatic segmentation of brain tumor and edema using Driclet priors in MR images. In the study, 3D segmentation method was also developed. Patino-Correa et al. [15] implemented conventional image processing techniques for segmentation of brain's white matter using MR images. In their study for brain tumor segmentation, Wu et al. [16] developed a method using model-aware affinity. Here, features were extracted Gabor filters and images segmented using support vector machines (SVM).

There are also various holistic CAD approaches for brain tumor classification and detection. Fractal-based method was proposed Zook et al. [17] for brain tumor detection. In another study, Iscan et al. [18] used a method for the detection of tumor on MR brain images by aid of ANN and wavelet transform. In their study, Garcia-Gomez et al. [19] presented a pattern recognition approach for diagnosis of benign and malignant brain tumors in MR images. For classification of brain MR images, Chaplot et al. [20] developed a method using self-organizing maps (SOM) and SVM. Georgiadis et al. [21] proposed a software system based on probabilistic neural network (PNN) for classification of metastatic and primary brain tumors, including meningioma and glioma, using texture features in T1-weighted MR images. Jensen and Schmainda [22] proposed computer-aided detection method for distinguishing of edema types for brain tumors. In another study, El-Dahshan et al. [23] used hybrid method for the classification of brain MR images using k-nearest neighbors (k-NN) and ANN. Arakeri and Reddy [24] proposed an automatic CAD system for brain tumor classification and detection using some state-of-the-art methods. In another study, Kaplan et al. [25] proposed a system using machine learning methods such as ANN, k-NN, random forest, and linear discriminant analysis for brain tumor classification.

In recent years, there are also the proposed hybrid approaches combined with meta-heuristic algorithms and machine learning methods for brain tumor classification and detection. In another study, Sharif et al. [26] designed hybrid machine learning approach for brain tumor detection. In the study, the skull removing on the images was achieved using particle swarm optimization (PSO).In addition, genetic algorithm (GA) was applied for feature selection. Moreover, ANN and other classifiers were used for classification of tumor grades. In their study [27], Dixit and Nanda improved a method using whale optimization algorithm and radial basis neural network for brain tumor classification. Besides, due to the development of deep learning models in recent years and their widespread use in many areas, it is seen that deep learning methods were proposed for brain tumor classification and detections [1, 28–31].

In this study, a hybrid machine learning approach is proposed for brain tumor classification on MR scans. The approach is designed for classification of benign and malignant brain tumors by the help of image processing techniques and machine learning methods. The contribution of this study as follow:

- Presenting the systematic review of studies proposed previously and state-of-the-art in brain tumor detection and classification,
- Comparison of state of the art classification method (Bayes, SVM, ANN and PSO-ANN) for brain tumor classification,
- Performing the detailed experiments on large MRI dataset,
- The high scores in accuracy, sensitivity and specifity for brain tumor classification.

The organization of the rest of the work is as follows. In Sect. 2, the materials used in proposed approach are explain and the design methodology is defines. In Sect. 3, experimental results of the proposed approach in MR images are presented comprehensively. Finally, Sect. 4 presents the discussion and conclusions on this study.

2 Material and Methods

The proposed approach is composed of two main stage called as Brain Tumor Segmentation and Brain Tumor Classification. Tumor segmentation step includes sub-steps such as pre-processing, eliminating the skull and segmentation of the brain tumor. Tumor Differentiation step also includes the sub-steps of feature extraction and selection and classification. Block diagram of the proposed approach is given in Fig. 2.

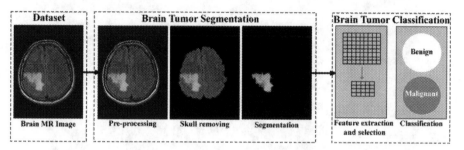

Fig. 2. Block diagram of proposed machine learning approach for brain tumor classification

2.1 Dataset

The proposed approach for brain tumor classification was evaluated using an original MR image dataset in this study. This dataset was composed a total of 188 T2-weighted MR images with axial view from 67 different patients for the proposed approach. According to pathological results, 124 of these images included malignant tumor, whereas 64 of them included benign tumor. All of MR images in dataset were obtained from Sincan Nafiz Körez State Hospital. The scans were performed on 1.5T Siemens Magnetom MR scanner in DICOM format. All images in dataset were converted to jpg for experimental studies. There were between 20 and 70 the MR slices for each patient in scans. Some benign and malignant brain images in this dataset are indicated in Fig. 3.

The most appropriate slice from each MR scan series of the dataset was selected with accompaniment of radiologist. In this dataset, the gold standard segmentations, delineated by experts, manually obtained for the each MR image by aid of the developed segmentation software, as seen in Fig. 4. The expert delineates edges of tumor by the developed software tool, giving to outline contours of the brain tumor boundaries.

2.2 Pre-processing and Enhancement

In this step, the MR images were enhanced by increasing the image quality and removing the noise. The noise and artifacts caused by the scanning error were removed for enhancing the images using median filter. Laplacian filter were used for precisification of tumor contours. Moreover, histogram equalization was also used to reduce contrast differences during scanning errors.

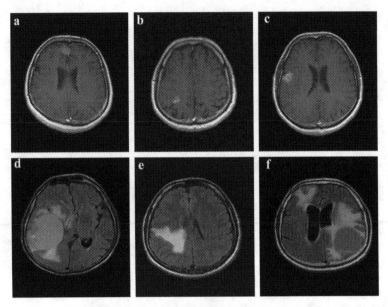

Fig. 3. Samples of benign (a, b, c) and malignant (d, e, f) tumors on MR images in dataset

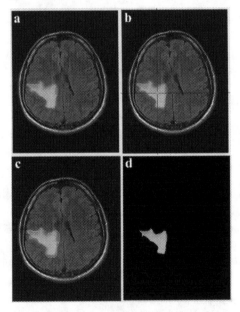

Fig. 4. The developed software tool for manual segmentation of tumor, (a) T2-weighted MR image with brain tumor, (b) outline of tumor region by the tool, (c) delineating brain tumor boundaries by experts, (d) segmentation of brain tumor by the developed software tool (reference segmentation)

2.3   Skull Removing

Skull removing process is a required step before any brain tumor segmentation. When not performed, it inclines to tumor segmentation errors. The goal of this step is to eliminate the skull completely from the full brain image to remove the unnecessary regions [32]. But, skull removing from MR image is especially challenging task. Therefore, an effective method was proposed in this study for skull stripping on MR images. In this method [33], firstly, an MR image in dataset was converted to double image. Consequently, binary values (true or false) for low and high threshold were assigned to each pixel of image. It is shown the samples for the skull removing step in Fig. 5 for brain tumor classification.

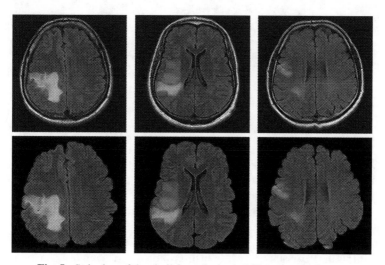

Fig. 5. Stripping of the skull from the brain MR images in dataset

2.4   Brain Tumor Segmentation

Automatic segmentation of images has been major research area to extract meaning from pixels. Brain MR images may contain noise, artifacts, inhomogeneity and divergence. In this step of the proposed approach, tumor segmentation process was applied on brain MR images. In this work, spatial-FCM (s-FCM) method was used for brain tumor segmentation [34]. Comparing to conventional FCM segmentation method, s-FCM indicates the spatial relating of a pixel with other pixels. In s-FCM, firstly, a local spatial similarity model is set up, and the initial clustering center and initial membership are decided adaptively. Afterwards, the fuzzy membership function is modified using the high inter-pixel correlation. Lately, the image segmentation is completed using the s- FCM method [35]. Thus, the clustering of set carries out more successful and are removed the artifacts and noise. The samples images segmented from skull-stripping MR slices using s-FCM are denoted in Fig. 6.

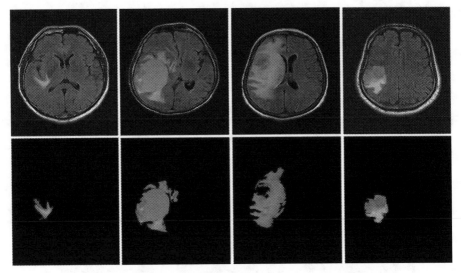

Fig. 6. Brain tumor segmentation in dataset using s-FCM method

2.5   Feature Extraction

After the brain tumor segmentation step is completed, it has to be determined if brain tumor type is benign or malignant. Even though the determination is accomplished with the help of classifier method, some important features on MR images can facilitate the decision process. Therefore, the characteristic features of brain tumors on segmented MR images have to be obtained. In MR images of dataset, benign tumors are rounder and have softer borders, while malignant tumors have grainier, sharper and harsher border features. This is the key point of successful discrimination of benign and malignant brain tumors from each other.

Brain tumors in MR images are generally determined according to geometric shapes, gray level statistical and energy level features of tumors. Therefore, important features were extracted from the segmented MR images for brain tumor classification, as benign or malignant. Firstly, shape features were extracted for analyzing tumor geometry. To obtain global statistic about tumor region, first-statistical features were used using histogram of the image. Moreover, texture features were utilized for gray level statistical values of tumor regions on MR images using gray-level co-occurrence matrix (GLCM). On the other hand, energy features were employed of brain tumors using wavelet decomposition transform. Finally, boundary features were extracted to get borders and edges of tumor using fractal dimensions.

Shape features allow extracting the features such as sharpness, circularity, convexity from an image using geometric parameters [36]. First-order statistical features are obtained from the histogram values with gray-level in image [37]. Statistical features of gray level textures on image are first derived with GLCM [38]. GLCM gives the relationship in image pixels of different gray level [33]. When the image is 2D, features from different angles can be obtained. The method calculates the pixel densities of i and j pixels in a specific angle and distance to create the GLCM matrix. Energy features can be obtained using wavelet decomposition for different regions. The segmented tumor region the brain MR image is divided into four sub-bands with 2-level wavelet decomposition. Using 2D wavelet decomposition, it is created three images for low frequencies and one image for high frequencies [39]. Boundary fractal features give information about the borders and edges of an image. With this method, characteristics features on an image can be extracted to discriminate the regions that are separate from the image [40]. It is based on obtaining binary images in fractal dimensions from border values of inside regions of the input image. In this study, 30 features were extracted in total with boundary features by using the border values of brain tumors. Table 1 denotes the details of feature extraction methods and the number of extracted features for each method in this study.

Table 1. The number of features extracted from segmented brain MR images in this study

Feature extraction method	The number of extracted features
Histogram features (HF)	7
Shape features (SF)	13
Texture features (TF)	88
Energy features (EF)	13
Boundary fractal features (BF)	30
Total	**151**

2.6 Feature Selection

In this paper, most appropriate features have be determined for selected MR images because more features occurs unnecessary processing overhead. Since 151 features are large for the classifier, they have to be selected in order to provide effective detection. In brain tumor classification, two fundamental approaches were utilized for feature selection. One of these is Sequential floating forward selection (SFFS) and the other one is relief-F (RF). 10 important features were selected both SFFS and RF.

SFFS is a general purpose feature selection method that allows the selection of the most effective features from among large pieces of information based on level of importance [41]. In this study, other feature selection technique used for brain tumor classification is relief-F (RF). RF denotes weight for each feature using relationship between a specific class and a feature to rank it [42].

A total of 151 different features are many to correct classification. Therefore, two different feature selection techniques are used in this paper for most suitable features. The top 10 features extracted from SFFS and RF are presented in Table 2.

Table 2. Detailed list of top 10 features selected with SFFS and RF feature selection methods

Rank	Feature Selection with RF			Feature Selection with SFFS		
	Order	Method	Selected Feature	Order	Method	Selected Feature
1	78	TF	Information measures of correlation (45^0)	33	TF	Entropy (0^0)
2	96	SF	Area	112	EF	WAV4
3	128	EF	WAV10	94	HF	Mean
4	94	HF	Mean	102	SF	Circularity
5	39	TF	Homogeneity (90^0)	38	TF	Homogeneity (45^0)
6	110	SF	Solidity	90	HF	Variance
7	43	TF	Energy (90^0)	42	TF	Energy (45^0)
8	125	BF	BOUN4	124	BF	BOUN3
9	143	BF	BOIN18	63	TF	Sum of entropy (90^0)
10	63	TF	Sum of entropy (90^0)	96	SF	Area

2.7   Brain Tumor Classification

Classification of brain tumors is important step for making a decision about differentiating of benign and malignant tumors. In the proposed approach, Bayes, ANN, SVM and PSO-ANN methods were used for brain tumor classifications and results were evaluated.

2.7.1   Hybrid Classifier Based on Particle Swarm Optimization and Artificial Neural Networks (PSO-ANN)

In this work, ANN is also trained by PSO, a heuristic algorithm, to acquire optimum parameters of weights in ANN. Thus, it is aimed to improve the classification performance of ANN.PSO is a swarm intelligence algorithm and belongs sub-field of computational intelligence [43]. PSO is also a heuristic optimization method based on population. It is inspired by behavior flocking some animals such as birds or fish while searching food. The goal of PSO algorithm is to arrange optimum location for all the particles in a multi-dimensional hyper-plane. The algorithm is initialized with random particles. Afterwards, it searches optimal solution by updating its velocity and position. The particles are updated for two specific particles in each generation. The first particle is *Pbest,* best known position for all particles. Second is *Gbest* which is the global best position to swarm [44]. Pseudocode for the PSO algorithm is denoted in Algorithm 1.

For PSO algorithm, velocity of a particle is updated using Eq. (1) in each iteration:

$$v_i(t+1) = v_i(t) + c_1 \times rand() \times (Pbest_i - p_i(t)) + c_2 \times rand() \times (Gbest - p_i(t))$$
$$(1)$$

where $v_i(t+1)$ is the new velocity for the ith particle, c_1 and c_2 are the weighting coefficients for the best and global best positions, respectively [45]. $p_i(t) and p_i(t+1)$

Are the ith particle's position and new position at time t respectively. The *rand* () functions generate a uniformly random variable between *0* and *1*. A particle's position is updated using Eq. (2):

$$p_i(t + 1) = p_i(t) + v_i(t + 1) \qquad (2)$$

Algorithm 1. Pseudocode of particle swarm optimization algorithm

1:	Initialize *P* particles with random
2:	Assign *Pbest* from *P*
3:	Assign *Gbest* from *P*
4:	**Repeat** (do not provided stop criteria)
5:	**while** (*i=1* to the number of particle in *P*)
6:	**If** (*Pbest$_i$* is better than *Pbest*)
7:	*Pbest* = *Pbest$_i$*
8:	**end if**
9:	**If** (*Pbest* is better than *Gbest*)
10:	*Gbest* = *Pbest*
11:	**end if**
12:	**end while**
13:	**while** (*i=1* to the number of particle in *P*)
14:	Update *Velocity$_i$*
15:	Update *Position$_i$*
16:	**end while**
17:	**Until**

In this study, it was used an ANN structure that identified the weights of feed-forward multi-layer network both back-propagation and PSO algorithm. As denoted in Fig. 7, there are three layers in ANN structure: input layer, hidden layer, and an output layer. Besides, ANN performance was assessed with mean square error (MSE) in Eq. (3) in the course of training.

$$MSE = \frac{1}{2N} \sum_{i=1}^{N} h_i^2 \qquad (3)$$

where, h is the error between actual and predicted output results after submitting the ith pattern to the network, and N is the number of data (pattern) in the training set.

In PSO-ANN, $[X_1, X_2, X_3,..., X_{10}]$ are inputs obtained from feature selection methods. Y is the output value. This value is *0* for benign and *1* for malignant brain tumor. The weights between network layers can be calculated using ANN learned by PSO (PSO-ANN). Here, the training of ANN is started with random weights and biases. In addition, c_1 and c_2 constants of PSO algorithm were determined as *2*. According to the experimental evaluations, the number of particles and maximum generation were determined by aid of test results and error of the network. These parameters were assigned as 30 and 200, respectively. The number of neurons in hidden layer was selected according to best performance of network ad determined as 3.

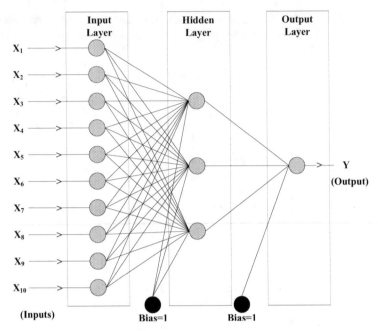

Fig. 7. Feed-forward multi-layer ANN structure used in this study for PSO-ANN

2.7.2 Other Classifiers

Artificial neural network (ANN) is an artificial intelligence method that practices the formation of a new system by copying the operations of the human brain [46]. Back-propagation (BP) is an orderly algorithm for training multi-layer ANN. BP learning rule allows obtaining the most optimum values for network weights by calculating the total error in the output at minimum level [47]. In ANN structure used in this study, BP algorithm was used to train the network and Levenberg-Marquardt algorithm was utilized as learning method for desired ANN model.

Support vector machines (SVM) are an important method used especially in the classification of two-category data. SVM classifier is an effective method for classifying data belonging to two different categories, finding the most appropriate linear equation separating the data. Nevertheless, proper kernel functions are used in SVM in separating classes with large data [6]. It is very important to determine the appropriate kernel function for the SVM to have a high classification success. In addition, sequential minimal optimization method is used in training of the SVM with kernel function to separate hyperplanes. Bayes classifier is based on Bayesian decision rule principles that approach classification problems statistically as displayed in Eq. (4). In Bayes classification, which example belongs to which class is probabilistically determined. The Bayes rule that shows the probability of X_k of belonging to class ($p(C_i|X_k)$). Here, (C_1, $C_2,...,C_n$) is the classes and $X_k = [x_1, x_2.........x_k]$ are k. feature vectors. In this study, Gauss distribution was used for classification of benign and malignant brain tumors in

Bayes classifier.

$$p(C_i|X_k) = \frac{p(X_k|C_i)p(C_i)}{p(X_k)} i = 1, 2, \ldots n \tag{4}$$

3   Experimental Results

In this study, a hybrid machine learning approach based on ANN and PSO was developed for the classification of benign and malignant brain tumors. In the proposed approach (PSO-ANN), all experimental studies were implemented in the Matlab software environment. Experimental studies were carried out with a computer with 64 bit Windows 10 operating system, Intel Core i7 with 2.8 GHz CPU, 16 GB RAM hardware.

Classification performance of the proposed approach was obtained with four types of classifiers such as Bayes, SVM, ANN and PSO-ANN. In order to assess the performance of these classifiers, we took into consideration leave-one-out (LOO) cross validation method. In LOO, each pattern in the dataset is included in both the training and testing steps of the classification process. In this study, LOO cross validation was performed using188 MR images in entire dataset. Firstly, each classifier was trained with 187 images and tested on the remaining one image. These operations were then repeated for all images in the dataset. The final results were obtained to evaluate the performance of the classifier after 188 iterations.

The classification performance of the proposed approach was evaluated using both a total of 151 features (first experiment) and the selected top 10 features (second experiment). In first experiment, the 151 features were used to assess the classification performance of the proposed approach for all classifiers. In second experiment, top 10 features were selected with both SFFS and RF. Classification accuracies were achieved with four classifiers using the top 10 selected features. Thus, the effects of feature selection were evaluated using the classification results.

The true and false scores denotes between the actual results and the predicted results in a confusion matrix. In this study, true positive (TP) stands for malignant tumors classified as malignant, false positive (FP) shows benign tumors classified as malignant, true negative (TN) presents benign tumors classified as benign and false negative (FN) shows malignant tumors classified as benign. To evaluate the classification performance of the proposed approach, many experiments were performed and measured the metrics such as accuracy, sensitivity, specifity, as seen Eq. (5), Eq. (6) and Eq. (7), respectively.

$$\text{Accuracy(ACC)} = \frac{\text{TP} + \text{TN}}{\text{TP} + \text{TN} + \text{FN} + \text{FP}} \times 100 \tag{5}$$

$$\text{Sensitivity(SEN)} = \frac{\text{TP}}{\text{TP} + \text{FN}} \times 100 \tag{6}$$

$$\text{Specifity(SPE)} = \frac{\text{TN}}{\text{FP} + \text{TN}} \times 100 \tag{7}$$

The obtained results of each classifier are indicated in Table 3 for the first experiment. In this experiment, whole 151 features were used for performance evaluating. Results show that best performance values are achieved by using PSO-ANN classifier for all features. As shown Table 3 for PSO-ANN, 164 of the 188 malignant and benign brain tumors were classified successfully. Therefore, it is clearly seen that best classifier is PSO-ANN compared to other classifiers. In addition, Table 4 presents the classification scores of the proposed approach according to results in Table 3. According to the table, best performance values were obtained with PSO-ANN classifier compared to other classifiers.

Table 3. Brain tumor classification results for classifiers without feature selection

Number of features	Classifier	Classification results			
		TP	FP	FN	TN
151	Bayes	101	20	23	44
151	ANN	105	18	19	46
151	SVM	109	16	15	48
151	PSO-ANN	113	13	11	51

Table 4. Performance measures of proposed approach using different classifiers for all features

Number of features	Metric	Performance scores of classifiers			
		Bayes	ANN	SVM	PSO-ANN
151	ACC (%)	77.13	80.32	83.51	87.23
151	SEN (%)	81.45	84.68	87.90	91.13
151	SPE (%)	68.75	71.88	75.0	79.69

Similarly, the results of each classifier are denoted in Table 5 for second experiment. In this experiment, top 10 features are selected with both SFFS and RF. As shown Table 5, the best classifier result was obtained with PSO-ANN method for both RF and SFFS. When RF is used in feature selection, 178 of the 188 brain tumors were identified successfully, only 10 tumors were misclassified. On the other hand, the results are as follows: TP = 121, FP = 4, FN = 3 and TN = 60 with PSO-ANN when SFFS is used for feature selection. Here, 181 of the 188 brain tumors were identified successfully. Therefore, it is clearly seen that best classifier is PSO-ANN with SFFS compared to other classifiers with SFFS and RF. Similarly, Table 6 presents the detection results of the proposed machine learning approach according to results of Table 5. ACC, SEN and SPE are a performance measure that evaluates the overall effectiveness of the approach. In Table 6, ACC, SEN and SPE obtained using the PSO-ANN classifier and SFFS feature selection method is 96.28%, 97.58% and 93.75% respectively which is higher than other classifier methods.

Table 5. Brain tumor classification results for each classifier with feature selection

Number of features	Classifier	Feature Selection with RF				Feature Selection with SFFS			
		TP	FP	FN	TN	TP	FP	FN	TN
10	Bayes	112	14	12	50	113	14	11	50
10	ANN	116	13	8	51	116	10	8	54
10	SVM	117	10	7	54	118	6	6	58
10	PSO-ANN	120	6	4	58	121	4	3	60

Table 6. Performance measures of proposed approach using different classifiers for feature selection methods

Number of features	Metric	Feature Selection with RF				Feature Selection with SFFS			
		Bayes	ANN	SVM	PSO-ANN	Bayes	ANN	SVM	PSO-ANN
10	ACC (%)	86.17	88.83	90.96	94.68	86.7	90.43	93.62	96.28
10	SEN (%)	90.32	93.55	94.35	96.77	91.13	93.55	95.16	97.58
10	SPE (%)	78.13	79.69	84.38	90.63	78.13	84.38	90.63	93.75

4 Conclusions

This study proposes an automated hybrid machine learning approach for classification of benign or malignant brain tumors on MR images. There are several contributions of this study. Firstly, in this work, a holistic machine learning approach was proposed for brain tumor classification, including steps such as pre-processing, skull removing, segmentation, feature extraction, feature selection and tumor classification. Secondly, the proposed approach using PSO-ANN and SFFS shows high ACC, SEN and SPE scores with 96.28%, 97.58%, 93.75%, respectively. In the third, this study compares the-state-of-the-art classifiers such as PSO-ANN, SVM, ANN, and Bayes for brain tumor classification. This approach may guide the researchers to selection of classifier method on brain tumor distinguishing. In the fourth, this work gives the review of the proposed methods and the-state-of-the-art methods in brain tumor classification. To sum up, the proposed approach can facilitate decision making of tumor distinctions and it can be used by the experts as a secondary tool for benign and malignant brain tumor classification. On the other hand, there are a number of limitations of the study. One of the most important limitations of the proposed approach is related to the fact that system performance was only tried on T2-weighted MR images with axial directions. Use of scans on coronal and sagittal planes along with other MR densities can contribute to system reliability. Moreover, proposed approach was performed by means of 2D slices. The MR slices, in which couldn't be exactly observed the tumors, were not included to our dataset. It is clear that using of technically suitable images instead of all volume can ensure more successful diagnosis. In addition, experts have to check all slices of volume for reference segmentation and they have to try to diagnose according to best suitable slices. Moreover, for more reliable assessment, proposed approach has to be evaluated in the clinic settings.

Acknowledgements. We thank to Sincan Nafiz Körez State Hospital in classification of brain tumor for providing MR image dataset.

References

1. Shaik, N.S., Cherukuri, T.K.: Multi-level attention network: application to brain tumor classification. SIViP **16**(3), 817–824 (2022)
2. Ayadi, W., Elhamzi, W., Charfi, I., et al.: Deep CNN for brain tumor classification. Neural. Process. Lett. **53**(1), 671–700 (2021)
3. Padma Nanthagopal, A., Sukanesh Rajamony, R.: Classification of benign and malignant brain tumor CT images using wavelet texture parameters and neural network classifier. J. Visualization **16**(1), 19–28 (2013)
4. Fan, Y., Zhang, X., Gao, C., et al.: Burden and trends of brain and central nervous system cancer from 1990 to 2019 at the global, regional, and country levels. Arch. Public Health **80**(1), 1–14 (2022)
5. Huo, J., Brown, M.S., Okada, K.: CADrx for GBM brain tumors: predicting treatment response from changes in diffusion-weighted MRI. In: Machine Learning in Computer-Aided Diagnosis: Medical Imaging Intelligence and Analysis, IGI global (2012)
6. Dandıl, E., Çakıroğlu, M., Ekşi, Z.: Computer-aided diagnosis of malign and benign brain tumors on MR images. In: Bogdanova, A.M., Gjorgjevikj, D. (eds.) ICT Innovations 2014. AISC, vol. 311, pp. 157–166. Springer, Cham (2015). https://doi.org/10.1007/978-3-319-09879-1_16
7. Doi, K.: Computer-aided diagnosis in medical imaging: historical review, current status and future potential. Comput. Med. Imaging Graph. **31**(4–5), 198–211 (2007)
8. Suzuki, K.: A review of computer-aided diagnosis in thoracic and colonic imaging. Quant. Imaging Med. Surg. **2**(3), 163 (2012)
9. Kolen, J.F., Hutcheson, T.: Reducing the time complexity of the fuzzy C-means algorithm. IEEE Trans. Fuzzy Syst. **10**(2), 263–267 (2002)
10. Murugavalli, S., Rajamani, V.: A high speed parallel fuzzy C-mean algorithm for brain tumor segmentation. BIME J. **6**(1), 29–33 (2006)
11. Fletcher-Heath, L.M., Hall, L.O., Goldgof, D.B., et al.: Automatic segmentation of non-enhancing brain tumors in magnetic resonance images. Artif. Intell. Med. **21**(1–3), 43–63 (2001)
12. Juang, L.-H., Wu, M.-N.: MRI brain lesion image detection based on color-converted K-means clustering segmentation. Measurement **43**(7), 941–949 (2010)
13. Ambrosini, R.D., Wang, P., O'Dell, W.G.: Computer-aided detection of metastatic brain tumors using automated three-dimensional template matching. J. Magn. Reson. Imaging **31**(1), 85–93 (2010)
14. Popuri, K., Cobzas, D., Murtha, A., et al.: 3D variational brain tumor segmentation using dirichlet priors on a clustered feature set. Int. J. Comput. Assist. Radiol. Surg. **7**(4), 493–506 (2012)
15. Patino-Correa, L.J., Pogrebnyak, O., Martinez-Castro, J.A., et al.: White matter hyper-intensities automatic identification and segmentation in magnetic resonance images. Expert Syst. Appl. **41**(16), 7114–7123 (2014)

16. Wu, W., Chen, A.Y., Zhao, L., et al.: Brain tumor detection and segmentation in a CRF (conditional random fields) framework with pixel-pairwise affinity and superpixel-level features. Int. J. Comput. Assist. Radiol. Surg. **9**(2), 241–253 (2014)
17. Zook, J.M., Iftekharuddin, K.M.: Statistical analysis of fractal-based brain tumor detection algorithms. Magn. Reson. Imaging **23**(5), 671–678 (2005)
18. Iscan, Z., Dokur, Z., Ölmez, T.: Tumor detection by using Zernike moments on segmented magnetic resonance brain images. Expert Syst. Appl. **37**(3), 2540–2549 (2010)
19. García-Gómez, J.M., Vidal, C., Martí-Bonmatí, D., et al.: Benign/Malignant classifier of soft tissue tumors using MR imaging. Magn. Reson. Mater. Phys., Biol. Med. **16**(4), 194–201 (2004)
20. Chaplot, S., Patnaik, L.M., Jagannathan, N.R.: Classification of magnetic resonance brain images using wavelets as input to support vector machine and neural network. Biomed. Signal Process. Control **1**(1), 86–92 (2006)
21. Georgiadis, P., Cavouras, D., Kalatzis, I., et al.: Improving brain tumor characterization on MRI by probabilistic neural networks and non-linear transformation of textural features. Comput. Methods Programs Biomed. **89**(1), 24–32 (2008)
22. Jensen, T.R., Schmainda, K.M.: Computer-aided detection of brain tumor invasion using multiparametric MRI. J. Magn. Resonance Imaging: Official J. Int. Soc. Magn. Resonance Med. **30**(3), 481–489 (2009)
23. El-Dahshan, E.-S.A., Hosny, T., Salem, A.-B.M.: Hybrid intelligent techniques for MRI brain images classification. Digit. Signal Process. **20**(2), 433–441 (2010)
24. Arakeri, M., Reddy, G.: Computer-aided diagnosis system for tissue characterization of brain tumor on magnetic resonance images. SIViP **9**(2), 409–425 (2015)
25. Kaplan, K., Kaya, Y., Kuncan, M., et al.: Brain tumor classification using modified local binary patterns (Lbp) feature extraction methods. Med. Hypotheses **139**, 109696 (2020)
26. Sharif, M., Amin, J., Raza, M., et al.: An integrated design of particle swarm optimization (PSO) with fusion of features for detection of brain tumor. Pattern Recogn. Lett. **129**, 150–157 (2020)
27. Dixit, A., Nanda, A.: An improved whale optimization algorithm-based radial neural network for multi-grade brain tumor classification. Vis. Comput. **38**(11), 3525–3540 (2022)
28. Cinar, N., Kaya, M., Kaya, B.: A novel convolutional neural network-based approach for brain tumor classification using magnetic resonance images. Int. J. Imaging Syst. Technol. **33**(3), 895–908 (2023)
29. Kang, J., Ullah, Z., Gwak, J.: MRI-based brain tumor classification using ensemble of deep features and machine learning classifiers. Sensors **21**(6), 2222 (2021)
30. Sharif, M.I., Khan, M.A., Alhussein, M., et al.: A decision support system for multimodal brain tumor classification using deep learning. Complex Intell. Syst. **8**(4), 3007–3020 (2022)
31. Swati, Z.N.K., Zhao, Q., Kabir, M., et al.: Brain tumor classification for mr images using transfer learning and fine-tuning. Comput. Med. Imaging Graph. **75**, 34–46 (2019)
32. Gambino, O., Daidone, E., Sciortino, M., et al.: Automatic skull stripping in MRI based on morphological filters and fuzzy c-means segmentation. Ann. Int. Conf. IEEE Eng. Med. Biol. Soc. **2011**, 5040–5043 (2011)
33. Dandıl, E.: A computer-aided pipeline for automatic lung cancer classification on computed tomography scans. J. Healthcare Eng. 2018 (2018)
34. Chuang, K.-S., Tzeng, H.-L., Chen, S., et al.: Fuzzy C-means clustering with spatial information for image segmentation. Comput. Med. Imaging Graph. **30**(1), 9–15 (2006)
35. Wang, X.-Y., Bu, J.: A fast and robust image segmentation using FCM with spatial information. Digit. Signal Process. **20**(4), 1173–1182 (2010)
36. Mingqiang, Y., Kidiyo, K., Joseph, R.: A survey of shape feature extraction techniques. Pattern Recogn. **15**(7), 43–90 (2008)

37. Saifullah, S., Suryotomo, A.P.: Identification of chicken egg fertility using SVM classifier based on first-order statistical feature extraction. arXiv preprint arXiv:2201.04063 (2022)
38. Haralick, R.M., Shanmugam, K., Dinstein, I.H.: Textural features for image classification. IEEE Trans. Syst. Man Cybern. **6**, 610–621 (1973)
39. Van de Wouwer, G., Scheunders, P., Van Dyck, D.: Statistical texture characterization from discrete wavelet representations. IEEE Trans. Image Process. **8**(4), 592–598 (1999)
40. Costa, A.F., Humpire-Mamani, G., Traina, A.J.M.: An efficient algorithm for fractal analysis of textures. In: 2012 25th SIBGRAPI Conference on Graphics, Patterns and Images, pp. 39–46 (2012)
41. Pudil, P., Novovičová, J., Kittler, J.: Floating search methods in feature selection. Pattern Recogn. Lett. **15**(11), 1119–1125 (1994)
42. Tahir, F., Fahiem, M.A.: A statistical-textural-features based approach for classification of solid drugs using surface microscopic images. Comput. Math. Methods Med. 2014 (2014)
43. Brownlee, J.: Clever Algorithms: Nature-Inspired Programming Recipes, Jason Brownlee (2011)
44. Liu, L., Liu, W., Cartes, D.A.: Particle swarm optimization-based parameter identification applied to permanent magnet synchronous motors. Eng. Appl. Artif. Intell. **21**(7), 1092–1100 (2008)
45. Castillo, O., Melin, P.: New Perspectives on Hybrid Intelligent System Design Based on Fuzzy Logic Neural Networks and Metaheuristics. . Springer, Cham (2022)
46. Kahramanli, H., Allahverdi, N.: Design of a hybrid system for the diabetes and heart diseases. Expert Syst. Appl. **35**(1–2), 82–89 (2008)
47. Haykin, S., Network, N.: A comprehensive foundation. Neural Netw. **2004**(2), 4 (2004)

Optimized KiU-Net: Lightweight Convolutional Neural Network for Retinal Vessel Segmentation in Medical Images

Hazrat Bilal[✉][iD] and Cem Direkoğlu[iD]

Intelligent Systems Group, Center for Sustainability (CFS), Electrical and Electronics Engineering (EEE) Department, Middle East Technical University, Northern Cyprus Campus, North Cyprus Mersin 10, 99738 Kalkanli, Guzelyurt, Turkey
{bilal.hazrat,cemdir}@metu.edu.tr

Abstract. Medical image segmentation helps with computer-assisted disease analysis, operations, and therapy. Blood vessel segmentation is very important for the diagnosis and treatment of different diseases. Lately, the U-Net and KiU-Net based vessel segmentation techniques have demonstrated reasonable achievements. The U-Net architecture belongs to the group of undercomplete autoencoders which ignores the semantic features of the thin and low contrast vessels. On the other hand, the KiU-Net uses a combination of undercomplete and overcomplete architectures to segment the small structure and fine edges better than U-Net. However, this solution is still not accurate enough and computationally complex. We propose an Optimized KiU-Net model to increase the segmentation accuracy of thin and low-contrast blood vessels and improve the computational efficiency of this lightweight network. The proposed model selects the ideal length of the encoder and the number of convolutional channels. Moreover, our proposed model has better convergence and uses a smaller number of parameters by combining the feature map at the final layer instead at each block. Our proposed network outperforms the KiU-Net on vessel segmentation in the RITE dataset. It obtained an overall enhancement of about 4% in terms of F1 score and 6% in terms of IoU compared to KiU-Net. Evaluation and comparison were also conducted on the GLASS dataset, and the results show that the proposed model is effective.

Keywords: Vessel Segmentation · Deep Learning · Optimized KiU-Net

1 Introduction

A vital component of the human body's circulatory system that maintains the internal organs functioning normally is the blood vessel. In medical practice, physicians diagnose illnesses (such as diabetic retinopathy, macular edema, and arteriosclerosis) and undertake clinical decisions and navigation based on the

A. Ortis et al. (Eds.): ICAETA 2023, CCIS 1983, pp. 373–383, 2024.
https://doi.org/10.1007/978-3-031-50920-9_29

shape and position of vessels [1,2]. For instance, it is possible to use the structural properties of retinal blood vessels like the diameter of the vessel, branching orientation, and length of the branch for initial detection and efficient retinal pathology monitoring [3]. The initial clinical images of blood vessels acquired cannot clearly reflect true morphological information because of imaging technology constraints and basic properties of human tissues. This requires an expert to manually segment blood vessel images, which would be labor-intensive and opinionated. As a result, the computerized artery segmentation method is crucial and has attracted the interest of many computer-assisted clinical imaging researchers.

Computer vision algorithms based on conventional clinical image analysis is time consuming and insufficient for segmenting vessels. These traditional techniques for segmenting vessels employ hand-crafted features [4,5], filtering based models [6], and statistical approaches [7]. Deep learning (DL) has also been extensively utilized in the area of clinical image analysis due to its superior representation learning capability, which benefits from the effect of data-driven and modern computer equipment. To increase the robustness and processing speed of segmentation, many researchers prefer to use DL-based approaches [8] since DL algorithms outperform the other methods. Among the DL-based medical image segmentation methods, U-Net [9] network has been one of the most widely used method [10–14].

Even though the U-Net [9] produces accurate medical image segmentation results in general, it does not perform well in the case of blood vessel segmentation. The U-Net [9] cannot identify all of the vessel pixels, which leads to poor vessel connectivity. A deep learning segmentation model developed on "encoder-decoder" design often uses layered convolution and downsampling to capture higher-level features and uses skip connections in conjunction with reversed processes to reconstruct the initial quality image features for pixel-level supervised methods. Whereas the feature map is shrunk through downsampling to broaden its receptive field and minimize computation, this operation does not suit blood vessel segmentation since it destroys high-resolution spatial data. The blood vessel images are very unique as compared to other medical images like heart or cells, the blood vessel images are generally narrow with poor pixel ratios and contrast resolution. As the downsampling operation reduces the amount of useful spatial information, learning semantic features of small-scale vessels is difficult for "encoder-decoder" based architecture. Particularly, fine low-level details are lost with these types of networks. The typical "encoder-decoder" design of U-Net falls towards the category of undercomplete convolutional autoencoders (AEs), in which the input data's dimension is decreased close to the end of an encoder. Therefore, under-complete designs are basically constrained in their capacity to collect high-resolution spatial information.

According to the above-mentioned problem with under-complete representations, the transfer of the image into a higher dimension is done through the introduction of an overcomplete network in [15], which is called Kite-Net (Ki-Net). In the research, it has been demonstrated that overcomplete models are more reliable and stable, particularly when noise is present [16]. They demonstrated that the Ki-Net architecture can acquire the features in the image, which benefits the capturing of high-resolution features as compared to the conventional undercomplete models. They also applied a unique bridge fusion technique to merge the advantages offered by the suggested Ki-Net and those from the conventional U-Net, which is named as KiU-Net [15] and outperformed the existing segmentation models. Moreover, the segmentation patterns still contain unidentified vessel pixels that lead to poor vessel connectivity graphically, irrespective of the fact that it obtained good segmentation performance on other types of medical images (other than retinal vessel images).

In this study, we present an approach to address the aforementioned problem and make it precise, reliable, and GPU-efficient. We improve and optimize the existing KiU-Net implementation by expanding the encoder layers and carefully choosing optimal convolutional channels. Moreover, we combined the feature map at the final layer instead of bridging at each block, which produces faster convergence and better accurate metrics in comparison to recently developed segmentation techniques. The proposed work captures high-resolution information and segments fine edges in the vessel images better than the traditional "encoder-decoder" structure as U-Net and KiU-Net.

The paper is organized as follows: In Sect. 2, we outline the proposed methodology. Section 3 presents our experiments, results, and comparisons on two different datasets. Finally, Sect. 4 is conclusions and suggestions for future research.

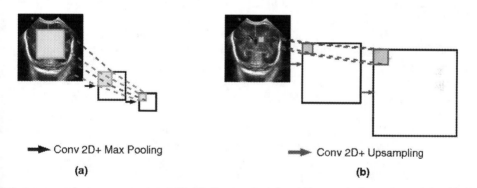

Conv 2D+ Max Pooling

(a)

Conv 2D+ Upsampling

(b)

Fig. 1. Type of design's impact on the receptive field. (a) U-Net: The outer layers concentrate on significantly bigger areas of the input data at each position. (b) Ki-Net: The outer layers concentrate on significantly smaller areas of the input data at each position [15].

2   Proposed Method

We have described U-Net [9] and KiU-Net [15] as DL based image segmentation techniques. The U-Net conventional encoder-decoder design falls under the category of undercomplete autoencoders (AEs), in which the input data's dimension is minimized close to the end of an encoder block and reduces the ability to acquire more fine features. As a result, as we move further into the network, all filters within a standard encoder-decoder structure have a larger receptive field as shown in Fig. 1. The reason for the growing receptive field becomes essential for Convnets to capture high-level information such as objects, patterns, or masses and it also decreases the effectiveness of the filters. On the contrary, the Ki-Net is a member of the overcomplete autoencoders (AEs) family [15]. This means that the geographical sizes at the middle layers are greater as compared to the input information. In contrast to the typical under-complete encoder-decoder design, this requires the overcomplete structure to behave differently [16]. Even as we move deeper into the encoder of the overcomplete architecture, the filters learn more precise low-level information because the receptive field is getting smaller as illustrated in Fig. 1.

In U-Net, the intermediate filters get smaller as we move down and capture high-level features, while in Ki-Net, the intermediate filters get larger as we move down and acquire fine low-level features across all levels with a higher resolution. So they combined Ki-Net and conventional U-Net to increase segmentation performance because Ki-Net alone will only extract edges. This integrated structure, known as KiU-Net, takes advantage of both the high-level structure acquiring image features of U-Net and the low-level finer edge capturing image features of Ki-Net.

Optimized KiU-Net: Even though KiU-Net produces remarkable segmentation performance in accordance with evaluation criteria on other datasets, it does not perform well in vessel segmentation since there are still unidentified vessel regions in segmented patterns. This leads to a lack of visual vascular connectivity in retinal vessel images. In contrast to many other clinical images of the heart, brain, or cells, vascular images are unusual because of thin vessel patterns, low pixel ratios, and contrast. We improve the KiU-Net by choosing the most effective convolution channels and changing the encoder length. A standard KiU-Net design from [15] is employed where the network's encoder depth was set to 3, and also the convolution channels at every encoder layer were 32, 64, and 128. Our investigation shows that raising the encoder's depth to 4 and changing the number of channels to 16, 32, 48, and 64, significantly improves the result for vessel segmentation.

We also implement a parallel connectivity design, with a U-Net and a Ki-Net on each side as shown in Fig. 2. The given image data is concurrently passed across two sides. We use four layers of convolution units within both the encoder and decoder for both branches. Every convolution unit in the encoder of the U-Net component comprises a 2D convolutional layer with kernel size 3 and stride 1, proceeded by a max-pooling with a pooling coefficient of 2 and coupled with instance normalization and LeakyReLU activation. Likewise, each convolution unit in the U-Net branch decoder is composed of a 2D convolutional layer with kernel size 3 and stride 1, proceeded by a bilinear interpolation with a scale factor of 2 and coupled with instance normalization and LeakyReLU activation.

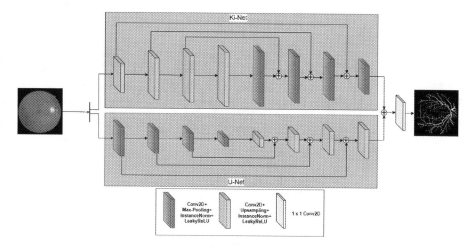

Fig. 2. Optimized KiU-Net Architecture

Throughout the U-Net section, we use an encoder-decoder design of U-Net. On the contrary, every convolution unit within the encoder of the Ki-Net component comprises a 2D convolutional layer with kernel 3 and stride 1, proceeded by a bilinear interpolation with a scale factor of 2 and coupled with instance normalization and LeakyReLU activation. Similar to this, every convolution unit in the Ki-Net branch decoder is made up of a 2D convolutional layer accompanied by a max-pooling with a pooling coefficient of 2 and coupled with instance normalization and LeakyReLU activation. In order to improve the localization, skip connections are used in both branches between the encoder and decoder sections as seen in Fig. 2.

The complementing features were collected from the two branches of the network and sent to each one simultaneously using a cross residual fusion block (CRFB) in KiU-Net [15]. However, based on our experiments, the cross residual fusion block did not enhance the segmentation of the fine low-level detail in the vessel images and made it computationally expensive. Consequently, we simply concatenate the features in the last layer. It is more efficient and optimal to use this method because it results in better convergence.

The prediction and label are compared using pixel-wise binary cross-entropy loss to train the model. The Eq. 1 describes the loss function between the prediction p and the label $\hat{p}$:

$$L_{CE(p,\hat{p})} = -\frac{1}{mn} \sum_{x=0}^{m-1} \sum_{y=0}^{n-1} (p(x,y)) \log(\hat{p}(x,y)) + (1-p(x,y)) \log(1-\hat{p}(x,y)) \quad (1)$$

In this equation, the m and n represent the size of the image, $p(x,y)$ and $\hat{p}(x,y)$ stand for the result of the prediction and true label at a given point (x,y), respectively.

3   Experiments and Results

Implementation: Optimized KiU-Net is trained using binary cross-entropy loss $L_{CE(p,\hat{p})}$, with a batch size of 1 and the Adam optimizer. The learning rate of 0.001 was used. The model was developed using the PyTorch framework and Google Colab for training. A total of 100 training epochs were used to develop the model.

3.1   RITE Dataset

A dataset called RITE (Retinal Images Vessel Tree Extraction) provides the segmentation of veins and arteries on the fundus images. This data consists of 40 sample images, 20 of which are used for learning and 20 of which are for testing. There are two reference standards included with the retinal fundus images, one is for vessels, and the other is for arteries and veins [17]. In our research, we use the fundus input images for the training of our model to forecast vessel segmentation. We reduce the size of each image to 128 × 128 in this work.

We evaluate the accuracy of our proposed Optimized KiU-Net while comparing with the most effective approaches such as Seg-Net [18], U-Net [9], and KiU-Net [15]. Experimental results show that our proposed Optimized KiU-Net outperforms existing approaches in terms of F1 score and IoU [19], are shown in Table 1. Compared to Seg-Net [18], our model showed a 27.57 and 27.16 point increase in F1 score and IoU respectively. When comparing with U-Net [9], our model showed a 24.56 and 35.19 point increase in F1 score and IoU. In comparison to KiU-Net [15], our model showed a 4.63 and 6.13 point increase in F1 score and IoU respectively. Our proposed approach yielded high-quality results as shown in Fig. 3. The images clearly show that our technique captures fine low-level details more effectively and provides a more accurate segmentation result as compared to the other methods.

Furthermore, Our proposed model boasts a computational advantage over existing approaches due to its smaller parameter count. This makes it more efficient and simpler to implement. The reduced number of parameters is a stand-out feature of our model that contributes to decrease computational complexity. These results demonstrate the superior performance of our proposed Optimized KiU-Net in retinal vessel segmentation.

Nevertheless, it is important to note that the Optimized KiU-Net still needs to be improved because it may neglect to segment the thin edges of the retinal vessels as compared to the ground truth.

3.2   GlaS Dataset

In this study, we also conducted experiments on the GlaS (gland segmentation in histology images) dataset [27] to assess the effectiveness of our approach. Clinical photographs of Hematoxylin and Eosin (H&E) stained slides are included in the Gland segmentation dataset, as well as the associated label annotations by experienced clinicians. This dataset has 165 total colon histology images, of

Table 1. Quantitative contrast of the proposed technique with other models.

Method	F1	IoU	Parameters
Seg-Net [18]	52.23	39.14	12.5M
U-Net [9]	55.24	31.11	3.1M
KiU-Net [15]	75.17	60.37	0.29M
Optimized KiU-Net (ours)	**79.80**	**66.30**	**0.18M**

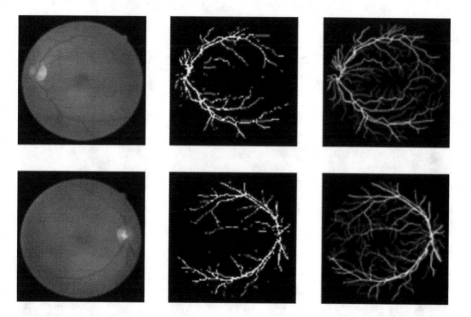

Fig. 3. The proposed model prediction on the retinal images. The left column is the actual image while the right column is the label and the center column is the prediction of the proposed model.

which 85 are used for training and 80 for testing. We reduce the size of each image to a resolution of 128×128 for all of our tests because the images in the data are various sizes. We evaluate the performance of our proposed Optimized KiU-Net as compared to the most effective approaches as shown in Table 2. Results show that the Optimized KiU-Net significantly outperforms other existing techniques except the HistoSeg [26]. The HistoSeg generated patches from the 85 images of training data using augmentation, and the augmented dataset contains 2347 images in total. Therefore, the performance of our proposed network is better than the other network on the actual size of the dataset. The qualitative results of our proposed method are shown in Fig. 4.

Table 2. Quantitative contrast of the proposed technique with other models.

Method	F1	IoU
FCN [20]	66.61	50.84
U-Net [9]	77.78	65.34
U-Net++ [21]	78.03	65.55
Res-UNet [22]	78.83	65.95
Deeplabv3+ [23]	76.01	67.04
Axial Attention U-Net [24]	76.26	63.03
MedT [25]	81.02	69.61
HistoSeg [26]	**98.07**	**76.73**
Optimized KiU-Net (ours)	82.21	71.03

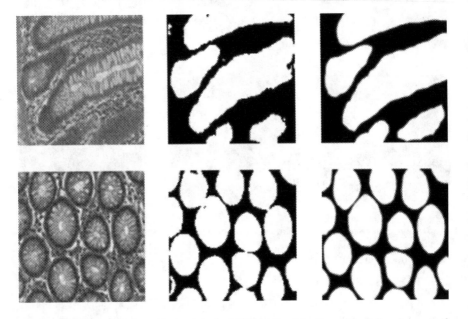

Fig. 4. proposed model prediction on the H&E stained images. The left column is the actual image while the right column is the label and the center column is the prediction of the proposed model.

4   Conclusion

We presented a new network named Optimized KiU-Net that is developed by optimizing the standard KiU-Net architecture. The purpose of this optimization is to capture low-level features and fine edges that are commonly missed by other approaches. Furthermore, we simply used concatenation at the last layer for features from the two networks instead of bridging the features at each block, which further improves the capturing of low-level features. Based on our experiments, we selected the optimal length of the encoder and convolutional channels, which gives a further advantage to the proposed design such as using less number of parameters and faster convergence time. The presented technique outperforms the existing techniques on a challenging dataset (RITE) because of the ability to extract both low-level and high-level features. However, the proposed model may fail to detect the very thin ends of retinal vessels, which can lead to incorrect interpretations and diagnostic outcomes. The accuracy of the model can be increased by training with a more extensive and diverse dataset of fundus images. Then the model will have more exposure to a wider range of vessel shapes and sizes, which will help to differentiate between the thin edges and the surrounding tissue.

References

1. Xu, W., Yang, H., Zhang, M., Pan, X., Liu, W., Yan, S.: DECNet: a dual-stream edge complementary network for retinal vessel segmentation. In: IEEE International Conference on Bioinformatics and Biomedicine (BIBM), vol. 2021, pp. 1595–1600 (2021). https://doi.org/10.1109/BIBM52615.2021.9669751
2. Li, R.Q., et al.: Real-time multi-guidewire endpoint localization in fluoroscopy images. IEEE Trans. Med. Imaging **40**(8), 2002–2014 (2021). Epub 2021 Jul 30. PMID: 33788685. https://doi.org/10.1109/TMI.2021.3069998
3. Soomro, T.A., et al.: Deep learning models for retinal blood vessels segmentation: a review. IEEE Access **7**, 71696–71717 (2019). https://doi.org/10.1109/ACCESS.2019.2920616
4. Javidi, M., Pourreza, H.R., Harati, A.: Vessel segmentation and microaneurysm detection using discriminative dictionary learning and sparse representation. Comput Methods Programs Biomed. **139**, 93–108 (2017). Epub 2016 Oct 26. PMID: 28187898. https://doi.org/10.1016/j.cmpb.2016.10.015
5. Javidi, M., Harati, A., Pourreza, H.: Retinal image assessment using bi-level adaptive morphological component analysis. Artif. Intell. Med. **99**, 101702 (2019). Epub 2019 Jul 30. PMID: 31606110. https://doi.org/10.1016/j.artmed.2019.07.010
6. Wang, C., et al.: Tensor-cut: a tensor-based graph-cut blood vessel segmentation method and its application to renal artery segmentation. Med. Image Anal. **60**, 101623 (2019). https://doi.org/10.1016/j.media.2019.101623
7. Kalaie, S., Gooya, A.: Vascular tree tracking and bifurcation points detection in retinal images using a hierarchical probabilistic model. Comput. Methods Programs Biomed. **151**, 139–149 (2017). https://doi.org/10.1016/j.cmpb.2017.08.018
8. Jia, D., Zhuang, X.: Learning-based algorithms for vessel tracking: a review. Comput. Med. Imaging Graph. **89**, 101840 (2021). https://doi.org/10.1016/j.compmedimag.2020.101840. Epub 2021 Jan 30. PMID: 33548822
9. Ronneberger, O., Fischer, P., Brox, T.: U-Net: convolutional networks for biomedical image segmentation. In: Navab, N., Hornegger, J., Wells, W.M., Frangi, A.F. (eds.) MICCAI 2015. LNCS, vol. 9351, pp. 234–241. Springer, Cham (2015). https://doi.org/10.1007/978-3-319-24574-4_28
10. Çiçek, Ö., Abdulkadir, A., Lienkamp, S.S., Brox, T., Ronneberger, O.: 3D U-Net: learning dense volumetric segmentation from sparse annotation. In: Ourselin, S., Joskowicz, L., Sabuncu, M.R., Unal, G., Wells, W. (eds.) MICCAI 2016. LNCS, vol. 9901, pp. 424–432. Springer, Cham (2016). https://doi.org/10.1007/978-3-319-46723-8_49

11. Zhao, N., Tong, N., Ruan, D., Sheng, K.: Fully automated pancreas segmentation with two-stage 3D convolutional neural networks. In: Shen, D., et al. (eds.) MICCAI 2019. LNCS, vol. 11765, pp. 201–209. Springer, Cham (2019). https://doi.org/10.1007/978-3-030-32245-8_23

12. Li, X., Chen, H., Qi, X., Dou, Q., Fu, C.-W., Heng, P.-A.: H-DenseUNet: hybrid densely connected UNet for liver and tumor segmentation from CT volumes. IEEE Trans. Med. Imaging **37**(12), 2663–2674 (2018). https://doi.org/10.1109/TMI.2018.2845918

13. Milletari, F., Navab, N., Ahmadi, S.: V-Net: fully convolutional neural networks for volumetric medical image segmentation. In: 2016 Fourth International Conference on 3D Vision (3DV), Stanford, CA, USA, pp. 565–571 (2016). https://doi.org/10.1109/3DV.2016.79

14. Zhou, Z., Siddiquee, M.M.R., Tajbakhsh, N., Liang, J.: UNet++: redesigning skip connections to exploit multiscale features in image segmentation. IEEE Trans. Med. Imaging. **39**(6), 1856–1867 (2020). https://doi.org/10.1109/TMI.2019.2959609. Epub 2019 Dec 13. PMID: 31841402; PMCID: PMC7357299

15. Valanarasu, J.M.J., Sindagi, V.A., Hacihaliloglu, I., Patel, V.M.: KiU-Net: towards accurate segmentation of biomedical images using over-complete representations. In: Martel, A.L., et al. (eds.) MICCAI 2020. LNCS, vol. 12264, pp. 363–373. Springer, Cham (2020). https://doi.org/10.1007/978-3-030-59719-1_36

16. Lewicki, M.S., Sejnowski, T.J.: Learning overcomplete representations. Neural Comput. **12**(2), 337–65 (2000). https://doi.org/10.1162/089976600300015826. PMID: 10636946

17. Hu, Q., Abràmoff, M.D., Garvin, M.K.: Automated separation of binary overlapping trees in low-contrast color retinal images. In: Mori, K., Sakuma, I., Sato, Y., Barillot, C., Navab, N. (eds.) MICCAI 2013. LNCS, vol. 8150, pp. 436–443. Springer, Heidelberg (2013). https://doi.org/10.1007/978-3-642-40763-5_54 PMID: 24579170

18. Badrinarayanan, V., Kendall, A., Cipolla, R.: SegNet: a deep convolutional encoder-decoder architecture for image segmentation. IEEE Trans. Pattern Anal. Mach. Intell. **39**(12), 2481–2495 (2017). https://doi.org/10.1109/TPAMI.2016.2644615

19. Taha, A.A., Hanbury, A.: Metrics for evaluating 3D medical image segmentation: analysis, selection, and tool. BMC Med. Imaging. **12**, 15–29 (2015). https://doi.org/10.1186/s12880-015-0068-x. PMID: 26263899; PMCID: PMC4533825

20. Long, J., Shelhamer, E., Darrell, T.: Fully convolutional networks for semantic segmentation. In: IEEE Conference on Computer Vision and Pattern Recognition (CVPR), vol. 2015, pp. 3431–3440 (2015). https://doi.org/10.1109/CVPR.2015.7298965

21. Zhou, Z., Siddiquee, M.M.R., Tajbakhsh, N., Liang, J.: UNet++: redesigning skip connections to exploit multiscale features in image segmentation. IEEE Trans. Med. Imaging **39**(6), 1856–1867 (2020). https://doi.org/10.1109/TMI.2019.2959609

22. Xiao, X., Lian, S., Luo, Z., Li, S.: Weighted res-unet for high-quality retina vessel segmentation. In: 2018 9th International Conference on Information Technology in Medicine and Education (ITME), pp. 327–331 (2018). https://doi.org/10.1109/ITME.2018.00080

23. Chen, L.C., Zhu, Y., Papandreou, G., Schroff, F., Adam, H.: Encoder-decoder with atrous separable convolution for semantic image segmentation. In: Proceedings of the European conference on computer vision (ECCV), p. 801818 (2018)

24. Wang, H., Zhu, Y., Green, B., Adam, H., Yuille, A., Chen, L.-C.: Axial-DeepLab: stand-alone axial-attention for panoptic segmentation. In: Vedaldi, A., Bischof, H., Brox, T., Frahm, J.-M. (eds.) ECCV 2020. LNCS, vol. 12349, pp. 108–126. Springer, Cham (2020). https://doi.org/10.1007/978-3-030-58548-8_7

25. Shen, X., Wang, L., Zhao, Y., Liu, R., Qian, W., Ma, H.: Dilated transformer: residual axial attention for breast ultrasound image segmentation. Quant. Imaging Med. Surg. **12**(9), 4512 (2022). https://doi.org/10.21037/qims-22-33

26. Wazir, S., Fraz, M.M.: HistoSeg: Quick attention with multi-loss function for multi-structure segmentation in digital histology images. In: 2022 12th International Conference on Pattern Recognition Systems (ICPRS), pp. 1–7 (2022). https://doi.org/10.1109/ICPRS54038.2022.9854067

27. Sirinukunwattana, K., et al.: Gland segmentation in colon histology images: the glas challenge contest. Med. Image Anal. **35**, 489–502 (2017). https://doi.org/10.1016/j.media.2016.08.008. Epub 2016 Sep 3. PMID: 27614792

On the Characteristic Functions in Listing Stable Arguments

Samer Nofal[✉], Amani Abu Jabal, Abdullah Alfarrarjeh, and Ismail Hababeh

Department of Computer Science, German Jordanian University, Amman, Jordan
{samer.nofal,amani.abujabal,abdullah.alfarrarjeh,
ismail.hababeh}@gju.edu.jo

Abstract. An abstract argumentation framework (AF) is viewed as a directed graph such that graph vertices represent abstract arguments while graph edges denote attacks between these arguments. We say that a set, S, of arguments, are conflict-free if and only if for every $(x, y) \in S \times S$, x does not attack y. A set, S, of arguments of a given AF, is called a stable extension in AF if S is conflict-free and such that every argument outside S is attacked by an argument inside S. To the best of our knowledge, a thorough mathematical analysis of the truth of what so-called characteristic functions (which are an essential component in generating all stable extensions of a given AF) is not previously addressed in the literature. We fill this gap; we rigorously analyze the verity of characteristic functions employed in listing all stable extensions in a given AF.

Keywords: Multiagent Systems · Nonmonotonic Reasoning · Computational Argumentation · Stable Semantics · Characteristic Functions

1 Introduction

An *abstract argumentation framework* (AF) is a pair (A, R) where A is a set of *abstract arguments* and $R \subseteq A \times A$ is the *attack* relation between them. Let $H = (A, R)$ be an AF, $S \subseteq A$ be a subset of arguments and $S^+ = \{y \mid \exists x \in S \text{ with } (x, y) \in R\}$. Then, S is a *stable extension* in H if and only if $S^+ = A \setminus S$. Since introduced in [6] as a formalism for nonmonotonic reasoning, AFs have attracted a substantial body of research due to their promising applications (see e.g. [3, 4, 12, 14, 15]). As a motivation example, we show an application of AFs for modeling negotiation between agents in multiagent systems. The following example is adapted from [13]). Consider two home-improvement agents: agent 1 needs a nail to hang a painting but has a screw, whereas agent 2 is trying to hang a mirror and has a nail to hang the mirror. Suppose the following dialog between the two agents.

A1. *Agent 1: Give me a nail to hang my painting.*
A2. *Agent 2: I need the nail for hanging my mirror.*

A3. *Agent 1: You can use a screw to hang the mirror.*
A4. *Agent 2: Then, I need to buy a screw for hanging my mirror.*
A5. *Agent 1: I can provide you with a screw.*

This dialog between the two agents can be modeled as a directed graph with a directed-edge set $\{(A5, A4), (A4, A3), (A4, A2), (A2, A4), (A3, A2), (A2, A1)\}$. Recall that a directed edge (x, y) represents x *attacking* y, which means that if we accept x, then y will be rejected. Back to the dialog between the two agents, the two agents must decide what to do next that dialog. Therefore, computing a set of acceptable arguments in the dialog between the two agents is necessary. Under stable semantics, we note that arguments A1, A3, and A5 are acceptable since they form a stable extension together. This means Agent 1 succeeded in persuading Agent 2 to exchange the nail with the screw.

It is known that the problem of generating all stable extensions of a given AF is NP-hard, see e.g. [7]. Thereby, in the literature, one can find different implemented methods for solving this problem, such as backtracking procedures, dynamic programming, and reduction-based methods, see e.g. [5] for a fuller review. Implementing backtracking procedures for generating stable extensions of a given AF might be accomplished using characteristic functions, see for example [1,9]. For generating stable extensions, characteristic functions had been validated experimentally in the literature; see the articles of [2,8] for example. In our previous work, [10], we investigated the verity of set-theoretic structures employed in implementing a backtracking process for generating stable extensions of an AF.

Since we do not see in the literature a thorough, rigorous validation of characteristic functions utilized for generating stable extensions in a given AF, we fill this gap; we present in Sect. 2 a complete, formal analysis of the verity of characteristic functions exploited in implementing a backtracking process for listing all stable extensions in a given AF. We conclude the paper in Sect. 3.

2   Our Analysis

Let $H = (A, R)$ be an AF and $T \subseteq A$ be a subset of arguments. Then,

$$T^+ \stackrel{\text{def}}{=} \{y \mid \exists x \in T \text{ with } (x, y) \in R\}, \text{ and } T^- \stackrel{\text{def}}{=} \{y \mid \exists x \in T \text{ with } (y, x) \in R\}.$$

To enumerate all stable extensions in H, let S denote an under-construction stable extension of H. Thus, we start with $S = \emptyset$ and then let S grow to a stable extension (if any exists) incrementally by choosing arguments from A to join S. For this, we denote by *choice* a set of arguments eligible to join S. More precisely, take $S \subseteq A$ such that $S \cap S^+ = \emptyset$, then $choice \subseteq A \setminus (S \cup S^- \cup S^+)$. Additionally, we denote by *tabu* the arguments that do not belong to $S \cup S^+ \cup choice$. More specifically, take $S \subseteq A$ such that $S \cap S^+ = \emptyset$, and $choice \subseteq A \setminus (S \cup S^- \cup S^+)$. Then, $tabu = A \setminus (S \cup S^+ \cup choice)$.

Now we define a characteristic function to indicate the status of an argument in a given AF concerning S, S^+, *choice*, and *tabu*.

Definition 1. *Let* $H = (A, R)$ *be an* AF, $S \subseteq A$, $S \cap S^+ = \emptyset$, *choice* $\subseteq A \setminus (S \cup S^+ \cup S^-)$, *tabu* $= A \setminus (S \cup S^+ \cup choice)$, *and* $\mu : A \to \{blank, in, out, must\text{-}out\}$ *be a total mapping. Then,* μ *is a characteristic function of* H *concerning* S *if and only if for all* $x \in A$ *it is the case that*

$$x \in S \iff \mu(x) = in,$$
$$x \in S^+ \iff \mu(x) = out,$$
$$x \in choice \iff \mu(x) = blank, \quad and$$
$$x \in tabu \iff \mu(x) = must\text{-}out.$$

We specify the conditions under which a characteristic function corresponds to a stable extension.

Proposition 1. *Let* $H = (A, R)$ *be an* AF, $S \subseteq A$, $S \cap S^+ = \emptyset$, *choice* $\subseteq A \setminus (S \cup S^+ \cup S^-)$, *tabu* $= A \setminus (S \cup S^+ \cup choice)$, *and* μ *be a characteristic function of* H *with respect to* S. *Then,* $\{x \mid \mu(x) = in\}$ *is a stable extension in* H *if and only if* $\{x \mid \mu(x) = blank\} = \emptyset$ *and* $\{x \mid \mu(x) = must\text{-}out\} = \emptyset$.

Proof. We prove $\implies$ firstly. Considering Definition 1, $S = \{x \mid \mu(x) = in\}$ and $S^+ = \{x \mid \mu(x) = out\}$. Suppose $\{x \mid \mu(x) = in\}$ is stable. Then, $\{x \mid \mu(x) = out\} = A \setminus \{x \mid \mu(x) = in\}$; and subsequently, $\{x \mid \mu(x) = in\} \cup \{x \mid \mu(x) = out\} = A$. By Definition 1, observe that

$$\begin{aligned} tabu &= \{x \mid \mu(x) = must\text{-}out\} \\ &= A \setminus (\{x \mid \mu(x) = in\} \cup \{x \mid \mu(x) = out\} \cup \{x \mid \mu(x) = blank\}) \\ &= A \setminus (A \cup \{x \mid \mu(x) = blank\}) \\ &= \emptyset \end{aligned}$$

Likewise,

$$\begin{aligned} choice &= \{x \mid \mu(x) = blank\} \subseteq A \setminus (\{x \mid \mu(x) = in\} \cup \{x \mid \mu(x) = out\} \cup S^-) \\ &= \{x \mid \mu(x) = blank\} \subseteq A \setminus (A \cup S^-) = \emptyset \end{aligned}$$

Now we prove $\impliedby$. Suppose $\{x \mid \mu(x) = blank\} = \emptyset$ and $\{x \mid \mu(x) = must\text{-}out\} = \emptyset$. Note that

$$\begin{aligned} tabu &= \{x \mid \mu(x) = must\text{-}out\} \\ &= A \setminus (\{x \mid \mu(x) = in\} \cup \{x \mid \mu(x) = out\} \cup \{x \mid \mu(x) = blank\}), \end{aligned}$$

which can be rewritten as

$$\emptyset = A \setminus (\{x \mid \mu(x) = in\} \cup \{x \mid \mu(x) = out\} \cup \emptyset).$$

And subsequently,

$$A = \{x \mid \mu(x) = in\} \cup \{x \mid \mu(x) = out\}.$$

Thus, and since $\{x \mid \mu(x) = in\} \cap \{x \mid \mu(x) = out\} = \emptyset$, we note that $\{x \mid \mu(x) = out\} = A \setminus \{x \mid \mu(x) = in\}$. Consequently, $\{x \mid \mu(x) = in\}$ is a stable extension in H. $\square$

Next, we discuss three propositions that speed up the process of stable extension generation.

Proposition 2. *Let* $H = (A, R)$ *be an* AF, $S \subseteq A$, $S \cap S^+ = \emptyset$, *choice* $\subseteq A \setminus (S \cup S^+ \cup S^-)$, *tabu* $= A \setminus (S \cup S^+ \cup choice)$, μ *be a characteristic function of* H *with respect to* S, *and* v *be an argument with* $\mu(v) = blank$ *such that for all* $y \in \{v\}^-$ *it holds that* $\mu(y) \in \{out, must\text{-}out\}$. *If there is a stable extension* $T \supseteq \{x \mid \mu(x) = in\}$ *such that* $T \setminus \{x \mid \mu(x) = in\} \subseteq \{x \mid \mu(x) = blank\}$, *then* $v \in T$.

Proof. Suppose $v \notin T$. Then, $v \in T^+$ because T is stable. Thus,

$$\exists y \in \{v\}^- \text{ such that } y \in T.$$

However, according to the premise of this proposition,

$$\forall y \in \{v\}^- \ \mu(y) \in \{out, must\text{-}out\}.$$

Therefore,

$$T \cap (\{x \mid \mu(x) = out\} \cup \{x \mid \mu(x) = must\text{-}out\}) \neq \emptyset \qquad (\star)$$

By the assumption of the premise of this proposition, we note that

$$S \cap (S^+ \cup tabu) = \emptyset \text{ and } choice \cap (S^+ \cup tabu) = \emptyset.$$

Applying Definition 1, this can be rewritten respectively as

$$\{x \mid \mu(x) = in\} \cap (\{x \mid \mu(x) = out\} \cup \{x \mid \mu(x) = must\text{-}out\}) = \emptyset, \text{ and}$$
$$\{x \mid \mu(x) = blank\} \cap (\{x \mid \mu(x) = out\} \cup \{x \mid \mu(x) = must\text{-}out\}) = \emptyset.$$

Again, given the premise of this proposition, note that

$$T \supseteq \{x \mid \mu(x) = in\}, \text{and } T \setminus \{x \mid \mu(x) = in\} \subseteq \{x \mid \mu(x) = blank\}.$$

Thus,

$$T \cap (\{x \mid \mu(x) = out\} \cup \{x \mid \mu(x) = must\text{-}out\}) = \emptyset. \qquad (\star\star)$$

Observe the contradiction between $(\star)$ and $(\star\star)$. $\square$

Proposition 3. *Let* $H = (A, R)$ *be an* AF, $S \subseteq A$, $S \cap S^+ = \emptyset$, *choice* $\subseteq A \setminus (S \cup S^+ \cup S^-)$, *tabu* $= A \setminus (S \cup S^+ \cup choice)$, μ *be a characteristic function of* H *with respect to* S, *and* v *be an argument with* $\mu(v) = blank$ *such that for some* y *with* $\mu(y) = must\text{-}out$ *it is the case that* $\{x \in \{y\}^- \mid \mu(x) = blank\} = \{v\}$. *If there is a stable extension* $T \supseteq \{x \mid \mu(x) = in\}$ *such that* $T \setminus \{x \mid \mu(x) = in\} \subseteq \{x \mid \mu(x) = blank\}$, *then* $v \in T$.

Proof. Suppose $v \notin T$. Then,

$$T \setminus \{x \mid \mu(x) = in\} \subseteq \{x \mid \mu(x) = blank\} \setminus \{v\}.$$

Due to $tabu = A \setminus (S \cup S^+ \cup choice)$, and according to Definition 1 it holds that

$$tabu = \{x \mid \mu(x) = must\text{-}out\}$$
$$= A \setminus (\{x \mid \mu(x) = in\} \cup \{x \mid \mu(x) = out\} \cup \{x \mid \mu(x) = blank\}).$$

Since $T \supseteq \{x \mid \mu(x) = in\}$ such that $T \setminus \{x \mid \mu(x) = in\} \subseteq \{x \mid \mu(x) = blank\}$,

$$T \subseteq \{x \mid \mu(x) = in\} \cup \{x \mid \mu(x) = blank\}.$$

Referring to the premise of this proposition, as $y \in \{x \mid \mu(x) = must\text{-}out\}$, $y \in tabu$ (recall Definition 1) and so $y \notin T$. However, since T is stable, $y \in T^+$. Thus,

$$\{y\}^- \cap T \neq \emptyset.$$

Because $y \in \{x \mid \mu(x) = must\text{-}out\}$, it holds that $y \notin \{x \mid \mu(x) = out\}$; hence $y \notin S^+$ consistently with Definition 1. Therefore,

$$\{y\}^- \cap \{x \mid \mu(x) = in\} = \emptyset.$$

Thereby,

$$\{y\}^- \cap (\{x \mid \mu(x) = blank\} \setminus \{v\}) \neq \emptyset.$$

From the premise of this proposition, it is the case that $\{x \in \{y\}^- \mid \mu(x) = blank\} = \{v\}$. Thus,

$$\{y\}^- \cap \{x \mid \mu(x) = blank\} = \{v\}.$$

Contradition. □

Proposition 4. *Let $H = (A, R)$ be an* AF, *$S \subseteq A$, $S \cap S^+ = \emptyset$, choice $\subseteq A \setminus (S \cup S^+ \cup S^-)$, $tabu = A \setminus (S \cup S^+ \cup choice)$, μ be a characteristic function of H with respect to S, and x be an argument with $\mu(x) = must\text{-}out$ such that for all $y \in \{x\}^-$ it is the case that $\mu(y) \in \{out, must\_out\}$. There does not exist a stable extension $Q \supseteq \{v \mid \mu(v) = in\}$ such that $Q \setminus \{v \mid \mu(v) = in\} \subseteq \{v \mid \mu(v) = blank\}$.*

Proof. Consider Definition 1 during the proof. As $x \in \{v \mid \mu(v) = must\text{-}out\}$ and $tabu = A \setminus (S \cup S^+ \cup choice)$,

$$x \notin \{v \mid \mu(v) = in\} \cup \{v \mid \mu(v) = out\} \cup \{v \mid \mu(v) = blank\}.$$

Subsequently,

$$\forall T \supseteq \{v \mid \mu(v) = in\} \text{ such that } T \setminus \{v \mid \mu(v) = in\} \subseteq \{v \mid \mu(v) = blank\}, \, x \notin T.$$

Since for all $y \in \{x\}^-$ it holds that $\mu(y) \in \{out, must\_out\}$,

$$\{x\}^- \subseteq \{v \mid \mu(v) = out\} \cup \{v \mid \mu(v) = must\text{-}out\}.$$

Hence, $\{x\}^- \cap (\{v \mid \mu(v) = in\} \cup \{v \mid \mu(v) = blank\}) = \emptyset$. Therefore,

$\forall T \supseteq \{v \mid \mu(v) = in\} : T \setminus \{v \mid \mu(v) = in\} \subseteq \{v \mid \mu(v) = blank\}, \{x\}^- \cap T = \emptyset$.

Thus,

$$\forall T \supseteq \{v \mid \mu(v) = in\} : T \setminus \{v \mid \mu(v) = in\} \subseteq \{v \mid \mu(v) = blank\}, x \notin T^+.$$

Subsequently,

$$\forall T \supseteq \{v \mid \mu(v) = in\} : T \setminus \{v \mid \mu(v) = in\} \subseteq \{v \mid \mu(v) = blank\}, x \notin T^+ \cup T.$$

This means,

$$\forall T \supseteq \{v \mid \mu(v) = in\} : T \setminus \{v \mid \mu(v) = in\} \subseteq \{v \mid \mu(v) = blank\}, T^+ \neq A \setminus T.$$

Consequently,

$$\forall T \supseteq \{v \mid \mu(v) = in\} : T \setminus \{v \mid \mu(v) = in\} \subseteq \{v \mid \mu(v) = blank\}, T \text{ is not stable}.$$

This completes the proof of this proposition. □

Now, we turn to lines 2, 3, and 7 in Algorithm 1. Recall, by applying these lines, it is required to search (respectively) for an argument x such that

$x \in tabu$ with $\{x\}^- \subseteq S^+ \cup tabu$, or $x \in choice$ with $\{x\}^- \subseteq S^+ \cup tabu$, or
$x \in tabu$ with $|\{x\}^- \cap choice| = 1$.

To implement these lines (i.e., 2, 3, and 7), we define the following construct (inspired by a hint given in [9] and already utilized in [11]).

Definition 2. Let $H = (A, R)$ be an AF, $S \subseteq A$, $S \cap S^+ = \emptyset$, $choice \subseteq A \setminus (S \cup S^+ \cup S^-)$, $tabu = A \setminus (S \cup S^+ \cup choice)$, and μ be a characteristic function of H with respect to S. For all $x \in A$, if $\mu(x) \in \{blank, must\text{-}out\}$, then $\pi(x) \overset{\text{def}}{=} |\{y \in \{x\}^- : \mu(y) = blank\}|$.

The intuition behind Definition 2 is that instead of checking (recurrently over and over) the whole set of attackers, $\{x\}^-$, of a given argument x, to see whether $\{x\}^-$ is contained in $S^+ \cup tabu$, one might hold a counter of the attackers of x that are currently in $choice$. Whenever an attacker of x is being moved from $choice$, we decrease the counter. So, to check whether $\{x\}^- \subseteq S^+ \cup tabu$, we need to do is to test if the counter is equal to zero. In the following proposition, we prove the usage of this notion (i.e., counting attackers).

Proposition 5. Let $H = (A, R)$ be an AF, $S \subseteq A$, $S \cap S^+ = \emptyset$, $choice \subseteq A \setminus (S \cup S^+ \cup S^-)$, $tabu = A \setminus (S \cup S^+ \cup choice)$, and μ be a characteristic function of H with respect to S. Then, for all $x \in A$ it holds that

(i) $x \in tabu \wedge \{x\}^- \subseteq S^+ \cup tabu \iff \mu(x) = must\text{-}out \wedge \pi(x) = 0$.

(ii) $x \in choice \wedge \{x\}^- \subseteq S^+ \cup tabu \iff \mu(x) = blank \wedge \pi(x) = 0$.

(iii) $x \in tabu \wedge |\{x\}^- \cap choice| = 1 \iff \mu(x) = must\text{-}out \wedge \pi(x) = 1$.

Proof. Consider definitions 1 and 2 throughout this proof. From the premise of this proposition, note that $S \cup S^+ \cup choice \cup tabu = A$. Additionally, the sets: S, S^+, $tabu$, and $choice$ are pairwise disjoint. Thus, for all x

$$x \in tabu \wedge \{x\}^- \subseteq S^+ \cup tabu \iff x \in tabu \wedge \{x\}^- \cap (S \cup choice) = \emptyset \iff$$
$$x \in tabu \wedge x \notin S^+ \wedge \{y \in \{x\}^- \mid y \in choice\} = \emptyset \iff$$
$$\mu(x) = must\text{-}out \wedge \{y \in \{x\}^- \mid \mu(y) = blank\} = \emptyset \iff$$
$$\mu(x) = must\text{-}out \wedge \pi(x) = 0.$$

That completes the proof of (i). Likewise, (ii) is true since for all x, it holds that

$$x \in choice \wedge \{x\}^- \subseteq S^+ \cup tabu \iff$$
$$x \in choice \wedge \{x\}^- \cap (S \cup choice) = \emptyset \iff$$
$$x \in choice \wedge x \notin S^+ \wedge \{y \in \{x\}^- \mid y \in choice\} = \emptyset \iff$$
$$\mu(x) = blank \wedge \{y \in \{x\}^- \mid \mu(y) = blank\} = \emptyset \iff \mu(x) = blank \wedge \pi(x) = 0.$$

As to (iii), for all x, it is the case that

$$x \in tabu \wedge |\{x\}^- \cap choice| = 1 \iff$$
$$\mu(x) = must\text{-}out \wedge |\{y \in \{x\}^- : y \in choice\}| = 1 \iff$$
$$\mu(x) = must\text{-}out \wedge |\{y \in \{x\}^- : \mu(y) = blank\}| = 1 \iff$$
$$\mu(x) = must\text{-}out \wedge \pi(x) = 1.$$

$\square$

Using the characteristic function μ and the counter function π specified in Definition 1 & 2, respectively, we formulate in Algorithm 1 a backtracking process for generating all stable extensions in an AF. Let $H = (A, R)$ be an AF, $\mu : A \to \{in, out, must\text{-}out, blank\}$ be a total mapping such that for all $x \in A$,

$$\mu(x) = \begin{cases} must\text{-}out, & if\ (x, x) \in R; \\ blank, & otherwise. \end{cases}$$

And let $\pi : A \to \{0, 1, 2, ..., |A|\}$ be a total mapping such that for all $x \in A$, $\pi(x) = |\{y \in \{x\}^- : \mu(y) = blank\}|$. By invoking Algorithm 1 with list-stb-ext(μ, π, $\{x \mid \mu(x) = blank \wedge \pi(x) = 0\}$), the algorithm computes all stable extensions in H.

Let us apply Algorithm 1 to generate the stable extensions of an AF with $A = \{a, b, c, d, e, f\}$ and

$$R = \{(a, b), (b, c), (b, d), (d, b), (d, e), (d, f), (e, c), (e, a), (e, f), (f, a)\}.$$

So, we start the algorithm with list-stb-ext(μ, π, γ) such that

$$\mu = \{(a, blank), (b, blank), (c, blank), (d, blank), (e, blank), (f, blank)\},$$
$$\pi = \{(a, 2), (b, 2), (c, 2), (d, 1), (e, 1), (f, 2)\}, \text{and } \gamma = \emptyset. \tag{1}$$

Referring to line 15 in Algorithm 1, let x be the argument a. Then, invoke list-stb-ext(μ, π, γ) with

$$\mu = \{(a, blank), (b, blank), (c, blank), (d, blank), (e, blank), (f, blank)\},$$
$$\pi = \{(a, 2), (b, 2), (c, 2), (d, 1), (e, 1), (f, 2)\}, \text{and } \gamma = \{a\}. \tag{2}$$

Algorithm 1: list-stb-ext(μ, π, γ)

1   **while** $\gamma \neq \emptyset$ **do**
2   For some $q \in \gamma$ **do** $\gamma \leftarrow \gamma \setminus \{q\}$; $\mu(q) \leftarrow in$;
3   **foreach** $z \in \{q\}^+$ with $\mu(z) = must\text{-}out$ **do** $\mu(z) \leftarrow out$;
4   **foreach** $z \in \{q\}^- \cup \{q\}^+$ with $\mu(z) = blank$ **do**
5   **if** $z \in \{q\}^+$ **then** $\mu(z) \leftarrow out$ **else** $\mu(z) \leftarrow must\text{-}out$;
6   **foreach** $x \in \{z\}^+$ **do**
7         $\pi(x) \leftarrow \pi(x) - 1$;
8   **if** $\mu(x) = must\text{-}out$ with $\pi(x) = 0$ **then** return;
9   **if** $\mu(x) = blank$ with $\pi(x) = 0$ **then** $\gamma \leftarrow \gamma \cup \{x\}$;
10   **if** $\mu(x) = must\text{-}out$ with $\pi(x) = 1$ **then**
11          $\gamma \leftarrow \gamma \cup \{y \in \{x\}^- \mid \mu(y) = blank\}$;
12 **if** $\{x \mid \mu(x) = blank\} = \emptyset$ **then**
13   **if** $\{x \mid \mu(x) = must\text{-}out\} = \emptyset$ **then** $\{x \mid \mu(x) = in\}$ is stable;
14   return;
15 list-stb-ext(μ, π, $\{x\}$); // for some x with $\mu(x) = blank$
16 $\mu(x) \leftarrow must\text{-}out$;
17 **foreach** $z \in \{x\}^+$ **do**
18     $\pi(z) \leftarrow \pi(z) - 1$;
19   **if** $\mu(z) = must\text{-}out$ with $\pi(z) = 0$ **then** return;
20   **if** $\mu(z) = blank$ with $\pi(z) = 0$ **then** $\gamma \leftarrow \gamma \cup \{z\}$;
21   **if** $\mu(z) = must\text{-}out$ and $\pi(z) = 1$ **then** $\gamma \leftarrow \gamma \cup \{y \in \{z\}^- \mid \mu(y) = blank\}$;
22 list-stb-ext(μ, π, γ);

Apply a first round of the *while* loop of the algorithm. Therefore,

$$\mu = \{(a, in), (b, out), (c, blank), (d, blank), (e, must\text{-}out), (f, must\text{-}out)\},$$
$$\pi = \{(a, 0), (b, 2), (c, 0), (d, 0), (e, 1), (f, 1)\}, \text{and } \gamma = \{c, d\}. \tag{3}$$

After a second and third round of the *while* loop,

$$\mu = \{(a, in), (b, out), (c, in), (d, in), (e, out), (f, out)\},$$
$$\pi = \{(a, 0), (b, 2), (c, 0), (d, 0), (e, 1), (f, 1)\}, \text{and } \gamma = \emptyset. \tag{4}$$

Now, $\{a, c, d\}$ is stable; see line 13 in Algorithm 1. Applying line 14, backtrack to state (3.1). Perform the actions at lines 16–21, and then invoke list-stb-ext(μ, π, γ) (line 22) such that

$$\mu = \{(a, must\text{-}out), (b, blank), (c, blank), (d, blank), (e, blank), (f, blank)\},$$
$$\pi = \{(a, 2), (b, 1), (c, 2), (d, 1), (e, 1), (f, 2)\}, \text{and } \gamma = \emptyset. \tag{5}$$

Referring to line 15, let x be the argument b. Invoke list-stb-ext(μ, π, γ) with

$$\mu = \{(a, must\text{-}out), (b, blank), (c, blank), (d, blank), (e, blank), (f, blank)\},$$
$$\pi = \{(a, 2), (b, 1), (c, 2), (d, 1), (e, 1), (f, 2)\}, \text{and } \gamma = \{b\}. \tag{6}$$

Apply a first round of the *while* loop. Thereby,

$$\mu = \{(a, must\text{-}out), (b, in), (c, out), (d, out), (e, blank), (f, blank)\},$$
$$\pi = \{(a, 2), (b, 0), (c, 2), (d, 1), (e, 0), (f, 1)\}, \text{and } \gamma = \{e\}. \tag{7}$$

As $\gamma \neq \emptyset$, apply a second round of the *while* loop. Thus,

$$\mu = \{(a, out), (b, in), (c, out), (d, out), (e, in), (f, out)\}, \\ \pi = \{(a, 1), (b, 0), (c, 2), (d, 1), (e, 0), (f, 1)\}, \text{and } \gamma = \emptyset. \tag{8}$$

Referring to line 13 in Algorithm 1, $\{b, e\}$ is stable. Applying line 14, backtrack to state (3.5). Afterwards, apply the lines 16–21 and then invoke list-stb-ext(μ, π, γ) (line 22) with

$$\mu = \{(a, must\text{-}out), (b, must\text{-}out), (c, blank), (d, blank), (e, blank), (f, blank)\}, \\ \pi = \{(a, 2), (b, 1), (c, 1), (d, 0), (e, 1), (f, 2)\}, \text{and } \gamma = \{d\}. \tag{9}$$

In performing the first round of the *while* loop, we get

$$\mu = \{(a, must\text{-}out), (b, out), (c, blank), (d, in), (e, out), (f, out)\}, \\ \pi = \{(a, 0), (b, 1), (c, 0), (d, 0), (e, 1), (f, 1)\}, \text{and } \gamma = \{c, f\}. \tag{10}$$

However, as $\mu(a) = must\text{-}out$ and $\pi(a) = 0$, by applying line 8, we return to a previous state and eventually terminate the procedure.

Next, we will give four more propositions that (along with the previous propositions) will establish the verity of Algorithm 1. To this end, we denote by T_i the elements of a set T at the algorithm's state i. Algorithm 1 enters a new state whenever line 2 or line 16 are executed. Note that lines 2 and 16 are sensible to be selected to designate a beginning of a new state of the algorithm since they include a decision to re-map an argument to *in* or *must_out* rather than an imposed re-mapping under some conditions, such as those conditions in the lines 3–5. Focusing on the arguments that are mapped to *in*, in the initial state of the algorithm, we let

$$\{x | \mu_1(x) = in\} = \emptyset,$$

and for all states i it holds that

$$\{x | \mu_{i+1}(x) = in\} = \{x | \mu_i(x) = in\} \text{ (see line 16) or} \\ \{x | \mu_{i+1}(x) = in\} = \{x | \mu_i(x) = in\} \cup \{q\}_i$$

such that $\{q\}_i$ is a one-element set containing an argument from

$$\{x \mid \mu_i(x) = blank\} \text{ (see line 15 in Algorithm1)}$$

or an argument from

$$\{x \text{ with } \mu_i(x) = blank \mid \pi_i(x) = 0\} \text{ (see lines 9 and 20 in Algorithm1)}$$

or an argument from

$$\{x \text{ with } \mu_i(x) = blank \mid \exists y \in \{x\}^+ : \mu_i(y) = must\text{-}out \wedge \pi_i(y) = 1\},$$

see lines 10, 11, and 21 in Algorithm 1.

Proposition 6. *Let $H = (A, R)$ be an* AF *and* $\pi : A \rightarrow \{0, 1, 2, ..., |A|\}$ *be a total mapping such that for all* $x \in A$, $\pi_1(x) = |\{x\}^- \setminus \{y : (y, y) \in R\}|$. *And let* $\mu : A \rightarrow \{in, out, must\text{-}out, blank\}$ *be a total mapping such that for all* $x \in A$,

$$\mu_1(x) = \begin{cases} must\text{-}out, & if\ (x, x) \in R; \\ blank, & otherwise. \end{cases}$$

And assume that Algorithm 1 is started with list-stb-ext(μ_1, π_1, $\{x \mid \mu_1(x) = blank \wedge \pi_1(x) = 0\}$). For every state i it holds that $\{x \mid \mu_i(x) = in\} \cap \{x \mid \mu_i(x) = in\}^+ = \emptyset$.

Proof. Since $\{x \mid \mu_1(x) = in\} = \emptyset$, $\{x \mid \mu_1(x) = in\} \cap \{x \mid \mu_1(x) = in\}^+ = \emptyset$. We now show that for every state i,

$$\{x \mid \mu_i(x) = in\} \cap \{x \mid \mu_i(x) = in\}^+ = \emptyset \Longrightarrow$$
$$\{x \mid \mu_{i+1}(x) = in\} \cap \{x \mid \mu_{i+1}(x) = in\}^+ = \emptyset.$$

Assume that the premise of this implication is true, then we need to show that

$$\{x \mid \mu_{i+1}(x) = in\} \cap \{x \mid \mu_{i+1}(x) = in\}^+ =$$
$$(\{x \mid \mu_i(x) = in\} \cup \{q\}_i) \cap (\{x \mid \mu_i(x) = in\}^+ \cup \{q\}_i^+) = \emptyset.$$

Subsequently, we need to show that for any state i,

$$\{x \mid \mu_i(x) = in\} \cap \{x \mid \mu_i(x) = in\}^+ = \emptyset, \tag{6.1}$$
$$\{x \mid \mu_i(x) = in\} \cap \{q\}_i^+ = \emptyset, \tag{6.2}$$
$$\{q\}_i \cap \{x \mid \mu_i(x) = in\}^+ = \emptyset, \tag{6.3}$$
$$and\ \{q\}_i \cap \{q\}_i^+ = \emptyset. \tag{6.4}$$

Assuming the premise of the above implication in this proof, (6.1) follows immediately.
As to (6.2), recall that $\mu_i(q) = blank$ for all states i. Now, suppose that (6.2) is false. Thus,

at some state i, $\exists x$ with $\mu_i(x) = in : (q, x) \in R$ and $\mu_i(q) = blank$.

This means

at some state i, $\exists x$ with $\mu_i(x) = in : \exists y \in \{x\}^-$ with $\mu_i(y) = blank$.

This contradicts the actions of Algorithm 1 (lines 2–5), which indicate

$\forall i, \forall x$ with $\mu_i(x) = in, \forall y \in \{x\}^-, \mu_i(y) \in \{out, must\_out\}$.

Therefore, (6.2) holds. Now we prove (6.3). Recall, again, for every state i, $\mu_i(q) = blank$. Assume that (6.3) is false. Thereby,

at some state i, $\exists x$ with $\mu_i(x) = in : (x, q) \in R$ and $\mu_i(q) = blank$.

Thus,

at some state i, $\exists x$ with $\mu_i(x) = in : \exists y \in \{x\}^+$ with $\mu_i(y) = blank$.

This contradicts the actions of Algorithm 1 (lines 2–5), which require

$$\forall i, \forall x \text{ with } \mu_i(x) = in, \forall y \in \{x\}^+, \ \mu_i(y) = out.$$

Therefore, (6.3) holds. Now we show (6.4). According to the premise of this proposition,

$$\forall x \text{ with } (x, x) \in R, \ \mu_1(x) = must\_out,$$

and so the actions of Algorithm 1 (including line 3) entail that

$$\forall x \text{ with } (x, x) \in R, \text{ for all states } i, \mu_i(x) \in \{out, must\_out\}.$$

Now, suppose that (6.4) is false. Thus, $(q, q) \in R$ and $\mu_i(q) = blank$ at some state i. Contradiction. □

Proposition 7. *Let $H = (A, R)$ be an* AF *and $\pi : A \to \{0, 1, 2, ..., |A|\}$ be a total mapping such that for all $x \in A$, $\pi_1(x) = |\{x\}^- \setminus \{y : (y, y) \in R\}|$. And let $\mu : A \to \{in, out, must\text{-}out, blank\}$ be a total mapping such that for all $x \in A$,*

$$\mu_1(x) = \begin{cases} must\text{-}out, & if \ (x, x) \in R; \\ blank, & otherwise. \end{cases}$$

And assume that Algorithm 1 is started with list-stb-ext(μ_1, π_1, $\{x \mid \mu_1(x) = blank \wedge \pi_1(x) = 0\}$). For every state i, it holds that

$$\{x \mid \mu_i(x) = blank\} \subseteq A \setminus (\{x \mid \mu_i(x) = in\} \cup \{x \mid \mu_i(x) = in\}^+ \cup \{x \mid \mu_i(x) = in\}^-).$$

Proof. Since μ is a total mapping from A to $\{in, out, must\text{-}out, blank\}$, for every state i it holds that $\{x \mid \mu_i(x) = blank\} \subseteq A$. Now, we need to check that for every state i,

$$\{x \mid \mu_i(x) = blank\} \cap (\{x \mid \mu_i(x) = in\} \cup \{x \mid \mu_i(x) = in\}^+ \cup \{x \mid \mu_i(x) = in\}^-) = \emptyset.$$

According to the algorithm's actions (lines 2–5), note that for every state i,

$$\forall y \in \{x \mid \mu_i(x) = in\}^+ \ \mu_i(y) = out,$$
$$\forall z \in \{x \mid \mu_i(x) = in\}^- \ \mu_i(z) \in \{out, must\text{-}out\}.$$

Hence,

$$\{x \mid \mu_i(x) = in\}^+ \subseteq \{x \mid \mu_i(x) = out\},$$
$$\{x \mid \mu_i(x) = in\}^- \subseteq \{x \mid \mu_i(x) \in \{out, must\text{-}out\}\}.$$

Thus,

$$\{x \mid \mu_i(x) = in\}^+ \cup \{x \mid \mu_i(x) = in\}^- \subseteq$$
$$\{x \mid \mu_i(x) = out\} \cup \{x \mid \mu_i(x) = must\text{-}out\}.$$

Since $\mu : A \rightarrow \{blank, in, out, must\text{-}out\}$ is a total mapping, for all i

$$\{x \mid \mu_i(x) = blank\} \cap$$
$$(\{x \mid \mu_i(x) = in\} \cup \{x \mid \mu_i(x) = out\} \cup \{x \mid \mu_i(x) = must\text{-}out\}) = \emptyset.$$

So, it holds that for every state i

$$\{x \mid \mu_i(x) = blank\} \cap$$
$$(\{x \mid \mu_i(x) = in\} \cup \{x \mid \mu_i(x) = in\}^+ \cup \{x \mid \mu_i(x) = in\}^-) = \emptyset.$$

This completes our proof. □

Proposition 8. *Let* $H = (A, R)$ *be an* AF *and* $\pi : A \rightarrow \{0, 1, 2, ..., |A|\}$ *be a total mapping such that for all* $x \in A$, $\pi_1(x) = |\{x\}^- \setminus \{y : (y, y) \in R\}|$. *And let* $\mu : A \rightarrow \{in, out, must\text{-}out, blank\}$ *be a total mapping such that for all* $x \in A$,

$$\mu_1(x) = \begin{cases} must\text{-}out, & if \ (x, x) \in R; \\ blank, & otherwise. \end{cases}$$

And assume that Algorithm 1 is started with list-stb-ext(μ_1, π_1, $\{x \mid \mu_1(x) = blank \wedge \pi_1(x) = 0\}$). For every state i it holds that

$$\{x \mid \mu_i(x) = must\text{-}out\} =$$
$$A \setminus (\{x \mid \mu_i(x) = in\} \cup \{x \mid \mu_i(x) = in\}^+ \cup \{x \mid \mu_i(x) = blank\}).$$

Proof. Note that according to the actions of Algorithm 1 (lines 2–5), for every state i it is the case that

$$\forall x \text{ with } \mu_i(x) = in, \forall y \in \{x\}^+ \ \mu_i(y) = out.$$

Subsequently, for every state i it holds that

$$\forall y \in \{x \mid \mu_i(x) = in\}^+ \ \mu_i(y) = out.$$

So our target is to show that for all i

$$\{x \mid \mu_i(x) = must\text{-}out\} = A \setminus \{x \mid \mu_i(x) \in \{in, out, blank\}\}.$$

However, this immediately follows from the following observation. Since

$$\mu : A \rightarrow \{in, out, blank, must\text{-}out\}$$

is a total mapping, we note that for all i,

$$\{x \mid \mu_i(x) = must\text{-}out\} \cap \{x \mid \mu_i(x) \in \{in, out, blank\}\} = \emptyset, \text{ and}$$
$$\{x \mid \mu_i(x) = must\text{-}out\} \cup \{x \mid \mu_i(x) \in \{in, out, blank\}\} = A.$$

This completes our proof. □

Proposition 9. *Let $H = (A, R)$ be an* AF *and $\pi : A \to \{0, 1, 2, ..., |A|\}$ be a total mapping such that for all $x \in A$ it holds that $\pi_1(x) = |\{x\}^- \setminus \{y : (y, y) \in R\}|$. And let $\mu : A \to \{in, out, must\text{-}out, blank\}$ be a total mapping such that for all $x \in A$,*

$$\mu_1(x) = \begin{cases} must\text{-}out, & if\ (x, x) \in R; \\ blank, & otherwise. \end{cases}$$

Assume that Algorithm 1 is started with list-stb-ext(μ_1, π_1, $\{x \mid \mu_1(x) = blank \land \pi_1(x) = 0\}$). Then, the algorithm computes exactly the stable extensions in H.

Proof. The proof is composed of two parts:

P1. At some state i, let $\{x \mid \mu_i(x) = in\}$ be the set reported by Algorithm 1 at line 13. Then, we need to prove that $\{x \mid \mu_i(x) = in\}$ is a stable extension in H.

P2. For all Q, if Q is a stable extension in H and, Algorithm 1 is sound (i.e., *P1* is established), then there is a stable extension $\{x \mid \mu_i(x) = in\}$, reported by the algorithm (line 13) at some state i, such that $\{x \mid \mu_i(x) = in\} = Q$.

To establish *P1*, it suffices to show that $\{x \mid \mu_i(x) = in\} = A \setminus \{x \mid \mu_i(x) = in\}^+$, which means it is required to prove that

$$\{x \mid \mu_i(x) = in\} \cap \{x \mid \mu_i(x) = in\}^+ = \emptyset, \text{ and}$$

$$\{x \mid \mu_i(x) = in\} \cup \{x \mid \mu_i(x) = in\}^+ = A.$$

Observe that

$$\{x \mid \mu_i(x) = in\} \cap \{x \mid \mu_i(x) = in\}^+ = \emptyset$$

is already established in Proposition 6. Now, we need to prove that

$$\{x \mid \mu_i(x) = in\} \cup \{x \mid \mu_i(x) = in\}^+ = A.$$

As $\mu : A \to \{in, out, must\text{-}out, blank\}$ is a total mapping,

$$\begin{aligned} &\{x \mid \mu_i(x) = in\} \cup \{x \mid \mu_i(x) = out\} \cup \\ &\{x \mid \mu_i(x) = blank\} \cup \{x \mid \mu_i(x) = must\text{-}out\} = A. \end{aligned}$$

According to the algorithm's actions (lines 2–5),

$$\forall x \text{ with } \mu_i(x) = in, \forall y \in \{x\}^+, \mu_i(y) = out$$

and

$$\forall y \text{ with } \mu_i(y) = out, \exists x \text{ with } \mu_i(x) = in \land y \in \{x\}^+.$$

Thus, for every state i it holds that $\{x \mid \mu_i(x) = in\}^+ = \{x \mid \mu_i(x) = out\}$. Observe, Algorithm 1 reports that $\{x \mid \mu_i(x) = in\}$ is stable if and only if $\{x \mid \mu_i(x) = blank\} = \emptyset$ and $\{x \mid \mu_i(x) = must\text{-}out\} = \emptyset$, see lines 12–13. Therefore, it holds that

$$\{x \mid \mu_i(x) = in\} \cup \{x \mid \mu_i(x) = in\}^+ \cup \emptyset \cup \emptyset = A,$$

which means that $\{x \mid \mu_i(x) = in\} \cup \{x \mid \mu_i(x) = in\}^+ = A$.

This completes the proof of *P1*. Now we turn to *P2*. By modifying its consequence part, we rewrite *P2* into *Ṕ2*.

Ṕ2: For all Q, if Q is a stable extension in H and Algorithm 1 is sound, then there is a stable extension $\{x \mid \mu_i(x) = in\}$, reported by the algorithm (line 13) at some state i, such that for all $a \in Q$ it holds that $\mu_i(a) = in$.

We establish *Ṕ2* by contradiction. Later, we show that the consequence of *Ṕ2* is equivalent to the consequence part of *P2*. Now, assume that *Ṕ2* is false.

Negation of Ṕ2: There is Q such that Q is a stable extension in H, Algorithm 1 is sound, and for every $\{x \mid \mu_i(x) = in\}$ reported by the algorithm (line 13) at some state i, there is $a \in Q$ such that $\mu_i(a) \neq in$. We identify four cases.

Case 1. For $\mu_1(a) = blank$, if the algorithm terminates (line 8) during the very first execution of the while block (but not necessarily from the first round), then, since the algorithm is sound, H has no stable extensions. This contradicts the assumption that $Q \supseteq \{a\}$ is a stable extension in H. Hence, *Ṕ2* holds.

Case 2. If $(a, a) \in R$, then $\mu_1(a) = must\_out$, which contradicts the assumption that $Q \supseteq \{a\}$ is a stable extension in H. Hence, *Ṕ2* holds.

Case 3. With $\mu_1(a) = blank$, assume that after the very first execution of the while block (i.e., including one or more rounds), $\mu_k(a) = blank$ for some state $k \geq 1$. Then, for a state $i \geq k$, let $x = a$ (see line 15 in Algorithm 1). If for all subsequent states $j > i$, the set $\{x \mid \mu_j(x) = in\} \supseteq \{a\}$ is not reported stable by the algorithm, then, since the algorithm is sound, a does not belong to any stable extension. This contradicts the assumption that $Q \supseteq \{a\}$ is a stable extension in H. Hence, *Ṕ2* holds.

Case 4. With $\mu_1(a) = blank$, assume that after the very first execution of the while block (i.e., including one or more rounds), $\mu_i(a) \neq blank$ for some state $i > 1$. According to the actions of the while loop, this implies that $\mu_i(a) \in \{in, out, must\_out\}$. For $\mu_i(a) = in$, if for all subsequent states $j > i$, the set $\{x \mid \mu_j(x) = in\} \supseteq \{a\}$ is not reported stable by the algorithm, then, since the algorithm is sound, a does not belong to any stable extension. This contradicts the assumption that $Q \supseteq \{a\}$ is a stable extension in H. Likewise, for $\mu_i(a) \in \{out, must\_out\}$, since the algorithm's actions are sound, this implies that a does not belong to any stable extension. This contradicts the assumption that $Q \supseteq \{a\}$ is a stable extension in H. Hence, *Ṕ2* holds. Now we rewrite the consequence of *Ṕ2*.

The Consequence of Ṕ2: There is a stable extension $\{x \mid \mu_i(x) = in\}$, reported by the algorithm (line 13) at some state i, such that $Q \subseteq \{x \mid \mu_i(x) = in\}$. Q being a proper subset of $\{x \mid \mu_i(x) = in\}$ is impossible because otherwise, it contradicts the algorithm being sound or that Q is stable. Therefore, the consequence of *Ṕ2* can be rewritten as next.

The Consequence of Ṕ2: There is a stable extension $\{x \mid \mu_i(x) = in\}$, reported by the algorithm (line 13) at some state i, such that $Q = \{x \mid \mu_i(x) = in\}$.

This is exactly the consequence of *P2*. $\square$

3   Conclusion

We analyzed mathematically the verity of characteristic functions prevalently utilized for implementing backtracking processes for listing stable extensions of AFs. Future work may continue this line of research to rigorously investigate the verity of characteristic functions for generating argument extensions under different semantics, such as preferred, complete, and stage argumentation. For further information on the definition and properties of argumentation semantics, we encourage the reader to consult, for instance, the excellent review of [1].

References

1. Baroni, P., Caminada, M., Giacomin, M.: An introduction to argumentation semantics. Knowl. Eng. Rev. **26**(4), 365–410 (2011)
2. Bistarelli, S., Rossi, F., Santini, F.: Not only size, but also shape counts: abstract argumentation solvers are benchmark-sensitive. J. Log. Comput. **28**(1), 85–117 (2018)
3. Black, E., Maudet, N., Parsons, S.: Argumentation-based dialogue. In: Handbook of Formal Argumentation, vol. 2, p. 511. College Publications (2021)
4. Calegari, R., Omicini, A., Pisano, G., Sartor, G.: Arg2p: an argumentation framework for explainable intelligent systems. J. Log. Comput. **32**(2), 369–401 (2022)
5. Cerutti, F., Gaggl, S.A., Thimm, M., Wallner, J.P.: Foundations of implementations for formal argumentation. FLAP **4**(8), 2623–2705 (2017)
6. Dung, P.M.: On the acceptability of arguments and its fundamental role in non-monotonic reasoning, logic programming and n-person games. Artif. Intell. **77**(2), 321–358 (1995)
7. Dunne, P.E.: Computational properties of argument systems satisfying graph-theoretic constraints. Artif. Intell. **171**(10), 701–729 (2007)
8. Gaggl, S.A., Linsbichler, T., Maratea, M., Woltran, S.: Design and results of the second international competition on computational models of argumentation. Artif. Intell. **279**, 103193 (2020)
9. Modgil, S., Caminada, M.: Proof theories and algorithms for abstract argumentation frameworks. In: Argumentation in Artificial Intelligence, pp. 105–129 (2009)
10. Nofal, S., Abu Jabal, A., Alfarrarjeh, A., Hababeh, I.: Validation of labelling algorithms for abstract argumentation frameworks: The case of listing stable extensions. In: Artificial Intelligence and Soft Computing, pp. 423–435 (2023)
11. Nofal, S., Atkinson, K., Dunne, P.E.: Computing grounded extensions of abstract argumentation frameworks. Comput. J. **64**(1), 54–63 (2021)
12. Panisson, A.R., Engelmann, D.C., Bordini, R.H.: Engineering explainable agents: an argumentation-based approach. In: 9th International Workshop on Engineering Multi-Agent Systems, pp. 273–291 (2022)
13. Parsons, S., Sierra, C., Jennings, N.R.: Agents that reason and negotiate by arguing. J. Log. Comput. **8**(3), 261–292 (1998)
14. Potyka, N.: Interpreting neural networks as quantitative argumentation frameworks. In: Proceedings of the AAAI Conference on Artificial Intelligence, vol. 35, pp. 6463–6470 (2021)
15. Rago, A., Cocarascu, O., Bechlivanidis, C., Lagnado, D., Toni, F.: Argumentative explanations for interactive recommendations. Artificial Intell. **296**, 103506 (2021)

Generating Sub-emotions from Social Media Data Using NLP to Ascertain Mental Illness

K. S. Srinath[1]([✉]) [iD], K. Kiran[1] [iD], P. Deepa Shenoy[1] [iD], and K. R. Venugopal[2] [iD]

[1] Department of Computer Science and Engineering, University of Visvesvaraya College of Engineering, Bangalore University, Bangalore, India
solansrinath@gmail.com
[2] Bangalore University, Bangalore, India

Abstract. It is predicted that mental illness will be one of the leading causes of death in 2030. Many people will not share their details of the illness detail with others, including family and friends. Also, many are unaware that their mental disorder is affecting their thinking and behavior. Early detection and medical intervention are necessary, otherwise it leads to severe problems. More than half of the world population, that is around 58.4% of the people use Social Media (SM) to express their thoughts and feelings. By fetching their timely thoughts and feelings expressed in social media we can analyze their emotions and sub-emotions. In this study, we developed a novel model to generate the sub-emotions of social media users from EmoLEX lexicon using the Affinity Propagation (AP) algorithm and word2vect conversion word2vec-google-news-300. The number of clusters and vocabulary obtained for ten emotions such as Anger, Anticipation, Disgust, Fear, Joy, Sadness, Surprise, Trust, Positive and Negative is evaluated based on sub-emotions generated. By using word2vect conversions and AP algorithm it is found that the consistency of Mean words (μW) per cluster are equally distributed in each cluster with respect to all emotions on an average of 19, 20, 21 and 22. The obtained sub-emotions is used to mask the SM user post and it could be further used for detecting the mental illness of SM users.

Keywords: Affinity Propagation (AP) · EmoLEX Lexicon · Machine Learning · Mental Disorder · Social Media (SM) · Word2vect

1 Introduction

Our emotions, feelings, and ability to think clearly are all orchestrated together to transmit information throughout the body via neurotransmitters to make physical sensations. Failure of it may lead to mental illness. One out of four suffers from mental illness around the world and experts predict that mental illness will become the reason for the world's major cause of death by 2030. Currently, millions of people are suffering from mental illness and due to the fear of ostracizing they may furtive until it gets worsened [1–3]. This problem complicates further as many are unaware about their illness at initial stage.

© The Author(s), under exclusive license to Springer Nature Switzerland AG 2024
A. Ortis et al. (Eds.): ICAETA 2023, CCIS 1983, pp. 399–409, 2024.
https://doi.org/10.1007/978-3-031-50920-9_31

Nowadays, more than half of the world that is around 4.62 billion people, use Social Media (SM) to express their thoughts and feelings. (Datareportal- "January 2022 global overview"). The second most populous country, India had 448 million SM users in January 2021. On an average two hours and twenty-seven minutes are spent by SM users daily on applications like WhatsApp, YouTube, Facebook, Instagram and Twitter. English is one of the major languages used to write or express their feelings, thoughts, intentions and ideas in SM [4, 5].

According to Oxford English Dictionary (OED), there are more than six hundred thousand words available in the English language, and hundreds of new words are added to it every year [6], to improve the sentences over time. Usage of a few words is faded which is no longer applicable to modern life.

Usage of lexicon to express themselves in SM opens up a great opportunity for researchers to understand and analyze the patterns of communications made by SM users [7]. People suffering from mental illness have a unique pattern of expressing their emotions in SM. In [8], Xiao Lei Huang *et. al.* Have performed sentiment analysis by assigning 'positive' and 'negative' sentiments to users' tweets and ascertained that posts by users under depression are lengthy posts compared to normal users.

Emotion is a psychic state experienced by humans in response to some situation or event. The three key elements like subjective experience, a physiological response and behavioral response are responsible to make up emotions [9] [10]. According to psychologists Paul Ekman and Richard Davidson [11] there are eight primary emotions (Anger, Anticipation, Disgust, Fear, Joy, Sadness, Surprise and Trust) and two sentiments (Positive and Negative) as shown in Fig. 1. These emotions are experienced and expressed universally.

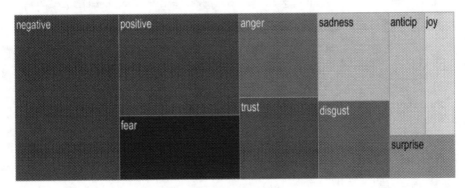

Fig. 1. Eight Primary Emotions

By understanding the unique pattern of expressing emotions in SM by mentally ill and normal users, we can analyze their mental state, but this is not sufficient to understand the level of emotion expressed by users [7, 12].

Using words like Surprise or Astonished, Anger or Furious, Happy or Excited and Sad or Sorrow clearly indicates the user is expressing different levels of emotions. The word furious indicates that the user is extremely angry than using the word anger. By taking this into consideration, the novelty of this research work is to build a model using word2vec-google-news-300 and AP algorithm that generates the sub-emotions which are used to measures the level of emotion expressed by SM users in their posts to analyze their mental health.

The rest of this article is organized as follows: Sect. 2 describes the technologies and methods used to detect mental illness via questionnaire and SM content. Section 3 gives clear idea about proposed method used to mitigate the current problem. Section 4 provides the results obtained from our proposed model. Section 5 concludes our proposed work and directs our next step for research.

2 Literature Survey

This section explores the previous works carried out on mental illness detection on textual content written by SM users. This section is broken into two main categories of research: a) Finding mental illness symptoms from questionnaire. b) Finding mental illness symptoms in social media.

Initially, hospitals and agencies were using questionnaires and interviews as one of the major tools to find mental illness. In our previous work, we have created a model to predict mental illness using DASS-42 questionnaire and ML algorithms such as a) Support Vector Machine (SVM) and Logistic Regression (LR) [1], b) Decision Tree (DT) and its ensemble XGBoost [13] and tuned it on dataset available in the following link: https://openpsychometrics.org/_rawdata/ [14].

Shuang Li, Yu Liu and Vijay Kumar proposed an optimized model that uses Convolutional Neural Network (CNN) along with Fully Connected Neural Network (FCNN) to recognize the mental disorder among Primary and Secondary School Students. Using this optimized model, the recall rate is improved by 0.19 and the accuracy was improved by 0.03, recall rate was improved by 0.19 and the F1-measure was improved by 0.05 [15].

Murat Acik, et. al. Used Mediterranean Diet Prevention (PREDIMED) dataset to predict the anxiety and psychological problems using DASS-42 dataset. The relation between the mental health and diet was analyzed using the Logistic regression model and found mental illness is inversely proportional to diet [16].

The limitation of using only questionnaire is that many users may be unaware of their illness and may not take up the assessment. Due to this it is not be possible to predict mental disorders. This has leaded the researchers to deviate from questionnaire to users SM data.

In [17], authors presented the Random Forest (RF) model for efficient classification to predict depression in twitter social media users. They have used random forest algorithm with wor2vec feature extracting technique and observed that the model resulted in less over fitting, fast performance, and can handle the data with high dimensions. However, it cannot balance the labelling of data leading to reduced accuracy and gradients reduction in case of long sentences.

To predict depression in twitter dataset, authors proposed a method to perform dual classification with the fusion of SVM and Naive Bayes algorithm. They tested and validated 2500 sentences, evaluated the performance and found that it produced better accuracy than other existing individual classifiers [18].

In recent years, NLP has drastically emerged in detecting mental disorders in SM users' emotions. But, working with emotions in SM users alone will not be sufficient to predict the mental illness accurately. Hence in our proposed work we generate sub-emotions as discussed in the following sections.

3   Bag of Sub-emotions Using Google-News-300

Our proposed model for generating sub-emotions is depicted in Fig. 2. It consists of two steps: (1) Generating sub-emotions and (2) mapping user text to sub emotions.

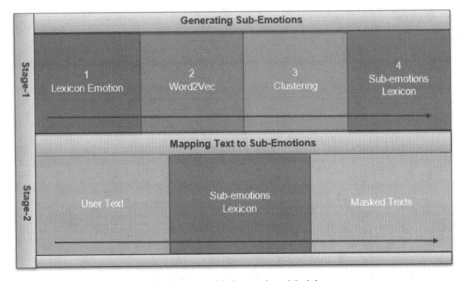

Fig. 2. Bag of Sub-emotions Model.

3.1   Generating sub-emotions

Generating sub-emotions involvers four stages namely: a) Lexicon Emotion b) Word2Vec conversion c) Clustering d) Sub-emotions. Flowchart for generating sub-emotions is shown in Fig. 3.

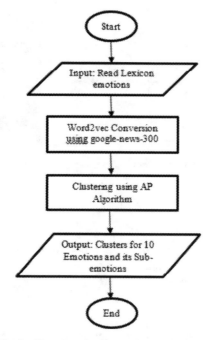

Fig. 3. Flowchart for Generating Sub-emotions

a) **Emotion Lexicon**

Every word represents or is attached to one or the other form of emotion to human beings. Lexicons are the collection of words available in the dictionary used for communicating with one another. Every individual word represents emotions attached to it, hence lexicon emotions are collection of eight basic emotions (anger, fear, anticipation, trust, surprise, sadness, joy, and disgust) and two sentiments (negative and positive) attached with each words.

The corpus of Emotion Lexicon used for our work is taken from online source which is created by Mohammad Saif M. and Turney Peter D in their research work [19, 20].

b) **Word2Vec Conversion**

To group the words into different clusters, words with emotions should be converted to its vector form. There are few word2vector conversion models available as open source like fastText used in [12].

In our proposed work, we have selected google-news-300 model to convert word to vector. Figure 4 depicts the converted vector values of a particular word ("anger") to its vector with dimension 300.

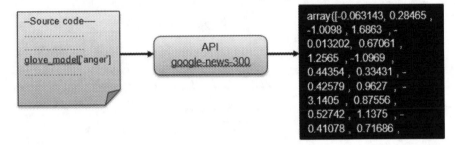

Fig. 4. Word to Vector Conversion using google-news-300 Model.

c) Clustering

All words which are converted into 300 dimensional vectors are grouped into different clusters using Affinity Propagation (AP) algorithm. AP algorithm checks the members of the input array that are representatives of clusters. After marking the centroid in the array, each centroid is represented as a different group of sub-emotions [21].

The code snippet of AP Algorithm in python language is shown.

```python
from sklearn.cluster import AffinityPropagation
clustering = AffinityPropagation().fit(X)
centers = clustering.cluster_centers_
labels = clustering.labels_
matrix = clustering.affinity_matrix_
```

d) Sub-emotions Lexicon

In this stage, the automatically generated clusters for the words in vector form are available as shown in Fig. 4 that contains sub-emotion vector arrays. The words corresponding to these vectors should be replaced back. After replacing the words we can find that every group of cluster contains the associated words with same emotion levels as shown in the example for anger0 (first cluster) and joy0.

```
************* anger0
alienate
odious
bother
chafe
bruising
```

```
************* joy0
abundance
abundant
bountiful
luxuriance
profusion
teemingness
```

3.2 Mapping User Text to Sub-emotions

In this second stage, we take users sentence or text as input and tokenize each word in the text or sentence with the help of python package "word_tokenize" and map each word to its corresponding bag of sub-emotions. Flowchart for mapping user text to sub-emotion is shown in Fig. 5. This process is also called as text masking or text embedding.

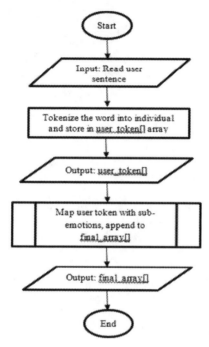

Fig. 5. Mapping user text to sub-emotion

4 Experimental Results

The main intention of this research work is to develop a novel model and evaluate the bag of sub-emotions generated from it. Further, the generated sub emotions can be used for detecting mental illness like depression, anxiety, stress etc. based on the vocabulary used by SM users. This section describes the results obtained by our proposed model.

4.1 Vocabulary and Clusters

After the experiment, the selected emotions obtained vocabulary and numbers of clusters created from AP algorithm for the vocabulary attached to emotions are tabulated. Mean words (μW) per cluster, standard deviation (σW) per cluster are calculated to check the consistency of words added per cluster by AP algorithm (Table 1).

Table 1. Emotions, Vocabulary and Clusters

Emotions	Vocabulary	Clusters	(μW)	σW)
Anger	5499	267	20.52	39.57
Anticipation	5306	260	20.33	44.69
Disgust	4706	216	21.69	40.70
Fear	6429	282	22.72	47.76
Joy	3975	192	20.60	42.91
Sadness	5306	260	20.33	44.69
Surprise	3448	174	19.71	40.70
Trust	5000	244	20.41	45.32
Positive	9879	436	22.61	47.17
Negative	11009	493	22.29	40.79

4.2 Sub-emotion Lexicon -Results

The words grouped in the clusters for particular emotions like anger, fear, joy and sadness are randomly selected from our result and tabulated in Table 2 for anger, Table 3 for fear, Table 4 for joy and Table 5 for sadness.

Table 2. Anger Clusters with Sub-Emotion

anger0	anger1	....	anger266
alienate	agitated		adversity
odious	angry		fee
bother	disgraceful		firearms
chafe	fuming		nuisance

Table 3. Fear Clusters with Sub-Emotion

fear0	fear1	....	fear286
abandon	abuse		adversity
abandoned	cruelly		batter
regulatory	mob		confidence
surprise	absent		contagious

Table 4. Joy Clusters with Sub-Emotion

joy0	joy1	....	joy192
abundance	admiration		closeness
abundant	adoration		confidence
bountiful	affection		excel
luxuriance	esteem		familiarity

Table 5. Sadness Clusters with Sub-Emotion

sadness0	sadness1	....	sadness260
abandon	abuse		dark
abandoned	molestation		darken
defy	lacking		debt
eschew	traveler		deserted

4.3   User Text to Sub-emotions Text

In the second stage of our proposed model, the "Actual sentence" read from user input is tokenized into each word and stored in "Tokenized sentence" array. Finally, each token is mapped to its corresponding sub-emotions from sub-emotions collection. Figures 6 and 7 display the two example results obtained for users input sentence. From the result it can be noticed that the stop words are not mapped with any emotions because in google-news-300 the sub-emotions for stop words are not defined.

```
Actual sentence:   I saw an accident for angry.

Tokenized sentence:   ['I', 'saw', 'an', 'accident', 'for', 'angry',
'.']

Sub Emotion for actual sentence:   I saw an fear85 for negative52.
```

Fig. 6. User Sentence to Sub-emotions Conversion Example-1

```
Actual sentence:   I was angry becauze he hit me.

Tokenized sentence: ['I', 'was', 'angry', 'becauze', 'he', 'hit',
'me', '.']

Sub Emotion for actual sentence:   I was negative52 becauze he
anger22 me.
```

Fig. 7. User Sentence to Sub-emotions Conversion Example-2

5 Conclusions and Future Work

In this research work, a novel approach is developed to create sub-emotions from emotion lexicon resource using glove model with google-news-300 and AP algorithm. Our proposed model is divided into two stages: a) Generating sub-emotions and b) Mapping text to sub-Emotions. We have selected google-news-300 model to convert word to vector. Hence, the emotions used for a particular word in Google news for 300 dimensions of associated words are considered for creating sub-emotions. Affinity propagation algorithm is used instead of K-nearest neighbor (KNN) for clustering emotions and we found that it creates consistent number of words in each cluster. Our model is ready to embed any user text and in our future work a) We intend to find other word to vector converting models and APIs to generate the sub-emotions and compare their vocabulary, b) Use these models to detect depression using standard dataset, and c) Detect stress and anxiety in social media users using sub-emotions along with machine learning algorithms.

References

1. Srinath, K.S., Kiran, K., Pranavi, S., Amrutha, M., Shenoy, P.D., Venugopal, K.R.: Prediction of depression, anxiety and stress levels using dass-42. In: 2022 IEEE 7th International conference for Convergence in Technology (I2CT), pp. 1–6 (2022). https://doi.org/10.1109/I2CT54291.2022.9824087
2. Mental Health: New Understanding New Hope in World Health Report, Geneva, p. 9 (2001)
3. Depression: a global crisis (2012). https://www.who.int/mental_health/management/depression/wfmh_paper_depression_wmhd_2012.pdf
4. https://datareportal.com/reports/digital-2022-global-overview-report
5. https://www.hootsuite.com/
6. https://www.oed.com/
7. Aragon, M.E., Lopez-Monroy, A.P., Gonzalez-Gurrola, L.C.G., Montes, M.: Detecting mental disorders in social media through emotional patterns - the case of anorexia and depression. In: IEEE Transactions on Affective Computing. https://doi.org/10.1109/TAFFC.2021.3075638
8. Huang, X., Zhang, L., Liu, T., Chiu, D., Zhu, T., Li, X.: Detecting suicidal ideation in Chinese microblogs with psychological lexicons. In: 2014 IEEE 11th Intl Conference on Ubiquitous Intelligence and Computing and 2014 IEEE 11th Intl Conference on Autonomic and Trusted Computing and 2014 IEEE 14th Intl Conference on Scalable Computing and Communications and Its Associated Workshops, pp. 844–849 (2014)
9. APA Dictionary of Psychology. https://dictionary.apa.org/emotion, American Psychological Association (2022)

10. Hockenbury, D., Hockenbury, S.E.: Discovering Psychology. Worth Publishers
11. Ekman, P.E.D., Davidson, R.J.: The Nature of Emotion: Fundamental Questions. Oxford University Press, New York, NY, US (1994)
12. Aragón, M.E., Monroy, A.P.L., González-Gurrola, L.C., Montes, M.: Detecting depression in social media using fine-grained emotions. In: Proceedings of the 2019 Conference of the North American Chapter of the Association for Computational Linguistics: Human Language Technologies, Volume 1 (Long and Short Papers), pp. 1481–1486 (2019). https://doi.org/10.18653/v1/N19-1151
13. Srinath, K.S., Kiran, K., Gagan, A.G., Jyothi, D.K., Shenoy, P.D., Venugopal, K.R.: Enhancing mental illness prediction using tree based machine learning approach. In: 2022 IEEE International Conference on Electronics, Computing and Communication Technologies (CONECCT), pp. 1–5 (2022). https://doi.org/10.1109/CONECCT55679.2022.9865689
14. https://openpsychometrics.org/_rawdata/
15. Li, S., Liu, Y., Kumar, V.: Deep learning-based mental health model on primary and secondary school students quality cultivation. Comput. Intell. Neurosci. **1687–5265** (2022). https://doi.org/10.1155/2022/7842304
16. Acik, M., Altan, M., Cakiroglu, F.P.: A cross-sectionally analysis of two dietary quality indices and the mental health profile in female adults. Current Psychol. 1–10 (2020)
17. Renaldi, A., Maharani, W.: Depression detection of user in media social Twitter using random forest. J. Inf. Syst. Res. (JOSH) **3**(4), 410–416 (2022)
18. Smys, S., Raj, J.S.: Analysis of deep learning techniques for early detection of depression on social media network-a comparative study. J. Trends Comput. Sci. Smart Technol. (TCSST), **03**, 24–39. (2021). https://doi.org/10.36548/jtcsst.2021.1.003
19. Mohammad, S.M., Turney, P.D.: Crowdsourcing a Word-Emotion Association Lexicon. Comput. Intell. **29**, 436–465 (2013)
20. Mohammad, S.M., Turney, P.D.: Emotions evoked by common words and phrases: using mechanical Turk to create an emotion lexicon. Comput. Linguist. pp. 26–34 (2010)
21. Thavikulwat, P.: Affinity propagation: a clustering algorithm for computer-assisted business simulation and experimental exercises. In: Developments in Business Simulation and Experiential Learning, vol. 35 (2008)

Non-Cryptographic Privacy Preserving Machine Learning Methods: A Review

Kevser Şahinbaş[1], Ferhat Ozgur Catak[2], Murat Kuzlu[3(✉)], Maliha Tabassum[3], and Salih Sarp[4]

[1] Istanbul Medipol University, Istanbul, Turkey
`ksahinbas@medipol.edu.tr`
[2] University of Stavanger, Rogaland, Norway
`f.ozgur.catak@uis.no`
[3] Old Dominion University, Norfolk, VA, USA
`{mkuzlu,mtaba006}@odu.edu`
[4] Virginia Commonwealth University, Richmond, VA, USA
`sarps@vcu.edu`

Abstract. In recent years, the use of Machine Learning (ML) techniques to exploit data and produce predictive models has become widespread in decision-making and problem-solving across various fields, including healthcare, energy, retail, transportation, and many more. Generally, a well-performing ML model requires large volumes of training data. However, collecting data and using it to predict behavior poses significant challenges to the privacy of individuals and organizations, such as data breaches, loss of privacy, and corresponding financial damage. Therefore, well-designed privacy-preserving ML (PPML) methods are significantly required for many emerging applications to mitigate these problems. This paper provides a comprehensive review of non-cryptographic privacy-preserving ML along with selected methods, such as differential privacy and federated learning. This paper aims to provide a roadmap for future research directions in the PPML field.

Keywords: Privacy-preserving · Machine Learning · Federated learning

1 Introduction

Privacy-Preserving Machine learning (PPML) is one of the most prominent application areas for data protection of computing operations [1]. This is especially crucial when the training sample contains sensitive or private data. Owners of such data may wish to use it to train a model but do not want to give up control over their data. PPML methods can help mitigate this risk by ensuring that the data used to train the model is not linked to personal identity. This can help protect the privacy of those predicted by the model.

This paper extensively reviews full-scale types of non-cryptographic privacy-preserving in the ML method, detailing what, where, and how privacy-preserving

A. Ortis et al. (Eds.): ICAETA 2023, CCIS 1983, pp. 410–421, 2024.
https://doi.org/10.1007/978-3-031-50920-9_32

can be provided. For this purpose, the concept of privacy-preserving ML is introduced and then reviewed in the literature, along with various methods for preserving data privacy while training ML models. The methodology is also briefly discussed to extend the literature review. Then, these methods are compared, and recommendations are provided for future work. The remaining sections of this paper are organized as follows. Section 2 presents non-cryptographic privacy-preserving machine learning methods. Section 3 provides opportunities and challenges. Section 4 examines the future research directions. Conclusions are drawn in Sect. 5.

1.1 Literature Review

This section provides the literature review on non-cryptographic privacy-preserving ML. Tables 1 and 2 summarize the current studies with their descriptions for differential privacy and federated learning methods, respectively.

Shokri et al. [2] developed a general system for learning from participants' data without disclosing private information. They created a neural network model using the "Distributed Stochastic Gradient Descent" optimization approach, training each participant independently. The method achieved success rates of 99.14% on MNIST data and 93.12% on SVHN data. To increase the security of the data and minimize the risks of leakage, a differential privacy approach has been applied by updating the parameters (adding noise). Firstly, the approach of applying differential privacy to Principal Component Analysis [3] was used for feature selection. A success rate of 73% was achieved on CIFAR-10 data and 97% on MNIST data. Chase et al. [4] developed a new method by using Secure Multi-Party Computation (SMPC) and differential privacy to protect the confidentiality of each sample in the training data used to create their neural network model.

Kotsogiannis et al. [5] offer One-Sided Differential Privacy (OSDP) that meets sensitivity masking. Their model assures that an attacker cannot considerably reduce the uncertainty about whether a record is sensitive using any technique. Bassily et al. [6] present a differential privacy type of the Stochastic Gradient Descent (SGD) method with enhanced composition and privacy amplification. For training models, Thakkar et al. [7] examine the adaptive gradient clip technique with user-level differential privacy, eliminating the requirement for comprehensive parameter tuning. Wang et al. [8] used non-IID (non-identically independently distributed) data to provide a new convergence analysis on local epoch size. In their study, a real-time control method that dynamically adjusts global aggregation frequency was developed. Yang et al. [9] present an extensive study of a secure federated learning framework in terms of definition, architecture, vertical FL, horizontal FL, and federated transfer learning. Chen et al. [10] check for inconsistencies between the global and lagged models by modifying the number of local periods to predict recession, expediting convergence, and avoiding straggler effect performance degradation. Konecny et al. [11] provide a technique of a communication-efficient FL model to decrease communication costs for methods of sketched updates and structured updates.

Table 1. Current studies in Differential Privacy

	Description
Differential Privacy	-Two levels of privacy protection
	-Doesn't share raw data during the learning process [12]
	-A refined analysis of privacy costs
	-Non-convex objectives, under a modest privacy budget [13]
	-Private Aggregation of Teacher Ensembles (PATE) is applied to preserve users' privacy
	-CT-Scan is affected by COVID-19 or not by comparing with CNN model [14]
	-Differentially private algorithms for convex empirical risk minimization
	-Optimal error rates are provided [6]
	-Applied the adaptive gradient clip method with user-level differential privacy [7]
	-One-sided differential privacy (OSDP) to protect sensitive records and for releasing count queries [5]
	-Local and Central Distinctive Privacy (LDP/CDP) techniques in FL
	-Decreases white-box membership inference attacks in FL [15]
	-PRECAD framework that provides both privacy and robustness for FL [16]

Hamm demonstrates the framework's performance with realistic tasks such as network intrusion detection, activity recognition, and malicious URL detection [27]. Choudhury et al. [12] illustrate the feasibility and usefulness of the federated learning framework in providing increased privacy while maintaining the global model's utility by applying 1 million patients' real-world electronic health data. Abadi et al. [13] developed a new method using stochastic gradient descent and differential privacy budget composition approaches. Noise is added to the gradient before updating the precision-limited network parameters of each training sample to preserve data. Chamikara et al. [18] propose a distributed perturbation algorithm called DISTPAB that achieves high accuracy, efficiency, attack resistance, and scalability for the privacy preserving of horizontally partitioned data. The privacy-preserving FedML demonstrates DISTPAB's perfect approach for preventing distributed machine learning privacy leaks while maintaining high data utility.

Tran et al. [28] propose a method for Privacy-Preserving ML models that can operate on a decentralized network setting without requiring a reliable third-party server and provide confidentiality of local data with low-cost communication bandwidth. They have designed a new method called a Decentralized Secure

Table 2. Current studies in Federated Learning

	Description
Federated Learning	-Find the best balance between local updating and global parameter aggregation for edge computing [17]
	-Sketched updates and structured updates [11]
	-A distributed perturbation algorithm using the asymmetry of resources of a distributed environment [18]
	-Two new summoning defense mechanisms, Krum and Trimmed Mean [19]
	-The LDP-FedSGD algorithm is used [20]
	-Privacy problems in composite learning [21]
	-Obtained small amounts of data from different sources from various hospitals and trains a global deep-learning model using blockchain-based FL [22]
	-FEDL outperforms vanilla FedAvg algorithm [23]
	-IoT data sharing, data offloading and caching, intrusion detection, localization, mobile audience detection, and IoT privacy and security [24]
	-A Distributed algorithm to develop an overall decentralized optimization framework [25]
	-Lightweight encryption protocol is performed [26]

Framework (SDTF) to protect the confidentiality of data in ML models. It aims to protect the data's privacy by supporting the parallel training process on a decentralized network without needing any third-party server. The Secure Sum Protocol is designed to safely calculate the sum of the participants' inputs in a large group. Randomization techniques and Secure Sum Protocol are combined to ensure the model-sharing process protects local models, even if two of them are confidential from honest but curious parties. This protocol aims to train a global model without leaking information about the local intermediate parameters and training inputs of the participants in the group. As a result of experiments on MNIST and UCI SMS spam datasets, the proposed method achieved a high success rate and efficiency for the created model.

Reich et al. [29] propose the method using Secure Multilateral Computing (SMC) to cover feature extraction from texts and classification with tree ensembles and logistic regression. They also make inferences about the reliability and accuracy of the solution. Ma et al. [30] present a new perspective on multilateral ML, which allows multiple neural networks to learn simultaneously and protects privacy in cloud computing, where huge volumes of training data are distributed among many parties. The authors conclude that the method meets the requirements for verifiability and confidentiality.

Kumar et al. [22] provide a model that obtains small amounts of data from different resources and uses blockchain-based federated learning for training a global ML method. Findings from the study show good performance. Liu et al. [31] suggest the use of federated learning for COVID-19 data training. They also compare the results of ResNet18, MobileNet, COVID-Net, and MoblieNet, four popular models with and without a federated learning method. Chaudhuri et al. focused on the classification problem of a deep neural network model where the training data consists of sensitive information [32]. They designed a method that aims to protect confidentiality in classifiers. The study used the approach of minimizing the average estimation error on the training data while determining the predictive value for each training sample by the classifier. They also used differential privacy methods on sensitive data to protect privacy. Other research on ML with differential privacy includes [33–35].

2   Non-Cryptographic Privacy-Preserving Methods

There are a variety of different methods that can be used to preserve the privacy of data while training ML models. These privacy-preserving methods are used to protect the privacy of individuals whose data is being used to train the model. These methods ensure that the data used to train the model is not linked to the person's identity. This section discusses widely used non-cryptographic privacy-preserving methods, which can be used to make ML algorithms more secure and protect sensitive data.

2.1   Differential Privacy

The concept of differential privacy is the core of privacy-preserving ML methods. Differential privacy (DP) was suggested by Dwork et al. [3], which establishes a sense of personal privacy and enables data analysis in ML. Then, it has become a prominent privacy protection technology. DP allows the extraction of useful information from a dataset without revealing any personally identifiable information about the individuals in the database, as illustrated in Fig. 1. DP is the foundation for ML and other encryption schemes that protect privacy. It is also an anonymization approach that can improve ML and mitigate privacy issues. DP can be used to generalize the ML process to mask the effects of specific input data and provide differential privacy concerning individuals, resulting in a verifiable guarantee of privacy [13]. A differential privacy method has been implemented to train data for ML algorithms based on the Stochastic Gradient Descent technique (SGD), an iterative process for incremental gradient updates to minimize a loss function.

DP is also applicable, especially for group SQL queries involving count, average, sum, maximum, minimum, and median. It can increase the privacy of the dataset by adding random noise to the query results.

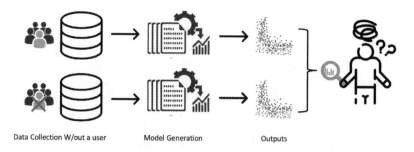

Fig. 1. The architecture of differential privacy (DP) overview

2.2 Federated Learning

Federated Learning (FedML or FL) was developed in 2016 as an efficient privacy-preserving ML technique. In this approach, many clients train their models cooperatively in a distributed environment managed by a central server while the training data is kept locally to protect privacy [9,36]. Figure 2 illustrates the general FL overview. FL enables the decentralization of ML processes by controlling the risk of compromising datasets and identity privacy as the participant limits the information exposed to datasets. Traditional centralized ML introduces system privacy issues and costs, which can be mitigated through FL. The convergence of non-IID data and communications in federated learning scenarios has been a common concern.

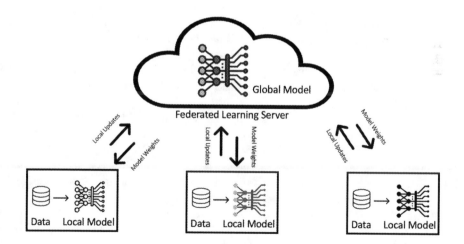

Fig. 2. General federated learning (FL) overview

FedML is branched into five types based on different aspects of federated learning as follows:

- **Federated averaging (FedAvg)**: In this method, each client trains its local model on its own data, then sends the model updates (not the data itself) to a central server. The server then aggregates the model updates from all clients and uses them to update the global model.
- **Split learning (Splitting)**: In this method, each client trains its local model on its local data and sends the output of its local model (not the data itself) to the central server. The central server then aggregates the outputs from all the clients to update the global model.
- **Federated Averaging with split learning (SplitFedAvg)**: In this method, each client sends both its local model and the output of its local model (not the data itself) to the central server. The central server then aggregates the models and outputs from all the clients to update the global model.
- **Federated Averaging with data sharing (ShareFedAvg)**: In this method, each client sends its local model and training data to the central server. The central server then aggregates the models and data from all the clients to update the global model.
- **Federated Averaging with data sharing and split learning (ShareSplitFedAvg)**: In this method, each client sends both its local data and local model to the central server. The central server then aggregates the models and the data from all the clients to update the global model.

3   Opportunities and Challenges

ML has become an integral part of many sectors, including image classification, speech recognition, natural language translation, and image analysis. These popular applications heavily rely on ML nowadays [37]. Amazon SageMaker [38], Microsoft Azure ML Studio [39], and Google Cloud ML Engine [40] are some of the known MLaaS (ML as a Service) providers. ML can be used to achieve various types of data privacy-related work. Human activity recognition (HAR) can generate massive data [41]. These datasets are from the synergy of communication [42–44] and the Medical Internet of Things (MIoT) [45,46]. These huge datasets are useful for the ML method because it enhances the study of the subject, such as the health diagnosis of patients. However, in the case of healthcare datasets, the privacy of the patient's information is sensitive. This kind of data needs to be protected from leakage. The two major algorithms for this purpose are homomorphic encryption [47,48] and differential privacy [49]. Applying these algorithms mentioned above allows a patient's data to be stored by providing privacy. Also, using Federated Learning, data sharing through ML is better and risk-free. There are remarkable outcomes in applying ML. The composition between the input and output consists of many layers. The training data consists of an individual's private information, which means the datasets can cause some risks if leaked. To prevent this, some privacy models have been adopted. Furthermore, financial companies can collect their users' information, transactions history, and other information. By applying these data to ML, it would be easier to detect fraud. As users' data plays a vital role in enhancing datasets' accuracy, this is one reason why large companies take their users' information to train and enhance these models. Those data can be used to recognize images, label photographs to objects, etc. [50].

During the process of ML, many challenges are present. These are some of the downsides of ML models. ML is a data-driven model, which means that the more data is present, the better the results. A large amount of data is needed to feed the ML model to achieve accurate results. ML models work with the maximum amount of input data from people and try to turn it into reality by providing accurate results. Recent studies have advanced and enabled vast knowledge to be learned from. Significant achievements include efficient storage, better processing, and computing on big datasets. However, collecting a large number of datasets for a particular project can sometimes become difficult due to the unavailability of data. A low number of data might give outputs that could be more accurate, which can ruin the outcome and provide false answers.

4   Future Research Directions

Despite ML's rapid development, it still has challenges and an ever-changing room for growth. This article reviewed privacy-preserving ML methods and their latest developments. We discussed FL and different kinds of privacy-preserving mechanisms. Some potential room for growth and future research directions are discussed below.

1. To provide data privacy, several FL frameworks have been developed. However, the quality and accuracy of the data tend to degrade those adapted FL frameworks. A basic framework of FL is provided for the privacy-preserving model. A good model can be collaborated using datasets, but privacy is not guaranteed [51,52].
2. In FL, data privacy for clients lays the most important part. Gradient communication between participants and the aggregator can reveal sensitive information about the participants' datasets [53,54]. Encryption techniques such as homomorphic encryption and secret sharing can be utilized to prevent this. However, computation and communication overhead is something that encryption-based FL faces. Therefore, it is necessary to find an efficient way to stop this from occurring. Also, there needs to be a balance between the trade-off. Perturbation techniques can be utilized to protect weight and gradient updates by adding noise, but this results in degraded model accuracy and increased computational overhead. A good balance between these two conflicting performances is necessary.
3. Sensitive information can be extracted from the final model if the query results are not protected properly. Efficient solutions are needed to protect the final model. Two possible directions are: a) utilizing encryption or perturbation techniques to protect the final model against external attacks, and b) utilizing the splitting technique to personalize the model for each participant by splitting the global model [52,55,56].

4. The cost of computation and effectiveness differs in terms of privacy-preserving mechanisms. Optimization of deployment of defense mechanisms or measurements is necessary. Studies [57] show a useful guide to conduct a comprehensive investigation on diverse metrics to measure data utility and data privacy. Most studies focus on frameworks with a central server. Future research is needed to determine whether privacy attacks against an FL framework without a central server work properly or not.

5   Conclusion

A large amount of data is used in developing ML models during training and estimation, and the data used may consist of personal data. These data may include sensitive data of individuals, such as hospital and bank databases. Using this data in ML models poses security and privacy risks for data owners. Tools applied to increase the confidentiality and security of data used in ML models are given in this study. These tools are typically based-on differential privacy and federated learning for non-cryptographic privacy-preserving ML. In addition, ML-based architectures created in the literature to increase data security and privacy using privacy-preserving tools are examined, along with how and at what stage these tools are applied to the models.

References

1. Çatak, F.Ö.: Secure multi-party computation based privacy preserving extreme learning machine algorithm over vertically distributed data. In: Arik, S., Huang, T., Lai, W.K., Liu, Q. (eds.) ICONIP 2015. LNCS, vol. 9490, pp. 337–345. Springer, Cham (2015). https://doi.org/10.1007/978-3-319-26535-3_39
2. Kubat, M.: The genetic algorithm. In: An Introduction to Machine Learning, pp. 309–329. Springer, Cham (2017). https://doi.org/10.1007/978-3-319-63913-0_16
3. Dwork, C., Roth, A., et al.: The algorithmic foundations of differential privacy. Found. Trends Theor. Comput. Sci. 9(3–4), 211–407 (2014)
4. Chase, M., Gilad-Bachrach, R., Laine, K., Lauter, K., Rindal, P.: Private collaborative neural network learning. Cryptology ePrint Archive (2017)
5. Kotsogiannis, I., Doudalis, S., Haney, S., Machanavajjhala, A., Mehrotra, S.: One-sided differential privacy. In: 2020 IEEE 36th International Conference on Data Engineering (ICDE), pp. 493–504. IEEE (2020)
6. Bassily, R., Smith, A., Thakurta, A.: Private empirical risk minimization: efficient algorithms and tight error bounds. In: 2014 IEEE 55th Annual Symposium on Foundations of Computer Science, pp. 464–473. IEEE (2014)
7. Thakkar, V., Gordon, K.: Privacy and policy implications for big data and health information technology for patients: a historical and legal analysis. Improving Usability, Safety and Patient Outcomes with Health Information Technology, pp. 413–417 (2019)
8. Wang, S., et al.: Adaptive federated learning in resource constrained edge computing systems. IEEE J. Select. Areas Commun. 37(6), 1205–1221 (2019)
9. Yang, Q., Liu, Y., Chen, T., Tong, Y.: Federated machine learning: concept and applications. ACM Trans. Intell. Syst. Technol. (TIST) 10(2), 1–19 (2019)

10. Ming Chen, Bingcheng Mao, and Tianyi Ma. Efficient and robust asynchronous federated learning with stragglers. In Submitted to International Conference on Learning Representations, 2019

11. Konečnỳ, J., McMahan, H.B., Yu, F.X., Richtárik, P., Suresh, A.T., Bacon, D.: Federated learning: Strategies for improving communication efficiency. arXiv preprint arXiv:1610.05492 (2016)

12. Choudhury, O., et al.: Differential privacy-enabled federated learning for sensitive health data. arXiv preprint arXiv:1910.02578 (2019)

13. Abadi, M., et al.: Deep learning with differential privacy. In: Proceedings of the 2016 ACM SIGSAC Conference on Computer and Communications Security, pp. 308–318 (2016)

14. Dewang, R.K., Raven, A., Mewada, A.: A machine learning-based privacy-preserving model for COVID-19 patient using differential privacy. In: 2021 19th OITS International Conference on Information Technology (OCIT), pp. 90–95. IEEE (2021)

15. Naseri, M., Hayes, J., De Cristofaro, E.: Local and central differential privacy for robustness and privacy in federated learning. arXiv preprint arXiv:2009.03561 (2020)

16. Gu, X., Li, M., Xiong, L.: PRECAD: privacy-preserving and robust federated learning via crypto-aided differential privacy. arXiv preprint arXiv:2110.11578 (2021)

17. Wang, X. S., Huang, Y., Zhao, Y., Tang, H., Wang, X., Bu, D.: Efficient genome-wide, privacy-preserving similar patient query based on private edit distance. In: Proceedings of the 22nd ACM SIGSAC Conference on Computer and Communications Security, pp. 492–503 (2015)

18. Chamikara, M.A.P., Bertok, P., Khalil, I., Liu, D., Camtepe, S.: Privacy preserving distributed machine learning with federated learning. Comput. Commun. **171**, 112–125 (2021)

19. Jiang, Y., Li, Y., Zhou, Y., Zheng, X.: Sybil attacks and defense on differential privacy based federated learning. In: 2021 IEEE 20th International Conference on Trust, Security and Privacy in Computing and Communications (TrustCom), pp. 355–362. IEEE (2021)

20. Zhao, Y., et al.: Local differential privacy-based federated learning for internet of things. IEEE Internet Things J. **8**(11), 8836–8853 (2020)

21. Xu, J., Glicksberg, B.S., Su, C., Walker, P., Bian, J., Wang, F.: Federated learning for healthcare informatics. J. Healthcare Inform. Res. **5**(1), 1–19 (2020)

22. Kumar, R., et al.: Blockchain-federated-learning and deep learning models for COVID-19 detection using CT imaging. IEEE Sens. J. **21**(14), 16301–16314 (2021)

23. Dinh, C.T., et al.: Federated learning over wireless networks: convergence analysis and resource allocation. IEEE/ACM Trans. Netw. **29**(1), 398–409 (2020)

24. Nguyen, D.C., Ding, M., Pathirana, P.N., Seneviratne, A., Li, J., Poor, H.V.: Federated learning for internet of things: a comprehensive survey. IEEE Commun. Surv. Tutor. (2021)

25. Brisimi, T.S., Chen, R., Mela, T., Olshevsky, A., Paschalidis, I.C., Shi, W.: Federated learning of predictive models from federated electronic health records. Int. J. Med. Inform. **112**, 59–67 (2018)

26. Fang, H., Qian, Q.: Privacy preserving machine learning with homomorphic encryption and federated learning. Future Internet **13**(4), 94 (2021)

27. Hamm, J., Cao, Y., Belkin, M.: Learning privately from multiparty data. In: International Conference on Machine Learning, pp. 555–563. PMLR (2016)

28. Tran, A.-T., Luong, T.-D., Karnjana, J., Huynh, V.-N.: An efficient approach for privacy preserving decentralized deep learning models based on secure multi-party computation. Neurocomputing **422**, 245–262 (2021)

29. Reich, D., et al. Privacy-preserving classification of personal text messages with secure multi-party computation. In: Advances in Neural Information Processing Systems, vol. 32 (2019)

30. Ma, X., Zhang, F., Chen, X., Shen, J.: Privacy preserving multi-party computation delegation for deep learning in cloud computing. Inf. Sci. **459**, 103–116 (2018)

31. Liu, B., Yan, B., Zhou, Y., Yang, Y., Zhang, Y.: Experiments of federated learning for COVID-19 chest x-ray images. arXiv preprint arXiv:2007.05592 (2020)

32. Chaudhuri, K., Monteleoni, C., Sarwate, A.D.: Differentially private empirical risk minimization. J. Mach. Learn. Res. **12**(3) (2011)

33. Kifer, D., Smith, A., Thakurta, A.: Private convex empirical risk minimization and high-dimensional regression. In: Conference on Learning Theory, pp. 25–1. JMLR Workshop and Conference Proceedings (2012)

34. Song, S., Chaudhuri, K., Sarwate, A.D.: Stochastic gradient descent with differentially private updates. In: 2013 IEEE Global Conference on Signal and Information Processing, pp. 245–248. IEEE (2013)

35. Wu, X., Kumar, A., Chaudhuri, K., Jha, S., Naughton, J.F.: Differentially private stochastic gradient descent for in-RDBMS analytics. CoRR, abs/1606.04722 (2016)

36. Thapa, C., Arachchige, P.C.M., Camtepe, S., Sun, L.: Splitfed: when federated learning meets split learning. arXiv preprint arXiv:2004.12088 (2020)

37. Zhang, T., He, Z., Lee, R.B.: Privacy-preserving machine learning through data obfuscation. arXiv preprint arXiv:1807.01860 (2018)

38. Rauschmayr, N., et al.: Amazon sagemaker debugger: a system for real-time insights into machine learning model training. Proc. Mach. Learn. Syst. **3**, 770–782 (2021)

39. Pliuhin, V., Pan, M., Yesina, V., Sukhonos, M.: Using azure machine learning cloud technology for electric machines optimization. In: 2018 International Scientific-Practical Conference Problems of Infocommunications. Science and Technology (PIC S&T), pp. 55–58. IEEE (2018)

40. Kuzlo, I., Strielkina, A., Tetskyi, A., Uzun, D.: Selecting cloud service for healthcare applications: from hardware to cloud across machine learning. In: PhD@ICTERI, pp. 26–34 (2018)

41. Owusu-Agyemeng, K., Qin, Z., Xiong, H., Liu, Y., Zhuang, T., Qin, Z.: MSDP: multi-scheme privacy-preserving deep learning via differential privacy. Pers. Ubiquit. Comput., 1–13 (2021)

42. Zhang, N., Peng Yang, J., Ren, D.C., Li, Yu., Shen, X.: Synergy of big data and 5G wireless networks: opportunities, approaches, and challenges. IEEE Wirel. Commun. **25**(1), 12–18 (2018)

43. Zhang, N., et al.: Software defined networking enabled wireless network virtualization: challenges and solutions. IEEE Netw. **31**(5), 42–49 (2017)

44. Chen, H., Guo, B., Zhiwen, Yu., Chen, L., Ma, X.: A generic framework for constraint-driven data selection in mobile crowd photographing. IEEE Internet Things J. **4**(1), 284–296 (2017)

45. Qin, Z., et al.: Learning-aided user identification using smartphone sensors for smart homes. IEEE Internet Things J. **6**(5), 7760–7772 (2019)

46. Qin, Z., Wang, Y., Cheng, H., Zhou, Y., Sheng, Z., Leung, V.C.: Demographic information prediction: a portrait of smartphone application users. IEEE Trans. Emerg. Topics Comput. **6**(3), 432–444 (2016)

47. Wagh, S., Gupta, D., Chandran, N.: SecureNN: efficient and private neural network training. Cryptology ePrint Archive (2018)
48. Mohassel, P., Zhang, Y.: SecureML: a system for scalable privacy-preserving machine learning. In: 2017 IEEE Symposium on Security and Privacy (SP), pp. 19–38. IEEE (2017)
49. Yin, C., Xi, J., Sun, R., Wang, J.: Location privacy protection based on differential privacy strategy for big data in industrial internet of things. IEEE Trans. Industr. Inf. 14(8), 3628–3636 (2017)
50. Ali, S., Irfan, M.M., Bomai, A., Zhao, C.: Towards privacy-preserving deep learning: opportunities and challenges. In: 2020 IEEE 7th International Conference on Data Science and Advanced Analytics (DSAA), pp. 673–682. IEEE (2020)
51. Geiping, J., Bauermeister, H., Dröge, H., Moeller, M.: Inverting gradients-how easy is it to break privacy in federated learning? In: Advances in Neural Information Processing Systems, vol. 33, 16937–16947 (2020)
52. Yin, X., Zhu, Y., Jiankun, H.: A comprehensive survey of privacy-preserving federated learning: a taxonomy, review, and future directions. ACM Comput. Surv. (CSUR) 54(6), 1–36 (2021)
53. Li, Q., Wen, Z., He, B.: Practical federated gradient boosting decision trees. In: Proceedings of the AAAI Conference on Artificial Intelligence, vol. 34, pp. 4642–4649 (2020)
54. Hao, Y., Yang, S., Zhu, S.: Parallel restarted SGD with faster convergence and less communication: demystifying why model averaging works for deep learning. In: Proceedings of the AAAI Conference on Artificial Intelligence, vol. 33, 5693–5700 (2019)
55. Fallah, A., Mokhtari, A., Ozdaglar, A.: Personalized federated learning with theoretical guarantees: a model-agnostic meta-learning approach. In: Advances in Neural Information Processing Systems, vol. 33, pp. 3557–3568 (2020)
56. Çatak, F.Ö., Mustacoglu, A.F.: CPP-ELM: cryptographically privacy-preserving extreme learning machine for cloud systems. Int. J. Comput. Intell. Syst. 11, 33–44 (2018)
57. Wagner, I., Eckhoff, D.: Technical privacy metrics: a systematic survey. ACM Comput. Surv. (CSUR) 51(3), 1–38 (2018)

Detection and Comparative Results of Plant Diseases Based on Deep Learning

Mübarek Mazhar Çakir[1] and Gökalp Çinarer[2]

[1] Department of Mechatronics Engineering, Yozgat Bozok University, Yozgat, Turkey
[2] Department of Computer Engineering, Yozgat Bozok University, Yozgat, Turkey
gokalp.cinarer@bozok.edu.tr

Abstract. Plant diseases are one of the problems that threaten crop health and yield in agriculture. Various diseases occurring in plants harm human health and economically producers and producer countries. Early diagnosis is very important in order to prevent the damage caused by diseases. For the early detection of these diseases in plants, continuous observation and examination of plants is required. In large agricultural areas, continuous monitoring of the plants by the producers or workers requires long periods of time and causes extra cost increase. In addition, the person who studies plant leaves must be an expert in plant science. A study was carried out to detect diseases by observing plants based on deep learning, which will be a technological solution to all these problems. Yolov5 and Yolov6 algorithms, one of the object recognition algorithms, was used for plant disease diagnosis. After comparing the two algorithms, the highest AP value with 58.4% belongs to the Yolov5-m model, the highest AR value with 69.3% belongs to the Yolov6-s model, and the highest F1 score with 62.4% belongs to the Yolov5-m. With the study, the comparative results of the models of the Yolo algorithms, together with the hyperparameter values, are given. According to the obtained values, it is seen that the small size models give the best performance. The higher performance of the small size models shows that deep learning models can be integrated into a mobile system, enabling rapid plant identification, sustainability in agriculture and cost reduction.

Keywords: Plant Diseases · Plant Leaf · Deep Learning · CNN · Yolo · Object Detection

1 Introduction

There are some difficulties with the automatic detection of plant diseases. The first of these is the complexity of the image background. The complexity of the image background makes it difficult to detect the plant disease, which is the target object, by distinguishing it from the background. Another difficulty is the problem of not being able to perceive the image clearly due to excessive exposure of plants to daylight under real environmental conditions. In addition, distinguishing the leaves of various plants, which are similar in color and shape, and the similarity of the spots, which are considered as the symptoms of plant diseases, are among the factors that make the detection of the disease difficult.

A. Ortis et al. (Eds.): ICAETA 2023, CCIS 1983, pp. 422–436, 2024.
https://doi.org/10.1007/978-3-031-50920-9_33

Developing sensing technologies and robot technology are widely used in agriculture because they provide solutions to many problems [1]. At the latest point of computer and image technologies, Artificial Intelligence (AI) and computer vision application areas are quite diverse. The decrease in the number of employees in the field of agriculture [2] requires the use of AI applications in agriculture more. With the developing deep learning (DL) methods, the possibility of artificial intelligence to make mistakes has been reduced and faster working structures have been created. Deep learning algorithms and architectural methods can be used for early detection of plant diseases. Early diagnosis of diseases with deep learning will be effective in ensuring the same high quality production of crops by making careful observation and control of plants periodically.

In this study, small models from deep learning algorithms were used. High performance has been achieved in small models of the mentioned object detection algorithms. The small size of the models means that the size of the weight file to be used in the detection of plant diseases is small as a result of training the models with the images in the dataset. The small size of the resulting file makes it possible to integrate the model results into mobile hardware such as mobile phones and android tablets. In this way, our study provides the opportunity to create a low-cost and fast-resulting system without the need for extra equipment to diagnose plant diseases.

For this study, which includes the detection and early diagnosis of plant diseases, first of all, studies on deep learning models used in object detection in the literature were examined. After the literature review, the details of the deep learning object detection algorithm Yolo and the Yolo models used were examined and a suitable dataset for plant diseases was found. As a result of the findings, the algorithm used was optimized, the hyperparameter values to be used were determined and the algorithms were run. The performances of the algorithms trained and tested with the dataset images were examined comparatively and the results were shared. Finally, it was compared with the studies in the literature.

2 Related Works

Studies on the subject in the literature are examined in Table 1 under three headings: citations, job details and algorithms.

Table 1. Related Works

Cites	Works Details	Algorithms
(Boudjid and Ramzan, 2021) [3]	Boudjid and Ramzan have studied the detection and tracking of mobile and stationary people for UAV applications. In their work, they used a dataset of 5000 images that they created themselves. The Yolov2 algorithm is run with this dataset. As a result, they obtained 98.2% accuracy from the Yolov2 algorithm	Yolov2

(*continued*)

Table 1. (*continued*)

Cites	Works Details	Algorithms
(Dharma et al. 2022) [4]	Dharma et al. studied a deep learning-based model for reading Batak Toba handwriting. The dataset used consists of 19 classes and 4674 images. 3739 of the images were reserved as trains and 935 as tests. They aimed to read Batak Toba handwritings by running Faster R-CNN and Yolov3 algorithms with the dataset they created. As a result of the training of its algorithms, Faster R-CNN has 97.3% and Yolov3 98.9% accuracy rates, while these values are 76.3% and 93.1%, respectively, when the algorithms are tested with letters. When tested with numbers, Faster R-CNN gives 73.6% accuracy and Yolov3 53% accuracy. When mixed letters and symbols are tested, Faster R-CNN 47.3% and Yolov3 35.5% results are obtained	Yolov3, Faster R-CNN
(Kasinathan and Uyyala, 2023) [5]	Kasinathan and Uyyala have worked on the deep learning-based detection of insects that harm food and crops in the field. In the study, the Fall Armyworm (FAW) dataset, which was created from the images of plant insects, was used. There are 190 images in the dataset and 70% of the images are reserved as train and 30% as test. Mask R-CNN, RetinaNet, R-CNN, SSD and Faster R-CNN algorithms were run with the FAW dataset. As a result of the studies, Mask R-CNN model with Resnet-101 backbone was found to be more successful than other models with a mAP value of 94.21%	Mask R-CNN, RetinaNet, R-CNN, SSD, Faster R-CNN
(Chethan Kumar et al. 2020) [6]	In the study, used a ready-made dataset of six classes for surveillance applications. The Yolov3 algorithm was run with the data consisting of images, and the Yolov4 algorithm was run with the data consisting of videos. As a result of the study, Yolov3 was 98% and Yolov4 was 99% accuracy rate	Yolov3, Yolov4
(Nepal et al. 2022) [7]	A dataset consisting of aerial images was used to determine the areas where the failed UAVs would land. Yolov3, Yolov4 and Yolov5-l algorithms were run with the data set consisting of aerial images. As a result of the study, the accuracy rate of the Yolov3 algorithm was 46%, and Yolov4 algorithm was 60%, and the accuracy rate of the Yolov5-l algorithm was 63%	Yolov3, Yolov4, Yolov5-l

3   Material and Method

3.1   Deep Learning and Object Detection

Artificial intelligence can be defined as machines that imitate humans by performing learning and problem-solving actions to perform specific tasks [8]. The term artificial intelligence was first used by a group of scientists in 1956, [9] and it has continued to develop until today and has become the focus of many research topics. With the continuation of the development, the concept of machine learning (ML) has emerged. ML is an AI approach that covers the process of creating algorithms that will enable machines to learn based on the data used [10]. Data given as input in machine learning for extract the features and the programmer creates the algorithm by examining the data. With the created algorithm, the properties of the data are extracted and the classification of the data is performed with machine learning.

With the development of artificial neural networks, an approach called deep learning (Deep Convolution Neural Network), in which multi-layered and each layer output is given as input to the next layer, has been presented [11]. With the developed neural network, analysis works can be performed by processing data such as audio, image, video and signal. In CNN structures, after some preliminary operations are performed on the input data, the necessary information in the data, namely the features, are extracted and classification processes are performed according to the obtained features [12].

Object detection with DL methods is very popular area of study by artificial intelligence researchers recently. In object recognition and object detection processes, images given as input to CNN structures are given with the information in which region the object to be recognized is. This pre-process is called the labeling process. CNN extracts the features from the labeled areas of the images given as input. These properties are the properties of the object such as color, texture, shape etc. Finally, CNN performs the classification. Classification is the process of separating the objects whose properties are extracted from each other, that is, categorizing the objects [13]. Figure 1 has a schematic representation of object detection with CNN.

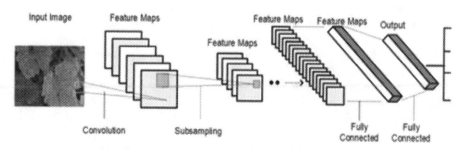

Fig. 1. Object Detection With CNN

There are multiple methods that perform object detection using the CNN structure. These methods are SSD, R-CNN, Yolo, Mask R-CNN, Fast and Faster R-CNN. Among these methods used in object detection, Yolo is more popular for real-time detection [14].

3.2 Yolo

Yolo is a one-step object recognition algorithm. The first version of Yolo was developed by Joseph Redmon [15]. It means You Only Look Once. It passes the pictures through the neuron structure in one go, not in two stages. That's why it works faster. It turns the objects it detects into boxes called bounding boxes. Figure 2 shows the bounding boxes used in Yolo.

Fig. 2. An Output From Yolo

In Yolo, each grid is in charge of detecting an object in it. There was a difficulty in detecting two objects in case of multiple objects in a grid.

To overcome this challenge, Joseph Redmon developed Yolov2, which uses the anchor box method [16]. With the anchor box method, two different objects in a single grid become detectable.

With the Yolov3 version [17], the understanding that an object can belong to more than one class has been introduced. In addition, Darknet 19 was used for feature extraction in the Yolov2 version, while Darknet-53 was used in the Yolov3 version. The Yolov4 version used the more advanced CSPDarknet53 for feature extraction. Instead of the FPN structure used in the neck layer of Yolov3, SAM, PAN and SSP are used in the neck layer of Yolov4 [18].

Yolov5 was developed by Glenn Jocher in 2020. In this version Fig. 3, CSPDarknet53 and Focus structure in the backbone layer, PANet in the neck layer and Pytorch as the framework are among the innovations made [7]. The biggest difference between Yolov4 and Yolov5 is the use of Pytorch in Yolov5. With the use of Pytorch framework, Yolov5 is easier to configure and more applicable than Yolov4 [19]. With the Focus structure used, the memory used in Cuda has been reduced. With the innovations made, the model has been accelerated and its performance increased.

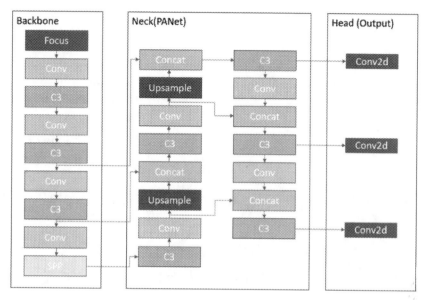

Fig. 3. Yolov5 Architecture [7]

Yolov6 was developed in 2022 by employees of the Meituan company. This version of Yolo on Fig. 4 shown has been developed mainly for industrial applications. By designing the backbone layer called EfficientRep, which consists of RepBlock, RepConv and CSPStackRep blocks, the developers have enabled more efficient use of hardware computing power and reduced the latency in feature extraction [20]. In the neck part, Rep-Pan was used by editing the PANet used in Yolov4 and Yolov5 in order to better balance the accuracy and speed of the model [20]. Finally, Efficient Decoupled Head, which solves the classification and detection processes separately, in a branched structure, is used in order to provide high accuracy and reduce the computational load in the head layer [20].

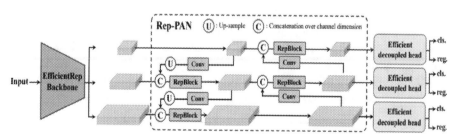

Fig. 4. Yolov6 Architecture [20]

3.3 Dataset

Diseased and disease-free leaf images are needed for the study. The data set named PlantDoc, created by Davinder Singh et al., was used in line with the need [21]. The PlantDoc dataset consists of 2598 images and 9216 labeled plant leaves divided into 27 classes (see Fig. 6). In Fig. 5, different images in the data set are given with their classes.

Fig. 5. Example Images From Different Classes Of PlantDoc Dataset

The performance rate of datasets created in the laboratory environment decreases in real life conditions. The images in the PlantDoc dataset were photographed in real environmental conditions (cultivated fields). For this reason, it is a suitable dataset for deep learning-based detection and diagnosis of plant diseases.

There are 13 plant species and 17 plant diseases in the dataset. The 17 disease types mentioned are given in Fig. 6.

In PlantDoc dataset, tag information is kept in XML files. In order to run Yolo algorithms with this data set, the file format in which the label information is stored with the Python software language is arranged as.txt.

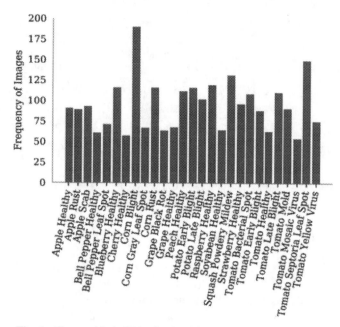

Fig. 6. Class and Label Distribution of The PlantDoc Dataset [21]

3.4   Running the Models

Yolov5 and Yolov6 models were trained on Google Colab. Colab is a free virtual computer platform designed by Google for artificial intelligence developers. In the Colab environment, the image data in the PlantDoc dataset were processed in Python software programe and Keras, Tensorflow, Pytorch libraries used in artificial intelligence. The PlantDoc dataset has been split into 3, as train, test and validation. 84%, 8% and 8% respectively.

The hyperparameters to be used in the algorithms were determined in accordance with the processing of the images (see Table 2).

The epoch value, which means how many times all the data in the data set will be shown to the neural network during the training, was determined as 100 for all models. The image is set to be 640x640. The batch size value, which can be defined as the training set, is determined as 16. Batch size value of 16 means that the images are processed in sets of 16 in each epoch and the parameter is updated. The batch size value has a great effect on the accuracy of the model. However, the small GPU memory limits the batch size. Not having a large enough GPU memory will cause the batch size to be small.

In the study, the learning rate was determined as 0.0032 for Yolov6 models and 0.01 for Yolov5 models. Learning rate is the value that defines the update rate of the parameters created in the neural network, which is connected to the optimization algorithm. Since it is the most effective parameter in the process of minimizing the error, a low value can slow down the learning process unnecessarily and reduce performance a lot. Choosing a large value may cause poor results in the accuracy performance of the model. Optimization algorithms increase the accuracy of results in complex learning processes of artificial

Table 2. Hyperparameters Used in Yolov5 and Yolov6 Models

Models	Learning Rate	Momentum	Activation Function	Optimization Algorithm	Epoch
Yolov5-n	0.01	0.937	SiLu	SGD	100
Yolov5-s	0.01	0.937	SiLu	SGD	100
Yolov5-m	0.01	0.937	SiLu	SGD	100
Yolov6-n	0.0032	0.843	ReLu - SiLu	SGD	100
Yolov6-s	0.0032	0.843	ReLu - SiLu	SGD	100
Yolov6-m	0.0032	0.843	ReLu - SiLu	SGD	100

intelligence. The Stochastic Gradient Descent (SGD) optimization algorithm used in all models. It continues to work until it reaches the desired minimum value. The momentum value, which enables the SGD optimization algorithm to run faster, was determined as 0.843 in Yolov6 models and 0.937 in Yolov5 models. The activation function is used to convert the outputs to nonlinear values in the learning process. Converting the outputs to non-linear values causes an increase in the performance of the models. ReLU activation functions are used in the neck layer and SiLu activation functions in the head layer of the three Yolov6 models. In the Yolov5 model, the SiLu activation function was used.

3.5 Evaluating the Results of Object Detection Algorithms

The basis for calculating the accuracy of an object detection model is the confusion matrix in Fig. 7. The confusion matrix shows the accuracy of the model based on the difference between the object detection model's prediction and the actual situation. TP and TN indicate that the model predicts correctly, while FP and FN indicate that the model predicts incorrectly. If there is a disease in the plant, the object detection model predicts that there is a disease in the plant, this situation is evaluated as TP. If there is no disease in the plant and the prediction of the object detection model is that there is no disease in the plant, this situation is evaluated as TN. If the plant has disease, but the object detection model predicts no disease, it is considered FN. If the plant has no disease, but the object detection model predicts disease, it is considered FP.

	Predicted	
Actual	True Positive (TP)	False Negative (FN)
	False Positive (FP)	True Negative (TN)

Fig. 7. Confusion Matrix

Accuracy is the ratio of correct predictions to all predictions (see Eq. 1). It gives the success of the model. However, it is not sufficient alone to determine the performance.

$$Accuracy = (TP + TN)/(TP + FP + TN + FN) \qquad (1)$$

The precision value given in Eq. 2 is the unit of measure that shows how many of the values predicted by the object detection model are actually correct.

$$Precision = TP/(TP + FP) \qquad (2)$$

Recall, on the other hand, shows how many of the actually correct values were predicted correctly by the object detection model (see Eq. 3).

$$Recall = TP/(TP + FN) \qquad (3)$$

The F1 score given in Eq. 4 is obtained by taking the harmonic mean of the two values to evaluate the precision and recall values over a single value. If the two values are at the extreme point, since the normal mean calculation will cause an incorrect evaluation, harmonic mean calculation has been made instead of the normal mean.

$$F_1 = 2 * (Precision * Recall)/(Precision + Recall) \qquad (4)$$

4   Result and Discussion

In the study 6 different models of Yolov5 and Yolov6 algorithms (Yolov5-n, Yolov5-s, Yolov5-m, Yolov6-n, Yolov6-s, Yolov6-m) were used with PlantDoc dataset. The results of the classification and detection of plant diseases were shared (see Table 3).

Table 3. Training Results of Yolov5 and Yolov6 Models

Models	Precision	Recall	F1	Train Time (hour)	mAP
Yolov5-n	0.544	0.528	0.535	2.443	0.563
Yolov5-s	0.575	0.539	0.556	2.663	0.574
Yolov5-m	0.611	0.555	0.581	2.838	0.603
Yolov6-n	0.54	0.513	0.486	2.928	0.530
Yolov6-s	0.586	0.528	0.520	3.259	0.560
Yolov6-m	0.547	0.572	0.522	3.806	0.547

Among the results of the PlantDoc dataset in the trained Yolo models, the highest precision value is 0.611, which belongs to the Yolov5-m model. The Yolov6-s and Yolov5-s models follow the Yolov5-m model in terms of precision. In the comparative analysis of the data obtained by the models in terms of the recall value, the highest recall value of 0.572 belongs to the Yolov6-m model. Yolov5-m and Yolov5-s models follow the Yolov6-m model as recall values, respectively. F1 value is checked to evaluate the precision and recall values together. Yolov5 models have a higher F1 value than Yolov6 models. Among the Yolov5 models, the Yolov5-m has the highest F1 value. The model with the highest mAP value is the Yolov5-m model. By comparing all values, the best model in terms of performance is seen as the Yolov5-m model. In addition to performance in object recognition, training time is also important. The model with the best performance for training time is Yolov5-n model. Yolov5-m model comes after the Yolov5-n model in terms of time performance. However, considering the object detection performances, Yolov5-m model gives a more balanced performance.

When the mAP-Epoch curves of the models are examined (see Fig. 8), an increase in the accuracy rate is observed in each model due to the increase in epoch and time. It is desired that the models used are stable as a result of training. The less fluctuations in the curves, the more stable the model will be. When the curves of Yolov6 models are examined, the mAP value is zero and suddenly increases until it reaches a certain epoch

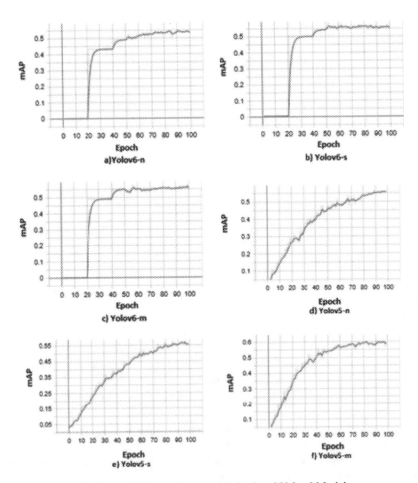

Fig. 8. mAP-Epoch Curves of Yolov5 and Yolov6 Models

value. Then it continues to follow a horizontal path. Although there is little fluctuation in the curves of Yolov6 models, sudden mAP changes disrupt the stability of the models. When the graphs of the Yolov5 models are examined, it is seen that the Yolov6 models are more stable than the graph curves. The Yolov5 model has a higher mAP value than other models, so its performance is better than other models.

After the models were trained, each model was tested with the test images in the dataset. When the test results in Table 4 are examined, the highest AP value is the Yolov5-m model with 0.584. Yolov5 models outperformed Yolov6 models in terms of AP value. The Yolov6-s model have the highest AR value. When all AR values were examined, Yolov6 models gave better results than Yolov5 models. However, when the F1 value, which is the joint evaluation of AP and AR values, is examined, Yolov5 models have higher performance than the Yolov6 models.

Table 4. Test Results of Yolov5 and Yolov6 Models

Models	Average Precision	Average Recall	F1
Yolov5-n	0.494	0.609	0.545
Yolov5-s	0.54	0.617	0.575
Yolov5-m	0.584	0.671	0.624
Yolov6-n	0.396	0.692	0.503
Yolov6-s	0.414	0.693	0.518
Yolov6-m	0.423	0.691	0.524

Another factor for performance evaluation of object detection algorithms is losses. Class loss is given in Fig. 9. Initially, there was a rapid decrease in class losses of all models. Afterwards, a more horizontal decrease was observed. When the curves are examined, it is seen that the losses are inversely proportional to the epoch increase. When the class losses are examined, at the end of 100 epochs, Yolov6-n fell below 1.0, Yolov6-s 0.8, Yolov6-m 0.7. Class losses of the Yolov5-n and Yolov5-s models are about 0.015, while the class losses of the Yolov5-m model are about 0.01. The class loss of the Yolov5-s model is lower than the Yolov5-n model. The class loss of the Yolov5-m model, on the other hand, has the lowest class loss among all models. In general, the class loss values of Yolov6 gave worse results than Yolov5.

Midhun P. Mathew et al. conducted a study for the detection of bacterial spot on bell pepper plant [22]. The dataset they prepared for disease detection consists of 2 classes, the healthy part and the bacterial stain. The images in the dataset are divided into two as 3000 trains and 1000 tests. With the data set, Yolov4 and Yolov5s algorithms were run and the researchers suggested the Yolov5s model for speed and accuracy. Achyut Morbekar et al. [23] used a readily available dataset for the detection of plant diseases. There are 24 classes in the dataset used. As a result of their studies, the researchers suggested the Yolov3 algorithm because it is fast and high performance. Abhishek Mohandas et al. created a special dataset to detect the disease of 5 different plant leaves [24]. They divided dataset 80% train and 20% test. As a result of the study, the Yolov4-tiny

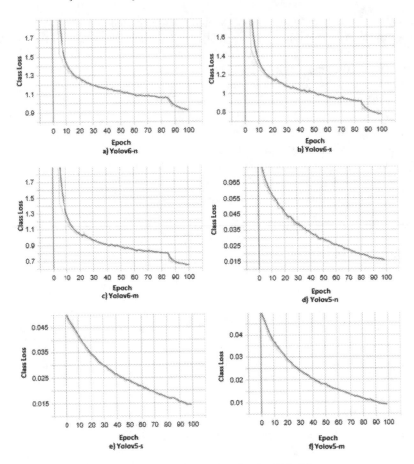

Fig. 9. Class Loss Curves of Yolov6 Models

model was successful in the classification and detection of plant diseases. Kazi Riad Uddin et al. [25] used the PlantDoc dataset for the identification and detection of plant diseases. Shamse Tasnim Cynthia et al. [26] created a special dataset for the detection of 5 different plant diseases. There are 236 images in the dataset. 80% of these images are split for train and 20% for testing. Faster R-CNN algorithm was proposed for plant disease detection with their work. It was stated that the model with the highest accuracy rate with a mAP value of 53.81% was EfficientD0.

As a result of the studies, the models of the Yolov6 algorithm performed well. Our research is to ensure sustainability and quality crop production in agriculture by providing early diagnosis of diseases on plants. According to the results obtained, the fact that the Yolov5-m model works faster than the other models and has a better performance with 65% mAP shows that it can be used in the detection of plant diseases.

In our study, the hyperparameter values used in the Yolov5 and Yolov6 algorithms were chosen so that the small models would perform well. The results obtained from small-sized object detection models are suitable for use in mobile systems. In addition, by training and testing 6 similar models with the same data set, a more homogeneous and comprehensive comparison was obtained compared to other studies. In this way, the sensibility rate of the results obtained is higher. In addition to all these, further studies, in order to obtain a higher accuracy rate, the number of images belonging to each class in the data set used in the training of the model should be increased and optimization studies should be carried out on the model.

References

1. King, A.: Technology: the future of agriculture. Nature **544**(7651), S21–S23 (2017). https://doi.org/10.1038/544s21a
2. "FAO: Ag employs 27% of world's workers, generates 4% of GDP," Regular Migration News. https://migration.ucdavis.edu/. Accessed 27 Dec 2022
3. Boudjit, K., Ramzan, N.: Human detection based on deep learning YOLO-v2 for real-time UAV applications, vol. 34, no. 3, pp. 527–544 (2021). https://doi.org/10.1080/0952813X.2021.1907793
4. Dharma, A.S., Kom, M., Tambunan, S., Naibaho, P.K.: Deteksi Objek Aksara Batak Toba Menggunakan Faster R-CNN dan YoloV3 (2022). https://www.academia.edu/68753590/Deteksi_Objek_Aksara_Batak_Toba_Menggunakan_Faster_R_CNN_dan_YoloV3. Accessed 16 Feb 2023
5. Kasinathan, T., Uyyala, S.R.: Detection of fall armyworm (spodoptera frugiperda) in field crops based on mask R-CNN. Signal Image Video Process, pp. 1–7 (2023). https://doi.org/10.1007/S11760-023-02485-3/FIGURES/10
6. Kumar, B.C., Punitha, R., Mohana.: YOLOv3 and YOLOv4: Multiple object detection for surveillance applications. In: Proceedings of the 3rd International Conference on Smart Systems and Inventive Technology, pp. 1316–1321 (2020). https://doi.org/10.1109/ICSSIT48917.2020.9214094
7. Nepal, U., Eslamiat, H.: Comparing YOLOv3, YOLOv4 and YOLOv5 for autonomous landing spot detection in faulty UAVs. Sensors **22**(2), 464 (2022). https://doi.org/10.3390/S22020464/S1
8. Ongsulee, P.: Artificial intelligence, machine learning and deep learning. In: International Conference on ICT and Knowledge Engineering, pp. 1–6 (2018). https://doi.org/10.1109/ICTKE.2017.8259629
9. Buchanan, B.G.: A (very) brief history of artificial intelligence. AI Mag. **26**(4), 53 (2005). https://doi.org/10.1609/AIMAG.V26I4.1848
10. Riedl, M.O.: Human-centered artificial intelligence and machine learning. Hum. Behav. Emerg. Technol. 33–36 (2019). https://doi.org/10.1002/HBE2.117
11. Hinton, G.E., Osindero, S., Teh, Y.W.: A fast learning algorithm for deep belief nets. Neural Comput. **18**(7), 1527–1554 (2006). https://doi.org/10.1162/NECO.2006.18.7.1527
12. Doğan, F., Türkoğlu, İ: Derin öğrenme modelleri ve uygulama alanlarina ilişkin bir derleme. Dicle Üniversitesi Mühendislik Fakültesi Mühendislik Dergisi **10**(2), 409–445 (2019). https://doi.org/10.24012/DUMF.411130
13. Zhao, Z.Q., Zheng, P., Xu, S.T., Wu, X.: Object detection with deep learning: a review. IEEE Trans. Neural. Netw. Learn. Syst. **30**(11), 3212–3232 (2019). https://doi.org/10.1109/TNNLS.2018.2876865
14. Tan, F.G., Yüksel, A.S., Aydemir, E., Ersoy, M.: Derin Öğrenme Teknikleri İle Nesne Tespiti Ve Takibi Üzerine Bir İnceleme. In: Avrupa Bilim ve Teknoloji Dergisi, no. 25, pp. 159–171 (2021). https://doi.org/10.31590/EJOSAT.878552

15. Redmon, J., Divvala, S., Girshick, R., Farhadi, A.: You only look once: unified, real-time object detection. In: Proceedings of the IEEE Computer Society Conference on Computer Vision and Pattern Recognition, vol. 2016, pp. 779–788 (2015).https://doi.org/10.48550/arxiv.1506.02640

16. Redmon, J., Farhadi, A.: YOLO9000: better, faster, stronger. In: Proceedings - 30th IEEE Conference on Computer Vision and Pattern Recognition, vol. 2017, pp. 6517–6525 (2016). https://doi.org/10.48550/arxiv.1612.08242

17. Redmon, J., Farhadi, A.: YOLOv3: an incremental improvement (2018). https://doi.org/10.48550/arxiv.1804.02767

18. Bochkovskiy, A., Wang, C.-Y., Liao, H.-Y.M.: YOLOv4: optimal speed and accuracy of object detection (2020). https://doi.org/10.48550/arxiv.2004.10934

19. Liu, Y., Zuo, X., Yun, H., Park, D.: Efficient object detection based on masking semantic segmentation region for lightweight embedded processors. Sensors **22**(22), 8890 (2022). https://doi.org/10.3390/S22228890

20. Li, C., et al.: YOLOv6: a single-stage object detection framework for industrial applications (2022). https://doi.org/10.48550/arxiv.2209.02976

21. Singh, D., Jain, N., Jain, P., Kayal, P., Kumawat, S., Batra, N.: PlantDoc: a dataset for visual plant disease detection. In: ACM International Conference Proceeding Series, pp. 249–253 (2020). https://doi.org/10.1145/3371158.3371196

22. Mathew, M.P., Mahesh, T.Y.: Leaf-based disease detection in bell pepper plant using YOLO v5. SIViP **16**(3), 841–847 (2022). https://doi.org/10.1007/S11760-021-02024-Y/METRICS

23. Morbekar, A., Parihar, A., Jadhav, R.: Crop disease detection using YOLO. In: 2020 International Conference for Emerging Technology (2020). https://doi.org/10.1109/INCET49848.2020.9153986

24. Mohandas, A., Anjali, M.S., Varma, U.R.: Real-time detection and identification of plant leaf diseases using YOLOv4-tiny. In: 2021 12th International Conference on Computing Communication and Networking Technologies (2021). https://doi.org/10.1109/ICCCNT51525.2021.9579783

25. Uddin, K., Khan, H.: Automated identification of plant disease using deep learning. In: 2nd Global Conference on Engineering Research GLOBCER'22 (2022). https://www.researchgate.net/publication/363487322_Automated_Identification_of_Plant_Disease_Using_Deep_Learning/citation/download. Accessed 29 Dec 2022

26. Cynthia, S.T., Hossain, K.M.S., Hasan, M.N., Asaduzzaman, M., Das, A.K.: Automated detection of plant diseases using image processing and faster R-CNN algorithm. In: 2019 International Conference on Sustainable Technologies for Industry 4.0 (2019). https://doi.org/10.1109/STI47673.2019.9068092

Triplet MAML for Few-Shot Classification Problems

Ayla Gülcü[1]([✉]) [iD], İsmail Taha Samed Özkan[2] [iD], Zeki Kuş[2] [iD],
and Osman Furkan Karakuş[3,4] [iD]

[1] Department of Software Engineering, Bahçeşehir University, Istanbul, Turkey
`ayla.gulcu@eng.bau.edu.tr`, `ismail.ozkan@bahcesehir.edu.tr`
[2] Department of Computer Engineering, Bahçeşehir University, Istanbul, Turkey
`zkus@fsm.edu.tr`
[3] Department of Computer Engineering, Fatih Sultan Mehmet Vakif University,
Istanbul, Turkey
[4] Department of Computer Engineering, Yıldız Technical University, Istanbul,
Turkey
`osman.karakus@yildiz.edu.tr`

Abstract. In this study, we propose a TripletMAML algorithm as an extension to Model-Agnostic Meta-Learning (MAML) which is the most widely-used optimization-based meta-learning algorithm. We approach MAML from a metric-learning perspective and train it using meta-learning tasks composed of triplets of images. The idea of meta-learning is preserved while generating the meta-learning tasks and training our novel meta-model. The experimental results obtained on four few-shot classification datasets show that TripletMAML that is trained using a combined loss yields in high quality results. We compared the performance of TripletMAML to several metric learning-based methods and a baseline method, in addition to MAML. For fair comparison, we used the reported results of those algorithms that were obtained using the same shallow backbone. The results show that TripletMAML improves MAML by a large margin, and yields better results than most of the compared algorithms in both 1-shot and 5-shot settings. Moreover, when we consider the classification performance of other meta-learning algorithms that use much deeper backbones, we conclude that TripletMAML is not only competitive in terms of the classification performance but also very efficient in terms of the complexity.

Keywords: Meta-learning · MAML · Triplet Networks · Few-Shot Image Classification · Metric Learning

1 Introduction

Machine learning approaches that aim to make learning more generalizable with the help of meta-knowledge obtained from previous tasks are known as *meta-*

This work was supported by the Scientific and Technological Research Council of Turkey (TÜBİTAK) under grant number 121E240.

learning [1,23]. The necessity for these methods stems from the fact that, in some circumstances, the amount of labelled data is too small to train a neural network model. For example, Convolutional Neural Networks (CNNs) can achieve excellent results for many vision-related tasks when there is huge data; however, their performance drops significantly under a limited data regime. The type of problems for which there are limited labeled training examples are available is referred to as *Few-Shot Learning* (FSL) [12,16,17] and there is ever-increasing interest among the machine learning researchers for these problems.

Lu et al. [19] state that most of the recent studies for computer vision-related FSL problems focus on deep metric learning and meta-learning methods. In [9], meta-learning techniques for FSL are categorized as *(i)* memory-based, *(ii)* metric-based and *(iii)* optimization-based methods. In memory-based algorithms, a meta-learner is trained by a memory element to be able to learn new classes. In metric-based approaches, a meta-learner learns a representation in the feature space. Finally, optimization-based methods perform an optimization process using a differentiable loss function over support set samples.

In this study, we design a new TripletMAML algorithm that takes MAML [13] algorithm as a basis and extends it by incorporating Triplet Networks. MAML is an optimization-based algorithm that focuses on finding a good initial state of a neural network which then can be adapted easily to a novel task using a few optimization iterations. We adopt the same optimization procedure used in MAML, but replace the network model with a Triplet Network. We could therefore utilize both the metric-based loss function and the classification-based loss function to train the meta-model parameters. From this perspective, Triplet-MAML is the first meta-learning algorithm that is both an optimization-based and metric-based algorithm, to the best of our knowledge. It is also important to note that the meta-learning idea is preserved while generating the meta tasks and training the meta-model.

The structure of this paper is organized as follows: we first discuss related studies in Sect. 2, then we explain our TripletMAML algorithm in Sect. 3. Next, in Sect. 4, we mention the experimental settings, and in Sect. 5, we provide experimental results. Finally, in Sect. 6, we conclude the paper.

2   Related Studies

MAML [13] is an optimization-based method which can be used with any model trained with gradient descent and, therefore, convenient for different types of problems. MAML has received considerable popularity due to its performance on different types of FSL problems. Different variants of the method have been introduced in the literature. Nichol et al. [21] introduce a variant of MAML called *Reptile*, which suggests joint training. The performance of Reptile is comparable to that of MAML, but it is less sensitive to the selection of hyper-parameters. Raghu et al. [22] study the effectiveness of MAML and present a new method called *ANIL* (Almost No Inner Loop) which simplifies MAML by removing the inner loop everywhere except the task-specific head. In another study, Arnold et al. [2] investigate the performance of MAML under different architectural settings. Fan et al. [11] propose a variant of MAML, called SignMAML, whose time complexity is less than MAML. Huisman et al. [15] investigate MAML, and a memory-based method, meta-learner LSTM, and they propose a new meta-learning algorithm, namely, *TURTLE*, based on those methods.

Metric-based methods have also shown to be very successful for FSL problems. For the purpose of predicting the unknown class labels, Matching Networks [31] use a k-nearest neighbours algorithm which is end-to-end trainable using the learned embedding of the few labelled examples. In an embedding space, these networks can also be considered a weighted nearest-neighbour classifier. Snell et al. [26] propose a simple but efficient metric-based algorithm called *Prototypical Networks*. Chen et al. [6] investigate the performance of baseline methods. These methods, in general, pre-train the given network by applying standard transfer learning procedures and then apply fine-tuning.

Li et al. [18] approach the FSL problem using a triplet network-based metric learning method to generate embedding that will help discriminate the classes better. They formulate a deep *K-tuplet Network* that compare K negative samples all at the same time in a given mini-batch. They first train an embedding network using the samples in the training dataset with K-tuplet loss in order to learn feature embedding. Then, a distance metric module with a non-linearity functionality is used to learn to discriminate the embeddings of novel classes. Although their technique looks similar to our TripletMAML method in terms of incorporating a Triplet Network, there are significant differences. As far as we are aware, our TripletMAML is the first meta-learning algorithm that is both an optimization-based and metric-based method. Moreover, the way the meta-model is trained conforms to the idea of meta-learning. Therefore, we assume that our TripletMAML is the first meta-learning method that uses Triplet Networks.

3   A New Meta-Learning Algorithm: TripletMAML

In our TripletMAML algorithm, we take *first-order MAML (fo-MAML)*, that does not compute second gradients as our basis. It is shown in [13] that fo-MAML is nearly as good as the second-order version, and therefore fo-MAML has become the default MAML due to its speed. We will be referring to fo-MAML when we say MAML in the rest of this paper.

3.1   MAML

MAML [13] is applicable in any model that can be trained with the Gradient Descent algorithm. The algorithm includes two optimization cycles, an outer loop and an inner loop. The inner loop tries to learn a new task consisting of a few labelled examples by optimizing the initial parameters. In contrast, the outer loop tries to find an appropriate starting point using meta-information obtained from those tasks.

In a typical machine learning setting, a dataset D is split as D_{train} and D_{test}, each containing a number of samples. However, in meta-learning, each batch contains a number *tasks* which is described as follows [23]: A meta-dataset D consists of a training meta-dataset, $\mathcal{D}_{train}$, and a test meta-dataset $\mathcal{D}_{test}$ which is used to see the performance of a trained meta-model on unseen tasks. $\mathcal{D}_{train}$ and $\mathcal{D}_{test}$ consist of multiple small datasets, $\mathcal{D}_i$, each of which is called a *task*. Each task is further split into a training,*support set*, and a test set, *query set*.

During training, MAML updates meta-model parameters, θ, with some gradient descent updates on the given task's, $\mathcal{T}_i$, support examples, and the model's parameters become θ_i' but this is temporary. The model with θ_i' is tested on the same task's query examples, and the loss is recorded. Then, the adaptation process for the next task starts with initial model parameters θ. Meta-model parameters θ are only updated during the meta-optimization stage, which starts upon completing all tasks in a given batch. So, θ is updated permanently using the error of the query set which was previously conditioned on the support set. After the meta-training phase, the meta-testing phase starts. The ability of the meta-learner to learn new tasks coming from novel classes is evaluated by temporarily training it on support set examples and testing on the query set examples.

3.2   Triplet Networks

Triplet networks inspired by Siamese networks are comprised of 3 instances of the same neural network with shared parameters. Scroff et al. [25] show that these networks are successful for numerous vision-related tasks. The network is fed with triplets, each containing three samples consisting of an anchor, a positive example coming from the same class as the anchor and a negative example coming from a different class. The network is then trained with the aim of bringing the feature vectors of the anchor and positive closer while pushing the feature vectors of the anchor and negative example further away.

The triplet loss for a given triplet is computed as in Eq. 1, where x_a is the anchor, x_p is a positive sample, x^- is a negative sample, f is the feature vector obtained for a given input, and $[]_+ = max(., 0)$ is the hinge function. The positive pair is separated from the negative pair at least by a distance margin, α. In [27], it is claimed that (N+1)-tuplet loss is a better alternative to triplet loss in which N-1 negative examples are all used simultaneously to identify a positive example. A positive example is being compared against the samples from multiple negative classes to yield a balanced embedding vector for the positive example, which is far from the rest of the classes. Given an (N+1)-tuplet of training examples $x, x^+, x_1, .., x_{N-1}$ where x^+ is a positive example to x and $x_i{}_{i=1}^{N-1}$ are negative, and f is the embedding function, the (N+1)-tuplet loss is given in Eq. 2. It is also shown that it is equivalent to triplet loss when N=2 if the embeddings have unit form as shown in Eq. 3.

$$\mathcal{L}(x_a, x_p, x^-) = [\|f(x_a) - f(x_p)\|_2^2 - \|f(x_a) - f(x^-)\|_2^2 + \alpha]_+ \qquad (1)$$

$$\mathcal{L}(x, x_p, x_{i\,i=1}^{N-1}) = log\left(1 + \sum_{i=1}^{N-1} exp(f(x).f(x_i) - f(x).f(x_p))\right) \qquad (2)$$

$$\mathcal{L}_{(2+1)-tuplet}(x, x_p, x^-) = log\left(1 + exp(f(x).f(x^-) - f(x).f(x_p))\right) \qquad (3)$$

3.3 Triplet Generation for TripletMAML

In our TripletMAML method, we have selected *(2+1)-tuplet loss* shown in Eq. 3 as the metric loss function, which is also combined with classification loss. In this section we explain our triplet task generation scheme that conforms to meta-learning paradigm.

According to Ravi and Larochelle [23], if 5-way 1-shot classification is going to be performed, then training episodes should be comprised of 5 classes taking one image from each. In our 5-way 1-shot triplets-based task setting, the same number of shots is used for both the support and the query set. 4 triplets form a task, and a combination of k tasks forms a mini-batch. In a given task support set, the same anchor and positive samples are used in all triplets, whereas a different negative sample is used in each triplet. Since the anchor example and the positive example belong to the same class, each class except the positive class is represented by one sample. As a result, there are 4 triplets in a given task with a total of 4+2 samples taken from 4+1 distinct classes. While forming the query set, different examples from the same positive and negative classes are used, conforming to the meta-learning task setting. Moreover, the idea of meta-learning, using the error of the query set examples conditioned previously on the support set to update the meta-learner, is preserved in our TripletMAML algorithm.

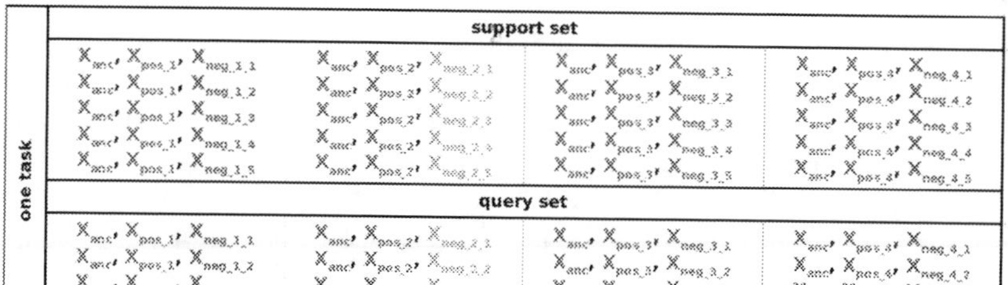

Fig. 1. Generation of triplets for 5-way 5-shot setting. Each support and query set contains five samples for each class, where the samples belonging to the same classes are denoted with the same colour (best viewed in colour).

In the 5-way 5-shot setting, each class is represented by five samples in both the support and the query sets. In Fig. 1, the examples from the same classes are denoted with the same colour. As can be seen in the figure, five images per class are used to form 20 triplets. As in the 1-shot setting, the same anchor image, X_{anc}, is used in all the triplets. X_{pos_1} to X_{pos_4} all belongs to the same class as X_{anc}. Therefore there are five samples from the same positive class. Remaining 4 classes are all denoted by X_{neg_i}, $i \in 1, .., 4$. For each negative class, again, there are five samples. $X_{neg\_1\_1}$ represents the first example belonging to the first negative classes. The same classes but different samples are used for the query set. It is also important to note here that while calculating the classification accuracy for TripletMAML to compare it to MAML and other methods, the accuracy contribution for a given sample is only calculated once.

Algorithm 1: Triplet-MAML Pseudo-code

    Input: α, β: Fast and meta-learning rate

1   Initialize model with θ

2   **for** Nbr_Outer **do**

3   Sample a batch of triplets: T

4   **for** $T_i \in \{T\}$ **do**

5   **for** Nbr_Inner **do**

6   Calculate loss on support: $\mathcal{L}_{T_i}^{support}(f_\theta)$

7   Calculate and record gradients: $\nabla_\theta \mathcal{L}_{T_i}^{support}(f_\theta)$

8   Adaptation: $\theta_i' = \theta - \alpha \nabla_\theta \mathcal{L}_{T_i}^{support}(f_\theta)$

9   Meta-update: $\theta \leftarrow \theta - \beta \nabla_\theta \sum_{T_i \in \{T\}} \mathcal{L}_{T_i}^{query}(f_{\theta_i'})$

TripletMAML algorithm is summarized in Algorithm 1. As can bee seen in the pseudo-code, the optimization process of MAML is strictly followed. The meta-model consisting of a triplet network is adapted for each task in the batch as shown in line 4. The model is adapted for a number of gradient descent steps which is shown as Nbr_Inner in the algorithm using the support set of the current task. The loss shown in line 6 contains both embedding and the classification components. The model which is adapted using the combined loss is tested on the query set. When all the tasks are complete in a given batch, meta-model parameters are updated using the gradients calculated on the query set.

4   Experimental Settings

4.1   Datasets for Few-Shot Classification

Omniglot dataset [16] contains 1623 characters belonging to 50 alphabets. For each character in a given alphabet, there are 20 examples in the dataset. Each of these examples is handwritten by a different writer. Most studies adopt the augmentation method of Vinyals et al. [31] which includes resizing the characters to 28 × 28 and rotating in multiples of 90°.

MiniImageNet was introduced by Vinyals et al. [31] by sampling from the ImageNet dataset. It is as complex as the ImageNet dataset but requires significantly fewer computational resources. This dataset contains 60K 84×84 coloured images from 100 classes. We use the splits from Ravi et al. [23] in this work.

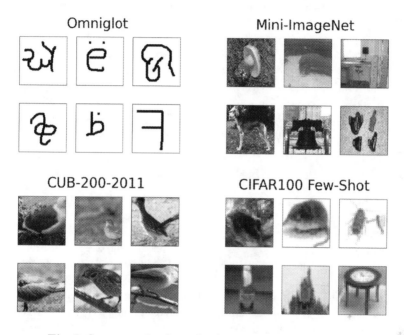

Fig. 2. Some samples from the datasets used in this study

CUB-200-2011 is an larger version of previous CUB-200 dataset [33] consisting of 6033 images over 200 bird species. CUB-200-2011 [32] doubles the number of images per category. There are 11,788 images across 200 classes in this dataset. Most studies follow Hilliard et al.'s [14] evaluation protocol. They also resize each image to 84 × 84 after some data augmentation process. The dataset provides bounding boxes to crop the images, but it is also possible to use the full bird images with the background, which provides a harder challenge. Triantafillou et al. [30] do not use bounding boxes to generate cropped images. Similarly, Chen et al. [6] do not use any bounding box information, and they first resize each image to 126 × 126 (84 × 1.5), then center crop to extract 84 × 84. We have followed the same approach to create images of 84 × 84 with no bounding box information. From now on, we will refer to this version as the *CUB*.

CIFAR Few-Shot (CIFAR-FS) was introduced by Bertinetto et al. [5] as a new few-shot learning dataset which is harder than Omniglot but easier than the MiniImagenet dataset. It consists of 60K colour images of sizes 32×32 pixels. The classes in this dataset are sampled from the CIFAR-100 dataset. In this work, we use the original splits from Bertinetto et al. [5].

4.2   Loss Function

Following [27], we decided to use (N+1)-tuplet loss with N = 2 as the metric loss in addition to the cross entropy loss which is used as the classification loss function. It is shown in the literature that multi-task learning approach performs well in various vision-related tasks. For example, in [4], the problem is modelled as a multi-task prediction task where the the classification loss and the embedding loss are combined. It is also shown in [29] that modelling the problem at hand as a multi-task learning problem is more effective in image retrieval tasks than relying only on a single embedding loss. Therefore, in this study, we combine these two loss functions using the weighted sum method. A parameter λ is used to control the contribution of the embedding loss as in [29].

4.3   Backbone and MAML Hyper-parameters

In our TripletMAML implementation, we adopted the architecture and the hyper-parameters from the CNN-4 model in [13,31] which consists of four blocks, each comprising 64 convolution filters of size 3 × 3 which is followed by a batch normalization layer. ReLU is used for the nonlinearity and a max-pooling layer is used for the dimentionality reduction. In our TripletMAML implementation, we have modified the final layer to consider both metric loss and classification loss. There are also MAML algorithm related hyper-parameters that need to be set such as fast learning rate, meta learning rate, train adaptation steps, test adaptation steps and the number of train iterations. For each dataset, we adopted the MAML hyper-parameters from the literature studies.

Table 1. The architectural and algorithm related hyper-parameters used in preliminary experiments

	MiniImageNet	CUB
Number of filters	32	32
Meta batch-size	4 (2 for 5-shot)	4
Fast learning rate	0.01	1e−2
Train adaptation steps	5	5
Test adaptation steps	10	10
Meta learning rate, optimizer	1e−3, Adam	1e−3

5 Results and Discussion

We have built our triplet meta-learning model taking the PyTorch MAML implementation in the learn2learn meta-learning library [3] as a basis. We have used
the same number of examples for each class for both the support and the query
set during the test phase. Experiments are carried out on a single RTX 3090
GPU with a memory of 24 GBs. For each dataset, the train iterations are set
to 60000, and the test performance averaged over 600 test episodes with 95%
confidence interval is shown.

Table 2. 5-shot preliminary experiments to select λ

	MiniImageNet 5-shot	CUB 5-shot
$\lambda = 0.5$	66.39 ± 0.88	81.50 ± 0.87
$\lambda = 1$	66.67 ± 0.87	81.69 ± 0.80
$\lambda = 1.5$	63.79 ± 0.87	81.77 ± 0.84

5.1 Preliminary Experiments

We first run a set of preliminary experiments to control the contribution of the
embedding loss, a λ. We have experimented with several λ values $\in \{0.5, 1, 1.5\}$
using MiniImageNet and CUB datasets. MiniImageNet dataset was already used
in the original MAML paper [13], so we have adopted their settings directly.
The CUB dataset was not used in the original MAML paper, so we adopted
these parameters from other studies that use the same CNN-4 architecture like
[15, 20, 24]. Selected settings for the two datasets are shown in Table 1. For each
of the two datasets, and each λ, TripletMAML is run for 60000 iterations and
tested for 600 test episodes (using validation set) and the results are given in
Table 2. Based on those, we decided to set $\lambda = 1$.

5.2 Few-Shot Classification Experiments

Omniglot was already used in the original MAML paper [13], so we adopted
their settings directly. For the CIFAR-FS dataset, we adopted the parameters
from the literature studies that propose a model similar to MAML [10, 11, 20].
These parameters are shown in Table 3. We compare our TripletMAML against
various baselines, including three metric learning-based methods, namely, Matching Networks (Matching Nets) [31], Prototypical Networks (ProtoNet) [26], and
Relation Networks (Relation Nets) [28], and a baseline method [6] in addition to
MAML. For a fair comparison, we have selected the literature results obtained
using CNN-4 backbone for each benchmark dataset.

Table 3. The architectural and algorithm related hyper-parameters used in Triplet-MAML for Omniglot and CIFAR-FS

	Omniglot	CIFAR-FS
Number of filters	64	64
Meta batch-size	32	4 (2 for 5-shot)
Fast learning rate	0.4	1e−2
Train adaptation steps	1	5
Test adaptation steps	3	10
Meta learning rate, optimizer	1e−3, Adam	1e−3, Adam

Table 4. 5-way accuracy values averaged over 600 test episodes on Omniglot and MiniImageNet datasets with 95% confidence interval

	Omniglot		MiniImageNet	
	1-shot	5-shot	1-shot	5-shot
MAML [13]	$98.7 \pm 0.4\%$	$99.9 \pm 0.1\%$	$48.70 \pm 1.84\%$	$63.11 \pm 0.92\%$
Matching Nets [31]	98.1%	98.9%	46.6%	60.0%
ProtoNet [26]	98.8%	99.7%	$49.42 \pm 0.78\%$	$68.20 \pm 0.66\%$
Relation Nets [28]	$99.6 \pm 0.2\%$	$99.8 \pm 0.1\%$	$50.44 \pm 0.82\%$	$65.32 \pm 0.70\%$
Baseline++ [6]	–	–	$48.24 \pm 0.75\%$	$66.43 \pm 0.63\%$
TripletMAML	$98.06 \pm 0.50\%$	$98.50 \pm 0.26\%$	$53.53 \pm 0.17\%$	$68.90 \pm 0.89\%$

Table 5. 5-way accuracy values averaged over 600 test episodes on CUB and CIFAR-FS datasets with 95% confidence interval

	CUB		CIFAR-FS	
	1-shot	5-shot	1-shot	5-shot
MAML [13]	$53.12 \pm 0.93\%^\star$	$70.90 \pm 0.75\%^\star$	$34.97 \pm 0.70\%^\star$	$47.41 \pm 0.73\%^\star$
Matching Nets [31]	$57.70 \pm 0.87\%^\star$	$71.42 \pm 0.71\%^\star$	$36.97 \pm 0.67\%^\star$	$49.44 \pm 0.71\%^\star$
ProtoNet [26]	$51.34 \pm 0.86\%^\star$	$67.56 \pm 0.76\%^\star$	$36.83 \pm 0.69\%^\star$	$51.21 \pm 0.74\%^\star$
Relation Nets [28]	$59.47 \pm 0.96\%^\star$	$73.88 \pm 0.74\%^\star$	$36.40 \pm 0.69\%^\star$	$51.35 \pm 0.69\%^\star$
Baseline++ [6]	$60.53 \pm 0.83\%$	$79.34 \pm 0.61\%$		
TripletMAML	$70.46 \pm 0.17\%$	$81.43 \pm 0.86\%$	$64.03 \pm 1.79\%$	$73.47 \pm 0.90\%$

$\star$ refers to results from [7]

The results of the experiments regarding Omniglot and MiniImageNet are given in Table 4. These suggest that TripletMAML yields high accuracy on the Omniglot with no data augmentation applied. In all the other baselines (MAML, Matching Nets, ProtoNet, Relation Nets), the dataset is augmented with rotations. In addition, ProtoNet [26] used different shots for the query set during training. Likewise, Sung et al. [28] used 19 and 15 query images in 1-shot and 5-shot settings, respectively during training. Omniglot was not used in [6], so we left its result empty.

For the MiniImageNet dataset, TripletMAML's results are superior to all other algorithms in both settings. It improves the accuracy of the closest baseline, Relation Nets, by 3% for 1-shot setting. As in the case with Omniglot, MAML, ProtoNet and Relation Nets use large number of query points. There are also other studies in the literature that result in higher accuracy values; however, those methods use a deeper feature backbone like ResNet-12 as in Meta-Baseline [8]. We did not include those results for a fair comparison.

The results regarding CUB and CIFAR-FS datasets are given in Table 5. For the CUB dataset, the results suggest that TripletMAML is better than all other methods in both settings. It improves Baseline++'s accuracy by a large margin of 10% in 1-shot setting. When it comes to 5-shot setting, the improvement is about 2%. MetaMix+MAML [7] algorithm achieves 73.04% and 86.10% accuracy for 1-shot and 5-shot settings, respectively. Our TripletMAML is no worse by 5% than MetaMix+MAML, which was pretrained using ResNet-12 and used 16 query points during training.

TripletMAML is again superior to all the baseline methods for the CIFAR-FS dataset as shown in the table. The CIFAR-FS dataset was first introduced by Bertinetto et al. [5]. Hence, we wanted to consider the accuracy of their closed formed solver. Similar to [26], they train using 20 classes for CIFAR-FS using a random number of training shots. They achieve 65.3% and 78.3% accuracy for 5-way 1-shot and 5-shot settings, respectively and these state that TripletMAML is no worse than that by 5%. On the other hand, MetaMix+MAML [7] algorithm performs poorly on the CIFAR-FS dataset. It yields 43.58% and 58.27% accuracy values for 1-shot and 5-shot configurations, respectively, which are clearly inferior to TripletMAML's results.

6 Conclusion

In this study, we propose a TripletMAML algorithm as an extension to MAML which is a well-known optimization algorithm for few-shot classification problems. We adopt the same optimization procedure used in MAML, but replace the network model with a Triplet Network. We could therefore utilize both the metric-based loss function and the classification-based loss function to train the meta-model parameters. The idea of meta-learning idea is preserved while generating the meta-learning tasks and training the meta-model. For our Triplet-MAML algorithm, we adopted the shallow CNN architecture used in the original MAML algorithm. Most meta-learning algorithms start to yield similar results when very deep architectures are used as the backbone, so we believe using a simple backbone would help to reveal the performance of a given algorithm better. We have performed experiments on four few-shot classification datasets. For each of those datasets, we compared the performance of TripletMAML to several metric learning-based methods and a baseline method, in addition to MAML. For fair comparison, we used the reported results of those algorithms that were obtained using the same shallow backbone. The experiments state that Triplet-MAML improves MAML by a large margin and also yields better results than most of the compared algorithms. Moreover, the results state that it yields competitive results when compared to much complex algorithms. As a future study, we would like to observe its performance for image retrieval problems.

References

1. Andrychowicz, M., et al.: Learning to learn by gradient descent by gradient descent. In: Advances in Neural Information Processing Systems, vol. 29 (2016)
2. Arnold, S., Iqbal, S., Sha, F.: When MAML can adapt fast and how to assist when it cannot. In: International Conference on Artificial Intelligence and Statistics, pp. 244–252. PMLR (2021)
3. Arnold, S.M.R., Mahajan, P., Datta, D., Bunner, I., Zarkias, K.S.: Learn2Learn: a library for meta-learning research (2020). http://arxiv.org/abs/2008.12284
4. Bell, S., Bala, K.: Learning visual similarity for product design with convolutional neural networks. ACM Trans. Graph. (TOG) **34**(4), 1–10 (2015)
5. Bertinetto, L., Henriques, J.F., Torr, P., Vedaldi, A.: Meta-learning with differentiable closed-form solvers. In: International Conference on Learning Representations (2019). http://openreview.net/forum?id=HyxnZh0ct7
6. Chen, W.Y., Liu, Y.C., Kira, Z., Wang, Y.C.F., Huang, J.B.: A closer look at few-shot classification. arXiv preprint arXiv:1904.04232 (2019)
7. Chen, Y., Ma, Y., Ko, T., Wang, J., Li, Q.: MetaMix: improved meta-learning with interpolation-based consistency regularization. In: 2020 25th International Conference on Pattern Recognition (ICPR). IEEE (2021). https://doi.org/10.1109/icpr48806.2021.9413158
8. Chen, Y., Liu, Z., Xu, H., Darrell, T., Wang, X.: Meta-baseline: exploring simple meta-learning for few-shot learning. In: Proceedings of the IEEE/CVF International Conference on Computer Vision, pp. 9062–9071 (2021)
9. Chen, Y., Wang, X., Liu, Z., Xu, H., Darrell, T.: A new meta-baseline for few-shot learning. arXiv preprint arXiv:2003.04390 (2020)
10. Devos, A., Chatel, S., Grossglauser, M.: Reproducing meta-learning with differentiable closed-form solvers. In: RML@ ICLR (2019)
11. Fan, C., Ram, P., Liu, S.: Sign-MAML: efficient model-agnostic meta-learning by SignSGD. CoRR abs/2109.07497 (2021). http://arxiv.org/abs/2109.07497
12. Fei-Fei, L., Fergus, R., Perona, P.: One-shot learning of object categories. IEEE Trans. Pattern Anal. Mach. Intell. **28**(4), 594–611 (2006)
13. Finn, C., Abbeel, P., Levine, S.: Model-agnostic meta-learning for fast adaptation of deep networks. In: International Conference on Machine Learning, pp. 1126–1135. PMLR (2017)
14. Hilliard, N., Phillips, L., Howland, S., Yankov, A., Corley, C.D., Hodas, N.O.: Few-shot learning with metric-agnostic conditional embeddings. arXiv preprint arXiv:1802.04376 (2018)
15. Huisman, M., Plaat, A., van Rijn, J.N.: Stateless neural meta-learning using second-order gradients. Machine Learning **111**(9), 3227–3244 (2022). https://doi.org/10.1007/s10994-022-06210-y
16. Lake, B., Salakhutdinov, R., Gross, J., Tenenbaum, J.: One shot learning of simple visual concepts. In: Proceedings of the Annual Meeting of the Cognitive Science Society, vol. 33 (2011)
17. Lake, B.M., Salakhutdinov, R., Tenenbaum, J.B.: Human-level concept learning through probabilistic program induction. Science **350**(6266), 1332–1338 (2015)

18. Li, X., Yu, L., Fu, C.W., Fang, M., Heng, P.A.: Revisiting metric learning for few-shot image classification. Neurocomputing **406**, 49–58 (2020)
19. Lu, J., Gong, P., Ye, J., Zhang, C.: Learning from very few samples: a survey. arXiv preprint arXiv:2009.02653 (2020)
20. Na, D., Lee, H., Kim, S., Park, M., Yang, E., Hwang, S.J.: Learning to balance: Bayesian meta-learning for imbalanced and out-of-distribution tasks. CoRR abs/1905.12917 (2019). https://arxiv.org/abs/1905.12917
21. Nichol, A., Achiam, J., Schulman, J.: On first-order meta-learning algorithms. arXiv preprint arXiv:1803.02999 (2018)
22. Raghu, A., Raghu, M., Bengio, S., Vinyals, O.: Rapid learning or feature reuse? Towards understanding the effectiveness of MAML. arXiv preprint arXiv:1909.09157 (2019)
23. Ravi, S., Larochelle, H.: Optimization as a model for few-shot learning (2016)
24. Ruan, X., Lin, G., Long, C., Lu, S.: Few-shot fine-grained classification with spatial attentive comparison. Knowledge-Based Systems **218**, 106840 (2021). https://doi.org/10.1016/j.knosys.2021.106840
25. Schroff, F., Kalenichenko, D., Philbin, J.: FaceNet: a unified embedding for face recognition and clustering. In: Proceedings of the IEEE Conference on Computer Vision and Pattern Recognition, pp. 815–823 (2015)
26. Snell, J., Swersky, K., Zemel, R.: Prototypical networks for few-shot learning. In: Advances in Neural Information Processing Systems, vol. 30 (2017)
27. Sohn, K.: Improved deep metric learning with multi-class n-pair loss objective. In: Advances in Neural Information Processing Systems, pp. 1857–1865 (2016)
28. Sung, F., Yang, Y., Zhang, L., Xiang, T., Torr, P.H., Hospedales, T.M.: Learning to compare: Relation network for few-shot learning. In: Proceedings of the IEEE Conference on Computer Vision and Pattern Recognition (CVPR) (2018)
29. Thong, W., Snoek, C.G., Smeulders, A.W.: Cooperative embeddings for instance, attribute and category retrieval. arXiv preprint arXiv:1904.01421 (2019)
30. Triantafillou, E., et al.: Meta-dataset: a dataset of datasets for learning to learn from few examples. arXiv preprint arXiv:1903.03096 (2020)
31. Vinyals, O., Blundell, C., Lillicrap, T., Wierstra, D., et al.: Matching networks for one shot learning. In: Advances in Neural Information Processing Systems, vol. 29, pp. 3630–3638 (2016)
32. Wah, C., Branson, S., Welinder, P., Perona, P., Belongie, S.: The Caltech-UCSD birds-200-2011 dataset. Technical report. CNS-TR-2011-001, California Institute of Technology (2011)
33. Welinder, P., et al.: Caltech-UCSD birds 200. Technical report. CNS-TR-2010-001, California Institute of Technology (2010)

Iterative Mask Filling: An Effective Text Augmentation Method Using Masked Language Modeling

Himmet Toprak Kesgin$^{(\boxtimes)}$![ORCID] and Mehmet Fatih Amasyali ![ORCID]

Department of Computer Engineering, Yildiz Technical University, Istanbul, Turkey
tkesgin@yildiz.edu.tr

Abstract. Data augmentation is an effective technique for improving the performance of machine learning models. However, it has not been explored as extensively in natural language processing (NLP) as it has in computer vision. In this paper, we propose a novel text augmentation method that leverages the Fill-Mask feature of the transformer-based BERT model. Our method involves iteratively masking words in a sentence and replacing them with language model predictions. We have tested our proposed method on various NLP tasks and found it to be effective in many cases. Our results are presented along with a comparison to existing augmentation methods. Experimental results show that our proposed method significantly improves performance, especially on topic classification datasets.

Keywords: text augmentation · data augmentation · mask filling · language modeling

1 Introduction

Training neural networks with larger amounts of data can improve their generalization ability and performance. In fact, increasing the amount of data often has a greater impact on model performance than using a more complex model [4]. However, obtaining large quantities of data can be expensive, especially in supervised learning, where each sample must be labeled.

Data augmentation is a way to increase the amount of training data available without collecting and labeling more data. In machine learning, data augmentation involves generating synthetic data from existing data in order to increase the amount of training data. Data augmentation can serve as a regularizer and improve the performance of machine learning models.

In image datasets, data augmentation involves applying transformations such as rotating, cropping, and changing the brightness of images. Data augmentation has been studied more extensively in computer vision than in natural language processing [23]. Numerous studies have demonstrated the remarkable success of data augmentation on image datasets [8,15,18]. One reason for this; transformations create new, valid images when applied to data. Applying transformations to

A. Ortis et al. (Eds.): ICAETA 2023, CCIS 1983, pp. 450–463, 2024.
https://doi.org/10.1007/978-3-031-50920-9_35

text datasets is not straightforward because they can disrupt syntax, grammatical correctness, and even alter the meaning of the original text. In addition, it is more challenging to preserve the label of an augmented sample in text than in images. These difficulties make it challenging to find effective data augmentation methods for text datasets.

Existing text augmentation methods include back-translation [9], synonym word substitution [20], easy data augmentation (EDA) [26] techniques such as random insertion, random swap, random deletion, or transformer-based text generation techniques such as; GPT-3 [5] or BERT [7].

In this paper, we propose a novel augmentation method that leverages the Fill-Mask feature of the transformer-based BERT model. We have tested our proposed method on a variety of natural language processing (NLP) tasks and found it to be effective in many cases. We compare our proposed method to existing text augmentation methods and present our results. The rest of this paper is organized as follows: Literature review in Sect. 2, methods in Sect. 3, experiments in Sect. 4, discussion and limitations in Sect. 5, and conclusions in Sect. 6.

2 Literature Review

Synonym replacement is one of the automated text augmentation techniques. Synonym replacement is the process of replacing, especially, nouns or verbs with their synonyms from a formal database source. Zhang et al. used the WordNet database for synonym replacement to increase the size of the training dataset [29].

The EDA offers four simple data augmentation operations: random swaps, random insertions, random deletions, and synonym replacements [26]. EDA methods are tested with RNNs and CNNs extensively in their experiments for text classification. Authors found that model performance is significantly improved by EDA techniques, especially for small datasets.

As an alternative to EDA, AEDA (An Easier Data Augmentation) only incorporates random punctuation marks into the original text [10]. The AEDA method is simpler to implement, and it preserves all the input sentence information, since it does not include deleting or replacing. AEDA has shown to improve performance on 5 text classification datasets.

Word embeddings can be used for a similar purpose as well. Instead of replacing words with synonyms from a specific source, With Word2Vec [16] words are replaced with their most similar counterparts [14]. The authors demonstrate that when a formal synonym model is not available, Word2vec-based augmentation can be beneficial.

Text augmentation can also be accomplished using back-translation. Back translation is the translation of a sentence from one language into another and then translating it back into the original. It is shown that the use of back translation resulted in a decrease in overfitting and an improvement over the BLUE score for IWSLT tasks [22].

There have been significant gains across different NLP tasks using transformer-based models, such as BERT [7], GPT-2 [19], and BART [12]. It is possible to use the text generation capabilities of these models for text augmentation [6,11]. The main challenge of these methods is preserving the label of the augmented sentence.

The augmentation of data does not have to be at the textual level. Once the text is represented as a vector, various techniques such as noise injection [27] or mix-up [24,28] can be applied for augmentation. These methods are not specific to text augmentation, but can be used for other data types. These augmentation methods can be applied both with and without text-level augmentation, yet they are not the main topic of our analysis.

We propose an augmentation method based on a transformer-based BERT model. As opposed to the previous methods, it uses the masked language modeling (Fill-Mask) feature instead of directly generating text. The Fill-Mask task involves masking and predicting which words to replace the masks. These models are useful for obtaining statistical understanding of the language in which the model was trained. A detailed description of our proposed augmentation method is can be found in the method section.

3   Methods

Increasing the number of samples in the training set can generally improve the performance of a machine learning model, but the extent of the improvement depends on how different the new samples are from the existing ones, whether they contain noise, and how accurate their labeling is. Adding new examples that are identical to existing ones may not have a significant impact on model training. On the other hand, if the new samples are sufficiently different from the existing ones, the model may be able to learn better decision boundaries. The performance of the model can also be negatively affected by a large amount of input or label noise in new samples.

In data augmentation, it is important that the generated instances are differentiated from the existing ones while still maintaining their labels. We propose Iterative Mask Filling Augmentation, a method that aims to replace existing sentences or paragraphs as much as possible while preserving their meaning and structure. This method uses masked-language modeling (MLM), which has been trained on a large corpus of text. MLMs predict words that have been intentionally hidden within a sentence and provide valuable information for understanding the statistical properties of language. Words can have different meanings in different contexts, so it is important to learn context-dependent representations of each word. In MLM, the context of a sentence is taken into account when generating mask predictions, allowing the model to make confident predictions about which words can replace masked words.

Iterative Mask Fill Augmentation is given in Algorithm 1. In this algorithm, each word in a sentence is replaced with the <Mask> symbol. The MLM model then determines which words can replace the masked word, along with their associated scores. The k hyperparameter of the algorithm determines how many words the new word is chosen from. There is a confidence score associated with each word that can replace the <Mask>, and these scores are normalized to probability values such that their sum is one. The word to be replaced is selected based on these probabilities.

Since the MLM model is trained on a large corpus of text, it suggests only plausible words for the given context. Taking into account the context of the entire sentence, if only one word seems plausible at a given point, it is most likely to be selected due to its high score in comparison to other words. Therefore, not all the words in the sentence will be changed; only those that are reasonable to change will be replaced. In cases where there is a chance of more than one word being plausible for a given point, probabilistic selection is performed, and the word is replaced. At the end of each iteration, an augmented version of the input sentence is created.

Algorithm 1. Iterative Mask Fill

Require:
1: MLM : Pretrained Mask Language Model
2: k : Number of top labels to be returned by MLM
3: $Sent$: Sentence to be augmented

4: **procedure** GENERATE_AUGMENTATION
5:     $tokenized\_sent \leftarrow$ WORD_TOKENIZE$(Sent)$
6:     $l \leftarrow$ LEN$(tokenized\_sent)$
7:     **for** $i \in \{1...l\}$ **do**
8:         $tokenized\_sent[i] \leftarrow$ " $< mask >$ "
9:         $scores, preds \leftarrow$ MLM$(tokenized\_sent[i])$
10:         $word \leftarrow$ SELECT_WORD$(scores, preds, k)$
11:         $tokenized\_sent[i] \leftarrow word$
12:     **return** JOIN$(tokenized\_sent)$
13: **procedure** SELECT_WORD(scores, preds, k)
14:     $sum =$ SUM$(scores)$
15:     $p = scores/sum$
16:     $j \leftarrow$ Choose j from $[1...k]$ with probability p
17:     **return** $preds[j]$

For example, the augmented version of the sentence "We introduce a new language representation model called BERT" produced by the algorithm is "they developed a natural language processing system called BERT". The steps of the algorithm can be seen in Fig. 1.

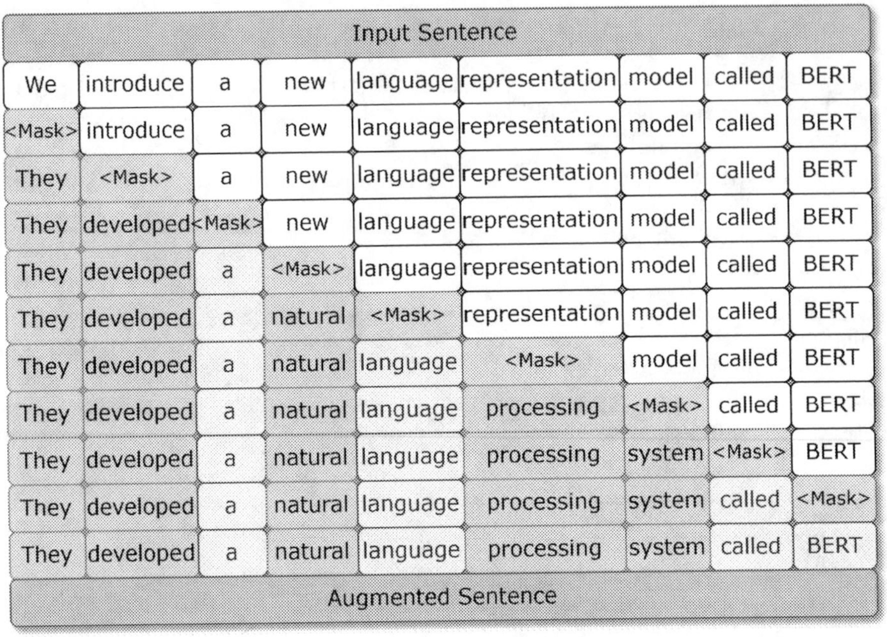

Input Sentence								
We	introduce	a	new	language	representation	model	called	BERT
<Mask>	introduce	a	new	language	representation	model	called	BERT
They	<Mask>	a	new	language	representation	model	called	BERT
They	developed	<Mask>	new	language	representation	model	called	BERT
They	developed	a	<Mask>	language	representation	model	called	BERT
They	developed	a	natural	<Mask>	representation	model	called	BERT
They	developed	a	natural	language	<Mask>	model	called	BERT
They	developed	a	natural	language	processing	<Mask>	called	BERT
They	developed	a	natural	language	processing	system	<Mask>	BERT
They	developed	a	natural	language	processing	system	called	<Mask>
They	developed	a	natural	language	processing	system	called	BERT
Augmented Sentence								

Fig. 1. Example Sentence Augmentation

4   Experiments

In this section, we will introduce the datasets used, describe the details of the experiments we performed, compare augmentation algorithms, and propose several improvements.

4.1   Datasets

We determined text data sets with various tasks for experiments. These datasets are the news category dataset (News) [3,17], financial sentiment analysis dataset (FinSent) [13], twitter sentiment analysis dataset (TwitSent) [2] and the New York Times news (New York). The New York dataset contains 800 news articles for training and 3000 news articles for testing. The News dataset has 10 classes, the New York dataset has 4 classes, and the FinSent and TwitSent datasets have 3 classes each. The datasets News, New York, TwitSent, and FinSent consist of 45000, 800, 74664, and 5843 samples, respectively, although we used subsets of these datasets in our experiments (as shown in Fig. 2).

We created the New York dataset with news articles from the year 2022, sourced from the New York Times website. We did this because existing language models may have been trained on older datasets, which could bias their predictions. However, since these news articles were written after the language models were created, they were not trained on these texts.

The datasets used in our experiments can be grouped into two broad categories: category determination and sentiment analysis. To examine the effects of real training examples on the training dataset, we split the training dataset into subsets of various sizes and tested the performance of the model for each subset size. For each dataset, we used the same test set in all experiments. Figure 2 displays the results of these analyses. By examining the differences between the model's performance on the real training samples and the augmented samples, we can calculate the accuracy gain that would be achieved by adding real training samples to the training dataset. The orange line in Fig. 2 indicates the number of training samples chosen for the experiments. Since augmented sentences cannot be as good as real training examples, these differences represent the potential maximum success of augmentation methods. These analyses allow us to compare different augmentation methods and see their potential gains.

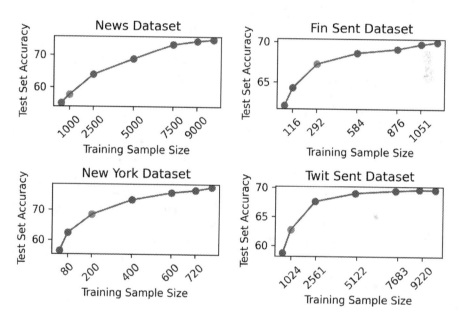

Fig. 2. Training Set Size - Test Set Accuracy Analysis (Color figure online)

4.2 Training Settings

For classification, all texts in the dataset were converted to lowercase and sentence vectors were created using the transformer-based model [1]. Throughout all experiments, neural networks were used as classifiers. First layer of the neural network consist of 384 neurons which are representations of texts. Three hidden layers, consisting of 64, 16, and 4 neurons respectively, are then connected to this layer. Tanh is the activation function of these layers. Output layer of neural network determined by datasets number of classes. The activation function of the last layer is softmax, which gives confidence scores for each class.

In all experiments, we used a hyperparameter k as 5 in the Iterative Mask Fill (IMF) algorithm. This parameter determines how many words are considered as candidates to replace the masked word at each iteration of the IMF algorithm. The value of k can affect the performance of the IMF algorithm, and choosing an appropriate value for k may require experimentation. In general, increasing the value of k may allow the IMF algorithm to consider a wider range of words as candidates to replace the masked word, potentially leading to more diverse augmented sentences. However, a larger value of k may also lead to slower execution of the algorithm and may result in augmented sentences that are less faithful to the original text. In our experiments, we found that a value of $k = 5$ produced good results for the datasets and tasks considered. It is possible that different values of k may be more suitable for different types of data and tasks.

4.3   Comparison of Text Augmentations

The purpose of this section is to test how much the Iterative Mask Fill (IMF) augmentation algorithm improves performance by adding new examples to the training set. We compare the performance of the IMF Augmentation algorithm to several other basic text augmentation algorithms, including random insertion, random swap, random deletion, synonym word substitution, back translation augmentation, and BERT replacement.

Random insertion (ri): This method involves inserting one of the synonyms of a randomly chosen word anywhere in the sentence. Random swap (rs): This method involves replacing two randomly chosen words in a sentence. Random delete (rd): This method involves randomly deleting each word in a sentence based on a probability p. Synonym word substitution: This method involves replacing n words with one of their synonyms among the non-stop words in a sentence. Back translation (bt): This method involves translating a sentence from one language to another and back to the original language. In our experiments, the data sets we used were in English, so we translated the sentences into Turkish and augmented them by translating them back into English. BERT replacement (br): This method involves masking some words in sentences and replacing them with BERT predictions. This method is similar to the non-iterative version of our proposed IMF method, as well as the synonym replacement method, where words are replaced with BERT predictions instead of their synonyms. In all of these methods, we used an alpha ratio of 0.1, which yielded the best performance in our EDA study. The alpha ratio indicates the percentage of words in the sentence that will be changed. Unless it is 0, it guarantees that at least one word will be changed.

In this section, we compare the performance of various methods of text augmentation. Based on the training size-accuracy analysis shown in Fig. 2, we used the training set sizes indicated by the orange line in the figure. For each sample, we included 1 and 4 augmentations using the existing methods. Therefore, the training set size increased by 100% and 400% in each set of experiments. As part of the performance comparisons, we also included the results when real sentences were added to the training dataset. The results of these experiments are shown in Table 1.

Table 1. Comparison of text augmentation methods

	New York	News	Fin Sent	Twit Sent
vanilla	69.46 ± 2.74	58.98 ± 2.19	67.23 ± 0.36	62.25 ± 0.64
real_sample	74.70 ± 1.71	63.35 ± 2.32	68.59 ± 0.53	65.16 ± 0.70
ri (100%)	71.13 ± 1.42	60.30 ± 1.56	64.94 ± 0.59	61.73 ± 0.54
rs (100%)	69.50 ± 2.37	59.03 ± 2.47	64.73 ± 0.75	61.53 ± 0.47
rd (100%)	70.61 ± 2.45	57.81 ± 3.23	64.66 ± 0.62	60.93 ± 0.75
sr (100%)	69.76 ± 2.06	58.17 ± 3.13	65.06 ± 0.59	61.46 ± 0.41
bt (100%)	69.58 ± 2.38	60.73 ± 3.50	66.81 ± 1.34	62.69 ± 0.35
br (100%)	68.58 ± 2.69	58.40 ± 1.47	67.12 ± 0.68	61.82 ± 0.46
imf (100%)	71.43 ± 1.68	60.67 ± 1.92	66.56 ± 0.82	61.75 ± 0.44
ri (400%)	69.05 ± 2.09	59.58 ± 1.73	64.90 ± 0.78	61.29 ± 0.36
rs (400%)	69.92 ± 2.35	60.13 ± 2.27	64.73 ± 0.66	61.52 ± 0.69
rd (400%)	70.41 ± 1.88	59.52 ± 1.68	64.63 ± 0.75	61.63 ± 0.38
sr (400%)	70.69 ± 1.38	60.45 ± 2.81	65.36 ± 0.68	61.99 ± 0.61
br (400%)	69.79 ± 1.53	**61.52 ± 1.71**	67.09 ± 0.33	61.76 ± 0.56
imf (400%)	**71.93 ± 1.58**	**61.25 ± 1.64**	64.47 ± 0.70	60.77 ± 0.78

These results show that the basic text augmentation methods generally improve the vanilla performance on category classification datasets, but not on sentiment analysis datasets. In sentiment analysis, the back translation method only improved performance on the TwitSent dataset. The EDA methods did not improve success much, but random insertion was the most promising among them. The IMF method improved performance on the two category classification datasets but decreased performance on the sentiment analysis datasets. Increasing the number of augmented sentences reduced overall accuracy, which suggests that the augmented sentences may contain label noise.

4.4 Improving Performance of Text Augmentations

While performing data augmentation, it is important to preserve the label of the instance. If the label changes during the augmentation process, it will mislead the model and reduce its performance. On the other hand, if the label is preserved without significantly altering the instance, the effect on the model's performance will be minimal, since the instance's representation in space will not be changed. Among the augmentation methods we used in the experiments, IMF is the one that changes the input sentence the most, since it has the potential to change each word in the sentence.

To visualize the impact of different augmentation methods on sentence representations, we plotted the vector representations of real and augmented texts using 2-dimensional TSNE. For each dataset, we randomly selected 100 representations and plotted the real and augmented sentences with different colors.

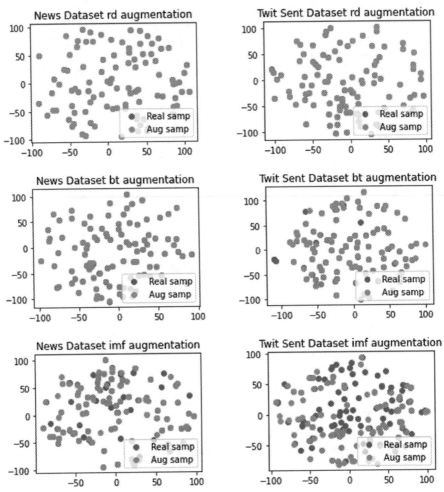

Fig. 3. Real and Augmented Sentences TSNE Representations

Figure 3 illustrates these plots. As shown in Fig. 3, the real and augmented sentences completely overlap for the other methods, but there are differences for the IMF method. These experiments support our claim that sentences augmented with IMF are more different from their original counterparts.

In this section, we also propose a method for filtering the texts generated by augmentation methods, selecting some of them, and including only the selected examples in the training set. To do this, we use a model trained with the vanilla method without augmentation for sample selection. With this model, we calculate the loss values of the augmented sentences and include only the k% of augmented sentences with the lowest loss values in the training set. This method

Table 2. Effect of filtering augmented sentences with low loss

vanilla real_sample	% smallest loss	New York	News	FinSent	TwitSent
		69.46 ± 2.74	58.98 ± 2.19	67.23 ± 0.36	62.25 ± 0.64
		74.70 ± 1.71	63.35 ± 2.32	68.59 ± 0.53	65.16 ± 0.70
ri	100%	71.13 ± 1.42	60.30 ± 1.56	64.94 ± 0.59	61.73 ± 0.54
rs	100%	69.50 ± 2.37	59.03 ± 2.47	64.73 ± 0.75	61.53 ± 0.47
rd	100%	70.61 ± 2.45	57.81 ± 3.23	64.66 ± 0.62	60.93 ± 0.75
sr	100%	69.76 ± 2.06	58.17 ± 3.13	65.06 ± 0.59	61.46 ± 0.41
bt	100%	69.58 ± 2.38	60.73 ± 3.50	66.81 ± 1.34	62.69 ± 0.35
br	100%	69.79 ± 1.53	61.52 ± 1.71	67.09 ± 0.33	61.76 ± 0.56
imf	100%	71.43 ± 1.68	60.67 ± 1.92	66.56 ± 0.82	61.75 ± 0.44
ri	80%	71.98 ± 2.62	61.42 ± 2.06	66.48 ± 0.21	61.81 ± 0.50
rs	80%	72.80 ± 0.51	62.12 ± 1.77	66.06 ± 0.38	61.93 ± 0.51
rd	80%	71.90 ± 1.58	60.74 ± 1.82	66.28 ± 0.35	61.91 ± 0.69
sr	80%	71.01 ± 2.29	60.10 ± 2.90	67.08 ± 0.29	62.05 ± 0.45
bt	80%	69.64 ± 2.29	58.76 ± 2.18	66.88 ± 0.54	63.15 ± 0.69
br	80%	68.44 ± 2.56	62.25 ± 1.21	67.42 ± 0.32	62.27 ± 0.40
imf	80%	**72.88 ± 1.29**	**62.87 ± 2.06**	67.49 ± 0.50	**63.24 ± 0.50**
ri	50%	69.44 ± 2.21	58.59 ± 1.75	66.67 ± 0.74	61.43 ± 0.51
rs	50%	66.71 ± 3.23	56.04 ± 4.37	66.97 ± 0.37	61.71 ± 0.71
rd	50%	68.19 ± 2.08	57.48 ± 2.78	66.89 ± 0.83	61.69 ± 0.90
sr	50%	70.20 ± 2.25	58.77 ± 1.97	67.27 ± 0.31	62.14 ± 0.33
bt	50%	71.04 ± 1.29	58.76 ± 2.09	67.08 ± 0.34	63.03 ± 0.65
br	50%	67.74 ± 1.31	55.77 ± 3.41	67.35 ± 0.27	62.11 ± 0.39
imf	50%	71.29 ± 1.83	62.07 ± 1.63	**67.80 ± 0.17**	62.03 ± 0.52
ri	25%	68.80 ± 2.59	58.18 ± 2.19	66.32 ± 0.96	61.53 ± 0.82
rs	25%	68.80 ± 3.67	56.40 ± 2.82	66.70 ± 0.47	61.61 ± 1.34
rd	25%	68.92 ± 2.40	57.34 ± 2.82	66.93 ± 0.48	61.93 ± 0.81
sr	25%	68.45 ± 3.55	58.07 ± 2.84	66.31 ± 1.54	62.38 ± 0.40
bt	25%	70.00 ± 3.05	59.40 ± 2.48	67.47 ± 0.37	62.72 ± 0.76
br	25%	66.99 ± 4.81	57.45 ± 1.45	67.23 ± 0.31	61.68 ± 0.84
imf	25%	70.26 ± 1.93	59.33 ± 1.44	66.94 ± 1.09	61.36 ± 0.69

allows us to choose only the most realistic and useful augmented examples for training the model.

Table 2 summarizes these results. In Table 2, four sentences were generated for each sentence in the dataset using every method except back translation, which only generates one augmented sentence for each sentence. For the lowest loss rates of 25%, 50%, and 80%, we included the corresponding percentage of augmented sentences in the training set. These results show that including low-loss augmented sentences in the training set significantly improves the performance of the augmentation methods. We believe that low-loss examples are less likely to contain label noise. Filtering low-loss samples, particularly improved the performance of the IMF algorithm, which performed very closely to including real samples in the training set for category classification datasets. Across the datasets and algorithms, including the 80% of augmented sentences with the lowest loss rates, except for FinSent, provided the best results.

4.5   Using Different Language Models

The IMF method we propose uses a masked language model to generate augmented sentences. For each sentence, it uses the predictions of the language model for the number of words in the sentence. As a result, the duration of the augmentation process depends on the number of words in the sentence and the prediction time of the model. Very large models, such as BERT, can take a long time to generate predictions because they have a large number of parameters. Therefore, in this section, we perform augmentation using smaller versions of BERT, namely distilBERT [21] and tinyBERT [25], and compare their performance to the original BERT model. Table 3 summarizes these results.

Table 3. Comparison of different language models for the IMF

MLM	param_count	time	New York	News	FinSent	TwitSent
Bert	110m	240 s	72.88 ± 1.29	62.87 ± 2.06	67.49 ± 0.50	63.24 ± 0.50
DistilBert	66m	156 s	72.36 ± 1.51	63.70 ± 1.51	66.85 ± 0.50	62.00 ± 0.39
TinyBert	4m	23 s	72.38 ± 1.63	61.25 ± 2.10	66.50 ± 0.38	61.88 ± 0.52

In Table 3, the param_count column indicates the number of parameters in the model, and the time column shows the time in seconds required for 100 sentence augmentations in the News dataset. In general, as the number of parameters in the language model decreases, the performance of the augmented sentences also decreases. However, reducing the size of the language model significantly speeds up mask estimation. This trade-off between performance and speed can be taken into consideration when selecting a language model for text augmentation. There is a close to linear relationship between the number of model parameters and the time required for mask estimation.

5   Discussion and Limitations

Previous methods proposed for text augmentation aim to enrich the dataset by making simple changes to the text. These methods are not very effective at improving model performance. The IMF method that we propose can increase the performance of augmented sentences more than other simple methods, but it requires a language model and can take a long time to generate augmented sentences, especially for longer texts such as paragraphs. As mentioned in the previous section, the algorithm can be accelerated by using a language model with fewer parameters. In our experiments, we found that the IMF method is more suitable for sentence-based augmentation and is more effective for news texts than for sentiment analysis datasets. This is because the sentiment of a sentence can change completely when a single word is changed, while the meaning of a news text is less sensitive to changes in individual words. Additionally, for all augmentation methods, as the size of the real dataset increases, the performance obtained with augmented texts decreases. This is because even when using real data, model performance changes very little as the dataset size increases. Therefore, augmentation methods are most effective when the dataset size is small.

6   Conclusions

In this paper, we propose a new text augmentation method using a MLM. We compare its performance with other augmentation methods on two news classification and two sentiment analysis datasets. Our results show that it can significantly improve the performance, especially in news classification datasets with a small number of training samples. We also propose a simple filtering process for augmented sentences to preserve their labels. We observe that existing augmentation methods do not significantly improve the performance of sentiment analysis tasks. One of the main challenges in text augmentation is to sufficiently modify the text while preserving its labeling. In future work, we aim to propose new augmentation methods using language models trained for different tasks that address these limitations. We also plan to study the effects of online text augmentation in future work.

Acknowledgment. This study was supported by the Scientific and Technological Research Council of Turkey (TUBITAK) Grant No: 120E100.

References

1. sentence-transformers/all-minilm-l6-v2 · hugging face (2022). https://huggingface.co/sentence-transformers/all-MiniLM-L6-v2
2. Twitter sentiment analysis | kaggle (2022). https://www.kaggle.com/datasets/jp797498e/twitter-entity-sentiment-analysis
3. News category dataset | kaggle (2023). https://www.kaggle.com/datasets/setseries/news-category-dataset
4. Brants, T., Popat, A.C., Xu, P., Och, F.J., Dean, J.: Large language models in machine translation (2007)
5. Brown, T., et al.: Language models are few-shot learners. In: Advance in Neural Information Processing System, vol. 33, pp. 1877–1901 (2020)
6. Coulombe, C.: Text data augmentation made simple by leveraging NLP cloud APIS. arXiv preprint arXiv:1812.04718 (2018)

7. Devlin, J., Chang, M.W., Lee, K., Toutanova, K.: Bert: Pre-training of deep bidirectional transformers for language understanding. arXiv preprint arXiv:1810.04805 (2018)
8. Fawzi, A., Samulowitz, H., Turaga, D., Frossard, P.: Adaptive data augmentation for image classification. In: 2016 IEEE International Conference on Image Processing (ICIP), pp. 3688–3692. IEEE (2016)
9. Hayashi, T., et al.: Back-translation-style data augmentation for end-to-end ASR. In: 2018 IEEE Spoken Language Technology Workshop (SLT), pp. 426–433. IEEE (2018)
10. Karimi, A., Rossi, L., Prati, A.: AEDA: an easier data augmentation technique for text classification. arXiv preprint arXiv:2108.13230 (2021)
11. Kumar, V., Choudhary, A., Cho, E.: Data augmentation using pre-trained transformer models. arXiv preprint arXiv:2003.02245 (2020)
12. Lewis, M., et al.: Bart: denoising sequence-to-sequence pre-training for natural language generation, translation, and comprehension. arXiv preprint arXiv:1910.13461 (2019)
13. Malo, P., Sinha, A., Korhonen, P., Wallenius, J., Takala, P.: Good debt or bad debt: detecting semantic orientations in economic texts. J. Am. Soc. Inf. Sci. **65**(4), 782–796 (2014)
14. Marivate, V., Sefara, T.: Improving short text classification through global augmentation methods. In: Holzinger, A., Kieseberg, P., Tjoa, A.M., Weippl, E. (eds.) CD-MAKE 2020. LNCS, vol. 12279, pp. 385–399. Springer, Cham (2020). https://doi.org/10.1007/978-3-030-57321-8_21
15. Mikołajczyk, A., Grochowski, M.: Data augmentation for improving deep learning in image classification problem. In: 2018 International Interdisciplinary PhD Workshop (IIPhDW), pp. 117–122. IEEE (2018)
16. Mikolov, T., Chen, K., Corrado, G., Dean, J.: Efficient estimation of word representations in vector space. arXiv preprint arXiv:1301.3781 (2013)
17. Misra, R.: News category dataset (2018). https://doi.org/10.13140/RG.2.2.20331.18729
18. Perez, L., Wang, J.: The effectiveness of data augmentation in image classification using deep learning. arXiv preprint arXiv:1712.04621 (2017)
19. Radford, A., Wu, J., Child, R., Luan, D., Amodei, D., Sutskever, I., et al.: Language models are unsupervised multitask learners. OpenAI Blog **1**(8), 9 (2019)
20. Rizos, G., Hemker, K., Schuller, B.: Augment to prevent: short-text data augmentation in deep learning for hate-speech classification. In: Proceedings of the 28th ACM International Conference on Information and Knowledge Management, pp. 991–1000 (2019)
21. Sanh, V., Debut, L., Chaumond, J., Wolf, T.: Distilbert, a distilled version of Bert: smaller, faster, cheaper and lighter. arXiv preprint arXiv:1910.01108 (2019)
22. Sennrich, R., Haddow, B., Birch, A.: Improving neural machine translation models with monolingual data. arXiv preprint arXiv:1511.06709 (2015)
23. Shorten, C., Khoshgoftaar, T.M., Furht, B.: Text data augmentation for deep learning. J. Big Data **8**(1), 1–34 (2021)
24. Sun, L., Xia, C., Yin, W., Liang, T., Yu, P.S., He, L.: Mixup-transformer: dynamic data augmentation for NLP tasks. arXiv preprint arXiv:2010.02394 (2020)
25. Turc, I., Chang, M.W., Lee, K., Toutanova, K.: Well-read students learn better: on the importance of pre-training compact models. arXiv preprint arXiv:1908.08962 (2019)
26. Wei, J., Zou, K.: Eda: Easy data augmentation techniques for boosting performance on text classification tasks. arXiv preprint arXiv:1901.11196 (2019)

27. Xie, Z., Wang, S.I., Li, J., Lévy, D., Nie, A., Jurafsky, D., Ng, A.Y.: Data noising as smoothing in neural network language models. arXiv preprint arXiv:1703.02573 (2017)
28. Zhang, H., Cisse, M., Dauphin, Y.N., Lopez-Paz, D.: Mixup: beyond empirical risk minimization. arXiv preprint arXiv:1710.09412 (2017)
29. Zhang, X., Zhao, J., LeCun, Y.: Character-level convolutional networks for text classification. In: Advances in Neural Information Processing Systems, vol. 28 (2015)

Sub Data Path Filtering Protocol for Subscription of Event Parts and Event Regeneration in Pub/Sub Pattern

Serif Inanir and Yilmaz Kemal Yuce[✉]

Alanya Alaaddin Keykubat University, Alanya, Antalya, Türkiye
yilmaz.yuce@alanya.edu.tr

Abstract. Pub/Sub is a common pattern allowing a producer to publish events to consumers. In types of Pub/Sub, structure of an event is either identified by publishers based-on static rules or by consumers based-on filtering approaches. In both scenarios, actors' total performance might get degraded due to required operations (e.g., filtering) impacting throughput. This study focuses on designing a filtering approach for both actors of Pub/Sub by reducing data size to be transmitted by producers and received and processed by consumers by creating a loosely coupled context, in which horizontal alterations to structure of any event can occur. Sub Data Path (SDP) approach presents a matching tree to separate an event with scope like JSON data, and each key in the relevant event act like a topic without being defined as a topic. Thereby, producers only must transmit part of a message through a path on the event structure to be located into a former event to create new event; consumers can subscribe to any subtopic (key for JSON format) to be able to receive data in terms of its own mechanism, not a producer's design. Therefore, creating an event can be completed with different producers which contribute a piece of the whole event; Bounded Context structure belongs to microservice architecture as a decomposition strategy can be handled by consumers in relation to their own business logic. To measure the proposed method, an experiment with gaze points collected by an eye tracker has been designed. By performing the filtering method for one, two and maximum SDP keys, filtering duration, event size reduction percent and transmission duration were revealed. The experimental results imply that the proposed method can send 7.5 events in average, instead of sending just one in the same period. Also, since worst case of the proposed method based-on events in the context can be calculated, an architecture can be prevented from bottlenecks. These benefits makes SDP advantageous over similar methods in terms of being both a fast and scalable alternative.

Keywords: Publisher/Subscriber · broker-system · loose-coupling · filtering · performance

A. Ortis et al. (Eds.): ICAETA 2023, CCIS 1983, pp. 464–481, 2024.
https://doi.org/10.1007/978-3-031-50920-9_36

1 Introduction

In today's world, with Industry 4.0, Internet of Things, smart city infrastructures and end-user applications becoming a necessity, devices with computing features need to communicate with each other efficiently and in accordance with defined rules [1]. Consequently, communication patterns that can create the necessary context for the harmonious operation of different systems have been defined and it is aimed to find solutions to the problems in the field of computing. Among these patterns, Publisher/Subscriber (Pub/Sub) is an accepted communication paradigm with the problems it focuses on solving [2, 3].

The Pub/Sub pattern acts as an intermediate layer for communication between producer (Pub) and consumer (Sub) clients, creating a dynamic, flexible, and loosely coupled asynchronous communication context [1, 4]. With the establishment of this structure, one-to-many or many-to-many communication can be implemented using intermediary systems (also called message processing software, broker, or Pub/Sub engine [5]) [6, 7].

Basically, by using Pub/Sub with intermediary systems (eg Apache Kafka [8]) or alternative architectures developed with software libraries (eg ZeroMQ [9]);

- Pub clients can send an event (message) to the communication context.
- Sub clients can subscribe to the type of event they want to receive and receive the events they subscribe to.

Additionally, Pub and Sub clients communicate anonymously: Sub clients cannot talk to the Pub client that creates the event they receive; Pub clients cannot see (identify) Sub clients accessing their events. Therefore, it is not possible for Pub clients to specifically send an event to a Sub client of their choice and set the condition for all Sub clients to receive the current event to create a new event. However, Pub/Sub is a flexible pattern that can be adapted to a particular part of an architecture and customized to suit requirements. In this respect, it works in a platform-dependent manner with the protocols of the systems used [5]. The following possibilities can be achieved that distinguish the Pub/Sub pattern from other design patterns [10–12]:

- Loose coupling between communication stakeholders in terms of resource, time, and concurrency
- Possibility of high and low level abstraction for different process groups
- Ability to replicate individual data structures across different processes
- Inter-actor division of work for failure recovery and system reconfiguration
- Generating the most stable syslogs before failure
- Distributed data structures and mechanisms to control them
- Filtering events based on individual consumer needs

Filtering events for on-demand consumption is about Sub clients subscribing to the type of event they want. Hence, the Sub client can get the data it is interested in instead of getting entire data in the context of communication. In this respect, event filtering prevents transmission of events deemed unnecessary and prevents clients from creating their own filtering mechanisms. Event filtering aims to reduce the use of network capacity and system resources at clients. Accordingly, performance compatibility between systems that do not know each other can be achieved. By filtering events, the subscription structure of a Pub/Sub context can be set up in three ways: topic-based (also known as group-based or channel-based), content-based, and type-based [6].

Topic-based filtering is a structure used for structured/semi-structured characterization and categorization of event content. Consumers can subscribe to a particular topic, thus, forming a group. In this way, the messages (publications) of the producers on a particular subject are distributed to the members of the consumer group created for the relevant subject [4, 6]. For example, let's say a group of consumers subscribe to a topic called "alive". Events shared by producers with the tag "alive" are to be received by the respective group consumers. Another possible model based on topic-based filtering is the establishment of a hierarchical structure among topics. This model is called hierarchy-based filtering and is accepted as a derived type of topic-based filtering [13]. For example, let the subject "human" be determined as the sub-topic of "alive". Events published with the theme "alive" are distributed among both "alive" and "human", while events with the topic "human" are going to be distributed only among the "human" consumer group [11]. Topic-based filtering (hence also hierarchy-based filtering) can be ideal for transmission, especially between systems that know each other. However, for systems that do not know each other or are loosely coupled, it may result in transmission of more than the desired information [14]. On the other hand, creating consumer groups in topic-based filtering is simple. Consumers who subscribe to a common topic all together represent a group [3, 4].

Content-based filtering is a structure used to categorize and query an event within itself, and events are not separated from each other by categorizing them in relation to their subject. Instead, transmission occurs when the content of an event satisfies a certain condition for the consumer [3, 6]. Therefore, conditional expressions can be defined by consumers using operators such as " $==$ ", " $>$ ", " $<$ ". Conditions and related operations in content-based filtering can be implemented in three different ways: with a string, with a template object, and with executable pieces of code [11]. With these methods, even if only a certain part of the event is meaningful to consumer (for example, if the "price" field in an event is greater than 90 or the "city" field begins, ends with the letter "i" in an expression written in SQL or a similar language), the entire event is transmitted to relevant consumer. There might cases where messages representing entire event may be unnecessary. Such messages are forwarded in topic-based filtering, whereas they are filtered, thus, not forwarded in content-based filtering. Therefore, it can help reduce network capacity use [14]. However, a formal language suitable for the subscription system is required to define equality in use of transmission capacity. On the other hand, with the use of formal languages, complex subscription operations can seriously slow down system. The fact that consumers can define an infinite or more than necessary number of equations and complex filtering requests might easily hinder the scalability of the system [14]. In addition, an effective solution for the creation of scalable and high-performance consumer groups in content-based filtering has not yet been designed [3]. The main reason behind it is that propositions to be made for the same data of interest can be expressed using many different conditional statements [4].

Type-based filtering is based on data and object type definitions and references class, interface, and similar concepts (abstract data type) of object-oriented languages (basically based on the principles of the Java language). Accordingly, the sub-parts that make up the event are objects [14]. A well-defined protocol should be used between systems that do not know each other, since the mechanisms used in the identification of events must be known. Considering the tight coupling between communication stakeholders, it can negatively affect the flexibility of certain architectures. On the other hand, when used with non-object-oriented languages, there may be problems in defining the class or similar structure that realizes subscription.

Hence, problems related to a Pub/Sub communication context include.

- Transmission of more than the necessary or desired information,
- Unnecessary transmission of the whole event rather than only the changed parts,
- Complex subscription operations causing the system to slow down and throughput to degrade,
- Tightly coupled communication stakeholders' potential negative impact on the flexibility of communication architecture.

In this study, a novel model called Sub Data Path (SDP) is proposed for Pub/Sub pattern. SDP tries to establish a standard for reducing the transaction and transmission cost of both parties during the group communication of producers and consumers. By defining a special protocol within the framework of the Pub/Sub pattern for the established communication context, SDP makes it possible to perform operations and indexing on its sub-parts instead of the whole event, thereby, making the transmission of only the necessary information and overcoming the issue of unnecessary transmission of the whole event. SDP also allows multiple producers to contribute to generating an event by producing a subset of event components/fields. Considering such abilities, here we report on SDP's impact on system performance based on throughput and other basic measures, i.e., Average Capacity Gain (ACG) and operational durations compared to types of Pub/Sub pattern in operation. To this purpose, experiments were conducted based on scenarios using an existing eye-tracking product, which employs Pub/Sub pattern for specific functioning.

The organization of this article is as follows; In the 2nd chapter named "Materials and Methods", the design of the SDP, its principles and architecture with an intermediary system are explained, while in the 3rd chapter named "Use Case and Performance Evaluation", the performance of SDP is calculated through a sample scenario. In the 4th section, "Conclusion", an assessment of the performance gave in results is investigated to reveal the benefits of SDP compared to other Pub/Sub models. In the last section, "Discussion", there are SDP properties based on obtained results, and a short overview for its comparison.

2 Literature

From its early studies to the present day, Pub/Sub methodology has increased its importance by diversifying both in Machine-to-Machine architecture and standards, as well as in the fields where it is used.

There exist many different implementations of Pub/Sub available, such as IBM Gryphon [15], Microsoft Scribe [16], Bayeux [17], Siena [18], Narada Brokering [19], XMessages [20], Echo [21] and others [22–25], dating back to years between 1990 and 2000. There are examples of early application software [4]. They used predominantly topic-based filtering, although they varied topologically (e.g., centralized and peer-to-peer). Studies using content-based filtering may have scalability and performance deficiencies, especially when there are devices that do not know each other in the architecture. In addition, in early applications, similar to the popular applications used today, various algorithms for fault tolerance can be found to maintain real-time communication [23, 25]. The popular systems and applications used today include Apache Kafka [8, 26], RabbitMQ [27, 28], Apache ActiveMQ [29], Apache RocketMQ [30], Amazon SQS [31], Google Cloud Pub/Sub [32], Eclipse Mosquitto [33], IBM MQ [34] and ZeroMQ [9] as a software library example.

In recent studies, distributed architecture designs that try to prioritize Pub/Sub pattern's fault tolerance and suitability against IoT have been revealed, by using different filtering-like approaches [35–40]. Architectures are generally developed in accordance with topological requirements to decide how to achieve minimum data loss, as well as some protocols defining the transmission between pub-to-broker-to-sub or decentralized clients to reduce data size [35–38]. As a different example, in a study with an innovative aspect, used its own distributed architecture to put events with only same topic into same broker [36]. In this structure, decentralized clients connect to another client which has got the flag showing the desired topic. Thereby, filtering operation to be done on the broker system can be reduced, because a broker houses just one topic together with their consumers, not all topics and their potential consumers. Different studies have small, many sub applications not housed by a broker design, and named as subsystem, by putting the heart of microservice architecture forth [40–43]. Hence, modules such as subscription registration, event storage, event transmission are expressed as different functioning components. While many studies focus all modules of a monolithic structure, there are some studies focus on a module such as storage module of a broker system supporting Pub/Sub. Rodriguez et.al also purposes to design an independent storage module working together with the broker [42]. In a similar effort to our study, Nasirifard et.al proposed a new methodology named as function-based matching, which they applied in microservice-like decentralized architecture [40]. In their design, subscription mechanism requires a source code, which is a filtering function and upon receiving the source code, the broker starts to filter using the filtering function. However, just like content-based filtering, we think this operation can only be effective on systems that know each other well. Otherwise, there is a potential for various bottlenecks and delays. Except for these, some researchers have also focused on decreasing security concerns in a communication context created with Pub/Sub [44]. Therefore, a privy and secure ground can be prepared for IoT, the topic of daily life of the future.

3   Methodology

In this section, the details of the SDP model, which tries to create a new standardization for the Pub/Sub pattern, and its structure on a central server topology are presented.

The focus of the SDP is to use filtering mechanisms not only for consumers, but to design filtering for both consumers and producers. Having this approach lets filtering to be used with a subscription strategy for consumers and a new event generation model/strategy for producers. Thus, it is aimed that both client types (producer and consumer) use less network capacity and resources during communication. In the model, Pub clients *(i)* can send the changed parts (fields/components) of the previous event instead of the whole new event and create a new event or *(ii)* can contribute to a piece of an event that will be created by multiple producers; instead of getting the entire event on a topic of interest, sub clients can subscribe to the relevant sub-segment of the event and receive only the fragment of topic of interest in accordance to their business logic design which defines the scope of an entity in the terms of Bounded Context as in microservice architecture [45]. Subscription and event creation primitives require that the fragment of an event be accessible to be able to read and write. Within this perspective, events in the context of communication are subjected to Boolean algebra operations through a tree structure and filtered based on content.

3.1 Sub Data Path in Pub/Sub Communication Pattern

By applying the SDP model to Pub-type actors, the parts of the scoped data (for example, JSON format or similar non-primitive [string, integer, etc.] non-data structure) previously transmitted to the intermediary system (for example, JSON format < key: The path information (key from value > pair) and the new data to be replaced with the previous part (the value from < key:value > of JSON format) must be received by the intermediary system [46].

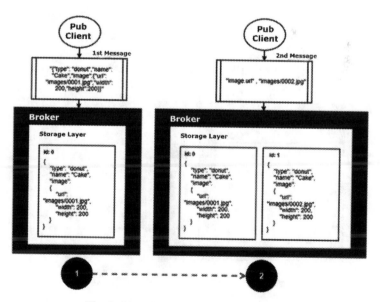

Fig. 1. Use of Sub Data Path in Pub clients.

An example scenario for using SDP for Pub client is given in Fig. 1. Accordingly, two messages were sent by the Pub client to the intermediary system [47].

After the transmission, the event information was kept in a queue structure in the storage layer in the intermediary system, i.e., the broker. The first message is a structured event in JSON format. It has the tag id:0 as the first element of the queue in the corresponding event storage layer. The second message is composed of two separate parts and contains the identifier and update value of the sub-part to be changed on the id:0 data. The value of "image.url" in the second message indicates that the "url" part (key) that is the sub-scope of "image", which is the sub-scope of the id:0 event, will be changed, and "images/0002.jpg" indicates the last value. The broker system matches the second message it received by creating a copy of the first message and adds it to the queue with the id:1 tag on the storage layer. Thus, a new event is created by simply performing a change request on the old data. Also, the concept of creating an event with multi-producers can be similarly implemented by removing the first message belongs to the corresponding structure. In this point, the event structure can be defined on the broker.

By the application of SDP model to Sub-type actors, an access identifier must be received by the intermediary system (broker) to reach the data pieces from which the events are desired to be filtered. The related mechanism, by incorporating topic-based filtering into content-based filtering, ensures that the topics are inferred from the events and aims to serve as a topic/group by identifying the sub-parts that make up the event. Moreover, topic-based filtering can also be used in this approach. In other words, events can be categorized by topic and sub-segments can be subscribed to.

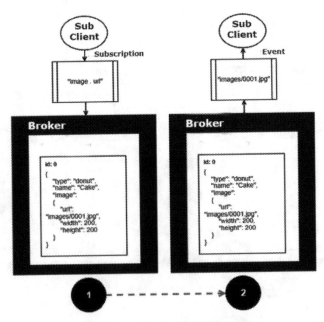

Fig. 2. Use of Sub Data Path in Sub clients.

An example scenario for using SDP for Sub client is given in Fig. 2. Given in the figure, there is an event configured using JSON with the id:0 in the storage layer on the intermediary system; a message regarding the subscription request has been sent from the sub client to the broker.

Subscription request forwarded is "image.url" as a single piece and it is a rule that the equivalent of the "url" subject is sent under the "image" subject in the event to be transmitted to subscriber by the intermediary system. The event message transmitted from the broker system to the Sub client was generated as an output of filtering the "image" and "url" fields in the "image" in the id:0 event.

3.2 Sub Data Path in Broker

The SDP model should be configured in compliance with the intermediary system to be used. Therefore, it must act in conjunction with existing protocol and API designs in a platform-dependent manner. In addition, SDP can be customized and adapted to the needs. There is an example broker system architecture in Fig. 3, in which SDP is used. The relevant architecture shows the minimum architectural requirements of SDP. The stated requirements are described below.,

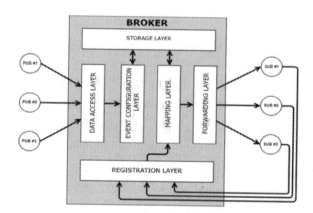

Fig. 3. Use of Sub Data Path in Sub clients.

Data Access Layer: All normal or SDP transmissions received from Pub clients are aggregated within this layer and transferred to the Event Configuration Layer.

Event Configuration Layer: It is checked whether the received message is created with SDP mechanisms. If the message is native SDP message, the requested old event of the Pub client that owns the message (for example, the previous or any subsequent event according to the designed protocol) is retrieved from the Storage Layer. The desired sub piece in the old event received is assigned the value in the new message just received. The new event created is sent to the Storage Layer. However, if the message is not native SDP, it can be sent directly to the Storage Layer and included in the relevant data model, saying that the message is already an event. Consequently, this layer can only output events.

Storage Layer. It is the event storage layer that ensures that any event data received from the Configuration Layer is kept in the specified data structure (e.g., queue) and shares it with other layers whenever it is asked for it.

Registration Layer. All native SDP messages received from sub clients are collected by the intermediary system over this layer and passed to the Mapping Layer. SDP transmissions can be of two different types. The first is that a native SDP message received is only valid for a single event. The second is that a native SDP received can be used continuously from the moment it is received or until a request declaring that message is not valid anymore is received. However, these and similar methods should be created according to the purposes of using SDP and the protocols of the platform to be developed and used.

Mapping Layer. Events received from the Storage Layer are fragmented in this layer using the SDP information of consumers and adapted to the form appropriate to the subscription topics. For the fragmentation process, it is necessary (i) to know the standard (for example, JSON) using which the event is created, and (ii) to create a hierarchical graph, as in Graph Theory, that will express its sub-parts. Through the hierarchical graph, by computing Boolean Algebra expressions (operations) nodes are traversed and visited, and filtering is realized to send the requested node or leaf to the relevant consumer. Filtered fields can be sent to the Forwarding Layer and shared with relevant consumers.

Forwarding Layer. Consumer identifiers and event messages to be sent to the relevant consumers passed from the Mapping Layer are transmitted to Sub clients via this layer using methods such as unacknowledged or reliable acknowledged transmission.

3.3 Event Filtering Based on Graph Computations

The graph structure is obtained by dividing the extensive events in the intermediary system into sub-parts. Each fragment has a unique identifier. Below is an example event in JSON format.

$$\text{JSON format: } \{\text{'T1'}:\{\text{'T2'}:\{\text{'T5'}:\text{'L1'}\},\text{'T3'}:\{\text{'T6'}:L2,\text{'T7'}:\text{Null}\}, \text{'T4'}:\text{'L3'}\}\} \qquad (1)$$

There is a graph representation in Fig. 4 using the related event. In relation with Fig. 4, the identifier might represent one of the following three: A new subpart: data block with scope, i.e., node (T); Atomic value: primitive data type (including string), i.e., leaf (L); Undefined message: null value.

Creating hierarchical layers as in Fig. 4 enables traversing and read/write operations on each entity (e.g., node, leaf). The SDP definitions received from Pub clients to over-write the event graph and from Sub clients to read from are used here. Typically, the SDP could be defined as follows:

- Processing Part: = Group1 ∧ Group2 ∧ Group3 ∧... ∧ GroupN

Group 1 corresponds to the traditional topic used to categorize topics. Although not required, it can be used as it can reduce the number of events for which the subscription can be specified. This approach forms a hierarchy and ranks groups as part of each other and requires that any group becomes a part of another group for an event piece to be received. Accordingly, the following condition must hold true:

- GroupN ∈ Group(N-1) - > GroupN
- Event Piece: = Group1 ∧ (Group1 ∈ Group2) ∧... ∧ (Group(N-1) ∈ GroupN)

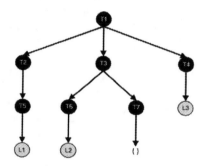

Fig. 4. Representation of events with their sub pieces as a graph.

ASCII characters can be preferred instead of Boolean Algebra operation symbols with the following structure since a formal expression is needed to use the corresponding logic rule in the implementation step. By reason of this, an expression that can be formed using the '.' instead of the AND symbol. With this structure, sample processing points (node or leaf to be used in read and write operations) can be extracted to receive a desired scope or atomic message over the hierarchical diagram in the given JSON format at (1):

- *T5:* = *T1.T2* and *L2:* = *T1.T3.T6* and *L3:* = *T1.T4*.

The designed SDP structures can have two forms as simple and compound. The transaction points T5, L2 and L3 given above are in simple form. The compound form is composed of simple forms to create more than one transaction point request. The composite structures are shown to reach the T5, L2 and L3 processing points given below:

- T5 ∧ L2 ∧ L3: = (T1 ∧ T2) ∧ (T1 ∧ T3 ∧ T6) ∧ (T1 ∧ T4)
- T5 ∧ L2 ∧ L3: = T1 ∧ (T2 ∧ (T3 ∧ T6) ∧ T4)

ASCII characters may be preferred over logic symbols with the following structure, since the compound structure needs a formal expression to be used in the implementation step:

- T5 ∧ L2 ∧ L3: = T1,T2,T3.T6;T4

The above notation replaces the symbols '∧' and '(' with ','; ')' and '∧' instead of ';'; it only uses the '.' character instead of the '∧' symbol and thus can express the route priorities and descriptions in a shorter way. By using 4 symbols in the second expression instead of 8 symbols in the first expression, the SDP message size was reduced by 50% for the relevant example within the scope of symbols. After this step, T1,T2,T3.T6;T4 message can be sent to the intermediary system for SDP that will act at T5, L2 and L3 points.

In order to detect the processing points in the SDP message by the intermediary system, the expression should be simplified according to the character equivalent used in the notation and the value of the last group point found should be obtained. The formula (2) shows the method of processing point detection after opening the compound SDP notation "T1,T2,T3.T6;T4" as T1 ∧ (T2 ∧ (T3 ∧ T6) ∧ T4).

$$\hspace{11cm} (2)$$

1. T1 ∧ (T2 ∧ (T3 ∧ T6) ∧ T4)
2. (T1 ∧ T2) ∧ (T1 ∧ (T3 ∧ T6)) ∧ (T1 ∧ T4)   // Distribution (1st)
3. T1 ∧ T2 -> T5   // Simplification (2nd)
4. T5   // Modus Ponens (3rd)
5. T1 ∧ T3 ∧ T6 -> L2   // Simplification (2nd)
6. L2   // Modus Ponens (5th)
7. T1 ∧ T4 -> L3   // Simplification (2nd)
8. L3   // Modus Ponens (7th)
   Output is T5 ∧ L2 ∧ L3

3.4   Broker and Library for SDP

For SDP to be implemented as a communication pattern, a library was developed in C# and Python programming languages. The Broker was implemented based on the given architecture in Python programming language and is available as an executable. The relevant projects can be reached via the following URL github.com/Serif-NNR/{DRI-Broker-Python, DRI-Lib-C-Sharp, DRI-Lib-Python}.

4   Use Case Performance Evaluation and Metrics

The event that the sub client is interested in and the size of the sub part it wants to retrieve may vary. At the same time, the current event, and the size of the sub part that the Pub client will use to update that event may change. For this reason, the data size that the SDP model requires, and the transmission performance of SDP may differ with respect to the configuration of the system used. Nevertheless, an example scenario and its results are discussed under this section. In the scenario, using a monocular eye tracking system called Pupil Lab Core, a heat map is to be created based on data collected from Pupil Capture v1.7.42 (PC) software [48]. PC is a multi-process architecture application and has a Pub/Sub communication backbone created using ZeroMQ [9] for processes to communicate. One of the processes collects eye tracking data (events) and publishes data in the Pub/Sub backbone for other processes collect and process. Even a separate process can subscribe and collect eye tracking data through this backbone for processing the data for its purpose. PC software contains different plug-ins, too, such as fixation detection (fix) and blink detection (blink) and can change an event content (keys and their hierarchy) according to plug-in activation status. In Fig. 5, an event collected with "surface" topic from PC is shown. The relevant event has been obtained without plug-ins which can increase the event size. Hence, the event can be classified and named as "base" event.

For analyses, we enable fixation and blink plug-ins sequentially and receive 1000 event for each event type in accordance with enabled plug-in combinations. Therefore, 4 event types were defined: base event, base with fixation, base with blink, and base with fixation & blink. The dataset has got 4000 events, 1000 event for each type. By using the obtained events, filtering duration and capacity gain (event size reduction percent) were measured as well as transmission duration with and without SDP event- event without SDP refers to the normal event for which no filtering operation was performed. However, since analyses for filtering and transmission are related with number of total subscriptions, not number of subscribers, measures based-on these subscriptions have been made: 1 SDP for each subscribable SDP keys singularly, Max SDP for the all subscribable SDP keys, and 2 SDP for creating a real-world project- a heat map. In a heat map, gaze points and their confidence amounts are significant and can be used to be able to perform such project as minimum requirements. Therefore, "gaze_on_srf.norm_pos" for coordination points on a surface such as a monitor, and "confidence" for their reliabilities were prepared as SDP phareses.

To calculate results correctly, we have indicated average amounts by calculating filtering durations 10.000 times, and transmission duration 100 times for each event. These operations were completed with a laptop which has Windows 10 OS, Nvidia GeForce RTX 3050 Laptop GPU and AMD Ryzen 7 5800H CPU, and with a program implemented in Python 3.8 and working as a single thread for broker side. As a note, in the experiments, consumer-side which just takes the event from broker works as a different process.

{"name": "unnamed", "uid": "1608920573.1084096", "m_to_screen": [[0.3808582575211248, -0.02668332410924973, 0.22680017352104195], [-0.07653411613944305, 0.4183003268912888, 0.1863123774528504], [-0.11989705918476559, -0.05684279974425613, 1.0]], "m_from_screen": [[2.7270824446361184, 0.08769176603849999, -0.6348408330633314], [0.34460153219525336, 2.59457283089543, -0.5615567198961999], [0.34655728115241313, 0.1579967687112137, 1.0]], "gaze_on_srf": [{"topic": "gaze.3d.0._on_surface", "norm_pos": [0.5431915603097689, 0.7655510955768282], "confidence": 0.997218687397818, "on_srf": "true", "base_data": {"topic": "gaze.3d.0.", "norm_pos": [0.46362095185488955, 0.5216428851201884], "eye_center_3d": [143.69992230572294, -175.93539680581722, 27.46961718359986], "gaze_normal_3d": [-0.3515501765528475, 0.3283953503539252, 0.8766806529355972], "gaze_point_3d": [-32.075165970700795, -11.737721628854601, 465.8099436513985], "confidence": 0.997218687397818, "timestamp": 125655.055173, "base_data": [{"topic": "pupil", "circle_3d": {"center": [-1.353598301472545, 9.199274426753385, 90.06375215327554], "normal": [-0.3454077923041379, 0.39413387588136567, -0.8516759623814176], "radius": 3.3091859004273294}, "confidence": 0.997218687397818, "timestamp": 125655.055173, "diameter_3d": 6.618371800854659, "ellipse": {"center": [86.92467705778508, 159.07034102538995], "axes": [36.70110230082387, 45.62527928325469], "angle": -52.61548801098084}, "norm_pos": [0.45273269300929725, 0.1715086404927607], "diameter": 45.62527928325469, "sphere": {"center": [2.7912952061771095, 4.469667916176998, 100.28386370185255], "radius": 12.0}, "projected_sphere": {"center": [113.25704379495141, 123.63349960536618], "axes": [148.3788064273108, 148.3788064273108], "angle": 90.0}, "model_confidence": 0.8078400038305622, "model_id": 29, "model_birth_timestamp": 125631.82925, "theta": 1.9759215779115222, "phi": -1.956088705267673, "method": "3d c++", "id": 0}]}, "timestamp": 125655.055173}], "fixations_on_srf": [], "timestamp": 125654.998853, "camera_pose_3d": "none"}

Fig. 5. Collected base eye tracking data in its original format.

4.1 Results

In Table 1, filtering duration and capacity percentages are shown. Clearly, there are about 50, 89, 76 and 84 total subscription SDP keys respectively for defined event types. Findings reveal that the filtering time increases as the number of subscribed SDPs increases. This difference becomes more pronounced as the number of SDPs subscribed increases. The same inference is true for capacity gain, too. Accordingly, for the worst-case scenario regarding eye tracking data; By filtering all SDPs (~89%) in the base with fixation event in about 0.0003 s and converting all the events in the base event into SDP events, the overall transmission size is reduced by ~ 93% on average.

Table 1. Filtering duration and capacity gain analyses for the purposes filtering approach

Event Type	Avg. SDP Count	1 SDP		Max SDP		Heat Map SDP	
		Avg. Duration (sec)	Avg. Capacity (%)	Avg. Duration (sec)	Avg. Capacity (%)	Avg. Duration (sec)	Avg. Capacity (%)
Base	50	1.93E-06	93.18%	9.10E-05	93.17%	1.59E-06	98.41%
Fix	89.074	3.27E-06	94.75%	2.95E-04	94.74%	1.58E-06	99.46%
Blink	76.172	2.85E-06	94.09%	2.38E-04	94.08%	1.62E-06	98.95%
Fix&Blink	84.008	3.09E-06	94.73%	2.78E-04	94.73%	1.58E-06	99.29%

In Table 2, transmission durations for SDP phrases and normal events are shown together with transmission gain calculated using 1 - (SDP transmission duration / normal transmission duration). For all four event types, the results indicate a decrease in the transmission duration starting from 88.27% and as high as 98.63%.

Table 2. Transmission duration analyses for events with and without SDP phrase

Event Type	Transmission for 1 SDP			Transmission for Max SDP			Transmission for Only Heat Map SDP		
	Avg. SDP (sec)	Avg. Event (sec)	Avg. Gain (%)	Avg. SDP (sec)	Avg. Event (sec)	Avg. Gain (%)	Avg. SDP (sec)	Avg. Event (sec)	Avg. Gain (%)
Base	1.80E-05	1.54E-04	88.27%	9.31E-04	7.72E-03	87.94%	2.09E-05	3.10E-04	93.27%
Fix	4.71E-05	8.00E-04	94.11%	4.41E-03	7.64E-02	94.22%	2.09E-05	1.61E-03	98.70%
Blink	3.72E-05	6.06E-04	93.86%	3.34E-03	5.70E-02	94.15%	2.13E-05	1.22E-03	98.25%
Fix &Blink	4.22E-05	7.44E-04	94.33%	4.10E-03	7.35E-02	94.43%	2.04E-05	1.49E-03	98.63%

5 Discussion

The experimental results imply that the filtering can be completed in about 0.0003 s in the worst case; and about 0.0000016 s in the best case for the given scenario. Thus, for the worst and best case scenarios, 3.333 and 625.000 events per second could be filtered, respectively, by just one thread, without other operations such as transmission. Moreover, results show that event size reduction was between 93% as a minimum and ~ 99% as a maximum. In relation to this reduction, transmission duration decreased between 87% and ~ 98%. In other words, instead of sending a single event to a consumer, we can send up to 50 events at the same time.

When compared with topic-based filtering model of Pub/Sub, proposed filtering requires a certain amount of time of execution. To understand that the proposed method can be faster than topic-based methodology, filtering and transmission durations should be compared, using max SDP analyses for base event. Computations show that transmission of a normal event can be completed in 0.00772 s, while transmission of an SDP event can be completed in 0.00093 s. When filtering duration is added to SDP transmission duration, we obtain 0.00102 (=0.00093 + 0.00009) second. In this case, with the proposed filtering approach, 7.5 events per second can be filtered and transmitted instead of 1 event, which results in a 650% increment.

Also, the proposed filtering can provide a load that can be foreseen while content-based can cause bottlenecks on the broker. Since we can calculate the total subscription of an event, we also can find the total filtering duration. In other words, regarding to the worst case of our eye tracking scenario using the computer configured as mentioned above, 3333 events can be sent per second. Thereby, designing the communication context in accordance with the worst case of a work to be done would be correct. However, same foresight may not be possible for content-based filtering.

6 Conclusion

In this study, a filtering protocol with a novel architecture for events with extensive structure in the Pub/Sub pattern is proposed and its minimum requirements are demonstrated through a broker system architecture. With the developed approach, the producer clients can transmit the changed parts of their previous events to the intermediary system and the consumer clients can receive only the data they need by subscribing to the sub-segments (fields) of the events. In this respect, the SDP model uses traditional topic-based and content-based filtering together and aims to increase performance for communication stakeholders (producer and consumer). The performance improvement is mainly about reducing the transmission size. Another aim of the study is to apply the filtering method to producers as well. Therefore, compared to similar filtering approaches, it uses filtering in two strategies: subscription actions and creating new events.

Using SDP, Sub clients can customize topic-based filtering towards sub-scopes of events. Therefore, the need for Subs to create filtering mechanisms can be addressed. Filtering for extensive event fragments can have four benefits for Sub clients.

- Less network capacity can be used as event sizes in transmission are reduced. Thus, the data transmission frequency can also be increased.
- Filtering operations of consumers who would apply similar filtering operations can be done in the intermediary system. Thus, a certain processing load required for filtering is required only in the intermediary system, and reoccurrence of the relevant load for each consumer can be avoided. This can lead to improved resource consumption, especially between systems that know each other or are loosely coupled (for example, only a group of consumers).
- Consumers can reduce their need to know the mechanisms for creating the events they need. Horizontal expansion of events for different environmental conditions may become insignificant for consumers. At this point, the only information that the consumer needs to know is the way to access the desired data. Thus, a loosely coupled environment can be gained in the description of the event. It can provide flexibility, especially in systems that do not know, or only know each other loosely.
- Bounded Context structure as a decomposition strategy can be defined among different Sub client groups [45]. Thereby, a consumer can't reach the whole event, they receive the event parts in accordance with their language of business logic. In other words, an event model defined by the producer can be depicted by the consumer as a model that has the same meaning but different content.

For pub clients, SDP is directly focused on less use of network capacity by transmitting only the changed parts of an event. But it can also be used for the cost of creating the event to be published. In an architecture where more than one producer is involved in event creation, reducing transmission between event stakeholders can also reduce other additional communication costs. Thereby, the mechanism of creating event by different producer stakeholders which alters the parts of the event in accordance with their existential purposes can be obtainable, therefore, an event's producer can be many different producers in the concept of Pub/Sub. However, the related mechanism containing multi producer wasn't implemented on the projects mentioned in subsection named Broker and Library for SDP. By reason of this, this will be a future work for us.

The SDP model can be updated according to the requirements of the system to be used. For example, instead of a process by which the event is generated and stored in the intermediary system using an SDP taken from Pub, a process where only the SDP is stored can be designed. Thus, only SDP is stored in the memory and events can be generated from the relevant SDP as requested. With this operation, storage requirements can also be reduced.

Scalability issues in content-based filtering will not be present in the SDP model. Because, as a maximum, the number of subscriptions is equal to the number of sub-parts (identifiers) of the event, and the number of consumers is not the main factor that increases the cost. Therefore, the minimum and maximum costs on the intermediary system can be calculated.

On the other hand, type-based filtering triggers tightly coupled transmission between producer and consumer. The types of data received and sent should be known to consumers. The horizontal expansion of the event structure may lead to the need for new type definitions for consumers. In SDP model, the hierarchical structure that forms the event or its sub-part does not have to be known. At this point, consumers should know exact location of the item they will use in the hierarchical structure. Consumers are not adversely affected if the event expands or changes except for the relevant bus.

As a future work, we plan to measure the proposed method in detail in a setting with more than 10 producers and more than 10 consumers at the same time. Additionally, after this end-to-end analysis, we can study a software design supporting high-scalability Pub/Sub communication context based on proposed architecture with filtering.

References

1. Huang, L., Liu, L., Chen, J., Lei, K.: An implementation of content-based pub/sub system via stream computation. In: Qiu, M. (eds.) Smart Computing and Communication, SmartCom 2017, vol. 10699, pp. 344–353. Springer, Cham (2018). https://doi.org/10.1007/978-3-319-73830-7_34
2. Chockler, G., Melamed, R., Tock, Y., Vitenberg, R: Constructing scalable overlays for pub-sub with many topics. In: Proceedings of the Twenty-Sixth Annual ACM Symposium on Principles of Distributed Computing, PODC 2007, Portland, Oregon, USA (2007)
3. Banavar, G., Chandra, T., Mukherjee, B., Nagarajarao, J., Strom, R., Sturman D.: An efficient multicast protocol for content-basedpublish-subscribe systems. In: Proceedings - International Conference on Distributed Computing Systems, pp. 262–272 (1999). https://doi.org/10.1109/ICDCS.1999.776528
4. Liu, Y., Plale, B.: Survey of publish subscribe event systems (2003)
5. Setty, V., Kreitz, G., Vitenberg, R., van Steen, M., Urdaneta, G., Gimåker, S.: The Hidden Pub/Sub of Spotify. In: DEBS'13, pp. 231–240 (2013).https://doi.org/10.1145/2488222.2488273
6. Aguilera, M., Strom, R., Sturman, D., Astley, M., Chandra, T.: Matching events in a content-based subscription system. In: Symposium on Principles of Distributed Computing, pp. 53–61 (1999)
7. Onica, E., Felber, P., Mercier, H., Piviére, E.: Confidentiality-preserving publish/subscribe: a survey. Assoc. Comput. Mach. **49**, 1–43 (2016). https://doi.org/10.1145/2940296
8. Apache Kafka A Brief Introduction.https://kafka.apache.org/intro. Accessed 2 Dec 2022
9. ZeroMQ Guide Chapter2- Sockets and Patterns.http://zguide.zeromq.org/page:chapter2. Accessed 22 Dec 2022

10. Tarkoma, S.: Publish/Subscribe Systems: Design and Principles, Section 1.2 Components of a Pub/Sub System. Wiley (2012)
11. Eugster, P., Felber, P., Guerraoui, R., Kermarrec, A.: The many faces of publish/subscribe. ACM Comput. Surv. **35**(2), 114–131 (2003)
12. Birman, K., Joseph, T.: Exploiting virtual synchrony in distributed systems. SIGOPS Oper. Syst. Rev. **21**(5), 123–138 (1987)
13. Eken, S.: A topic-based hierarchical publish/subscribe messaging middleware for COVID-19 detection in X-ray image and its metadata. Soft. Comput. **27**, 1–11 (2020). https://doi.org/10.1007/s00500-020-05387-5
14. Eugster, P., Guerraoui R., Sventek, J.: Type-Based Publish/Subscribe (2000). https://doi.org/10.5075/epfl-thesis-2503
15. Banavar, G., Chandra, T., Mukherjee, B., Nagarajarao, J., Storm, R., Sturman, D.: An efficient multicast protocol for content-based publish-subscribe systems. In: International Conference on Distributed Computing Systems (1999)
16. Castro M., Druschel, P., Kermarrec, A.M., Rowstron, A.: Scribe: a large-scale and decentralized publish-subscribe infrastructure - preliminary draft submitted for publication. In: 3rd International Workshop on Networked Group Communication (NGC2001), UCL, London, UK, pp. 100-110 (2002)
17. Zhuang, S.Q., Zhao, B.Y., Joseph, A.D., Katz, R.H., Kubiatowicz, J.D.: Bayeux: an architecture for scalable and fault-tolerant widearea data dissemination. Computer Science Division, Report No. UCB/CSD-2–1170, Technical report (2001)
18. Carzaniga, A., Rosenblum, D.S., Wolf, A.L.: Design and evaluation of a wide-area event notification service. ACM Trans. Comput. Syst. **19**(3), 332–383 (2001)
19. Fox, G., Pallickara, S.: An event service to support grid computational environments. J. Concurr. Comput.: Pract. Exp. Special Issue Grid Comput. Environ. (2002)
20. Slominski, A., Simmhan Y., Rossi, A.L., Farrellee, M., Gannon, D.: Xevents/xmessages: Application events and messaging framework for grid. Technical report, Indiana University (2001)
21. Eisenhauer, G., Bustamente, F., Schwan, K.: Event services for high performance computing. In: Proceedings of the 9th IEEE International High Performance Distributed Computing (HPDC), Los Alamitos, CA. IEEE Computer Society (2000)
22. Baldoni, R., Marchetti, C., Verde, L.: CORBA request portable interceptors: analysis and applications. Concurr. Comput.: Pract. Exp. **15**, 551–579 (2003)
23. Mishra, S., Peterson, L., Schlichting, R.: Consul: a communication substrate for fault-tolerant distributed programs. Department of Computer Science, The University of Arizona, TR pp. 91–32 (1991)
24. Powell, D.: Delta-4: A Generic Architecture for Dependable Computing. Springer, Cham (1991)
25. Kopetz, H., et al.: Distributed fault-tolerant real-time systems: the mars approach. IEEE Micro **9**, 25–40 (1989)
26. Kreps, J., Narkhede, N., Rao, J.: Kafka: a distributed messaging system for log processing. In: NetDB'11, Athens, Greece (2011)
27. RabbitMQ Homepage. https://www.rabbitmq.com/. Accessed 22 Dec 2022
28. Dobbelaere, P., Esmaili, K.S.: Kafka versus RabbitMQ: a comparative study of two industry reference publish/subscribe implementations: industry paper. In: 11th ACM International Conference (2017). https://doi.org/10.1145/3093742.3093908
29. Snyder, B., Bosanac, D., Davies, R.: Introduction to Apache ActiveMQ, ActiveMQ in Action, Green Paper ISBN: 1933988940 (2010). https://freecontent.manning.com/wp-content/uploads/introduction-to-apache-activemq.pdf
30. Fu, G., Zhang, Y., Yu, G.: A fair comparison of message queuing systems. IEEE Access **9**, 421–432 (2020). https://doi.org/10.1109/ACCESS.2020.3046503

31. Amazon SQS Homepage. https://aws.amazon.com/tr/sqs/. Accessed 22 Dec 2022
32. Google Cloud Pub/Sub Homepage. https://cloud.google.com/pubsub/architecture. Accessed 22 Dec 2022
33. Light, R.A.: Mosquitto: server and client implementation of the MQTT protocol. J. Open Source Softw. **2**(13), 265 (2017). https://doi.org/10.21105/joss.00265
34. Thyagaraj, L.,et al.: IBM MQ as a Service: A Practical Approach, Red Paper, First Edition. International Technical Support Organization (2016)
35. Longo, E., Redondi, A., Cesana, M., Arcia-Moret, A., Manzoni, P.: MQTT-ST: a spanning tree protocol for distributed MQTT brokers, pp. 1–6 (2020). https://doi.org/10.1109/ICC40277.2020.9149046
36. Longo, E., Redondi, A.: Design and implementation of an advanced MQTT broker for distributed pub/sub scenarios. Comput. Netw. **224**, 109601 (2023). https://doi.org/10.1016/j.comnet.2023.109601. ISSN 1389–1286
37. Guesmi, T., Kalghoum, A., Alshammari, B.M., Alsaif, H., Alzamil, A.: Leveraging software-defined networking approach for future information-centric networking enhancement. Symmetry **13**(3), 441 (2021). https://doi.org/10.3390/sym13030441
38. D'Ortona, C., Tarchi, D., Raffaelli, C.: Open-source MQTT-based end-to-end IoT system for smart city scenarios. Future Internet **14**(2), 57 (2022). https://doi.org/10.3390/fi14020057
39. Pu, C., Ding, X., Wang, P., Yang, Y.: Practical implementation of an OPC UA multi-server aggregation and management architecture for IIoT. In: 2022 IEEE International Conferences on Internet of Things (iThings) and IEEE Green Computing & Communications (GreenCom) and IEEE Cyber, Physical & Social Computing (CPSCom) and IEEE Smart Data (SmartData) and IEEE Congress on Cybermatics (Cybermatics), Espoo, Finland, pp. 476–481 (2022). https://doi.org/10.1109/iThings-GreenCom-CPSCom-SmartData-Cybermatics55523.2022.00099
40. Hafeez, F., Nasirifard, P., Jacobsen H.: A serverless approach to publish/subscribe systems. In: Proceedings of the 19th International Middleware Conference (Posters) (Middleware 2018), pp. 9–10. Association for Computing Machinery, New York (2018).https://doi.org/10.1145/3284014.3284019
41. Popovic, M., Kordic, B., Basicevic, I., & Popovic, M.: Distributed python software transactional memory supporting publish-subscribe pattern. In: 2022 30th Telecommunications Forum (TELFOR), Belgrade, Serbia, pp. 1–4 (2022). https://doi.org/10.1109/TELFOR56187.2022.9983693
42. Rodriguez, L., Bent, J., Shaffer, T., Rangaswami, R. Infusing pub-sub storage with transactions, pp. 23–30 (2022).https://doi.org/10.1145/3538643.3539739
43. Ahmed, N.: MPaS: a micro-services based publish/subscribe middleware system model for IoT. In: 2022 5th Conference on Cloud and Internet of Things (CIoT), Marrakech, Morocco, pp. 220–225 (2022). https://doi.org/10.1109/CIoT53061.2022.9766670
44. Hamad, M., Finkenzeller, A., Liu, H., Lauinger, J., Prevelakis, V., Steinhorst, S.: SEEMQTT: secure end-to-end MQTT-based communication for mobile IoT systems using secret sharing and trust delegation. In: IEEE Internet Things J. **10**(4), 3384–3406 (2023). https://doi.org/10.1109/JIOT.2022.3221857
45. Richardson, C.: Microservices Patterns with Examples in Java. Manning Publications Co., pp. 33–56 (2019). ISBN: 9781617294549
46. JSON Homepage. https://www.json.org/json-en.html. Accessed 22 Dec 2022
47. A Data Instance of Javascript Object Notation. https://www.w3schools.com/js/js_json_intro.asp. Accessed 22 Dec 2020
48. Pupil Capture Software v1.7.42 Repo. https://github.com/pupil-labs/pupil/releases/tag/v1.7. Accessed 22 Dec 2022

A Computerized Experimental Rig
for the Vibration Investigation of Cracked
Composite Materials Plate Structures

Muhannad Al-Waily[1]([✉]) and Muhsin J. Jweeg[2]

[1] Department of Mechanical Engineering, Faculty of Engineering, University of Kufa, Kufa,
Iraq
`muhaned1.alwaeli@uokufa.edu.iq`
[2] College of Technical Engineering, Al-Farahidi University, Baghdad, Iraq
`muhsin.jweeg@uoalfarahidi.edu.iq`

Abstract. A composite plate is strengthened by reinforcements in the form of
powder particles and short or long fibers, where the fiber glass was used as the
reinforcement fiber to modify the composite plate was investigated. The crack
effects on the plate natural frequencies were analyzed using different boundary
conditions (SSSS, SSCC, SSFF). The problem was solved numerically using the
Finite Element Method, adopting the ANSYS program, experimentally employing
the time variation and measuring the natural frequency. The comparison between
the finite element method and the experimental program has shown good agreement with a maximum discrepancy of not more than 8.5% for the tested cases
of the composite plates. It was noticed that the natural frequency decreases with
the presence of the crack. It was also noticed that the central damage effect has
a higher effect on the natural frequency than the other crack positions. Also, the
results show that the natural frequency decreases as the crack length or depth
increases due to the stiffness reduction in cracked samples. And the increase in
the aspect ratio of the plate means an increase in the mass of the plate, which
results in a decrease in the natural frequencies. In addition, the results indicate
a reduction in the natural frequency for the SSCC boundary condition was less
than the decrease in the natural frequency for other boundary conditions, SSSS
and SSFF. Finally, the results showed that the short fiber types modify the natural
frequency more than the natural frequency of other reinforcement fiber types.

Keywords: Crack · Composite Materials · Plate Vibration · Damage Effect ·
Reinforcements

1 Introduction

The composite plates have been used in many life applications, such as aircraft, auto-
mobiles, and industrial uses. The composite materials have proved their effectiveness
in manufacturing lightweight structures with high strength due to the fiber types such
as glass, boron, Kevlar, carbon and epoxy resin. These fibers have filament properties

© The Author(s), under exclusive license to Springer Nature Switzerland AG 2024
A. Ortis et al. (Eds.): ICAETA 2023, CCIS 1983, pp. 482–502, 2024.
https://doi.org/10.1007/978-3-031-50920-9_37

and excellent transverse shear properties. The composite materials used may suffer from different mode failures, such as delamination, cracks, and de-bonding, and the structure material may reach the plastic region.

Kessler et al. [1] used the wave equation for damage detection in sandwich beams for materials quasi-isotropic manufactured from graphite with epoxy. They investigated the damage patterns, delamination and transverse ply cracks through holes. A beam driven with piezo-electrical (PZT) was studied by Afshari and Inman [2]. The crack growth on the beam vibration characteristics was investigated, and the combination of the crack and PZT actuator on the beam simulation. Al-Waily [3] studied the cracked beams with different support conditions using a suggested analytical and numerical solution employing the finite element technique using the ANSYS package. The crack depth and position effects on the natural frequency were investigated. The equivalent stiffness EI. An exponential function was assumed for the bending stiffness with the crack depth and location. The results show that the natural frequency decrease in the central crack case is higher than when the crack is near beam ends. Many researchers modified the strength of composite materials structure by various techniques for different applications [4–14], such as reinforcement with Nanomaterials [15–18], reinvestment by natural fiber [19], and other methods [20–35]. In addition, many researchers investigated the effect of cracks on the natural frequency of plate structure with various parameter effects [36–40].

The novelty of the current work is to prove experimentally and verified numerically the effects of crack length and position on the natural frequencies of the composite plates, including the effects of the reinforcements types using short and long fibers using different boundary conditions of selected boundary conditions of the plates. The cases investigated in this work were not studied previously. The research methodology will cover the construction of an experimental program and manufacturing of the required samples to be tested to cover the required variables of the research points, crack length, position and type of reinforcements. The work methodology also includes the numerical verifications of the experimental work.

2 Experimental Work

The materials used here are short and long glass fibers, powders, and particle reinforcements with polyester resin. The natural frequency is obtained using vibration measurements for the samples with and without cracks. Different aspect ratios of the plates were used with variable thicknesses and other support conditions. The volume fraction of the resin and the reinforcement are calculated as follows,

$$\text{Weight of Fiber} = \rho f^* \forall t * \forall f, \text{ and, Weight of Resin} = \rho m * \forall t * \forall m \qquad (1)$$

The volume fraction of the reinforcement was ∀f = 30%. The parameters studied in calculating the natural frequency are the types of reinforcement, crack size and location, the aspect ratio of the plate (ba), the plate's thickness, and the plate's boundary conditions. The making of the sample requires the polyester resin and the reinforcement fiber type, as shown in Fig. 1. The dimensions of vibration plate samples used are, as shown in Fig. 2,

$$a_t = a + 10\,cm\ (Supported),\ b_t = b + 10\,cm\ (Supported),\ H = 5.5\,mm,\ a = 24\,cm$$
$$a_t = 24\,cm + 10\,cm\ (Supported)\ = 34\,cm \tag{2}$$

And difference aspect ratio, $\left(\frac{b}{a} = 1, 1.5, 2\right)$. Then,

$$b_t = b + 10\,cm\ (Supported)\ = (b/a).a + 10\,cm\ (Supported) \tag{3}$$

The composite plate made for short, long, powder, and particle reinforcement composite plates, and the weight required of plate samples for different aspect ratios of the plate and different thicknesses of volume fraction of reinforcement materials used 30%, as shown in Table 1 plate thickness t = 5.5 mm as and Table 2 presents plate samples with different thicknesses, aspect rations and type of fibers.

The vibration plate samples studied are supported with different situations, for different aspect ratios (b/a = 1, 1.5, 2) as follows,

1. Simply support along all edges and sides (SSSS).
2. It is supported along ζ = 0 and 1 edges and clamped along η = 0 and one advantage (SSCC).
3. It is supported along ζ = 0 and 1 edges and free reinforced along η = 0 and one advantage (SSFF).

Figure 3 shows the constructed rig used for the vibration measurements using different design parameters and support conditions. The rig is constructed from the following elements,

1. Supporting sample structure.
2. Impact hammer, mass = 0.16 kg, type (086C03), Piezo electronic (PCB) – vibration transducer. The measurement parameters are 2224 N, resonant frequency (≥22 kHz), excitation voltage (20 to 30 VDC), current excitation (2 to 20 mA), output voltage (8–14 VDC), discharge time (≥2000 s),
3. Accelerometer (weight = 11 g) of the following characteristics: model (4371), lower frequency decided by the user, Upper-Frequency limit + 10%, 12.6 kHz, and the resonant frequency = 42 kHz.
4. The Amplifier type 7749 is used to measure the accelerometer's response and send the output signal to the storage oscilloscope.
5. The digital storage oscilloscope, model GDS-810, with the following properties: maximum frequency = (100 MHz), with a rate of reading (25 GS/s)

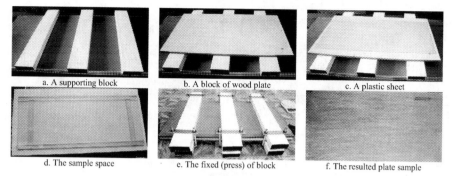

a. A supporting block   b. A block of wood plate   c. A plastic sheet

d. The sample space   e. The fixed (press) of block   f. The resulted plate sample

Fig. 1. Steps of manufacturing the composite plate

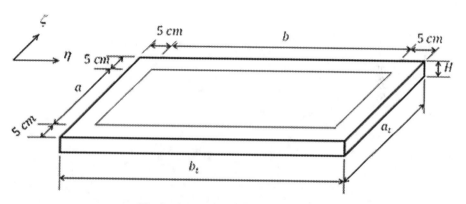

Fig. 2. Dimensions of the plate sample.

Table 1. Required Reinforcement and Resin Materials Weight (g) for Composite Sample of Volume fraction of Fiber 30% and H = 5.5 mm.

Materials		Aspect Ratio		
		2	1.5	1
Reinforcement Fiber Types	Powder	976.14	732.11	488.07
	Particle	780.92	585.69	390.46
	Short	637.75	478.32	318.88
	Long	894.80	671.10	447.40
Resin	Polyester	759.22	569.42	379.61

Table 2. The required reinforcement and resin materials weight (g) for composite sample of Volume fraction of fiber 30% and different plate thicknesses.

Aspect Ratio	Thickness (mm)	Reinforcement Fiber Types		Resin Materials
		Glass Short fiber	Glass Long fiber	Polyester
2	3.5	405.84	569.42	483.14
	9	1043.58	1464.21	1242.36
1.5	3.5	304.38	427.07	362.36
	9	782.69	1098.16	931.77
1	3.5	202.92	284.71	241.57
	9	521.79	732.12	621.18

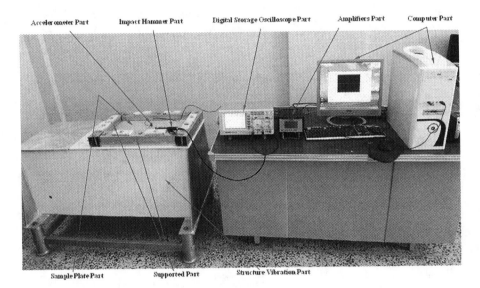

Fig. 3. The vibration measurement rig.

A sig-view program transfers the response signal from the digital storage. Oscilloscope to the FFT function generator reads the plate's natural frequency with the required geometry, support conditions and material properties [45–50]. The digital oscilloscope saves the response and the transient load as excel data. Figures 4 and 5 show the function of the sig view program in which the response and the load are measured from the vibration of the plate stored in the storage oscilloscope is transferred to the FFT to obtain the natural frequency. The crack ellipse shape was used as a crack model. The numerical investigation was achieved using Fenton power tools, using a power of 135 W, and the rotating speed is 3000–35000 rpm.

The composite plates in this work included the study of composite plate types (powder, particle, short, and long fibre) and the crack length as (10%, 15%, and 20%) from side parallel to z direction, then the crack length studied (24, 36, and 48 mm). Location middle (middle, side-1, and side-2), and examine the thickness effect of the plate and the depth of the crack through the thickness of the plate as (30%, 50%, and 70%) from the thickness of the plate (9 mm), and depth of crack for other thickness of the plate is (70%H), with different boundary conditions as (SSSS, SSCC, and SSFF) and aspect ratio of the plate (2, 1.5, and 1), as shown in Fig. 6. The information of the composite plate sample studied are,

1. The aspect ratio of the plate sample is 2, 1.5 and 1,
2. The boundary condition of the plate sample is SSSS, SSCC and SSFF.
3. Sample plate thickness is,
   a. 5.5 mm with crack depth (%H = 70%H = 3.85 mm) for,
      i. Crack length 2 middle, side-1 and side-2 crack location for powder, particle, short and long reinforcement composite plate.
      ii. Crack lengths 36 and 48 mm with the middle crack location for short and long reinforcement composite plates.
   b. 3.5 mm with crack depth (%H = 70%H = 2.45 mm) for crack length (24 mm) with the middle crack location for short and long reinforcement composite plate.
   c. 9 mm with crack depth (%H = 30%H = 2.7 mm, 50%H = 4.5 mm and 70%H = 6.3 mm) for crack length (24 mm) with the middle crack location for short and long reinforcement composite plate.

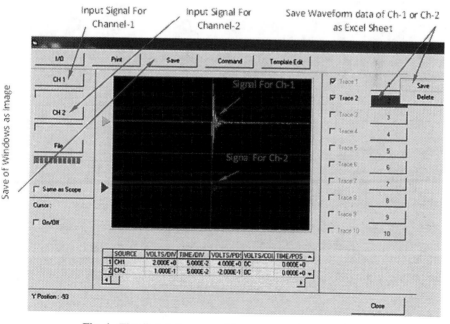

Fig. 4. The digital storage oscilloscope computer program.

However, the flow chart of the complete vibration test is shown in Fig. 7.

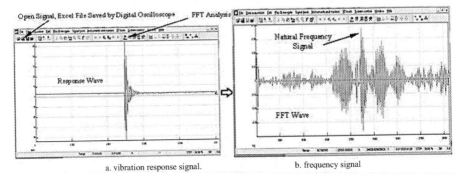

a. vibration response signal.   b. frequency signal

Fig. 5. The Sig-View Program to FFT Analysis Function.

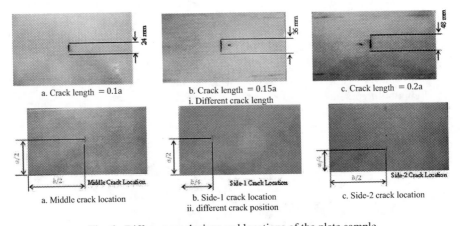

a. Crack length = 0.1a   b. Crack length = 0.15a   c. Crack length = 0.2a
                                 i. Different crack length

a. Middle crack location   b. Side-1 crack location   c. Side-2 crack location
                                 ii. different crack position

Fig. 6. Different crack sizes and locations of the plate sample.

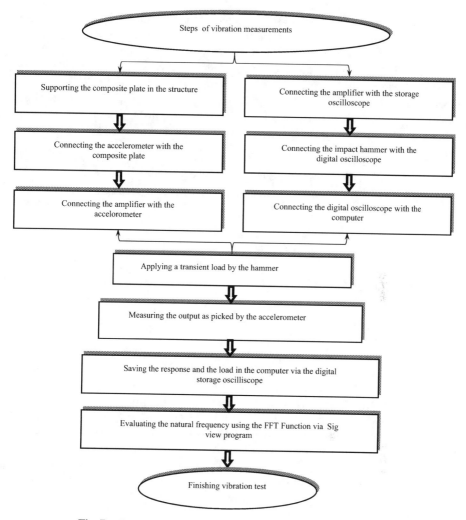

Fig. 7. The flow chart of vibration test composite plate structure.

3 Results and Discussions

The results of this work are the natural frequencies of the composite plates and an investigation of the effects of crack size and location. The following resin and reinforcements properties are:

$\forall_f = 30\%$, $G_{glass} = 30\,GPa$, $G_{Polyester} = 1.4\,GPa$, $\nu_{glass} = 0.25$, $\nu_{Polyester} = 0.4$, $E_{glass} = 95\,Ppa$, $E_{Polyester} = 3.8\,GPa$, $\rho_{glass} = 2600\,kg/m^3$, $\rho_{Polyester} = 1350\,kg/m^3$. However, the composite plate properties are listed in Table 3. The experimental program achieved in this work [25] was to evaluate the natural frequency of composite plate types with crack effect and validated by a numerical study using ANSYS Program [51–57]. The experimental and numerical studies were conducted for the composite plates for the reinforcements, powder, particles, short and long fibres, different crack sizes and positions (in ζ and η-directions), and different plate aspect rations, and thicknesses with different support conditions (simply supported in each edges SSSS, simply supported through $\zeta = 0$ and 1 edges and clamped supported along $\eta = 0$, and 1 edges SSCC, and simply supported through $\zeta = 0$ and 1 edges and Free supported along $\eta = 0$ and one edges SSFF). The numerical study using the ANSYS program [58–66] results are compared with those obtained experimentally, as shown in Figs. 8, 9, 10, 11, 12 and 13.

Table 3. Properties of Composite Materials of the plate, [25].

properties	Fiber Types		Short Fiber	Powder	Particle
	Long Fiber				
E (GPa)	$E_1 = 31.16$	$E_2 = 5.34$	15.86	7.1	4.66
G (GPa)	1.96		5.62	2.67	1.7
ρ (kg/m^3)	1525		1288	1600	1420
ν	0.355		0.411	0.375	0.375

Table 4 presents the natural frequencies. For different plate dimensions and boundary conditions for the composite plate thickness H = 5.5 mm, crack length 2C = 24 mm, crack depth ratio $\xi = 0.7$, with a volume fraction of fibre $\forall_f = 30\%$. The results indicate that the natural frequency for the plates (SSCC) is higher than that (SSSS) and (SSFF) for all the tested cases. Table 5 presents the natural frequency measurements with different crack lengths, type of fibres with a crack location in ζ and η-directions ($\zeta = 0.5$, $\eta = 0.5$), the volume fraction of fibre $\forall_f = 30\%$, and composite plate thickness H = 5.5 mm. Again, the natural frequency for the plate of the boundary condition (SSCC) is higher than those tested support conditions for the tested samples. Table 6 presents the natural frequencies measurements for the composite plate cases ($\zeta = 0.5$, $\eta = 0.5$), crack length 2C = 24 mm, H = 5.5 mm, and volume fraction of fibre $\forall_f = 30\%$. Similar conclusions for the results shown in Table 5 were noticed in Table 6. Table 7 presents results of the plate thickness effects for a central crack directions ($\zeta = 0.5$, $\eta = 0.5$), crack length 2C = 24 mm, crack depth ratio $\xi = 0.7$, and volume fraction of fibre $\forall_f = 30\%$. Figures 10 and 11 indicate the natural frequencies for the plates of aspect ratio = 1, 1.5, and 2 with different boundary conditions and crack position in ζ-direction, for the middle crack position in η-movement $\eta = 0.5$, plate thickness H = 5.5 mm, crack length 2C = 24 mm, crack depth ratio $\xi = 0.7$, the volume fraction of reinforcement fibre $\forall_f = 30\%$. The figures show a good agreement between the numerical and experimental results, where the percentage of discrepancy between numerical and experimental results is about (3 to 8%). Good agreement was obtained between the experimental and the numerical results with a percentage of discrepancy (3–8)%.

Figure 10, the results were for the short fibre reinforcement, while Fig. 11 was for the long fibre reinforcements. Figure 12 shows the natural frequency for short composite plate types with an aspect ratio (AR = 1, 1.5, and 2) for different boundary conditions

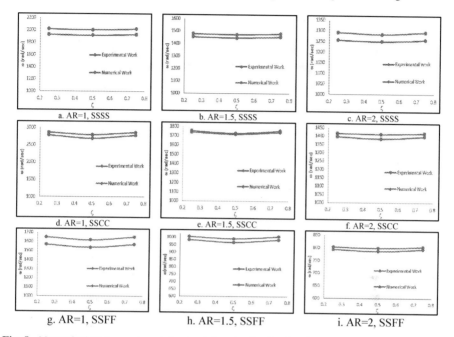

Fig. 8. Numerical and experimental natural frequencies for the short fiber composite plate with the variation of crack position in ζ-direction and AR = 1, 1.5, and 2 and different support conditions for, η = 0.5, 2C = 24 mm, ξ = 0.7.

(SSSS, SSCC and SSFF) for every kind of composite plate, respectively, the middle crack position in η and ζ-directions η = 0.5; ζ = 0.5, plate thickness H = 5.5 mm, crack depth ratio ξ = 0.7, the volume fraction of reinforcement fibre $\forall_f$ = 30%. The figures show a good agreement between the numerical and experimental results. The discrepancy between the numerical and experimental results is about (2 to 6%). The composite plates natural frequency results are obtained using short and long fibres respectively results are shown in Figs. 12 and 13 with the following design parameters: AR = 1, 1.5, and 2, with the support conditions SSSS, SSCC and SSFF, middle crack position in η and ζ-directions η = 0.5; ζ = 0.5, plate thickness H = 5.5 mm, crack depth ratio ξ = 0.7, fiber reinforcement volume fraction $\forall_f$ = 30%. A good agreement has been obtained between the numerical and experimental results with a percentage discrepancy between is about (2 to 5.5%).

Figures 8, 9, 10, 11, 12, 13 and Tables 4, 5, 6 and 7 present the effects of cracks location and size on the natural frequencies of the composite plates with the following main design parameters,

1. The crack position in (ζ and η-direction),

It was noticed that the natural frequency decrease as the position of the crack moves to the plate centre, as shown in all the studied cases of the AR = 1, 1.5, and 2 and the boundary conditions (SSSS, SSCC, and SSFF) and presented in Figs. 8, 9, 10, 11, 12, 13 and Tables 4, 5, 6 and 7.

2. The effects of the crack size length and depth.

It was noticed in Tables 5 and 6, and Figs. 12 and 13 using the same composite plates in (1), the natural frequency decreases as the crack length or depth increases due to the stiffness reduction in cracked samples.

3. Thickness of the plate effects,

The increase of the plate means an increase in the composite plate which results in an increase in the natural frequency for all the cases of the plates with different aspect ratios, as shown in Table 7.

4. Effect of the aspect ratio of the composite plate,

The increase in the aspect ratio of the plate means an increase in the mass of the plate, which results in a decrease in the natural frequencies shown in Tables 4, 5, 6, 7 and Figs. 8, 9, 10, 11, 12 and 13.

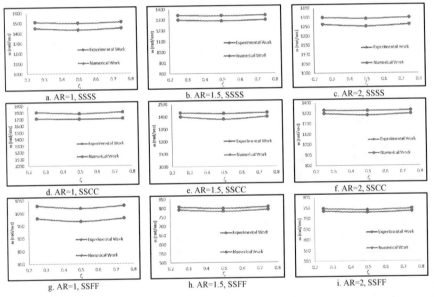

Fig. 9. Numerical and experimental natural frequencies for long Fiber Composite Plate with the variation of crack position in ζ-direction and AR = 1, 1.5, and 2 and different support conditions for, η = 0.5, 2C = 24 mm, ξ = 0.7.

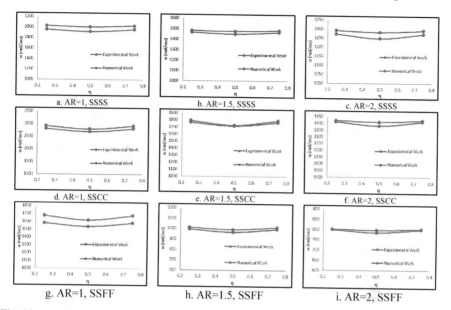

Fig. 10. Numerical and experimental of Natural Frequencies for short fiber composite Plate with the variation of crack position in η-direction and AR = 1, 1.5, and 2 and different support conditions for, ζ = 0.5, 2C = 24 mm, ξ = 0.7.

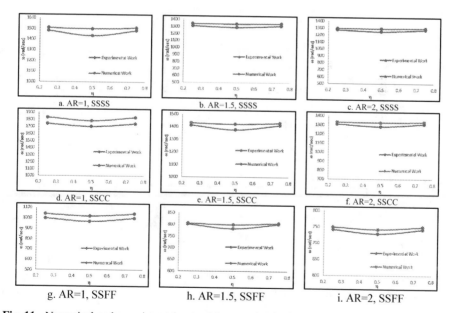

Fig. 11. Numerical and experimental natural frequencies for long fiber composite Plate with the variation of crack position in η-direction and AR = 1, 1.5, and 2 and different support conditions for, ζ = 0.5, 2C = 24 mm, ξ = 0.7.

494   M. Al-Waily and M. J. Jweeg

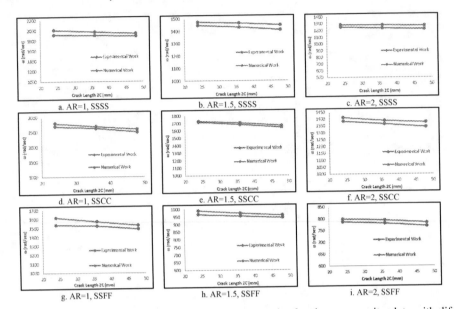

Fig. 12. Numerical and experimental of natural frequencies for short composite plate with different crack length effect with AR = 1, 1.5, and 2 and different support conditions for, ζ = 0.5, η = 0.5, ξ = 0.7.

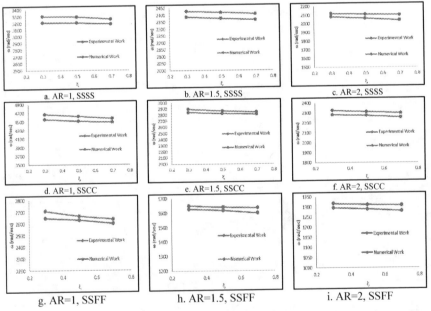

Fig. 13. Numerical and experimental of natural frequencies for short fiber different crack depth ratio with AR = 1, 1.5, and 2 and different support conditions Plate ζ = 0.5, η = 0.5, 2C = 24 mm, H = 9 mm.

Table 4. Frequency measurements for the plate (rad/sec) with different crack position in ζ and η directions, for H = 5.5 mm, 2C = 24 mm, ξ = 0.7.

Plate Type	Aspect Ratio	ζ	η	SSSS	SSCC	SSFF
Powder Reinforcement	1	0.5	0.5	1130.51	1550.12	914.06
		0.5	0.25	1165.12	1590.23	947.97
		0.25	0.5	1145.58	1555.23	925.59
	1.5	0.5	0.5	828.45	974.58	563.15
		0.5	0.25	850.47	1012.48	580.07
		0.25	0.5	836.14	996.14	571.55
	2	0.5	0.5	727.15	806.14	462.24
		0.5	0.25	742.13	829.36	469.01
		0.25	0.5	732.58	818.24	463.94
Short Reinforcement	1	0.5	0.5	1916.3	2690.36	1535.62
		0.5	0.25	1955.25	2810.44	1580.37
		0.25	0.5	1928.36	2780.24	1568.52
	1.5	0.5	0.5	1446.25	1716.49	965.90
		0.5	0.25	1463.25	1766.82	994.17
		0.25	0.5	1454.88	1738.34	983.74
	2	0.5	0.5	1248.36	1382.56	786.02
		0.5	0.25	1271.48	1411.68	800.23
		0.25	0.5	1254.87	1395.47	793.36
Particle Reinforcement	1	0.5	0.5	967.35	1356.48	781.58
		0.5	0.25	1002.59	1402.57	823.99
		0.25	0.5	995.48	1375.24	798.81
	1.5	0.5	0.5	708.79	861.35	497.81
		0.5	0.25	734.44	896.34	510.11
		0.25	0.5	724.55	874.35	503.79
	2	0.5	0.5	618.47	694.87	395.04
		0.5	0.25	631.42	715.49	405.88
		0.25	0.5	627.23	704.36	399.45
Long Reinforcement	1	0.5	0.5	1432.36	1702.26	967.09
		0.5	0.25	1482.34	1742.36	999.62
		0.25	0.5	1449.25	1712.24	978.89
	1.5	0.5	0.5	1288.4	1376.49	782.08

(*continued*)

Table 4. (*continued*)

Plate Type	Aspect Ratio	ζ	η	SSSS	SSCC	SSFF
		0.5	0.25	1310.55	1413.44	801.32
		0.25	0.5	1298.36	1397.35	791.89
	2	0.5	0.5	1251.36	1281.33	728.52
		0.5	0.25	1275.14	1309.22	741.50
		0.25	0.5	1261.36	1296.44	734.68

Table 5. Measured natural frequencies (rad/sec) with different crack length effect of short and long composite plate with, $\zeta = 0.5$, $\eta = 0.5$, $\forall_f = 30\%$, $\xi = 0.7$, H = 5.5 mm.

Plate Type	Aspect Ratio	2C (mm)	SSSS	SSCC	SSFF
Short Reinforcement	1	24	1916.3	2690.36	1535.618
		36	1902.48	2630.14	1523.88
		48	1892.36	2506.7	1490.492
	1.5	24	1446.25	1716.49	965.8957
		36	1431.62	1682.36	954.6085
		48	1409.48	1643.79	943.3947
	2	24	1248.36	1382.56	786.0231
		36	1232.44	1363.77	779.8126
		48	1221.47	1342.58	768.5424
Long Reinforcement	1	24	1432.36	1702.26	967.0915
		36	1422.36	1695.35	960.1985
		48	1410.45	1645.25	952.4368
	1.5	24	1288.4	1376.49	782.0746
		36	1273.48	1348.91	763.8887
		48	1264.89	1326.7	743.6384
	2	24	1251.36	1281.33	728.5212
		36	1243.87	1268.36	717.6966
		48	1228.66	1249.33	709.8559

Table 6. Measured Natural frequencies (rad/sec) with different crack depth effect of short and long composite plate for, $\zeta = 0.5$, $\eta = 0.5$, $\forall_f = 30\%$, $2C = 24$ mm, $H = 5.5$ mm.

Plate Type	Aspect Ratio	ξ	SSSS	SSCC	SSFF
Short Reinforcement	1	0.3	3210.58	4555.26	2646.972
		0.5	3205.14	4511.88	2631.533
		0.7	3190.25	4488.34	2602.777
	1.5	0.3	2386.36	2842.69	1625.859
		0.5	2375.12	2821.25	1615.254
		0.7	2366.14	2809.34	1597.114
	2	0.3	2080.35	2282.87	1294.12
		0.5	2061.87	2273.14	1288.553
		0.7	2041.43	2256.47	1279.46
Long Reinforcement	1	0.3	2402.33	2912.48	1661.909
		0.5	2385.47	2885.24	1642.843
		0.7	2363.4	2844.59	1628.234
	1.5	0.3	2170.49	2302.89	1305.904
		0.5	2138.41	2286.48	1293.505
		0.7	2121.35	2271.89	1288.553
	2	0.3	2087.42	2140.68	1210.648
		0.5	2064.23	2127.36	1204.24
		0.7	2048.36	2103.87	1195.745

Table 7. Natural Frequency (rad/sec) measurements with Different Plate Thickness for Middle Crack Location in ζ and η-Directions, $2C = 24$ mm, $\xi = 0.7$, $\forall_f = 30\%$.

Plate Type	Aspect Ratio	H (mm)	SSSS	SSCC	SSFF
Short Reinforcement	1	3.5	1213.55	1719.36	1016.228
		5.5	1916.3	2690.36	1535.618
		9	3190.25	4488.34	2602.777
	1.5	3.5	911.25	1093.14	623.9528
		5.5	1446.25	1716.49	965.8957
		9	2366.14	2809.34	1597.114

(*continued*)

Table 7. (*continued*)

Plate Type	Aspect Ratio	H (mm)	SSSS	SSCC	SSFF
Long Reinforcement	2	3.5	802.36	881.36	500.4709
		5.5	1248.36	1382.56	786.0231
		9	2041.43	2256.47	1279.46
	1	3.5	923.86	1113.79	632.5323
		5.5	1432.36	1702.26	967.0915
		9	2363.4	2844.59	1628.234
	1.5	3.5	829.34	891.24	505.6096
		5.5	1288.4	1376.49	782.0746
		9	2121.35	2271.89	1288.553
	2	3.5	804.87	831.74	469.9318
		5.5	1251.36	1281.33	728.5212
		9	2048.36	2103.87	1195.745

4 Conclusion

In this work, experimental and numerical investigation were achieved to investigate the effects of crack length and position on the natural frequencies of composite plates. The results showed that the central damage affects the plate stiffness more than the other crack positions, and the frequency reduction increases with the increase of the crack. It was also deduced that the increase in the aspect ratio results in a decrease in the plate natural frequency. The plate support end conditions affect the natural frequency keeping the design parameters constant, crack length, position and aspect ratio and result in a decrease in SSCC natural frequency, which was less than that in the SSSS and SSFF.

References

1. Kessler, S.S., Spearing, S.M., Soutis, C.: Structural health monitoring in composite materials using lamb wave methods. technology laboratory for advanced composites. Department of Aeronautics and Astronautics, Massachusetts Institute of Technology (2001)
2. Afshari, M., Inman, D.J.: Continuous crack modeling in piezoelectrically driven vibrations of Euler-Bernoulli beam. J. Vibr. Control **18**(10) (2012)
3. Al-Waily, M.: Theoretical and numerical vibration study of continuous beam with crack size and location effect. Int. J. Innov. Res. Sci. Eng. Technol. **2**(9) (2013)
4. Jweeg, M.J., Al-Waily, M., Muhammad, A.K., Resan, K.K.: Effects of temperature on the characterization of a new design for a non-articulated prosthetic foot. In: IOP Conference Series: Materials Science and Engineering, 2nd International Conference on Engineering Sciences, vol. 433 (2018)
5. Abbas, S.M., Resan, K.K., Muhammad, A.K., Al-Waily, M.: Mechanical and fatigue behaviors of prosthetic for partial foot amputation with various composite materials types effect. Int. J. Mech. Eng. Technol. **9**(9), 383–394 (2018)

6. Al-Waily, M., Al-Saffar, I.Q., Hussein, S.G., Al-Shammari, M.A.: Life enhancement of partial removable denture made by biomaterials reinforced by graphene nanoplates and hydroxyapatite with the aid of artificial neural network. J. Mech. Eng. Res. Dev. **43**(6), 269–285 (2020)
7. Abbas, E.N., Al-Waily, M., Hammza, T.M., Jweeg, M.J.: An investigation to the effects of impact strength on laminated notched composites used in prosthetic sockets manufacturing. In: IOP Conference Series: Materials Science and Engineering, 2nd International Scientific Conference of Al-Ayen University, vol. 928 (2020)
8. Al-Shammari, M.A., Bader, Q.H., Al-Waily, M., Hasson, A.M.: Fatigue behavior of steel beam coated with nanoparticles under high temperature. J. Mech. Eng. Res. Dev. **43**(4), 287–298 (2020)
9. Mechi, S.A., Al-Waily, M., Al-Khatat, A.: The mechanical properties of the lower limb socket material using natural fibers: a review. Mater. Sci. Forum **1039**, 473–492 (2021)
10. Abbas, E.N., Jweeg, M.J., Al-Waily, M.: Fatigue characterization of laminated composites used in prosthetic sockets manufacturing. J. Mech. Eng. Res. Dev. **43**(5), 384–399 (2020)
11. Al-Waily, M., Tolephih, M.H., Jweeg, M.J.: Fatigue characterization for composite materials used in artificial socket prostheses with the adding of nanoparticles. In: IOP Conference Series: Materials Science and Engineering, 2nd International Scientific Conference of Al-Ayen University, vol. 928 (2020)
12. Jebur, Q.H., Jweeg, M.J., Al-Waily, M.: Ogden model for characterising and simulation of PPHR Rubber under different strain rates. Aust. J. Mech. Eng. (2021)
13. Mechi, S.A., Al-Waily, M.: Impact and mechanical properties modifying for below knee prosthesis socket laminations by using natural kenaf fiber. In: 3rd International Scientific Conference of Engineering Sciences and Advances Technologies, Journal of Physics: Conference Series, vol. 1973 (2021)
14. Fahad, N.D., Kadhim, A.A., Al-Khayat, R.H., Al-Waily, M.: Effect of SiO2 and Al2O3 hybrid nano materials on fatigue behavior for laminated composite materials manufacture artificial socket prostheses. Mater. Sci. Forum **1039**, 493–509 (2021)
15. Haider, S.M.J., Takhakh, A.M., Al-Waily, M., Saadi, Y.: Simulation of gait cycle in sagittal plane for above-knee prosthesis. In: 3rd International Scientific Conference of Alkafeel University. AIP Conference Proceedings, vol. 2386 (2022)
16. Abbas, E.N., Jweeg, M.J., Al-Waily, M.: Analytical and numerical investigations for dynamic response of composite plates under various dynamic loading with the influence of carbon multi-wall tube nano materials. Int. J. Mech. Mechatron. Eng. **18**(6), 1–10 (2018)
17. Al-Waily, M., Al-Shammari, M.A., Jweeg, M.J.: An analytical investigation of thermal buckling behavior of composite plates reinforced by carbon nano particles. Eng. J. **24**(3) (2020)
18. Al-Waily, M., Jweeg, M.J., Al-Shammari, M.A., Resan, K.K., Takhakh, A.M.: Improvement of buckling behavior of composite plates reinforced with hybrids nanomaterials additives. Mater. Sci. Forum **1039**, 23–41 (2021)
19. Al-Waily, M., Deli, A.A., Al-Mawash, A.Z., Abud Ali, Z.A.A.: Effect of natural sisal fiber reinforcement on the composite plate buckling behavior. Int. J. Mech. Mechatron. Eng. **17**(1) (2017)
20. Al-Ansari, L.S., Al-Waily, M., Yusif, A.M.H.: Vibration analysis of hyper composite material beam utilizing shear deformation and rotary inertia effects. Int. J. Mech. Mechatron. Eng. **12**(4) (2012)
21. Jweeg, M.J., Hammood, A.S., Al-Waily, M.: Experimental and theoretical studies of mechanical properties for reinforcement fiber types of composite materials. Int. J. Mech. Mechatron. Eng. **12**(4) (2012)
22. Jweeg, M.J., Hammood, A.S., Al-Waily, M.: A suggested analytical solution of isotropic composite plate with crack effect. Int. J. Mech. Mechatron. Eng. **12**(5) (2012)

23. Al-Shammari, M.A., Al-Waily, M.: Theoretical and numerical vibration investigation study of orthotropic hyper composite plate structure. Int. J. Mech. Mechatron. Eng. **14**(6) (2014)

24. Jweeg, M.J., Al-Waily, M., Deli, A.A.: Theoretical and numerical investigation of buckling of orthotropic hyper composite plates. Int. J. Mech. Mechatron. Eng. **15**(4) (2015)

25. Al-Waily, M., Abud Ali, Z.A.A.: A suggested analytical solution of powder reinforcement effect on buckling load for isotropic mat and short hyper composite materials plate. Int. J. Mech. Mechatron. Eng. **15**(4) (2015)

26. Al-Waily, M., Resan, K.K., Al-Wazir, A.H., Abud Ali, Z.A.A.: Influences of glass and carbon powder reinforcement on the vibration response and characterization of an isotropic hyper composite materials plate structure. Int. J. Mech. Mechatron. Eng. **17**(6) (2017)

27. Kadhim, A.A., Al-Waily, M., Abud Ali, Z.A.A., Jweeg, M.J., Resan, K.K.: Improvement fatigue life and strength of isotropic hyper composite materials by reinforcement with different powder materials. Int. J. Mech. Mechatron. Eng. **18**(2) (2018)

28. Chiad, J.S., Al-Waily, M., Al-Shammari, M.A.: Buckling investigation of isotropic composite plate reinforced by different types of powders. Int. J. Mech. Eng. Technol. **9**(9), 305–317 (2018)

29. Ismail, M.R., Abud Ali, Z.A.A., Al-Waily, M.: Delamination damage effect on buckling behavior of woven reinforcement composite materials plate. Int. J. Mech. Mechatron. Eng. **18**(5), 83–93 (2018)

30. Abud Ali, Z.A.A., Kadhim, A.A., Al-Khayat, R.H., Al-Waily, M.: Review influence of loads upon delamination buckling in composite structures. J. Mech. Eng. Res. Dev. **44**(3), 392–406 (2021)

31. Al-Waily, M., Jaafar, A.M.: Energy balance modelling of high velocity impact effect on composite plate structures. Arch. Mater. Sci. Eng. **111**(1), 14–33 (2021)

32. Bakhy, S.H., Al-Waily, M., Al-Shammari, M.A.: Analytical and numerical investigation of the free vibration of functionally graded materials sandwich beams. Arch. Mater. Sci. Eng. **110**(2), 72–85 (2021)

33. Al-Shammari, M.A., Husain, M.A., Al-Waily, M.: Free vibration analysis of rectangular plates with cracked holes. In: 3rd International Scientific Conference of Alkafeel University, AIP Conference Proceedings, vol. 2386 (2022)

34. Jweeg, M.J.: A suggested analytical solution for vibration of honeycombs sandwich combined plate structure. Int. J. Mech. Mechatron. Eng. **16**(2) (2016)

35. Jweeg, M.J., Said, S.Z.: Effect of rotational and geometric stiffness matrices on dynamic stresses and deformations of rotating blades. J. Inst. Eng. (India) Mech. Eng. Div. **76**, 29–38 (1995)

36. Al-Shammari, M.A., Al-Waily, M.: Analytical investigation of buckling behavior of honeycombs sandwich combined plate structure. Int. J. Mech. Prod. Eng. Res. Dev. **8**(4), 771–786 (2018)

37. Njim, E.K., Al-Waily, M., Bakhy, S.H.: A review of the recent research on the experimental tests of functionally graded sandwich panels. J. Mech. Eng. Res. Dev. **44**(3), 420–441 (2021)

38. Njim, E.K., Bakhy, S.H., Al-Waily, M.: Analytical and numerical investigation of buckling behavior of functionally graded sandwich plate with porous core. J. Appl. Sci. Eng. **25**(2), 339–347 (2022)

39. Njim, E.K., Bakhy, S.H., Al-Waily, M.: Optimisation design of functionally graded sandwich plate with porous metal core for buckling characterisations. Pertanika J. Sci. Technol. **29**(4), 3113–3141 (2021)

40. Njim, E.K., Bakhy, S.H., Al-Waily, M.: Analytical and numerical flexural properties of polymeric porous functionally graded (PFGM) sandwich beams. J. Achiev. Mater. Manuf. Eng. **110**(1), 5–15 (2022)

41. Hussein, S.G., Al-Shammari, M.A., Takhakh, A.M., Al-Waily, M.: Effect of heat treatment on mechanical and vibration properties for 6061 and 2024 aluminum alloys. J. Mech. Eng. Res. Dev. **43**(1), 48–66 (2020)

42. Njim, E.K., Bakhy, S.H., Al-Waily, M.: Optimization design of vibration characterizations for functionally graded porous metal sandwich plate structure. Mater. Today Proc. (2021)

43. Njim, E.K., Bakhy, S.H., Al-Waily, M.: Analytical and numerical free vibration analysis of porous functionally graded materials (FGPMs) sandwich plate using Rayleigh-Ritz method. Arch. Mater. Sci. Eng. **110**(1), 27–41 (2021)

44. Njim, E.K., Bakhy, S.H., Al-Waily, M.: Analytical and numerical investigation of free vibration behavior for sandwich plate with functionally graded porous metal core. Pertanika J. Sci. Technol. **29**(3), 1655–1682 (2021)

45. Al-Baghdadi, M., Jweeg, M.J., Al-Waily, M.: Analytical and numerical investigations of mechanical vibration in the vertical direction of a human body in a driving vehicle using biomechanical vibration model. Pertanika J. Sci. Technol. **29**(4), 2791–2810 (2021)

46. Njim, E.K., Bakhy, S.H., Al-Waily, M.: Free vibration analysis of imperfect functionally graded sandwich plates: analytical and experimental investigation. Arch. Mater. Sci. Eng. **111**(2), 49–65 (2021)

47. Alhumdany, A.A., Al-Waily, M., Al-Jabery, M.H.K.: Theoretical and experimental investigation of using date palm nuts powder into mechanical properties and fundamental natural frequencies of hyper composite plate. Int. J. Mech. Mechatron. Eng. **16**(1) (2016)

48. Abbas, S.M., Takhakh, A.M., Al-Shammari, M.A., Al-Waily, M.: Manufacturing and analysis of ankle disarticulation prosthetic socket (SYMES). Int. J. Mech. Eng. Technol. **9**(7), 560–569 (2018)

49. Abbod, E.A., Al-Waily, M., Al-Hadrayi, Z.M.R., Resan, K.K., Abbas, S.M.: Numerical and experimental analysis to predict life of removable partial denture. In: IOP Conference Series: Materials Science and Engineering. 1st International Conference on Engineering and Advanced Technology, vol. 870 (2020)

50. Njim, E.K., Bakhy, S.H., Al-Waily, M.: Analytical and numerical investigation of buckling load of functionally graded materials with porous metal of sandwich plate. Mater. Today Proc. (2021)

51. Kadhim, A.A., Abbod, E.A., Muhammad, A.K., Resan, K.K., Al-Waily, M.: Manufacturing and analyzing a new prosthetic shank with adapters by 3D printer. J. Mech. Eng. Res. Dev. **44**(3), 383–391 (2021)

52. Al-Waily, M., Jweeg, M.J., Jebur, Q.H., Resan, K.K.: Creep characterization of various prosthetic and orthotics composite materials with nanoparticles using an experimental program and an artificial neural network. Mater. Today Proc. (2021)

53. Njim, E.K., Bakhy, S.H., Al-Waily, M.: Experimental and numerical flexural analysis of porous functionally graded beams reinforced by (Al/Al2O3) nanoparticles. Int. J. Nanoelectron. Mater. **152**, 91–106 (2022)

54. Al-Waily, M., Al-Baghdadi, M.A.R.S., Al-Khayat, R.H.: Flow velocity and crack angle effect on vibration and flow characterization for pipe induce vibration. Int. J. Mech. Mechatron. Eng. **17**(5), 19–27 (2017)

55. Resan, K.K., Alasadi, A.A., Al-Waily, M., Jweeg, M.J.: Influence of temperature on fatigue life for friction stir welding of aluminum alloy materials. Int. J. Mech. Mechatron. Eng. **18**(2) (2018)

56. Jweeg, M.J., Resan, K.K., Abbod, E.A., Al-Waily, M.: Dissimilar aluminium alloys welding by friction stir processing and reverse rotation friction stir processing. In: IOP Conference Series: Materials Science and Engineering, International Conference on Materials Engineering and Science, vol. 454 (2018)

57. Abdulridha, M.M., Fahad, N.D., Al-Waily, M., Resan, K.K.: Rubber creep behavior investigation with multi wall tube carbon nano particle material effect. Int. J. Mech. Eng. Technol. **9**(12), 729–746 (2018)

58. Abbas, H.J., Jweeg, M.J., Al-Waily, M., Diwan, A.A.: Experimental testing and theoretical prediction of fiber optical cable for fault detection and identification. J. Eng. Appl. Sci. **14**(2), 430–438 (2019)

59. Jebur, Q.H., Jweeg, M.J., Al-Waily, M., Ahmad, H.Y., Resan, K.K.: Hyperelastic models for the description and simulation of rubber subjected to large tensile loading. Arch. Mater. Sci. Eng. **108**(2), 75–85 (2021)

60. Aswad, T.S.N., Bin Razali, M.A., Al-Waily, M.: Numerical study of the shape obstacle effect on improving the efficiency of photovoltaic cell. J. Mech. Eng. Res. Dev. **44**(2), 209–224 (2021)

61. Jweeg, M.J., Mohammed, K.I., Tolephih, M.H., Al-Waily, M.: Investigation into the distribution of erosion-corrosion in the furnace tubes of oil refineries. Mater. Sci. Forum **1039**, 165–181 (2021)

62. Al-Khayat, R.H., Al-Fatlawi, A.W.A., Al-Baghdadi, M.A.R.S., Al-Waily, M.: Water hammer phenomenon in pumping stations: a stability investigation based on root locus. Open Eng. **12**, 254–262 (2022)

63. Abud Ali, Z.A.A., Takhakh, A.M., Al-Waily, M.: A review of use of nanoparticle additives in lubricants to improve its tribological properties. Mater. Today Proc. **52**(3), 1442–1450 (2022)

64. Al-Khayat, R.H., Kadhim, A.A., Al-Baghdadi, M.A.R.S., Al-Wail, M.: Flow parameters effect on water hammer stability in the hydraulic system using the state-space method. Open Eng. **12**, 215–226 (2022)

65. Jweeg, M.J., Alazawi, D.A., Jebur, Q.H., Al-Waily, M., Yasin, N.J.: Hyperelastic modelling of rubber with multi-walled carbon nanotubes subjected to tensile loading. Arch. Mater. Sci. Eng. **114**(2), 69–85 (2022)

66. Hakim, M., Omran, A.A.B., Ahmed, A.N., Al-Waily, M., Abdellatif, A.: A systematic review of rolling bearing fault diagnoses based on deep learning and transfer learning: taxonomy, overview, application, open challenges, weaknesses and recommendations. Ain Shams Eng. J. (2022)

Author Index